# Lecture Notes in Mechanical Engineering

**Lecture Notes in Mechanical Engineering (LNME)** publishes the latest developments in Mechanical Engineering—quickly, informally and with high quality. Original research reported in proceedings and post-proceedings represents the core of LNME. Volumes published in LNME embrace all aspects, subfields and new challenges of mechanical engineering. Topics in the series include:

- Engineering Design
- Machinery and Machine Elements
- Mechanical Structures and Stress Analysis
- Automotive Engineering
- Engine Technology
- Aerospace Technology and Astronautics
- Nanotechnology and Microengineering
- Control, Robotics, Mechatronics
- MEMS
- Theoretical and Applied Mechanics
- Dynamical Systems, Control
- Fluid Mechanics
- Engineering Thermodynamics, Heat and Mass Transfer
- Manufacturing
- Precision Engineering, Instrumentation, Measurement
- Materials Engineering
- Tribology and Surface Technology

To submit a proposal or request further information, please contact the Springer Editor of your location:

**China:** Ms. Ella Zhang at ella.zhang@springer.com
**India:** Priya Vyas at priya.vyas@springer.com
**Rest of Asia, Australia, New Zealand:** Swati Meherishi at swati.meherishi@springer.com
**All other countries:** Dr. Leontina Di Cecco at Leontina.dicecco@springer.com

To submit a proposal for a monograph, please check our Springer Tracts in Mechanical Engineering at https://link.springer.com/bookseries/11693 or contact Leontina.dicecco@springer.com

**Indexed by SCOPUS. All books published in the series are submitted for consideration in Web of Science.**

Suripeddi Srinivas · Badeti Satyanarayana ·
J. Prakash
Editors

# Recent Advances in Applied Mathematics and Applications to the Dynamics of Fluid Flows

## 5th International Conference on Applications of Fluid Dynamics (ICAFD) 2020

Springer

*Editors*
Suripeddi Srinivas
Department of Mathematics
VIT-AP University
Amaravati, Andhra Pradesh, India

Badeti Satyanarayana
Department of Mathematics
VIT-AP University
Amaravati, Andhra Pradesh, India

J. Prakash
Department of Mechanical Engineering
FEBE University of Johannesburg
Johannesburg, South Africa

ISSN 2195-4356 ISSN 2195-4364 (electronic)
Lecture Notes in Mechanical Engineering
ISBN 978-981-19-1928-2 ISBN 978-981-19-1929-9 (eBook)
https://doi.org/10.1007/978-981-19-1929-9



This Springer imprint is published by the registered company Springer Nature Singapore Pte Ltd.
The registered company address is: 152 Beach Road, #21-01/04 Gateway East, Singapore 189721, Singapore

# Contents

# About the Editors

**Prof. Suripeddi Srinivas** obtained his Masters (in Applied Mathematics) and Doctoral Degrees from the National Institute of Technology (Formerly Regional Engineering College) Warangal, India. Presently, he is a Professor and Dean (Research) at VIT-AP University, Amaravati, India. Earlier, he has served as Senior Professor, Department of Mathematics, School of Advanced Sciences, VIT University, Vellore. He has nearly 30 years of teaching and research experience. His thrust area of research is mathematical modelling and fluid dynamics with emphasis on heat and mass transfer, Nanofluids, Newtonian and Non-Newtonian fluid flows and Bio-fluid flows. He has published over 110 research papers in National and International journals and his publications have received more than 3,300 citations. His h-index is 26 and i10-Index is 50. As a Principal Investigator (PI), he has completed three major research projects sponsored by DRDO (Govt. of India), DST (Govt. of India) and NBHM (Govt. of India). He has also served as co-PI for several major research projects funded by various agencies. As of now, he has supervised nine Ph.Ds and four M.Phil. scholars. Several research students are actively working under his guidance. He has actively contributed to the research community in several capacities such as theses adjudicator for Ph.D/M.Phil. degrees, reviewer to several notable international journals. He has attended and organized several workshops, FDPs and conferences both in India and abroad.

**Dr. Badeti Satyanarayana** has over 17 years of teaching experience, having worked in well-known institutions. He joined VIT University, Vellore and under the guidance of Prof. V. Somaraju and Prof. S. Srinivas worked on his doctoral thesis on Geo Physical Fluid Dynamics. He has published several papers in National and International journals of repute. He has presented his work at various National and International conferences and symposiums both in India and abroad. Presently, he is working at VIT-AP University, Amaravati in the Department of Mathematics. His research interests are diverse include geophysical fluid dynamics, MHD boundary layers, perturbation theory and PDE's.

**Dr. J. Prakash** is currently a Professor of Mathematics in the Department of Mechanical Engineering Science, Faculty of Engineering and The Built Environment, University of Johannesburg, Auckland Park Kingsway Campus, South Africa. He obtained his B.Sc. and M.Sc. (Mathematics) from Agra University, Agra; M.Sc. (Statistics) from Meerut University, Meerut; D.Phil. (Applied Mathematics) from Garhwal University, Srinagar and Dip.Tchg. (Secondary) from Massey University, New Zealand. His major areas of research interests are in bio-fluid dynamics, fluid mechanics, computational fluid dynamics (CFD), mathematical modelling, mathematical methods. He has taught at the university level in India, Iraq, Zambia, Botswana and New Zealand for over 45 years. He has published over 175 research papers, attended over 60 international conferences and organized several national and international conferences. He is a recipient of three prestigious awards, namely: C. V. Raman International Fellowship, Global Visionary and an Honourable Scientist. He is a reviewer of over 70 international journals and on the editorial board of many accredited journals.

# Impacts of Variable Viscosity on Hydromagnetic Peristaltic Flow of a Bingham Fluid in a Vertical Channel

**R. Muthuraj, S. Srinivas, and D. Lourdu Immaculate**

## 1 Introduction

In the past few decades, studies on non-Newtonian fluid flows attract researchers because of its occurrence in many engineering field and it has been growing considerably in recent years also. Furthermore, traditional Newtonian fluid model cannot precisely describe the characteristics of most of the fluids involving in industrial applications. However, only limited studies are available in the field of non-Newtonian fluid flows with different flow configurations ([1–5] and several references therein). In view of these, Khalid et al. [1] have discussed influence of space porosity on hydromagnetic unsteady flow of a Casson fluid with oscillating vertical plate using Laplace transform method. Unsteady flow of a Williamson nanofluid in a channel bounded by two vertical walls was examined by Immaculate et al. [2]. Combined effects of wall slip and convective boundary conditions on non-Newtonian nanofluids in a channel with peristalsis were analyzed by Sayed et al. [3]. Chemical reaction effect on MHD flow of Powell-Eyring fluid in an inclined porous space was investigated by Muthuraj et al. [4]. More recently, impact of heat source on Powell-Eyring fluid flow with Dufour and Soret effects was studied by Muthuraj and Selvi [5]. Owing to occurrence of peristaltic mechanism in various industrial and physiological procedures, non-Newtonian fluid flows with peristalsis have received significant attention in the past. In view of its applications, several researchers investigate peristaltic transport of non-Newtonian fluids with different flow geometries [6–8]. Also, literature survey witnesses that most of the industrial materials can flow only with the help of

R. Muthuraj
Department of Mathematics, PSNA College of Engineering and Technology, Dindigul 624622, India

S. Srinivas
Department of Mathematics, VIT AP-University, Amaravati, India
e-mail: srinusuripeddi@hotmail.com

D. L. Immaculate (✉)
Department of Mathematics, The American College, Madurai 625002, India
e-mail: immaculate.jeeva@gmail.com



S. Srinivas et al. (eds.), *Recent Advances in Applied Mathematics and Applications to the Dynamics of Fluid Flows*, Lecture Notes in Mechanical Engineering,
https://doi.org/10.1007/978-981-19-1929-9_1

yield stress. These fluids exhibit non-Newtonian behavior. Such fluids are generally designated as Bingham fluids. Clay, printing ink, paint, and molten chocolate are some typical examples of Bingham plastics. For this reason, studies of Bingham fluid flows in various flow conditions have been made by several researchers ([9–12] and many worthy references therein). Later, Bingham plastic fluid flow with applied uniform magnetic field in a porous bed was examined by Misra and Adhikary [9]. Fusi and Farina [10] have investigated Bingham fluid flow in a channel with peristalsis. Later, they presented peristaltic transport of a Bingham fluid in a tube by lubrication approximation method [11]. More recently, pressure-driven flow of Bingham fluid in a curved channel was analyzed by Roberts and Cox [12]. Most of the previous investigations have considered all the physical properties are assumed to be a constant. However, it is obvious that some physical properties are functions of temperature. Also, existing studies witness that accurate prediction for the flow and heat transfer can be reached by taking some physical properties varying with temperature. Also, convective heat and mass transfer studies with chemical reaction effects have gotten significant attraction due to its importance in hydrometallurgical industries and chemical technology. Many studies can be found in this direction with different flow situations [13–20]. Notably, Eswaramoorthi et al. [16] examined convective flow of viscoelastic fluid with combined impacts of Soret and Dufour effect in the presence of radiation and chemical reaction. Muthuraj et al. [17] presented impacts of wall properties on dusty fluid flow with peristalsis and chemical reaction. The influence of wall slip condition and radiation on MHD convective stagnation flow with thermal diffusion and diffusion thermoeffects and chemical reaction was discussed by Niranjan et al. [18]. Sivasankaran et al. [19] demonstrate convective boundary condition effect on convective hydromagnetic convective stagnation flow with chemical reaction and wall slip. Bhuvaneswari et al. [20] illustrate MHD convective flow over a stretching surface in porous medium with the effects of cross diffusion and chemical reaction. Farooq et al. [21] presented viscosity variation on MHD flow of Jeffrey fluid in a symmetric channel with Dufour and Soret effect using Lubrication approach. Later, combined effects of diffusion thermo and thermal diffusion on MHD Jeffrey fluid flow in vertical porous walls with varying viscosity and viscous dissipation were analyzed by Selvi and Muthuraj [22]. Hassan [23] has studied Couette flow of a viscous fluid with the effects of variable viscosity and heat source using ADM. Ajibade and Tafida [24] have investigated channel flow of a viscous fluid with varying physical properties. Literature survey witness that most of the authors neglected coupled effects of variable viscosity and chemical reaction on peristaltic flow with heat and mass transfer effects. Therefore, here we study influence of varying viscosity and chemical reaction effects on peristaltic transport of a Bingham fluid with uniform magnetic field and heat source. Long wave length and low Reynolds number approximation are employed to reduce the governing system of partial differential equations into set of ordinary differential equations. Analytical expressions for velocity, temperature, and concentration fields are obtained and impacts of involved parameters are discussed graphically.

## 2 Formulation of the Problem

Consider a laminar, incompressible, hydromagnetic flow of a Bingham fluid in a vertical porous space bounded by two walls of width $d_1 + d_2$. The walls are maintained at different temperatures $T_1, T_2$ and concentrations $C_1, C_2$, respectively. The equations of the channel walls are described as

$$\begin{aligned} H_1 &= d_1 + a_1 cos\tfrac{2\pi}{\lambda}(\bar{X} - ct).......Right\ wall \\ H_2 &= -d_2 - b_1 cos(\tfrac{2\pi}{\lambda}(\bar{X} - ct) + \vartheta).......Left\ wall \end{aligned} \tag{1}$$

and $a_1, b_1, d_1, d_2$ and $\vartheta$ satisfies the condition

$$a_1^2 + b_1^2 + 2a_1 b_1 cos\vartheta \le (d_1 + d_2)^2. \tag{2}$$

The equations governing continuity, momentum, energy, and concentration are as follows:

$$\frac{\partial \bar{U}}{\partial \bar{X}} + \frac{\partial \bar{V}}{\partial \bar{Y}} = 0 \tag{3}$$

$$\begin{aligned} \rho\left(\frac{\partial \bar{U}}{\partial t} + \bar{U}\frac{\partial \bar{U}}{\partial \bar{X}} + \bar{V}\frac{\partial \bar{U}}{\partial \bar{Y}}\right) &= -\frac{\partial \bar{P}}{\partial \bar{X}} + \mu(T)\frac{\partial^2 \bar{U}}{\partial \bar{X}^2} - \frac{\partial}{\partial \bar{Y}}\left(\tau_0 - \mu(T)\frac{\partial \bar{U}}{\partial \bar{Y}}\right) \\ &\quad -\sigma B_0^2 \bar{U} - \frac{\mu(T)\phi^*}{k}\bar{U} + \rho g \beta_t (T - \bar{T}) + \rho g \beta_c (C - \bar{C}) \end{aligned} \tag{4}$$

$$\rho\left(\frac{\partial \bar{V}}{\partial t} + \bar{U}\frac{\partial \bar{V}}{\partial \bar{X}} + \bar{V}\frac{\partial \bar{V}}{\partial \bar{Y}}\right) = -\frac{\partial \bar{P}}{\partial \bar{Y}} + \mu(T)\left(\frac{\partial^2 \bar{V}}{\partial \bar{X}^2} + \frac{\partial^2 \bar{V}}{\partial \bar{Y}^2}\right) - \frac{\mu(T)\phi^*}{k}\bar{V} \tag{5}$$

$$\left(\frac{\partial T}{\partial t} + \bar{U}\frac{\partial T}{\partial \bar{X}} + \bar{V}\frac{\partial T}{\partial \bar{Y}}\right) = \frac{K}{\rho c_p}\left(\frac{\partial^2 T}{\partial \bar{X}^2} + \frac{\partial^2 T}{\partial \bar{Y}^2}\right) + Q(T - \bar{T}) \tag{6}$$

$$\left(\frac{\partial C}{\partial t} + \bar{U}\frac{\partial C}{\partial \bar{X}} + \bar{V}\frac{\partial C}{\partial \bar{Y}}\right) = D_m\left(\frac{\partial^2 C}{\partial \bar{X}^2} + \frac{\partial^2 C}{\partial \bar{Y}^2}\right) + \frac{D_m k_T}{\bar{T}}\left(\frac{\partial^2 T}{\partial \bar{X}^2} + \frac{\partial^2 T}{\partial \bar{Y}^2}\right) - k_1 C. \tag{7}$$

The coordinates and velocities in the fixed frame $(\bar{X}, \bar{Y})$ and the wave frame (x,y) are related by $x = \bar{X} - ct, y = \bar{Y}, u = \bar{U} - c, v = \bar{V}, p(x) = \bar{P}(x, t)$ and introducing the following non-dimensional quantities:

$$\begin{aligned} &\bar{x} = \frac{x}{\lambda}, \bar{y} = \frac{y}{d_1}, \bar{u} = \frac{u}{c}.\bar{v} = \frac{v}{c\delta}, \delta = \frac{d_1}{\lambda}, \bar{p} = \frac{d_1^2 p}{\mu_0 c \lambda}, \bar{t} = \frac{ct}{\lambda}, h_1 = \frac{H_1}{d_1}, h_2 = \frac{H_2}{d_2}, d = \frac{d_2}{d_1}, \\ &a = \frac{a_1}{b_1}, b = \frac{b_1}{d_1}, Re = \frac{\rho c d_1}{\mu}, \theta = \frac{T - \bar{T}}{T_1 - \bar{T}}, \phi = \frac{C - \bar{C}}{C_1 - \bar{C}}, n = \frac{T_2 - \bar{T}}{T_1 - \bar{T}}, m = \frac{C_2 - \bar{C}}{C_1 - \bar{C}}, \mu(\theta) = \frac{\mu(T)}{\mu_0}, \\ &\bar{\tau}_0 = \frac{d_1 \tau_0}{\mu_0 c}, P_r = \frac{\mu_0 c_p}{K}, S_c = \frac{\mu_0}{\rho D_m}, G_r = \frac{\rho g \beta_t (T_1 - \bar{T}) d_1^2}{\mu_0 c}, G_c = \frac{\rho g \beta_c (C_1 - \bar{C}) d_1^2}{\mu_0 c}, \gamma = \frac{k_1 d_1^2}{\nu}, \\ &D_a = \frac{k}{\phi^* d_1^2}, M = \sqrt{\frac{\sigma}{\mu_0}} B_0 d_1, \alpha = \frac{Q d_1^2}{K}, S_r = \frac{\rho D_m k_T (T_1 - \bar{T})}{\mu_0 \bar{T}(C_1 - \bar{C})}. \end{aligned} \tag{8}$$

Employing (8) into Eqs. (3)–(7) and dropping bars, the non-dimensional problem is

$$\frac{\partial u}{\partial x} + \frac{\partial v}{\partial y} = 0 \tag{9}$$

$$Re\delta\left(u\frac{\partial u}{\partial x} + v\frac{\partial u}{\partial y}\right) = -\frac{\partial P}{\partial x} + \delta^2\mu(\theta)\frac{\partial^2 u}{\partial x^2} - \frac{\partial}{\partial y}\left(\tau_0 - \mu(\theta)\frac{\partial u}{\partial y}\right) \\ H(u+1) + G_r\theta + G_c\phi \tag{10}$$

$$Re\delta^3\left(u\frac{\partial v}{\partial x} + v\frac{\partial v}{\partial y}\right) = -\frac{\partial p}{\partial y} + \mu(\theta)\delta^2\left(\delta^2\frac{\partial^2 v}{\partial x^2} + \frac{\partial^2 v}{\partial y^2}\right) - \mu(\theta)\frac{\delta^2}{D_a}v \tag{11}$$

$$ReP_r\delta\left(u\frac{\partial \theta}{\partial x} + v\frac{\partial \theta}{\partial y}\right) = \left(\delta^2\frac{\partial^2 \theta}{\partial x^2} + \frac{\partial^2 \theta}{\partial y^2}\right) + \alpha\theta \tag{12}$$

$$Re\delta(u\frac{\partial C}{\partial x} + v\frac{\partial C}{\partial y}) = \frac{1}{S_c}\left(\delta^2\frac{\partial^2 \phi}{\partial x^2} + \frac{\partial^2 \phi}{\partial y^2}\right) + S_r\left(\delta^2\frac{\partial^2 \theta}{\partial x^2} + \frac{\partial^2 \theta}{\partial y^2}\right) - \gamma\phi + c_1. \tag{13}$$

Using long wavelength procedure, we get

$$\frac{\partial}{\partial y}\left(\tau_0 - \mu(\theta)\frac{\partial u}{\partial y}\right) + H(u+1) - G_r\theta - G_c\phi = -\frac{\partial P}{\partial x} \tag{14}$$

$$\frac{\partial^2 \theta}{\partial y^2} + \alpha\theta = 0 \tag{15}$$

$$\frac{1}{S_c}\frac{\partial^2 \phi}{\partial y^2} + S_r\frac{\partial^2 \theta}{\partial y^2} - \gamma\phi + c_1 = 0 \tag{16}$$

with boundary conditions,

$$\begin{aligned} u = -1, \theta = 1, \phi = 1 \quad &\text{at} \quad y = h_1 \\ u = -1, \theta = n, \phi = m \quad &\text{at} \quad y = h_2, \end{aligned} \tag{17}$$

where $c_1 = -\frac{k_1 d_1^2 \bar{C}}{\nu(C_1 - \bar{C})}$, $H^2 = \left(M^2 + \frac{1}{D_a}\right)$, $h_1(x) = 1 + acos2\pi x$, $h_2(x) = -d - bcos(2\pi x + \vartheta)$, $a, b, d$, and $\vartheta$ satisfies the relation $a^2 + b^2 + 2abcos\vartheta \leq (1 + d)^2$.

## 3 Method of Solution

Solving Eq. (15) using (17), we obtain

$$\theta = A_1 cos\alpha_1 y + A_2 sin\alpha_1 y. \tag{18}$$

Plugging (18) into (16) and solving subject to the boundary condition (17), one can get

$$\phi = A_3 cosh\beta_1 y + A_4 sinh\beta_1 y + T_1 cos\alpha_1 y + T_2 sin\alpha_1 y + T_3, \tag{19}$$

where $\alpha_1 = \sqrt{\alpha}$, $\beta_1 = \sqrt{\gamma S_c}$, $T_1 = \frac{c_1}{\gamma}$, $T_2 = \frac{-\alpha A_1 S_r S_c}{\alpha + \gamma S_c}$, $A_1 = \frac{1 - A_2 sin\alpha_1 h_1}{cos\alpha_1 h_1}$, $A_2 = \frac{cos\alpha_1 h_2 - ncos\alpha_1 h_1}{sin\alpha_1(h_1 - h_2)}$, $A_3 = \frac{1 - A_4 sin\beta_1 h_1}{cosh\beta_1 h_1}$, $A_4 = \frac{s_1 cos\beta_1 h_2 - s_2 ncos\beta_1 h_1}{sinh\beta_1(h_1 - h_2)}$, $s_1 = 1 - T_1 cos\alpha_1 h_1 - T_2 sin\alpha_1 h_1 - T_3$, $s_2 = 1 - T_1 cos\alpha_1 h_2 - T_2 sin\alpha_1 h_2 - T_3$. In this analysis, viscosity variation can be taken in the following way (see Refs. [17, 18]), $\mu(\theta) = e^{-}\beta\theta = 1 - \beta\theta + O(\beta)^2$ where $\beta(<< 1)$ is the viscosity parameter. Therefore, Eq. (14) becomes

$$\tfrac{d}{dy}\left[(1-\beta\theta)\tfrac{du}{dy}\right] - H(u+1) + G_r\theta + G_c\phi = \tfrac{dp}{dx}. \tag{20}$$

Equation (20) is non-linear and it cannot be solved analytically. But, it can be solved analytically by perturbation method in terms of viscosity parameter. Accordingly, we can write

$$u = \beta^0 u_0 + \beta^1 u_1 + \beta^2 u_2 + \beta^3 u_3 + \ldots\ldots\ldots.. \tag{21}$$

Making use of Eq. (21) into Eqs. (17) and (20), we get the following.

### 3.1 Zeroth-Order System ($\beta^0$)

$$\frac{d^2u_0}{dy^2} - H(u_0+1) + G_r\theta + G_c\phi = \frac{dp_0}{dx} \tag{22}$$

with conditions

$$u_0 = -1 \quad \text{at} \quad y = h_1 \tag{23}$$

$$u_0 = -1 \quad \text{at} \quad y = h_2. \tag{24}$$

### 3.2 First-Order System ($\beta^1$)

$$\frac{d^2u_1}{dy^2} - \theta\frac{d^2u_0}{dy^2} - \frac{du_0}{dy}\frac{d\theta}{dy} - Hu_1 = \frac{dp_1}{dx} \tag{25}$$

with conditions

$$u_1 = 0 \quad \text{at} \quad y = h_1 \tag{26}$$

$$u_1 = 0 \quad \text{at} \quad y = h_2. \tag{27}$$

Solve the above system using appropriate boundary conditions finally we get the expression for velocity up to first order as

$$u_0 = A_5 cosh H_1 y + A_6 sinh H_1 y + T_{10} cos \alpha_1 y + T_{11} sin \alpha_1 y + T_{12} cosh \beta_1 y + T_{13} sinh \beta_1 y + T_0 \tag{28}$$

$$u_1 = A_7 cosh H_1 y + A_8 sinh H_1 y + T_{25} cosh(H_1 + i\alpha_1)y + T_{26} cosh(H_1 - i\alpha_1)y + T_{27} sinh(H_1 + i\alpha_1)y + T_{28} sinh(H_1 - i\alpha_1)y + T_{29} cosh(\beta_1 + i\alpha_1)y + T_{30} cosh(\beta_1 - i\alpha_1)y + T_{31} sinh(\beta_1 + i\alpha_1)y + T_{32} sinh(\beta_1 - i\alpha_1)y + T_{33} sin 2\alpha_1 y + T_{34} cos 2\alpha_1 y + T_{35}, \tag{29}$$

where $P_0 = \frac{dp_0}{dx}$, $P_1 = \frac{dp_1}{dx}$, $T_4 = T_3 + P_0 + H$, $T_5 = T_1 - G_r C_1$, $T_6 = T_2 - G_r C_2$, $T_7 = -G_c C_3$, $T_8 = -G_c C_4$, $T_9 = -\frac{T_4}{H}$, $T_{10} = -\frac{T_5}{\alpha_1^2 + H}$, $T_{11} = -\frac{T_6}{\alpha_1^2 + H}$, $T_{12} = -\frac{T_7}{\beta_1^2 - H}$, $T_{13} = -\frac{T_8}{\beta_1^2 - H}$, $T_{14} = C_1 C_5 H_1^2 + C_2 C_6 \alpha_1 h_1$, $T_{15} = C_1 C_6 H_1^2 + C_2 C_5 \alpha_1 h_1$, $T_{16} = -2\alpha_1^2 (C_2 T_{10} + C_1 T_{11}$, $T_{17} = C_1 T_{12} \beta_1^2 + C_2 T_{13} \alpha_1 \beta_1$, $T_{18} = C_1 T_{13} \beta_1^2 + C_2 T_{12} \alpha_1 \beta_1$, $T_{19} = C_2 C_5 H_1^2 + C_1 C_6 \alpha_1 \beta_1$, $T_{20} = C_2 C_6 H_1^2 - C_1 C_5 \alpha_1 H_1$, $T_{20} = C_2 C_6 H_1^2 - C_1 C_5 \alpha_1 H_1$, $T_{21} = C_2 T_{12} \beta_1^2 - C_1 T_{13} \alpha_1 \beta_1$, $T_{22} = C_2 T_{13} \beta_1^2 - C_1 T_{12} \alpha_1 \beta_1$, $T_{23} = \alpha_1^2 (C_2 T_{11} - C_1 T_{10})$, $T_{24} = -T_{23}$, $T_{25} = \frac{T_{14} - iT_{20}}{2[(H_1 + i\alpha_1)^2 - H]}$, $T_{26} = \frac{T_{14} + iT_{20}}{2[(H_1 - i\alpha_1)^2 - H]}$, $T_{27} = \frac{T_{15} - iT_{19}}{2[(H_1 + i\alpha_1)^2 - H]}$, $T_{28} = \frac{T_{15} + iT_{19}}{2[(H_1 - i\alpha_1)^2 - H]}$, $T_{29} = \frac{T_{17} - iT_{22}}{2[(\beta_1 + i\alpha_1)^2 - H]}$, $T_{30} = \frac{T_{17} + iT_{22}}{2[(\beta_1 - i\alpha_1)^2 - H]}$, $T_{31} = \frac{T_{18} - iT_{21}}{2[(\beta_1 + i\alpha_1)^2 - H]}$, $T_{32} = \frac{T_{18} + iT_{21}}{2[(\beta_1 - i\alpha_1)^2 - H]}$, $T_{33} = \frac{T_{23}}{2(4\alpha_1^2 - H^2)}$, $T_{34} = -\frac{T_{24}}{2(4\alpha_1^2 - H^2)}$, $T_{35} = -\frac{T_{23} + T_{24} + P_1}{2H}$, $T_{36} = -(1 + T_{10} cos \alpha_1 h_1 + T_{11} sin \alpha_1 h_1 + T_{12} cosh \beta_1 h_1 + T_{13} sinh \beta_1 h_1 + T_9)$, $T_{37} = -(1 + T_{10} cos \alpha_1 h_2 + T_{11} sin \alpha_1 h_2 + T_{12} cosh \beta_1 h_2 + T_{13} sinh \beta_1 h_2 + T_9)$, $T_{38} = -(T_{25} cosh(H + i\alpha_1) + T_{26} cosh(H - i\alpha_1) + T_{27} sinh(H + i\alpha_1) + T_{28} sinh(H - i\alpha_1) + T_{29} cosh(\beta + ih_1) + T_{30} cosh(\beta - ih_1) + T_{31} sinh(\beta + ih_1) + T_{32} sinh(\beta - ih_1) + T_{33} sin 2\alpha_1 h_1 + T_{34} cos 2\alpha_1 h_1 + T_{35})$, $T_{38} = -(T_{25} cosh(H + i\alpha_1)h_1 + T_{26} cosh(H - i\alpha_1)h_1 + T_{27} sinh(H + i\alpha_1)h_1 + T_{28} sinh(H - i\alpha_1)h_1 + T_{29} cosh(\beta + i\alpha_1)h_1 + T_{30} cosh(\beta - i\alpha_1)h_1 + T_{31} sinh (\beta + i\alpha_1)h_1 + T_{32} sinh(\beta - i\alpha_1)h_1 + T_{33} sin 2\alpha_1 h_1 + T_{34} cos 2\alpha_1 h_1 + T_{35})$, $T_{39} = -(T_{25} cosh(H + i\alpha_1) h_2 + T_{26} cosh(H - i\alpha_1)h_2 + T_{27} sinh(H + i\alpha_1)h_2 + T_{28} sinh(H - i\alpha_1)h_2 + T_{29} cosh(\beta + i\alpha_1)h_2 + T_{30} cosh(\beta - i\alpha_1)h_2 + T_{31} sinh(\beta + i\alpha_1)h_2 + T_{32} sinh(\beta - i\alpha_1)h_2 + T_{33} sin 2\alpha_1 h_2 + T_{34} cos 2\alpha_1 h_2 + T_{35})$, $A_6 = \frac{T_{36} cosh H_1 h_2 - T_{37} cosh H_1 h_1}{sinh H_1 (h_1 - h_2)}$, $A_5 = \frac{T_{36} - A_6 sinh 0 H_1 h_1}{sinh H_1}$, $A_8 = \frac{T_{38} cosh H_1 h_2 - T_{39} cosh H_1 h_1}{sinh H_1 (h_1 - h_2)}$, $A_5 = \frac{T_{36} - A_6 sinh H_1 h_1}{sinh H_1}$. We can find dimensionless stream function $\psi$ as follows:

$$\psi = \int u dy = \int (u_0 + \beta u_1) dy. \tag{30}$$

The dimensionless expression for the pressure rise per wavelength is written as follows:

$$\Lambda p_\lambda = \int_0^1 \frac{dp}{dx} dx. \tag{31}$$

The coefficient of heat transfer at the right wall is given by

$$Z = h_{1x} \theta_y. \tag{32}$$

The non-dimensional shear stress is given by

$$\tau_{yx} = -\tau_0 + \mu(\theta) \frac{du}{dy} = -\tau_0 + (1 - \beta\theta) \frac{du}{dy}. \tag{33}$$

## 4 Results and Discussion

In this section, our aim is to discuss the impacts of the pertinent parameters on velocity, temperature, and concentration fields. Throughout the calculations we employ M = 2, $D_a = 0.5$, $G_r = 50$, $G_c = 50$, x = 0.1, a = 0.4, b = 0.3, d = 1.2, $S_c = 0.5$, $S_r = 0.5$, $\gamma = 5$, $\vartheta = \pi/2$, $\alpha = 0.5$, m = 1, n = 1, $\beta = 0.1$, unless otherwise stated. Figure 1i–iii has been plotted to analyze the influence of heat source parameter ($\alpha$), chemical reaction parameter ($\gamma$), and phase difference ($\vartheta$) on velocity distribution. Figure 1i depicts that fluid velocity is increasing function when $\alpha$ is increased. Physically, an enhancement in $\alpha$ may lead to generate more heat inside the fluid which upsurges the fluid velocity throughout the channel. The influence of $\gamma$ on "u" can be seen in Fig. 1ii. As expected, the fluid velocity decreases by increasing $\gamma$. Figure 1iii shows the variation of geometric parameter phase angle $\vartheta$ on velocity. It illustrates that velocity of the fluid suppresses significantly by changing $\vartheta$. Figure 2 is plotted to show the influence of $G_r$, M, and $D_a$ on pressure gradient distribution. Figure 2i–iii depicts that dp/dx is decreasing function with changing $G_r$ and M whereas increasing $D_a$ lead to enhancement of the pressure gradient significantly. Further, it is noted that increasing $G_r$, M, and $D_a$ pressure gradient is small nearer to the boundary (i.e., x∈[0,0.2] and x∈[0.8,1]), whereas larger amount of pressure gradient is required to promote the flow in narrow part of the channel. Pressure rise against pumping is graphed in Fig. 3. Variation of pressure drop ($\Delta p$) against the time-average flux ($\Theta$)

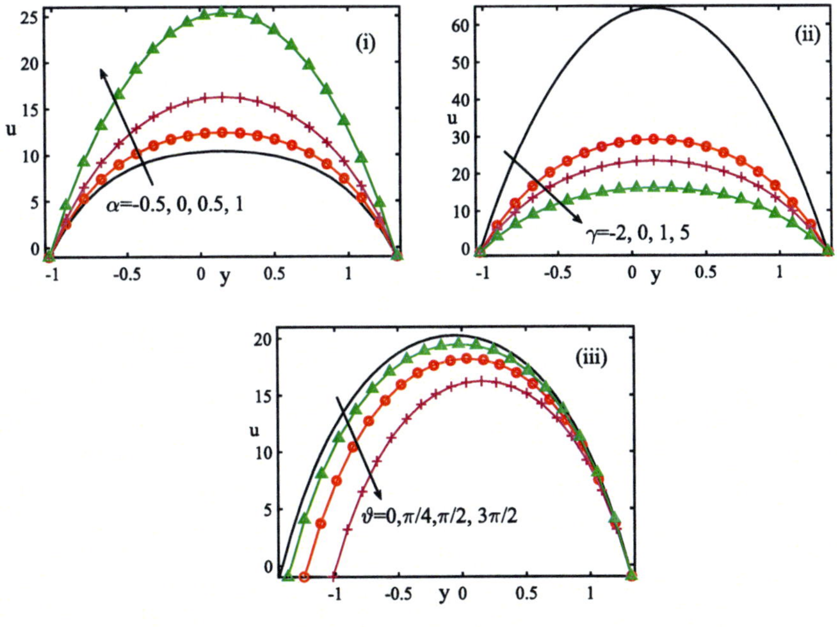


**Fig. 1** Effect of $\alpha$, $\gamma$, and $\vartheta$ on velocity distribution

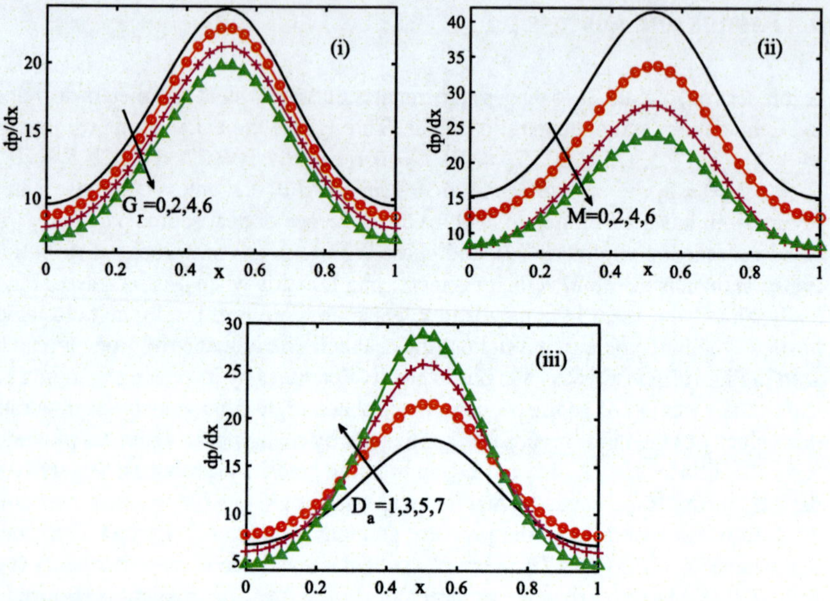


**Fig. 2** Influence of $G_r$, M, and $D_a$ on pressure gradient

with changing $D_a$, M, and $\gamma$ is presented in Fig. 3i–ii. It is observed that increasing $\gamma$ leads to increase in peristaltic pumping region, whereas mixed behavior is noted when $D_a$ and M are increased. It is well known that, in flow analysis shear stress is a very interesting phenomenon. Also, it is an essential characteristic that provides valuable information concerning the nature of dissipation at the boundaries. The effects of "d", $D_a$, and $\tau_0$ at the right wall on shear stress are illustrated in Fig. 4. Figure 4(i) is displayed to see the effect of "d" on $\tau$. It is found that enhancement of "d" leads to decrease of shear stress gradually. The reverse trend can be seen in Fig. 4ii, when "d" is replaced by $D_a$. Behavior of $\tau_0$ on shear stress is displayed in Fig. 4iii. From this figure, it is clear that the impact of this parameter on $\tau$ is quite similar to that of Fig. 4i.

Variation on temperature field with increasing $\alpha$ is shown in Fig. 5. It shows that temperature profiles are linear when $\alpha = 0$ while fluid temperature is enhanced as well as profiles become oscillatory in nature when $\alpha$ is increasing. Physically speaking, increasing heat source ($\alpha > 0$) leads to release of heat energy which causes fluid temperature to increase whereas opposite effect may exist in the presence of heat sink ($\alpha < 0$). Figure 6 serves to show heat transfer coefficient (Z) with different values of $\alpha$, "a", and "d". Influence of $\alpha$ is plotted in Fig. 6i. It shows that profiles are oscillatory in nature which may be due to peristalsis and also mixed behavior (increasing and decreasing) is noted by enhancing the values of $\alpha$. Similar impact can be observed in Fig. 6ii if $\alpha$ is replaced by "a". It is evident from these figures that Z is an increasing function with increasing "d" (see Fig. 6iii). Figure 7i–iii displays

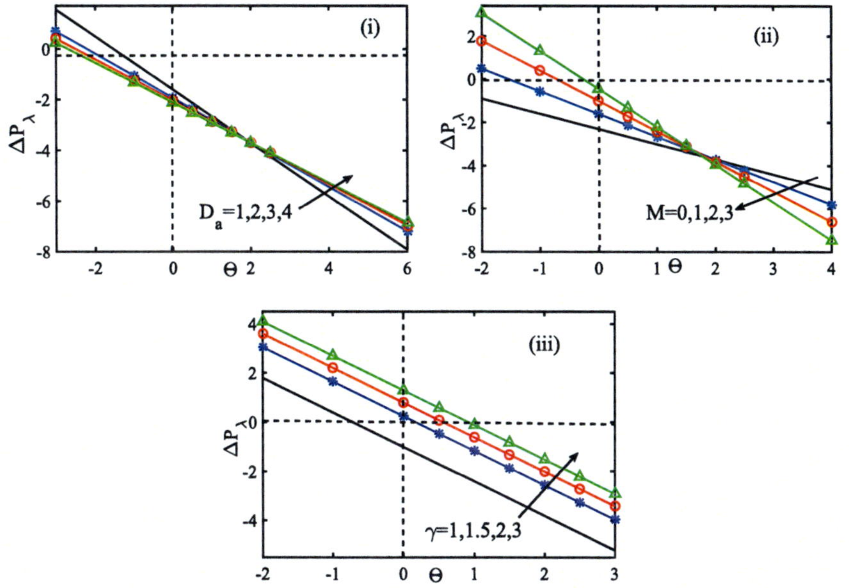


**Fig. 3** Variation on pressure drop with different values of $D_a$, M, and $\gamma$

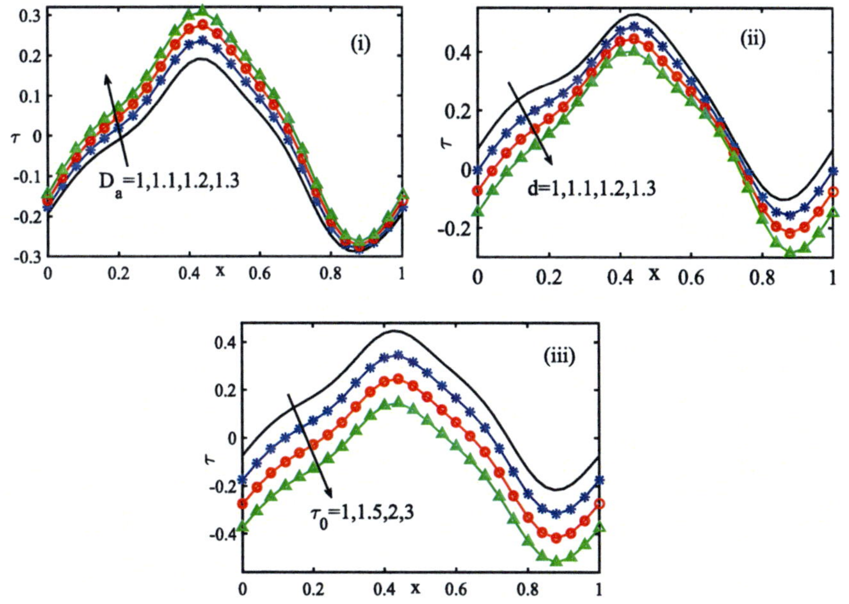


**Fig. 4** Variation on shear stress distribution with different values of $D_a$, d, and $\tau_0$

**Fig. 5** Variation on temperature distribution with changing $\alpha$

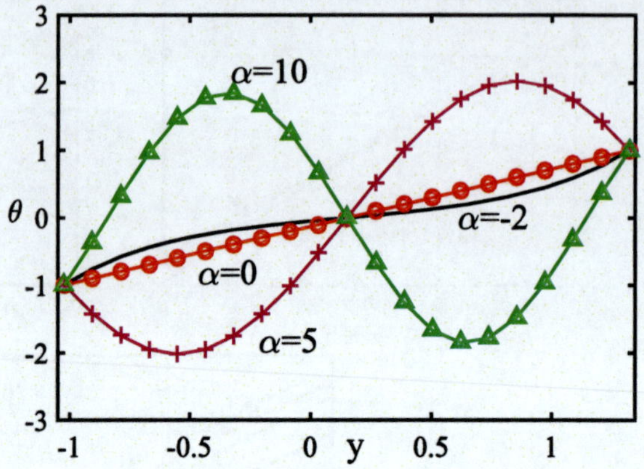


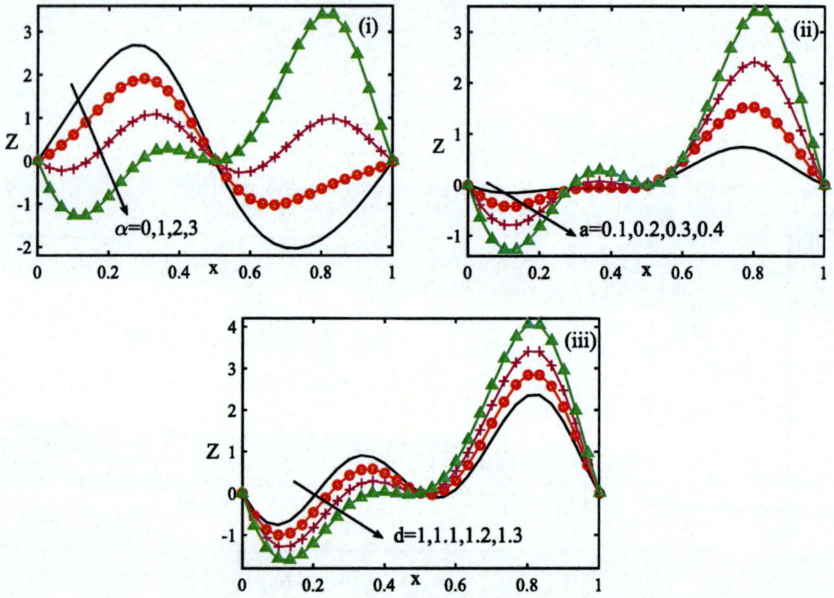


**Fig. 6** Variation on heat transfer coefficient with different values of $\alpha$, a, and d

the behavior of $\alpha$, $\gamma$, and $S_c$ on concentration distribution. From Fig. 7i, we noticed that enhancing $\alpha$ tends to increase fluid concentration in the first half of the channel while opposite is true in the other region. Impact of $\gamma$ on $\phi$ is shown in Fig. 7ii. It is noted that varying $\gamma$ tends to promote concentration of the fluid. The effect of $S_c$ on $\phi$ is examined in Fig. 7iii. It depicts that fluid concentration is increasing gradually with increasing $S_c$.

**Table 1** Nomenclature

| | |
|---|---|
| $a_1, b_1$ | Amplitudes of the waves |
| $B_0$. | Transverse magnetic field |
| c | Wave speed |
| $c_p$ | Specific heat |
| C | Dimensional concentration |
| $C_1, C_2$ | Wall concentrations |
| $\bar{C}$ | Mean value of $C_1$ and $C_2$ |
| $d_1 + d_2$ | Width of the channel |
| $D_a$ | Permeability parameter |
| $D_m$ | Coefficient of mass diffusivity |
| g | Acceleration due to gravity |
| $G_r$ | Grashof number |
| $G_c$ | Local mass Grashof number |
| K | Thermal conductivity |
| $k_1$ | Permeability of the medium |
| k | Dimensional chemical reaction parameter |
| $k_T$ | Thermal diffusion ratio |
| M | Magnetic parameter |
| $\bar{P}$ | Pressure in the fixed frame |
| p | Pressure in the wave frame |
| $P_r$ | Prandtl number |
| Re | Reynolds number |
| $S_c$ | Schmidt number |
| $S_r$ | Soret number |
| T | Dimensional temperature |
| $T_1, T_2$ | Wall temperatures |
| $\bar{T}$ | Mean value of $T_1$ and $T_2$ |
| $\bar{U}, \bar{V}$ | Dimensional velocity components in the fixed frame |
| u, v | Non-dimensional velocity component in the wave frame |
| | Greek symbols |
| $\alpha$ | Heat source/sink parameter |
| $\mu$ | Dynamic viscosity |
| $\rho$ | Density |
| $\beta_t$ | Coefficient of thermal expansion |
| $\beta_c$ | Coefficient of expansion with concentration |
| $\nu$ | Kinematic viscosity |
| $\phi^*$ | Porosity of the medium |
| $\tau_0$ | Yield stress |
| $\lambda$ | Wave length |
| $\sigma$ | Coefficient of electric conductivity |
| $\vartheta$ | Phase difference |
| $\gamma$ | Dimensionless chemical Reaction parameter |

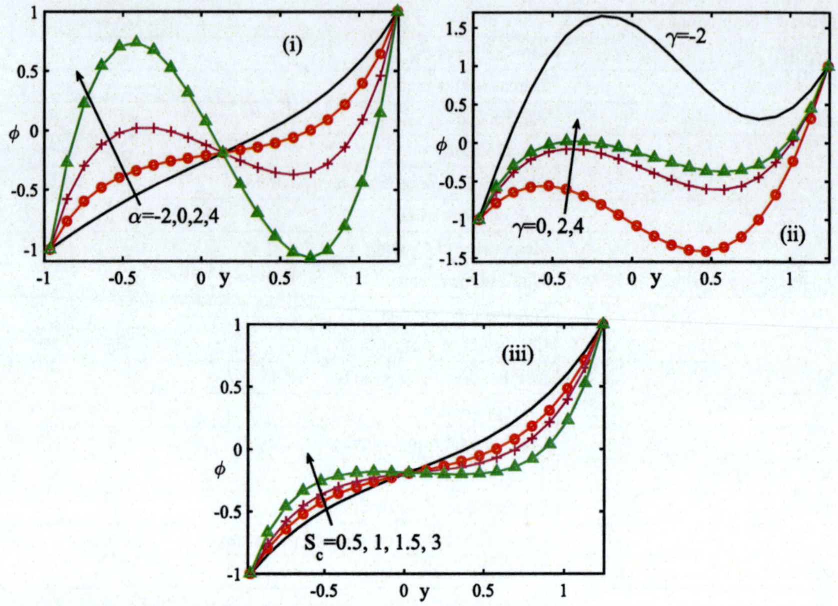


**Fig. 7** Effects of $\alpha$, $\gamma$, and $S_c$ on concentration distribution

## 5 Conclusion

In this article, MHD flow of Bingham fluid induced by peristaltic motion in a vertical asymmetric porous space by taking into account viscosity variation and chemical reaction was analyzed. The suitable approximation is employed to reduce the system of governing partial differential equations into non-linear ordinary differential equations. Analytical expressions for velocity, temperature, and concentration field are constructed. The main observations are listed below:

- Fluid velocity gets enhanced with increasing $\alpha$ while the opposite is true with changing $\gamma$ and $\vartheta$.
- In the absence of heat source temperature profile is linear whereas profiles become oscillatory in nature when increasing $\alpha$.
- Increasing $\alpha$, $S_r$, and $S_c$ lead to enhancement of fluid concentration in one region while the reverse effect is noted in the other region.
- The parameters M and $G_r$ tend to suppress pressure gradient gradually whereas $D_a$ produces the reverse impact.
- Increasing $\gamma$ enhanced pressure rise while mixed behavior can be noticed when $D_a$ and M are increased.
- Permeability parameter leads to enhancement of shear stress whereas mean width of the channel and yield stress produce the opposite effect.
- Heat transfer coefficient profiles become oscillatory in nature with increasing $\alpha$, a, and d.

## References


1. Khalid A, Khan I, Khan A, Shafie S (2015) Unsteady MHD free convection flow of Casson fluid past over an oscillating vertical plate embedded in a porous medium. Eng Sci Technol Eng Sci Technol Int J 18:309–317
2. LourduImmaculate D, Muthuraj R, Srinivas S (2016) MHD unsteady flow of a Williamson nanofluid in a vertical porous space with oscillating wall temperature. Frontiers Heat Mass Transf 12
3. Sayed HM, Aly EH, Vajravelu K (2016) Influence of slip and convective boundary conditions on peristaltic transport of non-Newtonian nanofluids in an inclined asymmetric channel. Alexandria Eng J 55:2209–2220
4. Muthuraj R, LourduImmaculate D, Srinivas S (2017) MHD Couette flow of Powell-Eyring fluid in an inclined porous space in the presence of temperature dependent heat source with chemical reaction. J Porous Media 20:559–575
5. Muthuraj R, Selvi RK (2020) Homotopy solution for MHD mixed convective unsteady flow of a Powell-Eyring fluid in a vertical porous space with consinusoidally fluctuating wall temperature. Int J Comput Sci Math 12:21–35
6. Mekheimer KhS (2002) Peristaltic transport of a couple stress fluid in a uniform and non-uniform channels. Biorheology 39:755–765
7. Srinivas S, Muthuraj R (2010) Peristaltic transport of a Jeffrey fluid under the effect of slip in an inclined asymmetric channel. Int J Appl Mech 2:1–19
8. Narahari M, Sreenadh S (2010) Peristaltic transport of a Bingham fluid in contact with a Newtonian fluid. Int J Appl Math Mech 6:41–54
9. Misra JC, Adhikary SD (2017) Flow of a Bingham fluid in a porous bed under the action of a magnetic field: application to magneto-hemorheology, Engg Sci Tech, an. Int J 20:973–981
10. Fusi L, Farina A (2017) Peristaltic flow of a Bingham fluid in a channel. Int J Non-linear Mech 97:78–88
11. Fusi L, Farina A (2018) Peristaltic axisymmetric flow of a Bingham fluid. Appl Math Comput 320:1–15
12. Roberts TG, Cox SJ (2020) An analytic velocity profile for pressure-driven flow of a Bingham fluid in a curved channel. J Non-Newtonian Fluid Mech 280:104278
13. Kameswaran PK, Narayana M, Sibanda P, Murthy PVSN (2012) Hydromagnetic nanofluid flow due to a stretching or shrinking sheet with viscous dissipation and chemical reaction effects. Int J Heat Mass Transf 55:7587–7595
14. Niranjan H, Sivasankaran S, Bhuvaneswari M (2016) Analytical and numerical study on magnetoconvection stagnation-point flow in a porous medium with chemical reaction. radiation and slip effects, mathematical problems in engineering, pp 1–12
15. Kasmani RM, Sivasankaran S, Bhuvaneswari M, Siri Z (2016) Effect of chemical reaction on convective heat transfer of boundary layer flow in nanofluid over a wedge with heat generation/absorption and suction. J Appl Fluid Mech 9:379–388
16. Eswaramoorthi S, Bhuvaneswari M, Sivasankaran S, Rajan S (2016) Soret and Dufour effects on viscoelastic boundary layer flow over a stretching surface with convective boundary condition with radiation and chemical reaction. Scientia Iranica - Trans B Mech Eng 23:2575–2586
17. Muthuraj R, Nirmala K, Srinivas S (2016) Influences of chemical reaction, wall properties on MHD Peristaltic transport of a Dusty Fluid with Heat and mass transfer. Alexandria Eng J 55:597–611
18. Niranjan H, Sivasankaran S, Bhuvaneswari M (2017) Chemical reaction, soret and dufour effects on MHD mixed convection stagnation point flow with radiation and slip condition. Scientia Iranica - Trans B Mech Eng 24:698–706
19. Sivasankaran S, Niranjan H, Bhuvaneswari M (2017) Chemical reaction, radiation and slip effects on MHD mixed convection stagnation-point flow in a porous medium with convective boundary condition. Int J Numerical Methods for Heat Fluid Flow 27:454–470

20. Bhuvaneswari M, Eswaramoorthi S, Sivasankaran S, Hussein A (2019) Cross-diffusion effects on MHD mixed convection over a stretching surface in a porous medium with chemical reaction and convective condition. Eng Trans 67(2019):3–19
21. Farooq S, Awais M, Naseem M, Hayat T, Ahmad B (2017) Magnetohydrodynamic peristalsis of variable viscosity Jeffrey liquid with heat and mass transfer. Nuclear Eng Technol 49:1396–1404
22. Selvi RK, Muthuraj R (2018) MHD oscillatory flow of a Jeffrey fluid in a vertical porous channel with viscous dissipation. Ain Shams Eng J 9:2503–2516
23. Hassan AR (2019) Thermodynamics analysis of an internal heat generating fluid of a variable viscosity reactive couette flow. J King Saud Univ Sci 31:506–510
24. Ajibade AO, Tafida MK (2020) The combined effect of variable viscosity and variable thermal conductivity on natural convection couette flow. Int J Thermofluids 5:100036

# Mathematical Modeling for Non-linear Wave Interaction of Submerged Body Using Hybrid Element Method

**Prashant Kumar, Prachi Priya, and Rajni**

## 1 Introduction

The study of wave and body dynamics deals with wave-body interactions for both submerged and floating bodies. The study of underwater navigation is important due to practical applications in many fields like submarine design, autonomous underwater vehicle, and also in under water drilling process. Non-linear wave interactions with the submerged body are useful for understanding the mechanism of submarine navigation and underwater activities. The construction of mooring lines of underwater pipe bridges and the precise prediction of forces over the submerged body are of obvious importance and special attention is also needed in terms of stability and protection.

In past years, various numerical methods are developed in which one method is known as high-order spectral method (HOSM) which includes non-linearity terms first established by Dommermuth and Yue [1] for modeling of non-linear gravity wave interactions which further Liu et al. [2] extended for the solution of interactions of non-linear wave with submerged body in time-domain analysis verified the experimental solutions of Miyata et al. [3]. Koo [4] developed a fully non-linear 2D Numerical Wave Tank (NWT) which is the combination of Mixed Eulerian–Lagrangian (MEL) and Boundary Element Method (BEM) and Bai (2014) developed the 3D NWT. A coupled FEM and BEM is developed by Wu et al. [5] to analyze the non-linear wave interactions with submerged bodies. The BEM is used in the

P. Kumar (✉) · P. Priya
Department of Applied Sciences, National Institute of Technology Delhi, Delhi 110040, India
e-mail: prashantkumar@nitdelhi.ac.in

Rajni
Jindal Global Business School, O P Jindal Global University, Sonipat, Haryana, India



S. Srinivas et al. (eds.), *Recent Advances in Applied Mathematics and Applications to the Dynamics of Fluid Flows*, Lecture Notes in Mechanical Engineering,
https://doi.org/10.1007/978-981-19-1929-9_2

region near the body surface while FEM is used in the region away from the body. Kent and Choi [6] studied the non-linear surface wave interaction with submerged body in an infinite depth using pseudo-spectral method and Desingularized Boundary Element Method (DBEM). Liu et al. [7] develop an efficient method for computing the radiation and diffraction in infinite depth of water waves with submerged sphere using multipole method for linear wave theory. A Desingularized Boundary Integral Method (DBIM) is used by Cao et al. [8, 9], Li [10] for the study of interaction of wave-body problems for both submerged and floating bodies. Bin et al. [11] studied the forces of wave on the submerged body in the viscous fluid. Recently, the interaction of solitary wave with moving cylinder using 2-D NWT in computational fluid dynamics (CFD).

In this article, a mathematical model based on BEM and HOSM is developed to calculate forces on submerged body due to the interactions of non-linear wave. The potential function is expressed as sum of spectral and body potential where spectral potential is represented by eigenfunction expansion in Fourier series and the coefficients are obtained by using HOSM and body potential is estimated from 2D BEM. The convergence analysis is discussed, and validation is conducted by comparing with existing study of Ogilvie [12]. In addition, the effect of various amplitudes incident wave at free surface is evaluated and horizontal and vertical drift force computed on submerged circular cylinder for various perturbation orders.

## 2 Mathematical Formulation

### 2.1 *Model Geometry*

The current model's geometry is defined in Fig. 1, the problem considers the interaction of non-linear wave by submerged body ($S_B(t)$) of arbitrary geometry where boundary $S$ consists of free surface boundary ($S_F(t)$), right ($S_R$) and left ($S_L$) pseudo-boundary having length of computational domain which is $2L$ and bottom boundary $S_0$. A global $x$–$z$ Cartesian co-coordinate system is considered. The flow is restricted to non-breaking waves and fluid is considered inviscid, irrotational, and incompressible. So, velocity potential $\Phi$ satisfies the Laplace equation

$$\nabla^2\Phi = 0 \tag{1}$$

The body surface's boundary conditions are defined as

$$\frac{\partial\Phi}{\partial n} = 0 \quad \text{onbottomsurface}(S_0) \tag{2}$$

$$\frac{\partial\Phi}{\partial n} = \vec{U}\cdot\hat{n} \quad \text{onbodysurface}(S_B(t)) \tag{3}$$

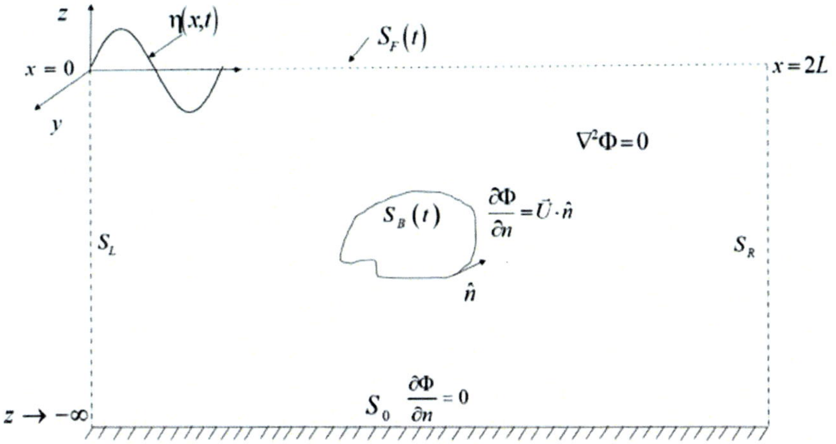


**Fig. 1** The geometry of the model, submerged body surface ($S_B(t)$) surrounded by free surface ($S_F(t)$), right ($S_R$) and left ($S_L$) pseudo-boundary and the bottom boundary ($S_O$)

$$\eta_t = (\nabla\Phi)\cdot(\hat{k}-\nabla\eta) \quad \text{onfreesurface}(S_F(t)) \tag{4}$$

$$-\frac{1}{\rho}(p-p_a) = \Phi_t + \frac{1}{2}\cdot(\nabla\Phi)^2 + gy\text{onz} = \eta \tag{5}$$

$$\nabla\Phi \rightarrow 0\,\text{when}\,z \rightarrow -\infty \tag{6}$$

where $\overrightarrow{U}$ represents velocity of body and $\eta(x, t)$ denotes the free surface. From [13], the potential function at free surface is written as $\Phi^s(x, t) \equiv \Phi(x, \eta, t)$, in terms of $\Phi^s(x, t)$ and the kinematic and dynamic boundary conditions given in Eqs. (4) and (5) are defined as

$$\eta_t(x,t) = \left(1+\eta_x^2(x,t)\right)\Phi_z(x,\eta,t) - \Phi_x^S(x,t)\cdot\eta_x(x,t) \tag{7}$$

$$\Phi_t^S(x,t) = -g\eta - \frac{1}{2}\left(\Phi_x^S(x,t)\right)^2 + \frac{1}{2}\left(1+\eta_x^2(x,t)\right)\Phi_z^2(x,t) \tag{8}$$

The free surface $\eta(x, 0)$ and surface potential $\Phi^s(x, 0)$ initially considering Stokes waves and to update free surface boundary conditions are integrated using conditionally stable fourth-order Runge–Kutta scheme.

## 2.2 High-Order Spectral Method (HOSM)

The velocity potential in HOSM is written in the perturbation series for certain order $M$

$$\Phi^s(x,t) = \sum_{m=1}^{M} \Phi^{(m)}(x,z,t) \tag{9}$$

where () $^{(m)}$ represents the $(kA)^{(m)}$ and $kA$ is wave steepness. Each $\Phi^{(m)}$ is expanded in Taylor series

$$\Phi^{(m)}(x,z,t) = \sum_{j=0}^{M-m} \frac{\eta^j}{j!}\frac{\partial^j}{\partial z^j}\Phi^{(m)}(x,0,t) \tag{10}$$

about z = 0. The radius of convergence of $\Phi$ cannot extend beyond the first singularity in the analytic continuation of $\Phi$ above z = $\eta$, which limits the validity and convergence of Eq. (10). From Eqs. (9) and (10), the order in which boundary conditions occur at mean free surface for unknown $\Phi^{(m)}$ also known as Dirichlet boundary condition is obtained and given as

$$\Phi^{(m)}(x,0,t) = \begin{cases} \Phi^S(x,t), & m = 1 \\ -\sum_{j=1}^{m-1} \frac{\eta^j}{j!}\frac{\partial^j}{\partial z^j}\Phi^{(m-j)}(x,0,t), & m = 2,3,4,\ldots\ldots\ldots\ldots M \end{cases} \tag{11}$$

## 2.3 Boundary Element Method (BEM)

The BEM is defined by using Green's identity formula and is given as

$$\int_{\Omega} \sigma(x_s, z_s, t) G(x, z, x_s, z_s) d\Omega = \Phi(x, z, t) \tag{12}$$

where $\sigma(x_s, y_s, t)$ represents the distribution of source on the surface of the body at time $t$ and $G$ is Green's function which represents periodic source potential

$$G(x, z; x_s, z_s) = \frac{1}{2}\log\left(2\cos h\frac{\pi(x-x_s)}{2L} - 2\cos\frac{\pi(z-z_s)}{2L}\right)$$
$$-\frac{1}{2}\log\left(2\cos h\frac{\pi(x-x_s)}{2L} - 2\cos\frac{\pi(z-2h_0+z_s)}{2L}\right) \tag{13}$$

where $h_0$ is the distance of mean free surface and center of body.

## 2.4 *Solution of the Problem*

The solution includes the combination of HOSM and BIEM, HOSM is used for free surface and on body surface BEM is imposed. The velocity potential is defined as total potential $(\Phi_{tp}(x, z, t))$, combination of body potential $(\Phi_B(x, z, t))$, and spectral potential $(\Phi_{sp}(x, z, t))$ and each potential is defined in perturbation series up to specified order. The spectral potential is defined as the Fourier series and each node coefficient is determined as

$$\Phi_{sp}^{(m)}(x, z, t) = \sum_{n=1}^{N} \Phi_{n_{sp}}^{(m)}(t)(\exp(ik_n x).\exp(\mathrm{i}|\mathrm{k_n}|\mathrm{z})) \tag{14}$$

In non-lifting body, the potential function is represented as sources on surface of body and its negative image is thus contribution of body potential which is zero at mean free surface using Eq. (11),

$$\Phi_{sp}^{(m)}(x, 0, t) = \begin{cases} \Phi_{sp}^{(1)}(x, 0, t) = \Phi^S(x, t) & m = 1, \\ \Phi_{sp}^{(1)}(x, 0, t) = -\sum_{j=1}^{m-1} \frac{\eta^j}{j!}\frac{\partial^j}{\partial y^j}\left(\Phi_{sp}^{(m-j)}(x, 0, t)\right. & m = 2, .M \\ \qquad \left. +\Phi_B^{(m-j)}(x, 0, t)\right) & \end{cases} \tag{15}$$

Since body potential is zero at mean free surface but not vertical derivative, the Fourier series coefficient of spectral potential is calculated from Eq. (15) and now using Eq. (12) in Eq. (3),

$$\frac{\partial}{\partial n}\left[\int_{S_B} \sigma(x_s, z_s)G(x, z; x_s, z_s)dS_\xi\right] = \begin{cases} -\frac{\partial\Phi_{sp}^{(m)}}{\partial n}|_{S_B} + \vec{U}\cdot\hat{n} & m = 1 \\ -\frac{\partial\Phi_{sp}^{(m)}}{\partial n}|_{S_B} & m = 2, 3, .....M \end{cases} \tag{16}$$

Thus, spectral and body potential will be calculated for all perturbation orders $M$, and required terms are known to measure the pressure and force on the surface of body. The kinematic and dynamic equations were used by spectral and body potential to update the free surface shape.

### 2.5 Diffraction Force

The force on the submerged body is defined as

$$F = \iint_{S_B} P n dS \tag{17}$$

where $P$ is the dynamic pressure on the surface of body. The diffraction force on the fully submerged stationary body due to interaction of non-linear waves is computed as

$$F(t) = -\rho \int_{S_B} \left[ \frac{\partial \Phi_{tp}(x, z, t)}{\partial t} + \frac{1}{2} \nabla \Phi_{tp}(x, z, t) \cdot \nabla \Phi_{tp}(x, z, t) \right] \cdot \hat{n} dS \tag{18}$$

where $\frac{\partial \Phi_{tp}}{\partial t}$ computed by temporal difference and $\nabla \Phi_{tp}$, by spatial central differencing.

## 3 Numerical Validation and Convergence Analysis

### 3.1 Numerical Validation

The time-series graph of pressure over the circular cylinder for time t = 0 s to t = 8 s is calculated from present numerical scheme for the length of domain 2L of free surface which is discretized uniformly Nx = 256 and body is discretized into NB = 128 number of segments. The numerical simulation results are validated with the available analytical solution for linear wave theory of [12]. In Fig. 2, the results acquired utilizing the present numerical scheme show good agreement with the previous study conducted by [12].

### 3.2 Convergence Analysis

The stable numerical method needs to be consistent and convergent, the consistency is demonstrated by conservation of volume and conservation of total energy that

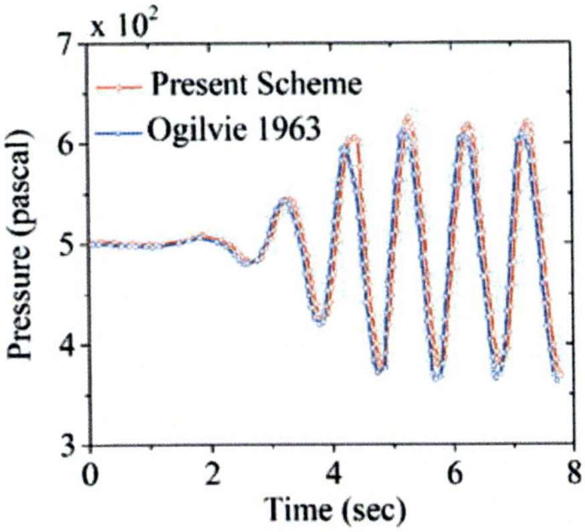


**Fig. 2** Pressure is calculated at the head of the circular cylinder for the present numerical scheme and Ogilvie [14] from time $t = 0\,s$ to $t = 8\,s$

**Table 1** The computation of horizontal drift force $F_h$ for order $M$ with $kA = 0.04$, $H/R = 2$, $kR = 0.4$ and $N_x = 1024$, $ST = 5\,T$

| $N_B = 2$ | $M = 2$ | $M = 3$ | $M = 4$ |
|---|---|---|---|
| 64 | −1.055 | −1.0782 | −1.0769 |
| 128 | −1.0656 | −1.802 | −1.0854 |
| 256 | −1.0698 | −1.0882 | −1.0896 |

requires $\int_{S_F(t)} \eta dx = \text{Constant}$ and $\int_{S_F(t)} P\eta_t dx = -\frac{1}{2}\frac{d}{dt}\left[\int_{S_F(t)} \Phi^S \eta_t dx + \int_{S_F(t)} \eta^2 dx\right]$, respectively where $P$ represents the pressure. Further for convergence analysis, the horizontal drift force ($F_h$) is calculated for increasing segments of body surface ($N_B$) of submerged circular cylinder which is given in Table 1. The horizontal drift force shows convergence for increasing $N_B$ with fixed mean free surface discretization $N_x$, wave steepness (*kA*), wave period (*T*), simulation time (*ST*), and *H/R*, where *A* is the amplitude of incident wave, *k* is wave number, *R* is radius, and *H* denotes depth of the circular cylinder up to three significant digit. The rate of convergence for increasing body segments is better in comparison with linear rate and exponential for increasing number of perturbation order. Similarly, convergence analysis with respect to number of free surface modes $N_x$ with order $M$ and convergence of numerical integration with $\Delta t$ can be obtained.

## 4 Simulation Results

The effect of various incident wave amplitude on free surface elevation in the presence of submerged body is described in Fig. 3 for wave amplitude A = 0.005 m, A = 0.015 m, and A = 0.02 m. The distortion of wave profile behind the submerged circular cylinder is low for small wave amplitude (A = 0.005 m) and became narrower

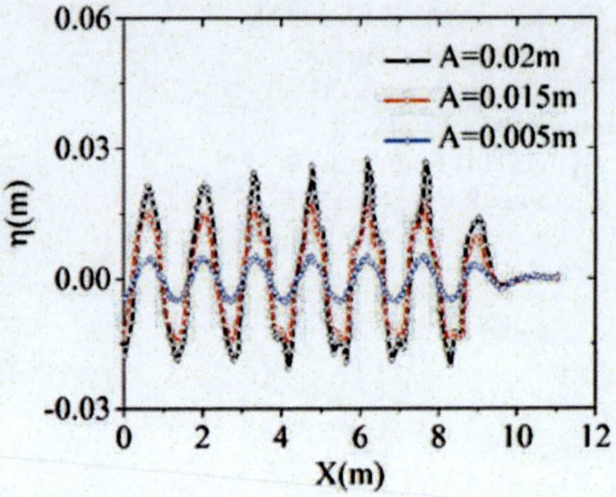


**Fig. 3** The free surface wave elevation at incident wave amplitudes $A = 0.02$ m, $A = 0.015$ m, and $A = 0.005$ m in the presence of circular submerged cylinder

and higher for wave amplitude A = 0.015 m and A = 0.02 m (see Fig. 3). The distortion and non-linearity are significantly noticeable due to high amplitude incident wave due to which higher harmonic components of wave are produced as waves pass over the cylinder. The conservation of energy is verified in all cases.

The drift forces on the submerged circular cylinder are analyzed by the graph for different perturbation orders $M = 2$, 3, and 4 as the function of wave steepness. It is observed in Fig. 4 that the negative horizontal drift force is obtained due to breaking of waves and the variations of drift forces when perturbation order increases (from $M = 2$ to 4) for the small wave steepness ($kA$). It shows significant differences for different perturbation orders. The variation of vertical drift force is constant for small values of $kA$ and shows slight variations for high wave steepness. The horizontal and vertical drift forces on the submerged circular cylinder are induced due to higher order effect.

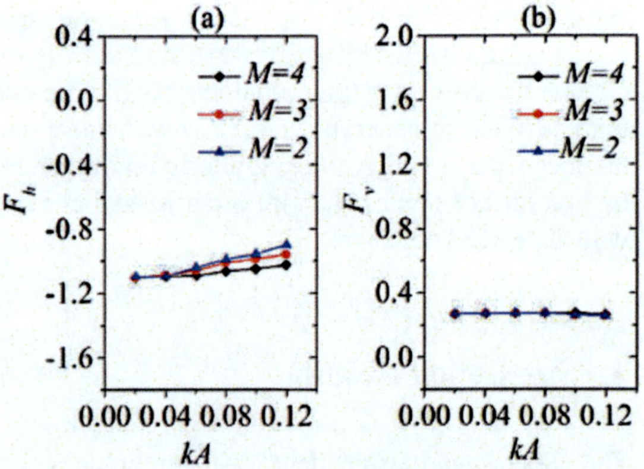


**Fig. 4** Variation of (**a**) horizontal drift force $F_h$ and (**b**) vertical drift force $F_v$ with respect to wave steepness ($kA$) for $M = 2, 3,$ and $4$ with $kR = 0.4$, $N_x = 512$, $N_B = 128$, and $H/R = 2$

## 5 Conclusions

A mathematical method based on the combination of boundary element method and high-order spectral method is developed for the analysis of the fully non-linear interactions of wave with submerged body in time domain. The convergence test considers the method shows better than linear rate for the increasing body segments and the exponential rate for the increase in order of perturbation. The pressure on submerged circular cylinder is validated with previous result of [12].

Further, the method is utilized to study the distortion of free surface profile for various amplitude incident waves to understand the effect of different wave amplitudes on the submerged body. The diffraction forces were demonstrated through vertical and horizontal drift forces on circular submerged cylinder. The negative horizontal drift force is observed as a consequences of decelerated fluid particles which exerts high pressure on the submerged cylinder. The method allows precise calculation of second and higher order forces and it can be extended to more general shape of the body.

**Acknowledgements** This work was supported by Department of Applied Science (Mathematics division), National Institute of Technology, Delhi.

## References

1. Dommermuth DG, Yue DKP (1987) A high-order spectral method for the study of nonlinear gravity waves. J Fluid Mech 184:267–288
2. Liu Y, Dommermuth DG, Yue DKP (1992) A high-order spectral method for nonlinear wave-body interactions. J Fluid Mech 245:115–136
3. Miyata JH, Khalil G, Lee YG, Kanai M (1988) An experimental study of the nonlinear forces on horizontal cylinders. J Kansai Soc Naval Archit 209:11–23
4. Bai W, Hannan MA, Ang KK (2014) Numerical simulation of fully nonlinear wave interaction with submerged structures: Fixed or subjected to constrained motion. J Fluids Struct 49:534–553. https://doi.org/10.1016/j.jfluidstructs.2014.05.011
5. Wu GX, Taylor RE (2003) The coupled finite element and boundary element analysis of nonlinear interactions between waves and bodies. Ocean Eng 30(3):387–400. https://doi.org/10.1016/S0029-8018(02)00037-9
6. Kent CP, Choi W (2007) An explicit formulation for the evolution of nonlinear surface waves interacting with a submerged body. Int J Numer Meth Fluids 55(11):1019–1038. https://doi.org/10.1002/fld.1504
7. Liu Y, Teng B, Cong P, Liu C, Gou Y (2012) Analytical study of wave diffraction and radiation by a submerged sphere in infinite water depth. Ocean Eng 51:129–141
8. Cao Y, Beck RF, Schultz WW (1993) An three-dimensional desingularised boundary integral methods for potential problems. Int J Numer Methods Fluid 12:785–803. https://doi.org/10.1002/fld.1650120807
9. Cao Y, Beck RF (2016) Desingularized boundary integral equations and their applications in wave dynamics and wave—body interaction problems. J Ocean Eng Sci 1(1):11–29. https://doi.org/10.1016/j.joes.2016.01.001
10. Li A (2019) New analytical solutions to water wave radiation by vertical truncated cylinders through multi-term Galerkin method New analytical solutions to water wave radiation by

vertical truncated cylinders through multi-term Galerkin method. Meccanica 54(3):429–450. https://doi.org/10.1007/s11012-019-00964-x
11. Bin T, Hong-fei M, Lin L (2018) Viscous effects on wave forces on a submerged horizontal circular cylinder. Chin Ocean Eng 32(3):245–255
12. Ogilvie TF (1963) First- and second-order forces on a cylinder submerged under a free surface. J Fluid Mech 16(3):451–472. https://doi.org/10.1017/S0022112063000896
13. Zakharov VE (1968) Stability of periodic waves of finite amplitude on the surface of a deep fluid. J Appl Mech Tech Phys 9:19–194
14. Koo W, Kim MH, Tavassoli A (2004) Fully nonlinear wave-body interactions with fully submerged dual cylinders. Int J Offshore Polar Eng 14(3):210–217

# Effect of Activation Energy and Slip Velocity on Convective Heat and Mass Transfer of Chemically Reacting Fluids Under Convective Surface Condition

**Poosappan Yesodha, Marimuthu Bhuvaneswari, Sivanandam Sivasankaran, and Kaliannan Saravanan**

## 1 Introduction

In the field of engineering and in the process of manufacture of materials, the final products' quality depends on many factors. In the existence of chemically reacting fluids, mass and energy transfer phenomena under convective boundary condition along with activation energy is useful to get the ultimate products with characteristics according to the expectations desired. This has motivated the authors of this paper to proceed with their present investigation.

The chemical reaction rate is enhanced when the activation energy is lowered. The effect of energy gained through the activated molecules (activation energy) on chemical reaction has been studied by several researchers. Maleque [1] after his research with fluid flow all through a flat permeable plate concluded that the rate constant of a chemical reaction for an exothermic reaction increases with an increase in temperature profile and velocity profile, but for endothermic reaction, the effects are opposite. Groppi and Spiga [2] presented the molecular behavior in a chemical reaction and transfer of mass and energy. The convective mass transfer in a pipe that was placed vertically in porous media was first considered by Bestman

P. Yesodha
Department of Chemistry, Emerald Heights College for Women, Ooty, Tamilnadu, India

M. Bhuvaneswari
Department of Mathematics, Kongunadu Polytechnic College, D. Gudalur, Dindigul, Tamilnadu, India

S. Sivasankaran (✉)
Department of Mathematics, King Abdulaziz University, Jeddah 21589, Saudi Arabia
e-mail: sd.siva@yahoo.com

K. Saravanan
Department of Chemistry, Thiruvalluvar Government Arts College, Rasipuram, Tamilnadu, India
e-mail: yesodhachemistry@gmail.com

 
S. Srinivas et al. (eds.), *Recent Advances in Applied Mathematics and Applications to the Dynamics of Fluid Flows*, Lecture Notes in Mechanical Engineering,
https://doi.org/10.1007/978-981-19-1929-9_3

[3] by activation energy and chemical reaction. Same way inside a vertical surface Mustafa et al. [4] passed magneto-nanofluid and discussed buoyancy effects. In a frame that is rotating with elastic surface, Shafique et al. [5] experimented with the non-Newtonian Maxwell fluid flow by means of activation energy. In a horizontal channel, Zeeshan et al. [6] showed on Couette-Poiseuille flow the effect of activation energy. When the liquid is rotating, the time-dependent flow with energy of activation was considered through Awad et al. [7]. Trapeznikova et al. [8] have cited the behavior of compressible fluid between the phases where there is a transfer of mass and energy in a permeable medium. Pekman and Tezer-Sezgin [9] studied about viscous fluid, which is incompressible in a permeable medium.

If the fluid is in the form of emulsion, suspension, foam or polymer solution, slip effects may occur along the boundary layer. This has attracted many researchers to work on the flow of fluid under slip condition. The non-Newtonian fluid flow above a plate at a position of stagnation with the slip effect was made clear by Labropulu and Li [10]. They suggested that the fluid encroaches diagonally or at an angle on the wall. Similarly, the convection flow with the condition of slip velocity was interpreted by Bhattacharyya et al. [11] and Harris et al. [12] over a perpendicular surface in a medium that is permeable. The MHD flow of an uncompressible fluid along with slip effects in a sheet that can be stretched or shrinked linearly was analyzed by Aman et al. [13] and Seini and Makinde [14]. A sheet that can be stretched or shrinked vertically with convection transfer of heat and steady flow using slip velocity was explored by Rosca and Pop [15]. Singh and Chamka [16] using second-order slip condition, observed flow of fluid and transport of heat near the sheet, which can be shrinked vertically and linearly. It was noticed by them that skin friction force diminishes on slip parameter increase. Niranjan et al. [17] numerically acquired the solution of a fluid along with slip condition for MHD convection flow for Dufour and Soret effects. Nandeppanavar et al. [18] found that on increasing the magnetic effect and slip parameters, increases the thickness of the layer of thermal boundary. The fluid flow along a plate that is moving with the slip effect, force of buoyancy and convective heating was investigated by Singh and Makinde [19]. Inside a carbon nanotube containing the permeable wall, Chan et al. [20] considered the behavior of fluid flow along with slip boundary state. Along with the influence of different factors, the effect of slip condition was studied by researchers Shateyl and Mabood [21], Abbas et al. [22] and Akbar and Khan [23].

During manufacturing processes in industries, conceded with the flow of fluid involves the generation of heat. This is identified as internal heat generation. This internal heat generation led to changes in the temperature distribution, forced convection, phase change process, heat transfer process and surface heat flux. The flow of thermo-micro polar fluid in the existence of generation of heat along a porous plate that was placed vertical was elucidated by Rahman et al. [24]. In the presence of perpendicular plate, flow of fluid in a permeable medium with radiation, magnetic field and internal heat generation was observed by Makinde [25]. In a sheet that was stretched linearly, Makinde and Sibanda [26] viewed the flow of fluid with internal heat generation in the boundary layer. In literature, the study of generation or absorption of heat with reactive fluid flow in MHD boundary layer has been shown by Salem and Abd El-Aziz [27], Patil and Kulkarni [28].

Inspired by the above works, the intention in this current study is to examine slip velocity, internal heat generation under convective boundary condition with energy of activation in a fluid that can undergo chemical reaction and flow in a frame with porous space that is in rotation.

## 2 Mathematical Model

Figure 1 illustrates the model of the present work. The representation describes a stretched sheet in a medium that is porous with boundary layer of three-dimensional *x*, *y* and *z* coordinate system. The velocities along these axes are *u*, *v* and *w*, respectively. Here, we consider the fluid to flow on the sheet only along the *x* axis with velocity $u = ax + L\frac{\partial u}{\partial z}$ and therefore along the *y* and *z* direction, the velocities *v* and *w* are zero. In addition, the fluid is subjected to rotation with angular velocity Ω. Influence of activation energy, slip velocity, internal heat generation effects has been carried out.

The flow is induced by a stretching level surface, which coincides as z ≥ **0** by means of the plane. With the above mentioned assumptions considering the velocity V = [*u(x, y, z), v(x, y, z), w(x, y, z)*], temperature T = T*(x, y, z)* and concentration C = C*(x, y, z)*, the governing equations can be written as (Makinde et al. [29]):

$$\frac{\partial u}{\partial x} + \frac{\partial v}{\partial y} + \frac{\partial w}{\partial z} = 0 \tag{1}$$

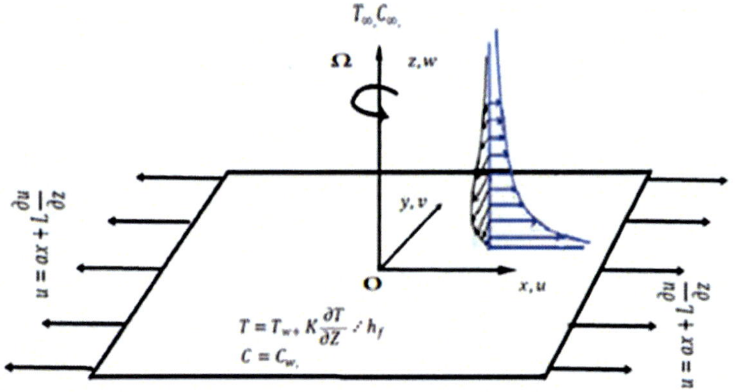


**Fig. 1** Physical model

$$u\frac{\partial u}{\partial x}+v\frac{\partial u}{\partial y}+w\frac{\partial u}{\partial z}-2\Omega v=\nu\frac{\partial^2 u}{\partial z^2}-\frac{\nu}{K}u-Fu^2 \tag{2}$$

$$u\frac{\partial v}{\partial x}+v\frac{\partial v}{\partial y}+w\frac{\partial v}{\partial z}+2\Omega u=\nu\frac{\partial^2 v}{\partial z^2}-\frac{\nu}{K}v-Fv^2 \tag{3}$$

$$u\frac{\partial T}{\partial x}+v\frac{\partial T}{\partial y}+w\frac{\partial T}{\partial z}=\alpha\frac{\partial^2 T}{\partial z^2}+\frac{Q}{\rho c_p}(T-T_\infty) \tag{4}$$

$$u\frac{\partial C}{\partial x}+v\frac{\partial C}{\partial y}+w\frac{\partial C}{\partial z}=D\frac{\partial^2 C}{\partial z^2}-k_r^2\left(\frac{T}{T_\infty}\right)^n e^{\frac{-E_a}{kT}}(C-C_\infty) \tag{5}$$

Here, u, $v$ and $w$ specify the velocity components. $F=\left(\frac{C_b}{K^{1/2}}\right)$ express non-uniform inertia co-efficient.$\nu=\left(\frac{\mu}{\rho}\right)$ represent kinematic viscosity, $\Omega$ symbolize angular velocity, $\rho$ stand for density, $D$ denotes the solute diffusibility and $K$ is porous medium permeability, $\alpha=\left(\frac{k}{\rho C_p}\right)$ indicates thermal diffusivity and $k$ thermal conductivity, $C_b$ implies drag co-efficient and $c_p$ specifies the specific heat. Modified Arrhenius function is characterized by the term $k_r^2\left(\frac{T}{T_\infty}\right)^n e^{\frac{-E_a}{\kappa T}}(C-C_\infty)$, $E_a$ means activation energy and $k_r^2$ is the rate constant of chemical reaction, exponent fitted rate constant is $n$ that typically lie in range $1<n<1$ and $\kappa=8.61\times10^{-5}eV/K$ is the Boltzmann constant.

The boundary conditions are:
$u=ax+L\frac{\partial u}{\partial z}$, $v=0$, $w=0$, $-k\frac{\partial T}{\partial z}=h_f(T_W-T)$, $C=C_w$, when $z=0$

$$u\to 0,\ v\to 0,\ C\to C_\infty,\ T\to T_\infty,\ as\ z\to\infty \tag{6}$$

Here, $T_w$ and $T_\infty$ are constant temperature and ambient temperature, respectively. $C_w$ and $C_\infty$ are surface concentration and ambient concentration, and $a$ is the stretching rate (> 0).

The following non-dimensional variables are introduced

$$\eta=z\sqrt{\frac{a}{\nu}},\ u=axf'(\eta),\ v=axg(\eta),\ w=-(a\nu)^{\frac{1}{2}}f(\eta),$$
$$\theta(\eta)=\frac{T-T_\infty}{T_W-T_\infty},\ \varphi(\eta)=\frac{C-C_\infty}{C_W-C_\infty} \tag{7}$$

Continuity Eq. (1) is automatically satisfied. Equations (2) to (6) become, in view of Eq. (7)

$$f'''+ff''-\lambda f'+2\beta g-(1+F_r)f'^2=0 \tag{8}$$

$$g''+fg'-f'g-2\beta f'-\lambda g-F_r g^2=0 \tag{9}$$

$$\left(\frac{1}{\Pr}\right)\theta'' + f\theta' + Hg\theta = 0 \tag{10}$$

$$\varphi'' + Scf\varphi' - Sc\sigma[1+\delta\theta]^n \exp\left[-\frac{E}{1+\delta\theta}\right]\varphi = 0 \tag{11}$$

With conditions

$$f' = 1 + df'',\ f' = 1,\ g = 0,\ \theta' = -Bi(1-\theta),\ \phi = 1\ at\ \eta = 0 \tag{12}$$

$$f' \to 0, g \to 0, \theta \to 0, \phi \to 0 \text{ as } \eta \to \infty.$$

where $\lambda = \frac{\nu}{Ka}$ represents porosity parameter, $\beta = \frac{\Omega}{a}$ rotational parameter, $Fr = \left(\frac{C_b}{K^{1/2}}\right)$ inertia coefficient, $\Pr = \frac{\nu}{\alpha}$ Prandtl number, $d = L\sqrt{\frac{a}{\nu}}$ slip parameter, $Hg = \frac{Q}{a\rho c_p}$ heat generation parameter, $\sigma = \frac{k^2}{a}$ reaction rate, $Sc = \frac{\nu}{D}$ Schmidt number and $Bi = -\frac{h_f}{k}\sqrt{\frac{\nu}{a}}$ Biot number.

The Nusselt number (*Nu*) is specified by the transference of heat in a fluid flowing through a porous rotating frame as a result of convection. In conditions of Sherwood number (*Sh*) and skin friction coefficient (*Cf*) are defined by the force and mass transport along the surface as follows:

$$Nu_x = \frac{xq_w}{k(T_w - T_\infty)},\ Sh_x = \frac{xj_w}{D(C_w - C_\infty)},\ C_f = \frac{T_w}{\rho U_w^2} \tag{13}$$

with

$$T_w = \left(\frac{\partial u}{\partial z}\right),\ q_w = -k_T\frac{\partial T}{\partial z}|_{z=0},\ j_w = -D\frac{\partial C}{\partial z}|_{z=0} \tag{14}$$

Finally, we have:

$$C_f\sqrt{Re_x} = f''(0),\ \frac{Nu_x}{\sqrt{Re_x}} = -\theta'(0),\ \frac{Sh_x}{\sqrt{Re_x}} = -\phi'(0), \tag{15}$$

Here, $Re_x = \frac{ax^2}{\upsilon}$ show local Reynolds number. For the governing model, the equations are non-linear; therefore, we have to seek a numerical solution. Hence, Runge–Kutta method is used to solve the equations along with shooting technique.

## 3 Results and Discussion

In this section of study, an examination of the influence of fluid velocity f'(η), fluid concentration ϕ(η) and fluid temperature θ(η) has been done. The results have been analyzed for various parameters as slip velocity (*d*), porosity parameter (λ), rotational

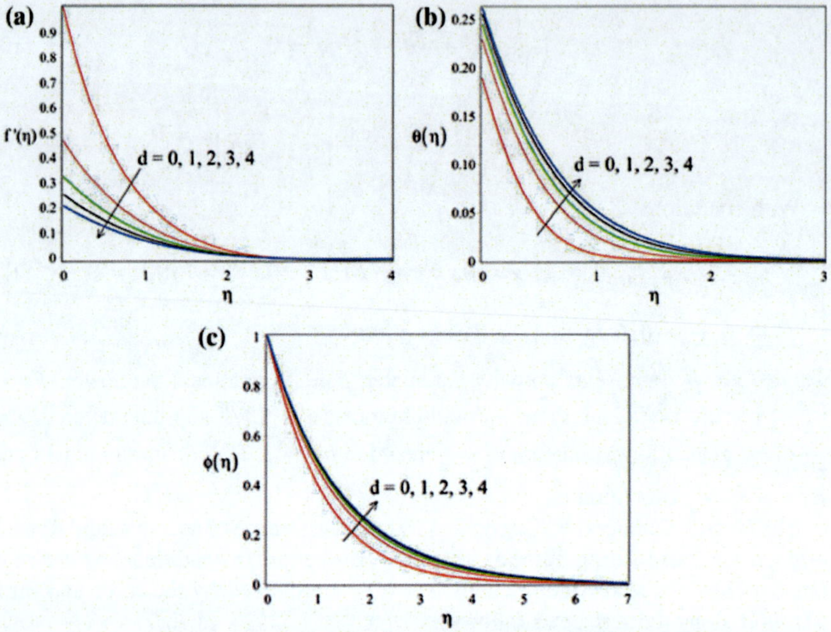


**Fig. 2** Plot of **a** velocity, **b** temperature and **c** concentration for various values of d with Pr = 7, Sc = 1, λ = 0.2, E = 1,n = 0.5, Fr = 1, σ =1, β = 0.5, Bi = 0.5,Hg = −0.2, δ = 1

parameter ($\beta$), temperature difference parameter ($\delta$), activation energy ($E$), inertia coefficient ($Fr$), Biot Number ($Bi$), heat generation parameter ($Hg$), reaction rate ($\sigma$), constant for fitted rate ($n$).

Figure 2a is plotted to analyze the profile of fluid velocity along the boundary layer for $Pr = 7$, $Sc = 1$, $\lambda = 0.2$, $E = 1$, $n = 0.5$, $Fr = 1$, $\sigma = 1$, $\beta = 0.5$, $Bi = 0.5$, $Hg = 0.2$, $\delta = 1$ and slip parameter with different values. The velocity drops down with the raise in slip parameter. Figure 2b,c is the representation of the effect of slip parameter over temperature and concentration, respectively. According to the result shown by the two graphs, both the profiles increase with slip parameter raise. Figure 3a–c indicates the graphical representation for a range of values of λ. In Fig. 3a, for the gradual increment in values of parameter λ, we see a reduction in fluid velocity, this is due to the resistance in flow of fluid. The outcome of temperature behavior for the increase in λ value is revealed in Fig. 3b. Fluid flows slowly because of increase in viscosity and therefore linking is higher and extra heat is produced and θ(η) enhances simultaneously. The effect of λ on fluid concentration is seen in Fig. 3c. The distribution of concentration increases with increase in λ.

In Fig. 4, we observe the graph of rotational parameter $\beta$ on profile of velocity, concentration and temperature. Figure 4a is a representation of $\beta$ on velocity. The graph reveals that flow of the fluid will reduce for addition in the value of $\beta$. The

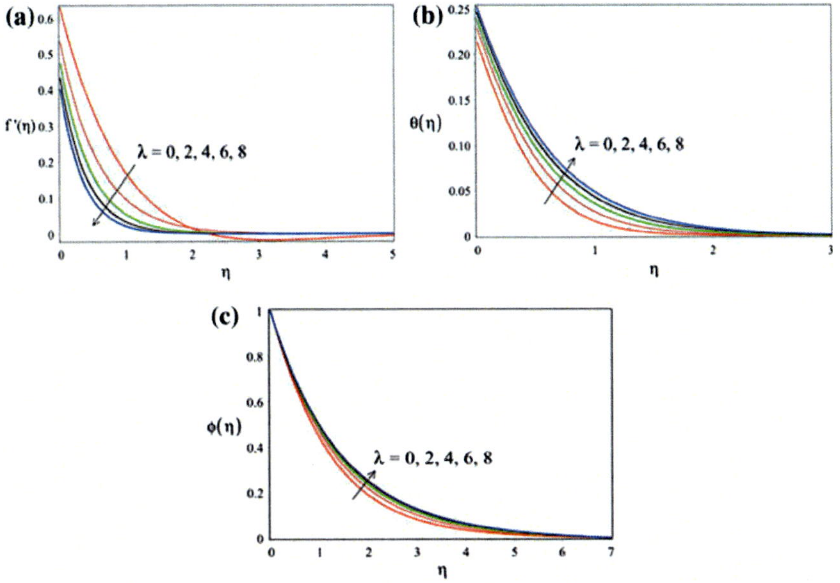


**Fig. 3** Plot of **a** velocity, **b** temperature and **c** concentration for various values of λ with Pr = 7, Sc = 1, $\beta = 0.5$, E = 1, n = 0.5, Fr = 1, $\sigma = 1$, d = 0.5, Bi = 0.5, Hg = −0.2, $\delta = 1$

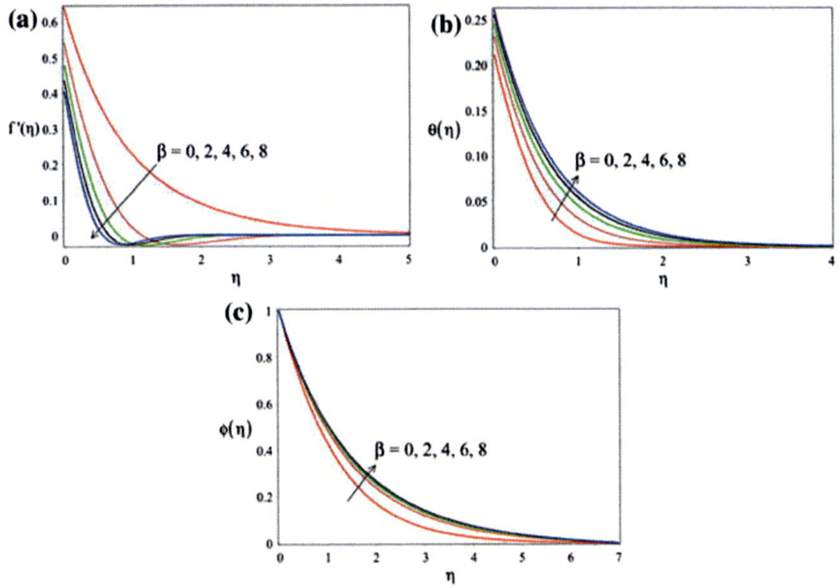


**Fig. 4** Plot of **a** velocity, **b** temperature and **c** concentration plots for various values of $\beta$ with Pr = 7, Sc = 1, λ = 0.2, E = 1, n = 0.5, Fr = 1, $\sigma = 1$, d = 0.5, Bi = 0.5, Hg = −0.2, $\delta = 1$

reduction in the velocity of the fluid is due to the reduction in the rate of stretching of the rotating frame. Figure 4b,c display the influence of $\beta$ on the profile of $\theta(\eta)$ and $\phi(\eta)$. The impact of Fig. 4b is that temperature of fluid and layer of the thermal boundary raises with value increment in $\beta$. Similarly, the survey from Fig. 4c tells the fact that increasing value of $\beta$, the $\phi(\eta)$ is elevated. Figure 5 reveals the decrease caused on fluid velocity for raise in the value of inertia coefficient *Fr*. We observed that fluid velocity is a decreasing function for slip parameter ($d = 0, 1, 2, 3, 4$), porosity parameter ($\lambda = 0, 2, 4, 6, 8$), rotational parameter ($\beta = 0, 2, 4, 6, 8$), and inertia coefficient ($Fr = 0, 1, 2, 3, 4$). We have to come to a conclusion that to all the above said four parameters, fluid velocity is inversely proportional.

The impact of Biot number variation on profile of temperature is revealed in Fig. 6. The elevation in the value of *Bi* brought about raise in the fluid temperature, this is due to the fact that inside a body, the heat transfer resistance is higher. Figure 7 displays the outcome of heat generation parameter, *Hg*, on temperature profile. It is found, fluid temperature increases with raise of *Hg*. This shows that boundary layer

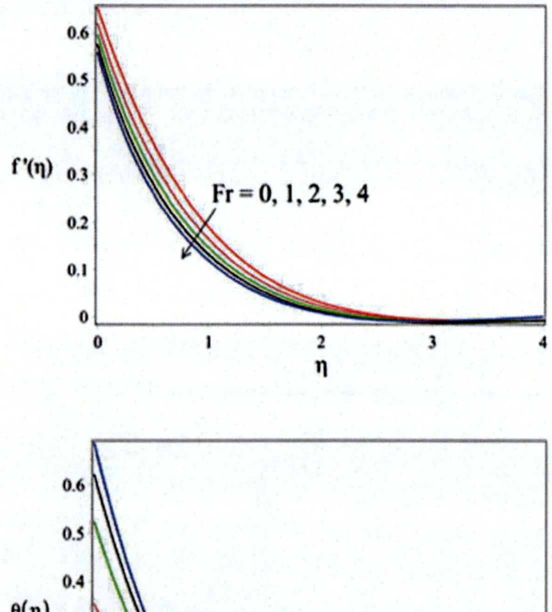


**Fig. 5** Velocity plot for various values of Fr with $Pr = 7$, $Sc = 1$, $\lambda = 0.2$, $E = 1$, $n = 0.5$, $d = 0.5$, $\sigma = 1$, $\beta = 0.5$, $Bi = 0.5$, $Hg = -0.2$, $\delta = 1$

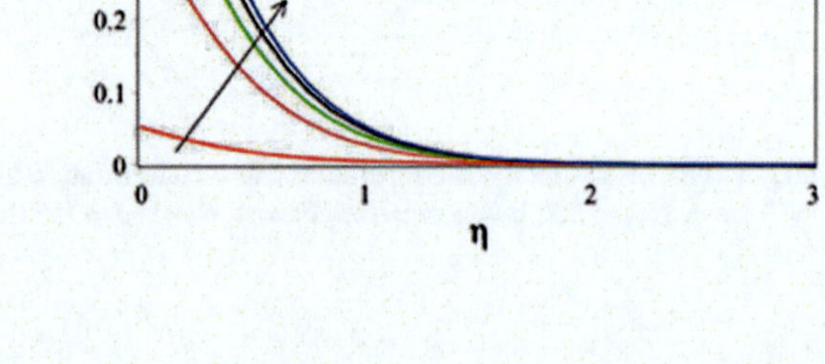


**Fig. 6** Temperature plot for different values of Bi with $Pr = 7$, $Sc = 1$, $\lambda = 0.2$, $E = 1$, $n = 0.5$, $d = 0.5$, $\sigma = 1$, $\beta = 0.5$, $Fr = 1.0$, $Hg = -0.2$, $\delta = 1$

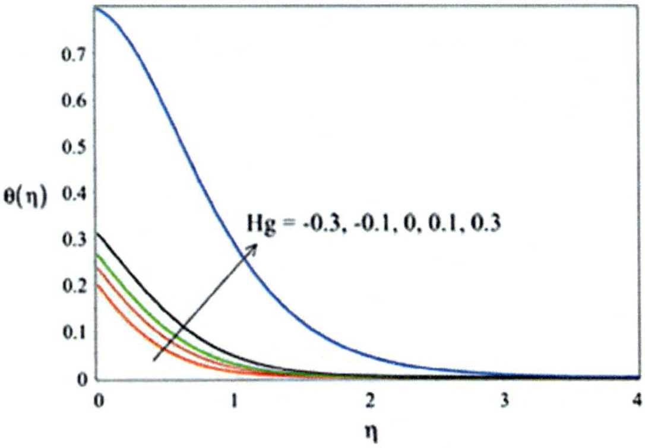


**Fig. 7** Temperature plot for different values of Hg with Pr = 7, Sc = 1, λ = 0.2, E = 1, n = 0.5, d = 0.5, $\sigma$ =1, $\beta$ = 0.5, Bi = 0.5, Fr = 1.0, $\delta$ = 1

becomes thicker by increase of heat generation and therefore the temperature of fluid rises and the rate of the heat transport increases.

A chemical reaction can take place at a specified rate owing to the collision of atoms or molecules but it depends on the factor of the highly energized intermediate activated molecules decomposition. The impact of energy of activation, *E,* on the profile of concentration is shown in Fig. 8. We are revealed that when *E* raises $\phi(\eta)$ also increases. For chemical reaction to be initiated, the enhancement of the fluid concentration is essential and throughout the flow of the fluid, which stimulates the chemical reaction rate ultimately.

The effect of $\sigma$, $\delta$ and $n$ on concentration of fluid is analyzed by Figs. 9, 10 and 11, respectively. The enhancement in the values of all the three parameters causes a decrease in concentration profile for $\sigma$, $n$ and $\delta$. According to Fig. 9, when $\sigma$ is increased, due to the speed up of chemical reaction of the fluid, the property of the fluid changes and as a result $\phi(\eta)$ reduces. From Fig. 10, the impact of enlarging the fitted rate constant values shows a decline in the concentration of the fluid as a result of decrease in stretching rate. For dissimilar values of $\delta$, the concentration of fluid

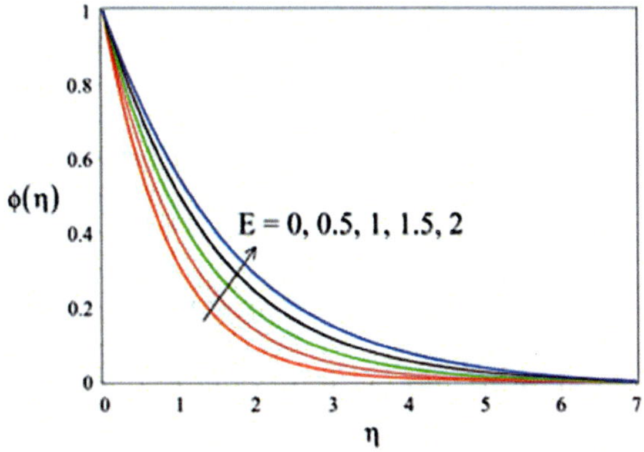


**Fig. 8** Concentration plot for different values of E with Pr = 7, Sc = 1, λ = 0.2, Fr = 1, n = 0.5, d = 0.5, $\sigma$ =1, $\beta$ = 0.5, Bi = 0.5, Hg = − 0.2, $\delta$ = 1

**Fig. 9** Concentration plot for different values of $\sigma$ with Pr = 7, Sc = 1, $\lambda = 0.2$, E = 1, n = 0.5, d = 0.5, Fr = 1, $\beta = 0.5$, Bi = 0.5, Hg = −0.2, $\delta = 1$

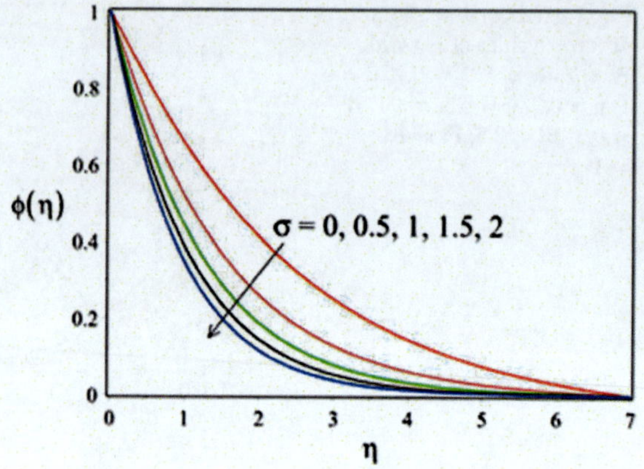


**Fig. 10** Concentration plot for different values of n with Pr = 1, Sc = 1, $\lambda = 0.2$, E = 1, Fr = 1, d = 0.5, $\sigma = 1$, $\beta = 0.5$, Bi = 0.5, Hg = −0.2, $\delta = 1$

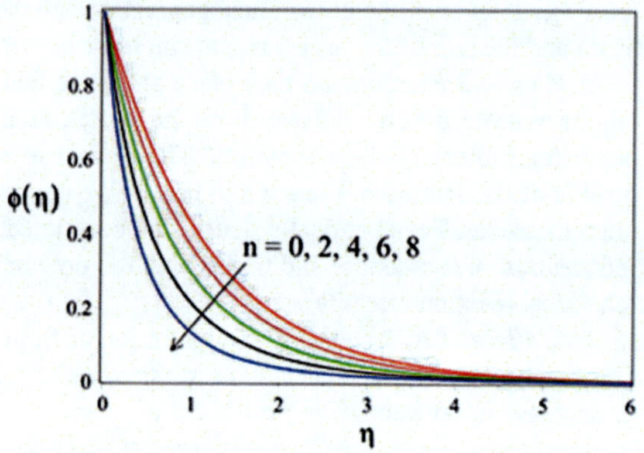


**Fig. 11** Concentration plot for different values of $\delta$ with Pr = 1, Sc = 1, $\lambda$=0.2, E = 1, n = 0.5, d = 0.5, $\sigma$=1, $\beta = 0.5$, Bi = 0.5, Hg = −0.2, Fr = 1

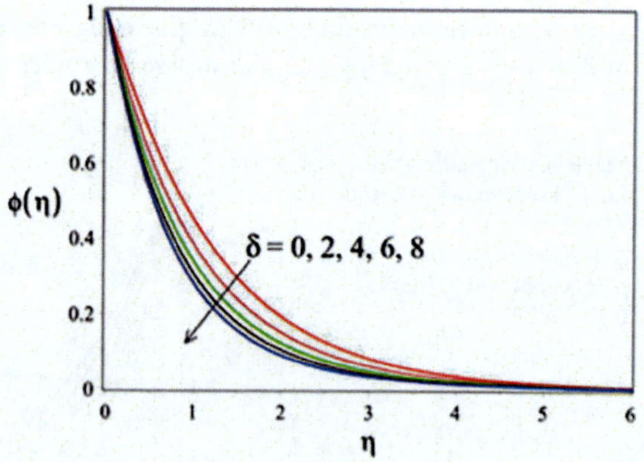

**Table 1** For different d, Fr, λ, $\beta$, the values of $-f''(0)$, $-g''(0)$, $-\theta'(0)$, $-\phi'(0)$ with Pr = 7, Sc = 1, E = 1.0, n = 0.5, Hg = -–0.2, $\sigma = 1$, $\delta = 1$ and Bi = 0.5

| β | λ | Fr | d | $-f''(0)$ | $-g''(0)$ | $-\theta'(0)$ | $-\phi'(0)$ |
|---|---|---|---|---|---|---|---|
| 0.5 | 0.2 | 0 | 0.5 | 0.696910 | 0.338881 | 0.394560 | 0.782882 |
| | | 1 | | 0.761053 | 0.318544 | 0.392345 | 0.770889 |
| | | 2 | | 0.810335 | 0.304055 | 0.390560 | 0.762019 |
| | | 3 | | 0.850433 | 0.293812 | 0.389049 | 0.756831 |
| | | 4 | | 0.884120 | 0.288282 | 0.387743 | 0.756559 |
| 0.5 | 0 | 1 | 0.5 | 0.740382 | 0.352203 | 0.393057 | 0.773719 |
| | 2 | | | 0.927499 | 0.168051 | 0.385606 | 0.740884 |
| | 4 | | | 1.048062 | 0.112634 | 0.380096 | 0.721871 |
| | 6 | | | 1.130544 | 0.085812 | 0.376207 | 0.711265 |
| | 8 | | | 1.191928 | 0.069775 | 0.373325 | 0.704559 |
| 0 | 0.2 | 1 | 0.5 | 0.819210 | 0.498652 | 0.389320 | 0.750736 |
| 1 | | | | 0.985666 | 0.832017 | 0.379927 | 0.714840 |
| 3 | | | | 1.088397 | 0.988372 | 0.374216 | 0.701930 |
| 5 | | | | 1.160748 | 1.087280 | 0.370477 | 0.695279 |
| 7 | | | | 1.215787 | 1.158124 | 0.367841 | 0.691192 |
| 0.5 | 0.2 | 1 | 0 | 1.445389 | 0.464183 | 0.405539 | 0.836592 |
| | | | 1 | 0.531214 | 0.252368 | 0.385276 | 0.745016 |
| | | | 2 | 0.337228 | 0.183733 | 0.377109 | 0.72109 |
| | | | 3 | 0.248970 | 0.146532 | 0.372312 | 0.709433 |
| | | | 4 | 0.197931 | 0.122582 | 0.369092 | 0.702431 |

is plotted as graph in Fig. 11. When the values of $\delta$ are increased, $\phi(\eta)$ showed a diminishing trend.

The values of $-f''(0)$, $-g''(0)$, $-\theta''(0)$ and $-\phi''(0)$ have been tabulated in Table 1, for dissimilar values of *Fr*, λ, $\beta$ and *d*. The impact of the different parameters on Nusselt number $Nu_x$(heat transfer rate), skin friction coefficient $C_f$ and Sherwood number $Sh_x$(mass transfer rate) has been shown. The skin friction shows an increase for the parameters *Fr*, λ and $\beta$ value increase and decrease for d value. The raise in the value of the parameters *Fr*, λ, $\beta$ and *d* shows a decrease in Nusselt number and Sherwood number.

From Table 2, we understand that with the increase in Biot number, the $Nu_x$ and $Sh_x$ increase. The behavior of heat generation parameter, *S*, differs. The $Nu_x$ decreases and $Sh_x$ increases, for the increase in *S* value. The $Sh_x$ values are tabulated in Table 3 for different values of the parameters *E*, $\sigma$, *N* and $\delta$ with fixed values such as $Sc = 1$, $\beta = 0.5$, $Fr = 1$, $\lambda = 0.2$, $Hg = 0.2$, d = 0.5, $Bi = 0.5$. The increase in activation energy decreases $-\phi'(0)$ but for $\sigma$, *N*, $\delta$ parameters it is a rising function.

**Table 2** For different Bi, S, the values of $-\theta'(0)$, $-\phi'(0)$ with Pr = 7, Sc = 1, d = 0.5, $\beta = 0.5$, Fr = 1, E = 1, $\delta = 1$, $\sigma = 1$, n = 0.5, $\lambda = 0.2$

| Bi | S | -θ'(0) | $-\phi'(0)$ |
|---|---|---|---|
| 0.1 | −0.2 | 0.094797 | 0.746864 |
| 1 | | 0.645672 | 0.790749 |
| 2 | | 0.953494 | 0.814120 |
| 3 | | 1.133648 | 0.827411 |
| 4 | | 1.251918 | 0.835985 |
| 0.5 | −0.3 | 0.400227 | 0.768941 |
| | 0.1 | 0.381976 | 0.777346 |
| | 0 | 0.367420 | 0.784333 |
| | 0.1 | 0.344562 | 0.795645 |
| | 0.3 | 0.102281 | 0.916072 |

**Table 3** The values of $-\phi'(0)$ for different Pr, $\delta$, $\sigma$, E, n with Sc = 1, $\beta = 0.5$, Fr = 1, $\lambda = 0.2$, Hg = −0.2, d = 0.5, Bi = 0.5

| Pr | $\delta$ | E | n | $\sigma$ | $-\phi'(0)$ |
|---|---|---|---|---|---|
| 7 | 1 | 0 | 0.5 | 1 | 1.121952 |
| | | 0.5 | | | 0.925229 |
| | | 1.0 | | | 0.770889 |
| | | 1.5 | | | 0.652092 |
| | | 2 | | | 0.563040 |
| 7 | 1 | 1 | 0.5 | 0 | 0.367710 |
| | | | | 0.5 | 0.605584 |
| | | | | 1.0 | 0.770889 |
| | | | | 1.5 | 0.905641 |
| | | | | 2 | 1.022640 |
| 1 | 1 | 1 | 0 | 1 | 0.806179 |
| | | | 2 | | 1.024866 |
| | | | 4 | | 1.375098 |
| | | | 6 | | 1.929882 |
| | | | 8 | | 2.793135 |
| 1 | 0 | 1 | 0.5 | 1 | 0.739063 |
| | 2 | | | | 0.946760 |
| | 4 | | | | 1.098606 |
| | 6 | | | | 1.216199 |
| | 8 | | | | 1.311939 |

## 4 Conclusions

In the current study, activation energy and slip velocity effects took place in permeable medium by internal heat generation in a stretched sheet along with rotating flow of chemically reacting fluid. The study concludes with the main outcome as below:

- Enhance in slip parameter value diminishes the velocity of fluid and uplifts the temperature and concentration of fluid.
- Influence of $\lambda$, $\beta$ and $Fr$ on fluid velocity is qualitatively alike.
- The fluid temperature is increased for both $Bi$ and $Hg$.
- Increment in activation energy shows a raise in fluid concentration.
- Behavior of $\beta$, $d$ and $\lambda$ on concentration are contradictory to $\sigma$, $\delta$ and $n$.
- Increment of $Fr$, $\lambda$ and $\beta$ presents diminish in transfer rate of heat and mass, which is evident from tabulation values.
- Development in activation energy reduces rate of mass transfer.
- Increase in slip parameter reveals the decline in rate of heat transfer, skin friction and rate of mass transfer.

## References


1. Maleque K (2013) Effects of exothermic/endothermic chemical reactions with Arrhenius activation energy on MHD free convection and mass transfer flow in presence of thermal radiation. J Thermodyn 11. Article ID 692516
2. Groppi M, Spiga G (1999) Kinetic approach to chemical reactions and inelastic transitions in a rarefied gas. J Math Chem 26:197–219
3. Bestman AR (1990) Natural convection boundary layer with suction and mass transfer in a porous medium. Int J Energ Res 14:389–396
4. Mustafaa M, Hayat T, Obaidat S (2013) Boundary layer flow of a nanofluid over an exponentially stretching sheet with convective boundary conditions. Int J Numer Method H 23:945–959
5. Shafique Z, Mustafa M, Mushtaq A (2016) Boundary layer flow of Maxwell fluid in rotating frame with binary chemical reaction and activation energy. Results Phys 6:627–633
6. Zeeshan A, Shehzad N, Ellahi R (2018) Analysis of activation energy in Couette-Poiseuille flow of nanofluid in the presence of chemical reaction and convective boundary conditions. Results Phys 8:502–512
7. Awad FG, Motsa S, Khumalo M (2014) Heat and mass transfer in unsteady rotating fluid flow with binary chemical reaction and activation energy. PLoS ONE 9(9):107622
8. Trapeznikova M, Churbanova N, Lyupa A (2020) CMMSE 2019: an explicit algorithm for the simulation of non isothermal multiphase multicomponent flow in a porous medium. J Math Chem 58:595–611
9. Pekmen B, Tezer-Sezqin M (2015) DRBEM solution of natural convective heat transfer with a non-Darcy model in a porous medium. J Math Chem 53:911–924
10. Labropulu F, Li D (2008) Stagnation-point flow of a second grade fluid with slip. Int J Nonlin Mech 43(9):941–947
11. Bhattacharyya K, Mukhopadhyay S, Layek GC (2011) Slip effects on boundary layer stagnation-point flow and heat transfer towards a shrinking sheet. Int J Heat Mass Transf 54(1–3):308–313

12. Harris SD, Ingham DB, Pop I (2009) Mixed convection boundary-layer flow near the stagnation point on a vertical surface in a porous medium: Brinkman model with slip. Transp Porous Media 77(2):267–285
13. Aman F, Ishak A, Pop I (2013) Magnetohydrodynamic stagnation-point flow towards a stretching/shrinking sheet with slip effects. Int Commun Heat Mass Transf 47:68–72
14. Seini IY, Makinde DO (2014) Boundary layer flow near stagnation-points on a vertical surface with slip in the presence of transverse magnetic field. Int J Numer Method H 24(3):643–653
15. Rosca AV, Pop I (2013) Flow and heat transfer over a vertical permeable stretching/shrinking sheet with a second order slip. Int J Heat Mass Transf 60(1):355–364
16. Singh G, Chamkha AJ (2013) Dual solutions for second-order slip flow and heat transfer on a vertical permeable shrinking sheet. Ain Shams Eng J 4(4):911–917
17. Niranjan H, Sivasankaranan S, Bhuvaneswari M (2017) Chemical reaction, Soret and Dufour effects on MHD mixed convection stagnation point with radiation and slip condition. Sci Iran 24(2):698–706
18. Nandeppanavar MM, Vajravelu K, Subhas Abel M, Siddalingappa MN (2013) MHD flow and heat transfer over a stretching surface with variable thermal conductivity and partial slip. Meccanica 48(6):1451–1464
19. Singh G, Makinde OD (2015) Mixed convection slip flow with temperature jump along a moving plate in presence of free stream. Therm Sci 19(1):119–128
20. Chan Y, Wu B, Ren Y, Xia L (2018) Analytical solution for Newtonian flow inside carbon nanotube taking into consideration vander waals forces. J Math Chem 56:158–169
21. Shateyi S, Mabood F (2017) MHD mixed convection slip flow near a stagnation-point on a non-linearly vertical stretching sheet in the presence of viscous dissipation. Therm Sci 21(6):2731–2745
22. Abbas Z, Wang Y, Hayat T, Oberlack M (2009) Slip effects and heat transfer analysis in a viscous fluid over an oscillatory stretching surface. Int J Numer Meth Fluids 59(4):443–458
23. Akbar NS, Khan ZH (2014) Heat transfer analysis of the peristaltic instinct of bi viscosity fluid with the impact of thermal and velocity slips. Int Commun Heat Mass Transf 58:193–199
24. Rahman MM, Eltayeb IA, Rahman SM (2009) Thermo-micropolar fluid flow along a vertical permeable plate with uniform surface heat flux in the presence of heat generation. Int J Thermal Sci 13(1):23–26
25. Makinde OD (2012) Heat and mass transfer by MHD mixed convection stagnation point flow toward a vertical plate embedded in a highly porous medium with radiation and internal heat generation. Meccanica 47(5):1173–1184
26. Makinde OD, Sibanda P (2011) Effects of chemical reaction on boundary layer flow past a vertical stretching surface in the presence of internal heat generation. Int J Numer Method H 21(6):779–792
27. Salem AM, Abd El-Aziz M (2008) Effect of Hall currents and chemical reaction on hydro magnetic flow of a stretching vertical surface with internal heat generation/absorption. Appl Math Model 32(7):1236–1254
28. Patil PM, Kulkarni PS (2008) Effects of chemical reaction on free convective flow of a polar fluid through a porous medium in the presence of internal heat generation. Int J Therm Sci 47(8):1043–1054
29. Makinde OD, Olanrewaju PO, Charles WM (2011) Unsteady convection with chemical reaction and radiative heat transfer past aflat porous plate moving through abinary mixture. Africka Mathematika 22(1):65–78

# Mixed Convective Flow of a Radiative Maxwell Liquid Past a Porous Riga Plate with Heat Consumption and Dual Stratifications

**S. Eswaramoorthi and S. Sivasankaran**

## 1 Introduction

The utility of non-Newtonian fluids is mounting because of their incredible needs in many industries like petroleum drilling, food production, optical fibers, metallic plates cooling, cosmetics products and many others. These liquids have the non-linear relationship between shear rate and shear stress, and each liquid has distinctive constitutive equation. One of these types of liquid is ML, and this illustrates the stress relaxation. The unsteady flow of ML on a SS with heat sink/source inspected by Swati [1]. They proved that the liquid speed upgrades with more presence of Maxwell parameter. Shafiq and Khalique [2] scrutinized the MHD flow of ML with heat generation/absorption. They discovered that the HT gradient upturns with escalating the Maxwell parameter. Heat generating/absorbing Maxwell nanoliquid flow over a heated SS was explained by Ramesh and Gireesha [3]. They proved that the MT gradient is low for VL compared to ML. Nadeem et al. [4] inspected the HT phenomenon of MHD ML on a porous plate. Their outcomes show that the LNN boosted up with rising the ML parameter. The problem of MHD flow of Maxwell nanoliquid towards a SS was numerically solved by Nadeem et al. [5]. Their results clearly show that the LT falling off due to the more presence of ML parameter. Few paramount investigation for this area is found in Refs. [6–11].

Radiative flow comes under the variations of plate temperature and ambient temperature is large. In plentiful industrial processes, radiation leads to altered the thickness of thermal boundary layer, such as gas turbines, missiles technology, crystal growing, nuclear reactors, etc. Kumar et al. [12] numerically addressed the thermally radiative pulsatile flow of Casson nanofluid on a porous surface with heat consumption/generation. Muthuraj et al. [13] demonstrated the radiative flow of two

S. Eswaramoorthi (✉)
Department of Mathematics, Dr. N.G.P. Arts and Science College, Coimbatore, Tamil Nadu, India
e-mail: eswaran.bharathiar@gmail.com

S. Sivasankaran
Department of Mathematics, King Abdulaziz University, Jeddah, Saudi Arabia
e-mail: sd.siva@yahoo.com

 
S. Srinivas et al. (eds.), *Recent Advances in Applied Mathematics and Applications to the Dynamics of Fluid Flows*, Lecture Notes in Mechanical Engineering,
https://doi.org/10.1007/978-981-19-1929-9_4

immiscible liquids on a porous space. They obtained that the LT enhances with raising the radiation parameter. The problem of pulsating flow of Casson liquid with radiation and heat sink/source was studied by Srinivas et al. [14]. Recent developments of this effect under various situations are seen in Refs. [15–20].

The major reason of this communication is to examine the consequences of ML flow along a porous Riga plate with radiation, chemical reaction and internal heat consumption. Because Riga plate is used to control the plate friction and the pressure drag of submarines, which averting the boundary layer separation and chemical reaction are used in many technological and industrial processes, see Kasmani et al. [21].

## 2 Mathematical Formulation

We take 2D ML flow along a porous RP. Let $\hat{T}_\infty$ and $\hat{C}_\infty$ be the ambient temperature and concentration, which is lesser than the LT $\hat{T}$ and LC $\hat{C}$, respectively. The HT phenomena are measured under the presence of radiation and solutal stratification. Under these assumption, the governing mathematical model is expressed as

$$\hat{u}_{\hat{x}} + \hat{v}_{\hat{y}} = 0, \tag{1}$$

$$\hat{u}\hat{u}_{\hat{x}} + \hat{v}\hat{u}_{\hat{y}} = \nu\hat{u}_{\hat{y}\hat{y}} - k_0\left[\hat{u}^2\hat{u}_{\hat{x}\hat{x}} + \hat{v}^2\hat{u}_{\hat{y}\hat{y}} + 2\hat{u}\hat{v}\hat{u}_{\hat{x}\hat{y}}\right]$$
$$+g[\beta_{\hat{T}}(\hat{T} - \hat{T_\infty}) + \beta_{\hat{C}}(\hat{C} - \hat{C_\infty})] - \frac{\hat{\nu}}{k_1}\hat{u} + \frac{\pi J_0 M_0}{8\rho} exp\left(-\frac{\pi}{a_1}\hat{y}\right), \tag{2}$$

$$\hat{u}\hat{T}_{\hat{x}} + \hat{v}\hat{T}_{\hat{y}} = \alpha\hat{T}_{\hat{y}\hat{y}} + \frac{1}{\rho C_p}\frac{16\sigma^*\hat{T}_\infty^3}{3k^*}\hat{T}_{\hat{y}\hat{y}} + \frac{\hat{Q}}{\rho C_p}\left(\hat{T} - \hat{T}_\infty\right), \tag{3}$$

$$\hat{u}\hat{C}_{\hat{x}} + \hat{v}\hat{C}_{\hat{y}} = D_B\hat{C}_{\hat{y}\hat{y}} - k_2(\hat{C} - \hat{C}_\infty), \tag{4}$$

where $\hat{u}$ & $\hat{v}$—velocity in $\hat{x}$ & $\hat{y}$ directions, $\nu$—kinematic viscosity, $k_0$—relaxation time of the UCM fluid, $J_0$—applied current density of the electrodes, $M_0$—magnetization of the permanent magnets, $\rho$—fluid density, $a_1$—width of magnets and electrodes, $k_1$—permeability of porous medium $\hat{T}$—fluid temperature, $\alpha$—thermal diffusivity, $C_p$—specific heat, $\sigma^*$—Stefan Boltzmann constant, $k^*$—mean absorption coefficient, $\hat{Q}$—heat generation/absorption coefficient, $D_B$—mass diffusivity and $k_2$—coefficient of chemical reaction.

The associated conditions are

$$\hat{u} = U_w(\hat{x}),\ \hat{v} = V_w(\hat{x}),\ \hat{T} = \hat{T}_w = \hat{T}_0 + b\hat{x}, \hat{C} = \hat{C}_w = \hat{C}_0 + c\hat{x}\ \ at\ \hat{y} = 0,$$
$$\hat{u} \to 0, \hat{u}_{\hat{y}} \to 0,\ \hat{T} \to \hat{T}_\infty = \hat{T}_0 + d\hat{x},\ \hat{C} \to \hat{C}_\infty = \hat{C}_0 + e\hat{x},\ \ as\ \hat{y} \to \infty \tag{5}$$

Define:

$$\psi = \hat{y}\sqrt{\frac{a}{\nu}},\quad \hat{u} = \hat{x}aF'(\psi),\quad \hat{v} = -\sqrt{\nu a}F(\psi),\ \Theta(\psi) = \frac{\hat{T}-\hat{T}_{\infty}}{\hat{T}_w-\hat{T}_{\infty}},\ \Phi(\psi) = \frac{\hat{C}-\hat{C}_{\infty}}{\hat{C}_w-\hat{C}_{\infty}}. \tag{6}$$

Substituting Eq. (6) in Eqs. (2)–(4), we have

$$F'''(\psi) - F'^2(\psi) + F(\psi)F''(\psi) + \lambda\left[2F(\psi)F'(\psi)F'''(\psi) - F^2(\psi)F'''(\psi)\right]$$
$$-\gamma F'(\psi) + Ri[\Theta(\psi) + \Lambda\Phi(\psi)] + Qe^{-\beta\psi} = 0 \tag{7}$$

$$\frac{1}{Pr}\left(1+\frac{4}{3}R\right)\Theta''(\psi) + F(\psi)\Theta'(\psi) - F'(\psi)\Theta(\psi) - S_1F'(\psi) + Hg\Theta(\psi) = 0 \tag{8}$$

$$\frac{1}{Sc}\Phi''(\psi) + F(\psi)\Phi'(\psi) - F'(\psi)\Phi(\psi) - S_2F'(\psi) - Cr\Phi(\psi) = 0 \tag{9}$$

where $\lambda$—Maxwell parameter, $\gamma$—porosity parameter, $Ri$—Richardson number, $\Lambda$—buoyancy ratio parameter, $Q$—modified Hartmann number, $Pr$—Prandtl number, $R$—radiation parameter, $S_1$—thermal stratification parameter, $Hg$—heat consumption/generation parameter, $Sc$—Schemidt number, $S_2$—solutal stratification parameter, $Cr$—chemical reaction parameter, $fw$—injection/suction parameter.

The interacted conditions are

$$F(\psi) = Fw,\ \ F'(\psi) = 1,\ \ \Theta'(\psi) = 1 - S_1,\ \Phi(\psi) = 1 - S_2,\ \text{at}\ \psi = 0$$
$$F'(\psi) = 0,\ \ \Theta(\psi) = 0,\ \ \Phi(\psi) = 0\ \text{as}\ \psi \to \infty. \tag{10}$$

The dimensional form of SFC, LNN and LSN is

$$c_F = \frac{\tau_w}{\rho U_w^2/2},\quad Nu = \frac{\hat{x}(q_w)}{k(\hat{T}_w - \hat{T}_{\infty})}\quad \text{and}\ Sh = \frac{\hat{x}j_w}{D_e(\hat{C}_w - \hat{C}_{\infty})},$$

The wall shear stress, heat flux and mass flux are

$$\tau_w = \mu\left[\hat{u}_{\hat{y}}\right]_{\hat{y}=0},\quad q_w = -\left[k\hat{T}_{\hat{y}} + \frac{4\sigma^*}{3k^*}\hat{T}^4{}_{\hat{y}}\right]_{\hat{y}=0}\quad \text{and}\ j_w = -k\left[\hat{C}_{\hat{y}}\right]_{\hat{y}=0}$$

The non-dimensional form of the SFC, LNN and LSN

$$\frac{1}{2}Cf\sqrt{Re} = (1+\lambda)F''(0),\quad Nu/\sqrt{Re} = -\left(1+\frac{4}{3}R\right)\Theta'(0),\ \ \text{and}\ Sh/\sqrt{Re} = -\Phi'(0).$$

## 3 HAM Solutions

The obtained equations (7)–(9) with (10) are solved with HAM technique. Let $F_0(\psi) = Fw + 1 - e^{-\psi}$, $\Theta_0(\psi) = (1 - S_1)e^{-\psi}$ and $\Phi_0(\psi) = (1 - S_2)e^{-\psi}$ are the initial approximations. The linear operators are $L_F = \frac{d^3F}{d\psi^3} - \frac{dF}{d\psi}$, $L_\Theta = \frac{d^2\Theta}{d\psi^2} - \Theta$ and $L_\Phi = \frac{d^2\Phi}{d\psi^2} - \Phi$ with $L_F\left[A_1 + A_2e^{\psi} + \frac{A_3}{e^{\psi}}\right] = L_\Theta\left[A_4e^{\psi} + \frac{A_5}{e^{\psi}}\right] = L_\Phi\left[A_6e^{\psi} + \frac{A_7}{e^{\psi}}\right] = 0$ where $A_j (j = 1 - 7)$ are constants. After executing the M$^{\text{th}}$ order HAM equations, we have the upcoming form, $F_M(\psi) = F_M^o(\psi) + A_1 + A_2Ee^{\psi} + \frac{A_3}{e^{\psi}}$; $\Theta_M(\psi) = \Theta_M^o(\psi) + A_4e^{\psi} + \frac{A_5}{e^{\psi}}$; $\Phi_M(\psi) = \Phi_M^o(\psi) + A_6e^{\psi} + \frac{A_7}{e^{\psi}}$ where $F_M^o(\psi)$, $\Theta_M^o(\psi)$ and $\Phi_M^o(\psi)$ are the particular solutions.

These solutions encompass with $h_F$, $h_\Theta$ and $h_\Phi$ parameters, and these parameters are responsible for solution convergency, see [23–25]. The range values are $-1.2 \leq h_F \leq -0.3$, and $-1.3 \leq h_\Theta$ & $h_\Phi \leq -0.3$. We fix $h$ value for all part of $\psi$ is $-0.8$, see Fig. 1. Table 1 indicates the HAM order, and it is seen that 18th is agreeable. Table 2 exhibits the comparison of our results with Abel and Tawade [22] and found in excellent agreement. Table 3 provides the abbreviations of our study.

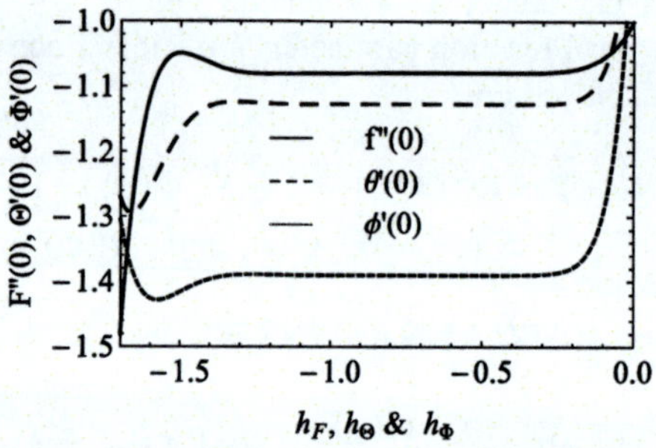


**Fig. 1** The *h*-curves

**Table 1** HAM order

| Order | $-F''(0)$ | $-\Theta(0)$ | $-\Phi(0)$ | Order | $-F''(0)$ | $-\Theta(0)$ | $-\Phi(0)$ |
|---|---|---|---|---|---|---|---|
| 1 | 1.04760 | 1.08267 | 1.29600 | 20 | 1.07944 | 1.12705 | 1.39015 |
| 5 | 1.08109 | 1.12649 | 1.38903 | 25 | 1.07944 | 1.12705 | 1.39015 |
| 10 | 1.07956 | 1.12717 | 1.39015 | 30 | 1.07944 | 1.12705 | 1.39015 |
| 15 | 1.07943 | 1.12707 | 1.39017 | 35 | 1.07944 | 1.12705 | 1.39015 |
| 18 | 1.07944 | 1.12705 | 1.39015 | 40 | 1.07944 | 1.12705 | 1.39015 |

**Table 2** Comparison of $-F''(0)$ for different values of $\lambda$ with Abel and Tawade [22]

| $\lambda$ | Present Study | Abel and Tawade [22] | $\lambda$ | Present Study | Abel and Tawade [22] |
|---|---|---|---|---|---|
| 0.0 | 1.0000 | 0.9999 | 1.2 | 1.2854 | 1.2853 |
| 0.4 | 1.1019 | 1.1019 | 16 | 1.3688 | 1.3686 |
| 0.8 | 1.1967 | 1.1967 | 2.0 | 1.4475 | 14476 |

**Table 3** Abbreviations

| | | | |
|---|---|---|---|
| HAM | Homotopy analysis method | ML | Maxwell liquid |
| HT | Heat transfer | ODE | Ordinary differential equations |
| LC | Liquid concentration | PDE | Partial differential equations |
| LNN | Local Nuselt number | RP | Riga plate |
| LSN | Local Sherwood number | SFC | Skin friction coefficient |
| LT | Liquid temperature | SS | Stretching sheet/surface |
| LV | Liquid velocity | VL | Viscous liquid |
| MHN | Modified Hartmann number | | |

# 4 Results and Discussions

In this segment, we inspect the changes of LF, LT, LC, SFC, LNN and LSN for variegate values of distinct physical parameters. Figure 2a, b gives the significance of $fw$, and $Q$ on LV profile. We achieved that the liquid speed becomes low for more presence of $fw$ and $Q$. The MHN generates a resistive force and this force control the liquid motion. The LT profile for variegate values of $Hg$ and $R$ and was shown in Fig. 3a, b. We found that the LT enriches due to the presence of $Hg$ and $R$. Physically, the energy transport rate of the liquid upturns due to the availability of radiation parameters, and this causes to developing the LT. Figure 4a, b portrays the

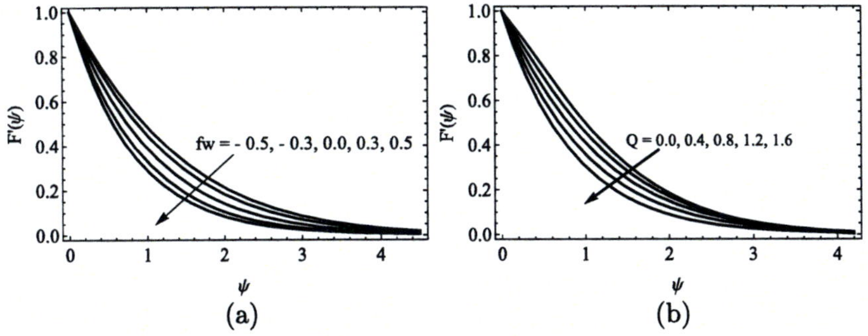


**Fig. 2** Velocity changes for variegate magnitude of $fw$ (**a**) and $Q$ (**b**)

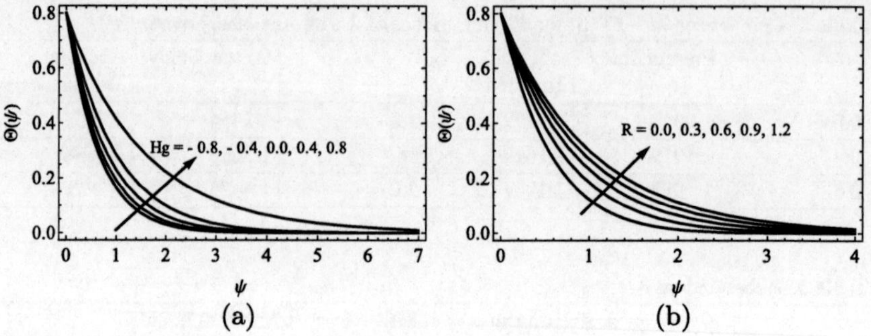


**Fig. 3** Temperature changes for variegate magnitude of $Hg$(**a**) and $R$ (**b**)

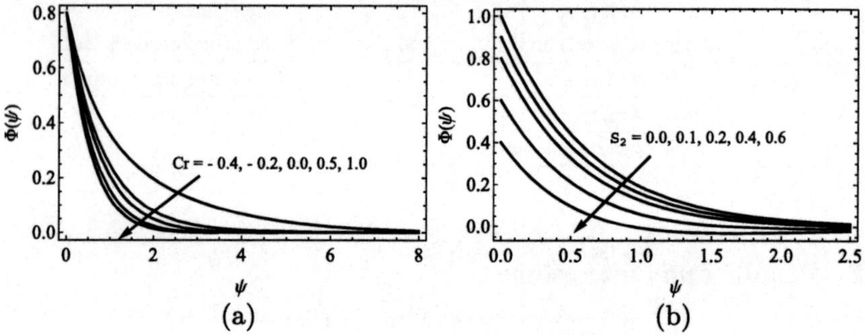


**Fig. 4** Concentration changes for variegate magnitude of $Cr$ (**a**) and $S_2$ (**b**)

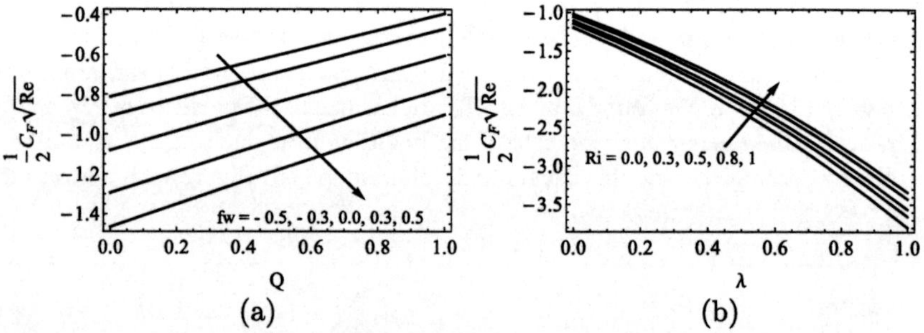


**Fig. 5** Skin friction coefficient for variegate magnitude of $\lambda$, $fw$, $Q$ and $Ri$

prominence of $Cr$ and $S_2$. We revealed that the LC is a non-elaborating function of $Cr$ and $S_2$.

The SFC for variegate values of $\lambda$, $fw$, $Q$, and $Ri$ is shown in Fig. 5a, b. We proved that the surface drag force enriches for enhancing the values of $Q$ and $Ri$ and it declines for higher values of $\lambda$ and $fw$. Figure 6a, b represents the effectuates of $fw$, $Q$, $Hg$ and $R$ on LNN. We noticed that the HT gradient enlarges for developing

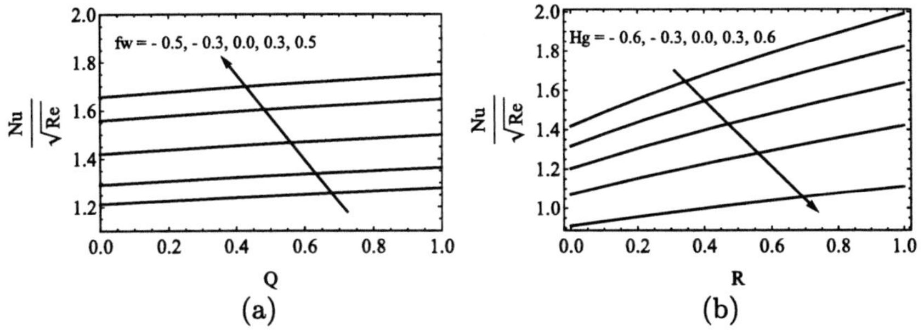


**Fig. 6** Local Nusselt number for variegate magnitude of $fw$, $Q$, $Hg$ and $R$

the values of $fw$, $Q$ and $R$. On the other hand, it downturns for more quantity of $Hg$. The variations of LSN for distinct values of $Cr$, $S_2$, $Q$ and $fw$ is plotted in Fig. 7a, b. We noted that MT gradient upgrades for strengthening $Cr$ $Q$ and $fw$ values. The LSN increases for higher quantity of $S_2$ in $Cr < -0.3$ and it suppresses at $Cr > -0.3$.

## 5 Conclusions

The present study focuses on the mixed convective flow of a radiative ML flow on a porous RP with heat consumption and dual stratifications. We enforce the HAM technique to find series solutions. In our analysis, we get the liquid velocity downturns when increasing the suction/injection and MHN. The liquid temperature enhances enriching heat consumption/generation parameter. The liquidic concentration becomes small for larger quantity of chemical reaction and solutal stratification parameters. The surface drag force upturns for rising of Richardson number. The HT and MT gradients are boosted up when enhancing the modified Hartmann number.

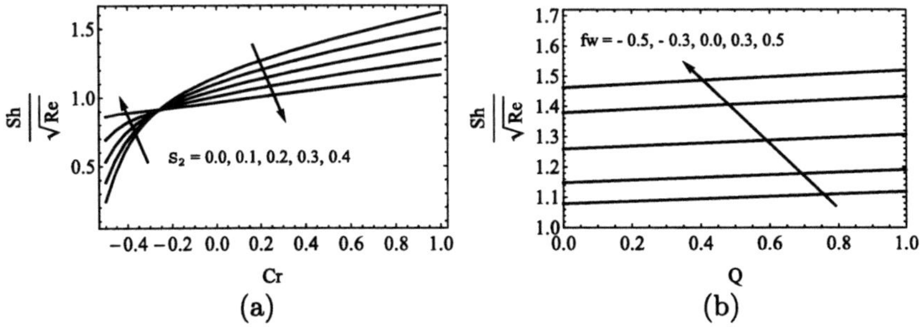


**Fig. 7** Local Sherwood number for variegate magnitude of $Q$, $fw$, $Cr$ and $S_2$

## References


1. Swati M (2012) Heat transfer analysis of the unsteady flow of a Maxwell fluid over a stretching surface in the presence of a heat source/sink. Chin Phys Lett 29(5):1–4
2. Shafiq A, Khalique CM (2020) Lie group analysis of upper convected Maxwell fluid flow along stretching surface. Alex Eng J 59(4):2533–2541
3. Ramesh GK, Gireesha BJ (2014) Influence of heat source/sink on a Maxwell fluid over a stretching surface with convective boundary condition in the presence of nanoparticles. Ain Shams Eng J 5(3):991–998
4. Nadeem S, Akhtar S, Abbas N (2020) Heat transfer of Maxwell base fluid flow of nanomaterial with MHD over a vertical moving surface. Alex Eng J 59(3):1847–1856
5. Nadeem S, Haq RU, Khan ZH (2014) Numerical study of MHD boundary layer flow of a Maxwell fluid past a stretching sheet in the presence of nanoparticles. J Taiwan Inst Chem Eng 45(1):121–126
6. Yin Y, Zhu K-Q (2006) Oscillating flow of a viscoelastic fluid in a pipe with the fractional Maxwell model. Appl Math Comput 173(1):231–242
7. Zhao JH, Zheng LC, Zhang XX, Liu FW (2016) Unsteady natural convection boundary layer heat transfer of fractional Maxwell viscoelastic fluid over a vertical plate. Int J Heat Mass Transf 97:760–766
8. Liu L, Liu F (2018) Boundary layer flow of fractional Maxwell fluid over a stretching sheet with variable thickness. Appl Math Lett 79:92–99
9. Shen M, Chen S, Liu F (2018) Unsteady MHD flow and heat transfer of fractional Maxwell viscoelastic nanofluid with Cattaneo heat flux and different particle shapes. Chin J Phys 56(3):1199–1211
10. Zhang Y, Zhang Y, Zhao Y, Liu H, Bai FY (2018) Analytical and numerical solutions of the unsteady 2D flow of MHD fractional Maxwell fluid induced by variable pressure gradient. Comput Math Appl 75(3):965–980
11. Chen X, Yang W, Zhang X, Liu F (2019) Unsteady boundary layer flow of viscoelastic MHD fluid with a double fractional Maxwell model. Appl Math Lett 95:143–149
12. Kumar CK, Srinivas S, Reddy AS (2020) MHD pulsating flow of Casson nanofluid in a vertical porous space with thermal radiation and Joule heating. J Mechani 36(4):535–549
13. Muthuraj R, Srinivas S, Shukla AK, Lourdu Immaculate D (2019) Effects of non-Darcian and temperature dependent heat source on two-phase flow in a vertical porous space with thermal radiation. Heat Transf Asian Res 48(1):5–23
14. Srinivas S, Kumar CK, Reddy AS (2018) Pulsating flow of Casson fluid in a porous channel with thermal radiation, chemical reaction and applied magnetic field. Nonlin Anal Model Contr 23(2):213–233
15. Eswaramoorthi S, Bhuvaneswari M, Sivasankaran S, Rajan S (2015) Effect of radiation on MHD convective flow and heat transfer of a viscoelastic fluid over a stretching surface. Procedia Eng 127:916–923
16. Eswaramoorthi S, Bhuvaneswari M, Sivasankaran S, Rajan S (2016) Soret and Dufour effects on viscoelastic boundary layer flow over a stretching surface with convective boundary condition with radiation and chemical reaction. Sci Iran Trans B Mech Eng 23(6):2575–2586
17. Niranjan H, Sivasankaran S, Bhuvaneswari M (2016) Analytical and numerical study on magnetoconvection stagnation-point flow in a porous medium with chemical reaction, radiation, and slip effects. Math Probl Eng Article ID 4017076 1–12
18. Sivasankaran S, Niranjan H, Bhuvaneswari M (2017) Chemical reaction, radiation and slip effects on MHD mixed convection stagnation-point flow in a porous medium with convective boundary condition. Int J Numer Methods Heat Fluid Flow 27(2):454–470
19. Hosseinzadeh K, Gholinia M, Jafari B, Ghanbarpour A, Olfian H, Ganji DD (2019) Nonlinear thermal radiation and chemical reaction effects on Maxwell fluid flow with convectively heated plate in a porous medium. Heat Transf Asian Res 48(2):744–759

20. Mallawi FOM, Bhuvaneswari M, Sivasankaran S, Eswaramoorthi S (2021) Impact of double-stratification on convective flow of a non-Newtonian liquid in a Riga plate with Cattaneo-Christov double-flux and thermal radiation. Ain Shams Eng J 12(1):969–981
21. Kasmani RM, Sivasankaran S, Bhuvaneswari M, Siri Z (2016) Effect of chemical reaction on convective heat transfer of boundary layer flow in nanofluid over a wedge with heat generation/absorption and suction. J Appl Fluid Mechan 9(1):379–388
22. Abel MS, Tawade JV, Shinde JN (2012) The effects of MHD flow and heat transfer for the UCM fluid over a stretching surface in presence of thermal radiation. Adv Math Phys 2012:1–21
23. Bhuvaneswari M, Eswaramoorthi S, Sivasankaran S, Ali Saleh A (2019) Effect of viscous dissipation and convective heating on convection flow of a second grade fluid over a stretching surface: analytical and numerical study. Sci Iran Trans B Mech Eng 26(3):1350–1357
24. Bhuvaneswari M, Eswaramoorthi S, Sivasankaran S, Hussein AK (2019) Cross-diffusion effects on MHD mixed convection over a stretching surface in a porous medium with chemical reaction and convective condition. Eng Trans 67(1):3–19
25. Srinivas S, Muthuraj R (2010) Effects of thermal radiation and space porosity on MHD mixed convection flow in a vertical channel using homotopy analysis method. Commun Nonlinear Sci Numer Simul 15(8):2098–2108

# Non-linear Convection with Time-Periodic Boundary Temperature Effect in Viscoelastic Ferromagnetic Fluids

**G. Jayalatha and Nivya Muchikel**

## 1 Introduction

The magnetic fluid research field is a multi-disciplinary area. For Physicists it is to study physical properties and to come out with theories explaining them, for a Chemists interest is to examine synthesis and production of ferrofluids, for Physicians and Biologists is to know possible biomedical applications, and for an Engineer is to come up with technological products. Major magnetic fluid applications depend on strength of the magnetic field, electromagnetic energy absorption at suitable frequencies, and change in physical properties with applied magnetic field. Magnetic fluids are made useful with these properties for many biomedical, material science, technological and engineering research. Ferrohydrodynamics was firstly studied by [1] and stability was by [2]. Rayleigh-Bénard situation with ferromagnetic fluids is studied extensively. Control mechanisms for convection with ferromagnetic fluid has many applications. Many authors addressed ferrofluid convection extensively (see in [3–7]), Mahajanan and Parashar [8] and references there in).

Fluids having elastic and also viscous components are viscoelastic fluids. They are non-Newtonian fluids that are mixture of some polymer and solvent. DNA suspensions, fluids from chemical industry, Biological fluids and paints fall into this category. Describing strain rate, stress tensor relation by constitutive equation, fluid properties are studied. They are working media for many practical problems (see in [9]). Pertaining to viscoelastic fluid convection many works are available in literature (see in [10–12] and references therein).

Convection in a Hele-shaw cell with external mechanisms such as through flow, magnetic field and variable gravity in a layer of nanofluid/couple stress fluid/ viscoelastic fluid is studied by Yadav (see in [13–18]).

G. Jayalatha (✉) · N. Muchikel
RV College of Engineering, Bengaluru-560059. An Autonomous Institution Affiliated to Visvesvaraya Technological University, Belgavi, Karnataka, India
e-mail: jayalathag@rvce.edu.in

N. Muchikel
e-mail: nivyamuchikel@rvce.edu.in



S. Srinivas et al. (eds.), *Recent Advances in Applied Mathematics and Applications to the Dynamics of Fluid Flows*, Lecture Notes in Mechanical Engineering,
https://doi.org/10.1007/978-981-19-1929-9_5

VFF are gel-based fluids. It is a type of smart fluid which on application of a magnetic field greatly increases its apparent viscosity, to the point of becoming a viscoelastic solid. For both oscillatory and stationary modes, convection in VFF is studied by Laroze et al. and Pérez et al. (see in [19–21]).

Motor cooling, noiseless printers, medicine, machine tools with variable speed are few application areas for heat transfer and convection in ferromagnetic fluid. Non-isothermal application of ferromagnetic/visco elastic fluids by regulating thermal convection onset and also transport of heat is considered to be important field, studied by Laroze et al. and Perez et al. (see in [22–24]), [25, 26] and references there in.

Much attention have been received recently for convection in modulated temperature concerning temperature gradients that are space dependent. Origin of this is transient cooling or heating at the boundaries. This makes basic temperature profile dependent on time and position. This leads to problem of thermal modulation involving energy equation solution, considering boundary conditions (BCs) that are temperature dependent. Works have been done on convection with modulated temperature/ gravity effect considering magnetic/viscoelastic/VFF see in [27–31]. In the present work considering the above literature VFF instability with temperature modulation is studied.

## 2 Mathematical Formulation

Magnetic fluid layer extended infinitely, confined between two infinite horizontal walls of thickness $d$ with viscoelastic carrier fluid and externally imposed wall temperatures is considered. Applied magnetic field $\mathbf{H_0}$ is considered in vertical direction. Ambient temperature is taken as $T_0$ by maintaining temperature at $z = d$ (upper plate) as $T_0 - \frac{\Delta T}{2}[1 - \epsilon cos(\Phi + \omega t)]$ and lower plate at $z = 0$ as $T_0 + \frac{\Delta T}{2}[1 + \epsilon cos(\omega t)]$, $\Delta T$ as difference in temperature between upper and lower surfaces in no modulation case, temperature modulation amplitude is $\epsilon$, frequency is $\omega$ and phase angle as $\Phi$.

Considering heating from below, flow configuration is given in Fig. 1. With Boussinesq approximation, balancing equations are:

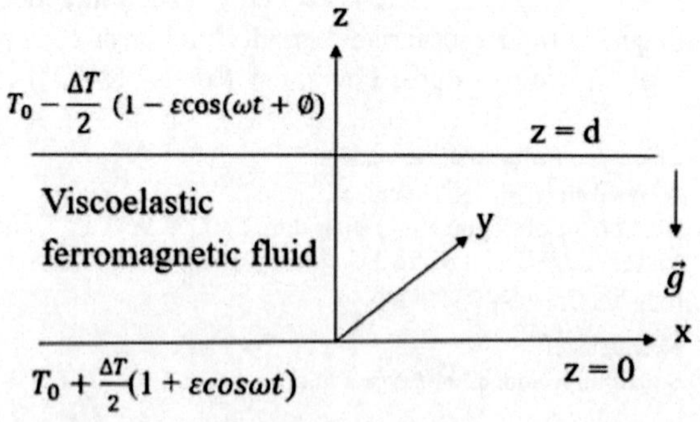


**Fig. 1** Physical configuration

Equation of Continuity:

$$q_{i,i} = 0. \tag{1}$$

Conservation equation for linear momentum:

$$\rho_0 \frac{\partial q_i}{\partial t} = \mu_0 \left(\mathbf{M}.\nabla\mathbf{H}\right) - \nabla p + \tau'_{ij,j} + \rho g_i. \tag{2}$$

Constitutive Equation for viscoelastic fluid: $\left[1 + \lambda_1 \frac{\partial}{\partial t}\right] \tau'_{ij} = \left[1 + \lambda_2 \frac{\partial}{\partial t}\right] [(q_{i,j} + q_{j,i})\mu]$.

Energy Conservation:

$$\frac{\partial T}{\partial t} + q_j T_{,j} = \kappa \left[T_{,j}\right]_{,\,j}, \tag{3}$$

where $z = d$, the upper wall and $z = 0$, the lower wall are subjected to the temperatures

$$T = T_0 + \frac{\Delta T}{2}\left[1 + \epsilon cos(\omega t)\right],\ T = T_0 - \frac{\Delta T}{2}\left[1 - \epsilon cos(\Phi + \omega t)\right]. \tag{4}$$

Total density for linear state of equation is:

$$\rho = -\rho_0[(T - T_0)\alpha - 1]. \tag{5}$$

Maxwell's equations:

$$\nabla \times \mathbf{H} = 0 \quad \text{and} \quad \nabla.\mathbf{B} = 0. \tag{6}$$

Between the fields assuming linear relationship,

$$\mathbf{B} = \mu_0[\mathbf{H} + \mathbf{M}], \tag{7}$$

and to fulfill Eq. (6) introduced scalar magnetic potential $\mathbf{H} = -\nabla\phi$. Assuming the magnetization field to instantaneously follow $\mathbf{M} = \mathrm{M(T, H)}\mathbf{H}$, the external field with equation of state:

$$M = M_0 + [T - T_0]k_l + [H - H_0]\chi_m, \tag{8}$$

where the physical variables are given in Table 1.

**Table 1** Nomenclature

| | | | Greek symbols |
|---|---|---|---|
| $k$ | Non-dimensional wave number ($m^{-1}$) | $\delta$ | Small positive constants |
| $d$ | Thickness of liquid layer (m) | $\alpha$ | Thermal expansion coefficient ($K^{-1}$) |
| $g_i$ | Gravitational acceleration (0,0,−g) ($m/s^2$) | $\kappa$ | Thermal diffusivity ($m^2/s$) |
| $q_i$ | Components of velocty (u,v,w)(m/s) | $\mu$ | Viscosity ($kgm^{-1}s^{-1}$) |
| $p$ | Effective pressure | $\rho$ | Density ($kg/m^3$) |
| $Pr$ | Prandtl number | $\rho_0$ | Reference density ($kg/m^3$) |
| $H_0$ | Applied magnetic field | $\epsilon$ | Amplitude of modulation |
| $Nu$ | Nusselt number | $\omega$ | Frequency ($s^{-1}$) |
| $M_1$ | Buoyancy magnetic number | $\lambda_2$ | Coefficient of strain retardation (s) |
| $M_3$ | Non-buoyancy magnetic number | $\lambda_1$ | Coefficient of stress relaxation (s) |
| $R$ | Rayleigh number | $\Lambda$ | Ratio of elasticity |
| $t$ | Time (s) | $\Lambda_1$ | Scaled stress relaxation parameter (Deborah number) |
| $T$ | Temperature (K) | $\Lambda_2$ | Strain retardation parameter (scaled) |
| $T_0$ | Constant temperature of the boundary (K) | $\tau_{ij}$ | Stress components ($N/m^2$) |
| $k_l$ | Pyromagnetic coefficient | $\chi_m$ | Magnetic susceptibility |
| | **Subscripts** | | **Superscripts** |
| $b$ | Basic state | $'$ | Perturbed quantities |

## 3 Steady State

The basic quiescent state quantities are:

$$\left.\begin{aligned} p = p_b(z) = -\rho_0 g d\left[\frac{z}{d} - \left(\frac{z}{d} - \frac{z^2}{2d^2}\right)\alpha\Delta T\right],\ q_{ib} = (0,0,0),\\ \rho_b(z) = -\rho_0\left[\left(\frac{z}{d} - 1\right)\alpha\Delta T - 1\right],\\ M_b(z) = M_0 + \left[\frac{k_l}{1+\chi_m}\left(\frac{z}{d} - 1\right)\right],\ H_b(z) = H_0 + \left[\frac{k_l z}{1+\chi_m}\left(1 - \frac{z}{d}\right)\right]. \end{aligned}\right\} \quad (9)$$

In the steady state, energy equation (3) is $\frac{\partial T_b}{\partial t} = \kappa\frac{\partial^2 T_b}{\partial z^2}$ and has the solution following Venezian, satisfying BCs (4) with two parts, viz., oscillating temperature $\epsilon T_1(z, t)$ and steady temperature field $T_s(z)$ as:

$$T_b(z,t) = \epsilon T_1(z,t) + T_s(z), \tag{10}$$

where

$$\left. \begin{aligned} T_s(z) = T_0 + \frac{\Delta T}{2d}(d-2z),\ T_1(z,t) = Re\left[b(-\lambda)e^{\frac{-\lambda z}{d}} + b(\lambda)e^{\frac{\lambda z}{d}}\right]e^{-i\omega t}, \\ b(\lambda) = \frac{\Delta T}{2}\left(\frac{e^{-\lambda} - e^{-i\Phi}}{e^{-\lambda - e^{\lambda}}}\right) \text{and } \lambda = \left(\frac{\omega d^2}{2\kappa}\right)^{\frac{1}{2}}(1-i). \end{aligned} \right\} \tag{11}$$

Super imposing amplitude perturbations which are finite, on the basic state, $T = T_b + T'$, $\mathbf{M} = M' + \mathbf{M_b}$, $q_i = q_{ib} + q_i'$, $\rho = \rho' + \rho_b$, $p = p' + p_b$, and $\mathbf{H} = H' + \mathbf{H_b}$. Here primes refer to quantities that are perturbed. Using classical procedure, taking the characteristic time as $d^2/\kappa$, characteristic length as $d$ and characteristic temperature as $\Delta T$, dimensionless equations by dropping primes are:

$$\begin{aligned} &\left(1+\Lambda_1\frac{\partial}{\partial t}\right)RM_1 J\left(\frac{\partial \phi}{\partial z}, T\right) - \left(1+\Lambda_1\frac{\partial}{\partial t}\right)R\frac{\partial T}{\partial x} + \frac{1}{Pr}\left(1+\Lambda_1\frac{\partial}{\partial t}\right)\frac{\partial}{\partial t}\left(\frac{\partial u}{\partial z} - \frac{\partial w}{\partial x}\right) \\ &- \left(1+\Lambda_2\frac{\partial}{\partial t}\right)\nabla^4\left(\frac{\partial u}{\partial z} - \frac{\partial w}{\partial x}\right) - RM_1\left(1+\Lambda_1\frac{\partial}{\partial t}\right)\left(\frac{\partial^2 \phi}{\partial x \partial z} - \frac{\partial T}{\partial x}\right)\frac{\partial T_b}{\partial z} = 0, \end{aligned} \tag{12}$$

$$\frac{\partial T}{\partial t} - \frac{\partial T_b}{\partial z}w - u\frac{\partial T}{\partial x} - \nabla^2 T + w = 0, \tag{13}$$

$$M_3\left(\frac{\partial^2 \phi}{\partial x^2}\right) + \frac{\partial^2 \phi}{\partial z^2} - \frac{\partial T}{\partial z} = 0, \tag{14}$$

where $\nabla^2 = \frac{\partial^2}{\partial x^2} + \frac{\partial^2}{\partial z^2}$.

In Eqs. (13) and (14), $\frac{\partial T_b}{\partial z}$ in the non-dimensional form is given below (Kanchana et al. [32]):
$\frac{\partial T_b}{\partial z} = -1 + \epsilon f$, where $f = Re\left\{\left[B(-\lambda)e^{-\lambda z} + B(\lambda)e^{\lambda z}\right]e^{-i\omega t}\right\}$, $B(\lambda) = \frac{\lambda}{2}\left[\frac{e^{-i\Phi} - e^{-\lambda}}{e^{\lambda} - e^{-\lambda}}\right]$ and $\lambda = \left[\frac{\omega}{2}\right]^{\frac{1}{2}}(1-i)$.

$B(\lambda)$ is non-dimensional representation for $b(\lambda)$ as mentioned in Eq. (11).

The non-dimensional parameters are:

$M_1 = \frac{\mu_0 k_1^2 \Delta T}{\alpha g \rho_0 (1+\chi_m) d}$ (Buoyancy magnetization number), $\Lambda = \frac{\lambda_2}{\lambda_1} = \frac{\Lambda_2}{\Lambda_1}$ (Elastic ratio), $\Lambda_2 = \frac{\lambda_2 \kappa}{d^2}$ (Scaled strain retardation parameter), $\Lambda_1 = \frac{\lambda_1 \kappa}{d^2}$ (Deborah number), $Pr = \frac{\mu_0}{\rho\kappa}$ (Prandtl number), $M_3 = \frac{1+\frac{M_0}{H_0}}{1+\chi_m}$ (Non-Buoyancy magnetization number), $R = \frac{\alpha \rho_0 g d^3 \Delta T}{\mu_0 \kappa}$ (Rayleigh number).

## 4 Derivation of Model of Khayat–Lorenz

Employing non-linear analysis, amount of transfer of heat is quantified by considering ferromagnetic and viscoelastic parameters. Along the y axis it is assumed that physical quantities are not varied, which gives stream function as:

$$w = \left(\frac{\partial \psi}{\partial x}\right), u = -\left(\frac{\partial \psi}{\partial z}\right). \tag{15}$$

Following first order equations are obtained after rearranging (11)–(13).

$$\Lambda_1 \frac{\partial M}{\partial t} = (1-\Lambda)\nabla^4 \psi - M, \tag{16}$$

where $M$ simplifies from

$$\frac{1}{Pr}\frac{\partial}{\partial t}\nabla^2\psi = \Lambda\nabla^4\psi + R\frac{\partial T}{\partial x} + RM_1(-1+\epsilon f)\left[\frac{\partial^2 \phi}{\partial x \partial z} - \frac{\partial T}{\partial x}\right] + RM_1 J\left(\frac{\partial \phi}{\partial z}, T\right) + M, \tag{17}$$

$$\frac{\partial T}{\partial t} = (-1+\epsilon f)\frac{\partial \psi}{\partial x} - \frac{\partial(\psi, T)}{\partial(x,z)} + \nabla^2 T, \tag{18}$$

$$M_3 \frac{\partial^2 \phi}{\partial x^2} + \frac{\partial^2 \phi}{\partial z^2} - \frac{\partial T}{\partial z} = 0. \tag{19}$$

Subjected to following BCs, Eqs. (16) to (19) to be solved.

$$At\ z = 0,\ 1,\ \psi = D^2\psi = D\phi = T = M = 0. \tag{20}$$

Representing solution as Fourier series, we get

$$\begin{aligned} \psi &= A(t)sin(\pi z)sin(kx), T = C(t)sin(2\pi z) + B(t)cos(kx)sin(\pi z), \\ \phi &= E(t)cos(2\pi z) + D(t)cos(kx)cos(\pi z), M = F(t)sin(\pi z)sin(kx). \end{aligned} \tag{21}$$

Here amplitudes are $A(t),\ B(t),\ C(t),\ D(t),\ E(t)$ and $F(t)$.

Choice of number of modes in truncated Fourier series is explained in references [30–35]. Substitution of Eq. (21) in Eqs. (16)–(19) and integrating after multiplying with correct eigen functions results in equations which are analogous with generalized Lorenz model [22] after scaling as follows:

$$\frac{dX}{d\tau} = Pr\left[\left(1 - \frac{M13}{\pi r}(z + \epsilon r N_1)\right)Y - \Lambda X - (1-\Lambda)H\right] \tag{22}$$

$$\frac{dY}{d\tau} = (1 - \epsilon N_1) r X - XZ - Y, \tag{23}$$

$$\frac{dZ}{d\tau} = XY - \frac{4\pi^2}{\delta^2} Z, \tag{24}$$

$$\frac{dH}{d\tau} = \frac{1}{\Lambda_1 \delta^2}(X - H). \tag{25}$$

where $\delta^2 = k^2 + \pi^2$, $r = \frac{R}{R_s}$, $M_{13} = \frac{M_1 M_3 \pi k^2}{k^2(1+M_1)M_3+\pi^2}$, $\tau = \delta^2 t$, $N_1 = \frac{(4\pi^2)Re(B(\lambda)(e^{\lambda}-1)+B(-\lambda)(1-e^{-}\lambda))e^{-i\omega^* t}}{\lambda(\lambda^2+4\pi^2)}$, $R_s = \frac{\delta^6[\pi^2+M_3k^2]}{k^2[k^2(1+M_1)M_3+\pi^2]}$ and $\omega^* =$ scaled frequency.

Note that if $\epsilon = 0$ and $M_{13} \to 0$ as $M_1 \to 0$ or $M_3 \to 0$, recovering Lorentz system. Solving (22)–(25) with initial conditions(ICs) $Y(0) = 1$; $X(0) = 1$; $H(0) = 1$ and $Z(0) = 1$, using RK45 solution is obtained.

## 5 Transfer of Heat

Heat transfer denoted by $Nu(\tau)$ is defined as:

$$Nu(\tau) = \frac{\text{Heat transport by convection}}{\text{Heat transport by conduction}} + 1,$$

$$= \frac{\frac{k}{2\pi}\int_0^{\frac{2\pi}{k}}\left[\frac{\partial T_b}{\partial z} + \frac{\partial T}{\partial z}dx\right]_{z=0}}{\frac{k}{2\pi}\int_0^{\frac{2\pi}{k}}\left[\frac{\partial T_b}{\partial z}dx\right]_{z=0}} + 1. \tag{26}$$

Simplifying equation (26), the Nusselt number is

$$Nu(\tau) = -\frac{2}{rQ(\tau)} Z(\tau) + 1, \tag{27}$$

where $Q(\tau) = \epsilon \text{Re}\left\{[B(\lambda) + B(-\lambda)]\, e^{i\omega^*\tau}\right\} - 1$.

The results are discussed considering mean Nusselt number $\overline{\text{Nu }(\tau)}$ vs various parameters in the next section.

## 6 Discussion of Result

RBC in a VFF is studied by considering finite amplitude perturbations. The viscoelastic parameters appearing in the problem are $\Lambda$ and $\Lambda_1$ while $M_1$ and $M_3$ are ferromagnetic parameters. The Prandtl number, $Pr$, is the ratio of viscous diffusion to thermal diffusion. In the case of viscoelastic fluids due to the presence of long molecules, the viscosity of the fluid is much more than that of a Newtonian fluid. In view of this we have assumed a larger value of $Pr$ than that of most Newtonian fluids. The stress relaxation and strain retardation due to elasticity appear in the form of Deborah number $\Lambda_1$ and another viscoelastic parameter $\Lambda_2$. In the case of a viscoelastic fluid, as discussed by Joseph [9], the values of $\Lambda_1$ and $\Lambda_2$ are both less than unity. Further it is now well known that oscillatory convection is possible in viscoelastic fluids only when $\Lambda < 1$. In view of the above observation the values of $\Lambda_1$ and $\Lambda_2$ have been taken to be $\Lambda_1 = 0.2$ and $\Lambda_2 = 0.1$ there by yielding an acceptable value of $\Lambda = 0.5$. It is important to note here that in the case of the Rivlin-Ericksen fluid $\Lambda_1 = 0$ and hence $\Lambda$ assumes a large value so in that case we have taken $\Lambda_1 = 0$ and $\Lambda = 50$. The values of $M_1$ and $M_3$ are also dictated by mainly magnetic susceptibility $\chi_m$ and the pyromagnetic coefficient $k_l$. The parameter $M_3$ does not involve dynamic coefficient of viscosity, $\mu_0$, and hence its value was taken to be that of a Newtonian Ferromagnetic fluid. However, $M_1$ has $\mu_0$ in the numerator and due to enhanced viscosity, $M_1$ should now take a slightly larger value than that of a Newtonian Ferromagnetic fluid. In the light of the above observation we have chosen $M_1$ to be either 10 or 20 and $M_3$ to be 1.1. With the above facts on VFF, all the computations have been done. A non-linear analysis of RBC in a VFF is studied. The influence of $Pr$, $\Lambda$, $\Lambda_1$, $M_1$, $M_3$ on $\overline{Nu(\tau)}$ is examined. The fluid elasticity effect on RBC with modulation of temperature in ferromagnetic fluids is examined for the resulting Khayat–Lorenz model(generalized) (23)–(26), which is solved by employing RKF45 (adaptive step size technique) with $H(0) = Z(0) = X(0) = Y(0) = 1$. Here convection control mechanism is modulation and its effect on transfer of heat through $\overline{Nu(\tau)}$ is considered.

Plot of $\overline{Nu}$ versus $\Lambda$ in Fig. 2 yields the result that with increasing elastic ratio heat transfer, diminishes without or with modulation. Plot of $\overline{Nu}$ versus $\Lambda_1$ in Fig. 3, yields the result as increasing elastic ratio increases the heat transfer with modulation being absent or present. Figures 4 and 5 reveal that the heat-transfer reduces with the enhancement in $M_3$ and $M_1$ with/without modulation. Plot of $\overline{Nu}$ versus $Pr$ in Fig. 6, gives us the result that increasing $Pr$ results in increasing $\overline{Nu}$ both in the absence/presence of modulated temperature. Following inequalities hold good for four different ferromagnetic fluids namely Rivlin-Ericksen, Maxwell, Oldroyd fluid B and Newtonian (see Fig. 7a, b).

$$\overline{Nu}^{Rivlin-Ericksen} = \overline{Nu}^{Newtonian} < \overline{Nu}^{Oldroyd\ fluid B} < \overline{Nu}^{Maxwell}.$$

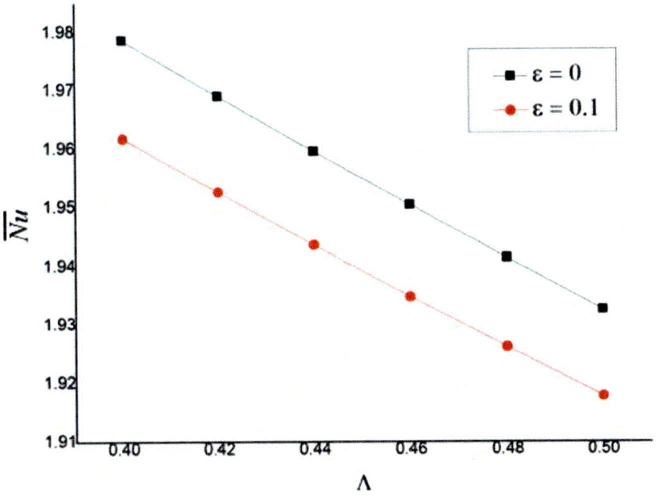


**Fig. 2** $\overline{Nu}$ against $\Lambda$ plot for $M_3 = 1.1$, $\omega^* = 5$, $M_1 = 10$, $r = 5$, $\phi = \pi/6$, $\Lambda_1 = 0.2$, $Pr = 10$ and for different modulation para- meter values

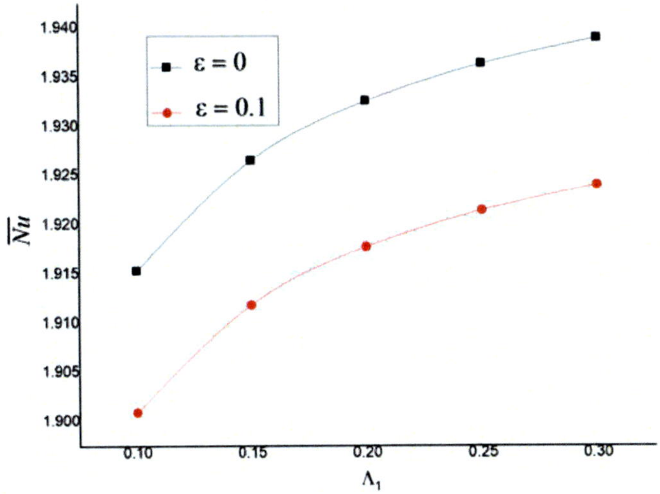


**Fig. 3** $\overline{Nu}$ against $\Lambda_1$ plot with $M_3 = 1.1$, $M_1 = 10$, $\phi = \pi/6$, $\Lambda = 0.5$, $\omega^* = 5$, $Pr = 10$, $r = 5$ and for different modulation para-meter values

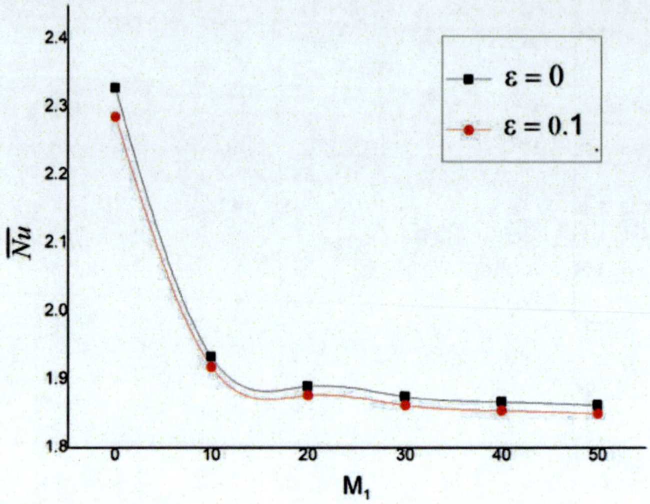


**Fig. 4** $\overline{Nu}$ against $M_1$ plot for $M_3 = 1.1$, $r = 5$, $\Lambda_1 = 0.2$, $\omega^* = 5$, $\Lambda = 0.5$, $Pr = 10$, $\phi = \pi/6$ and for different modulation para-meter values

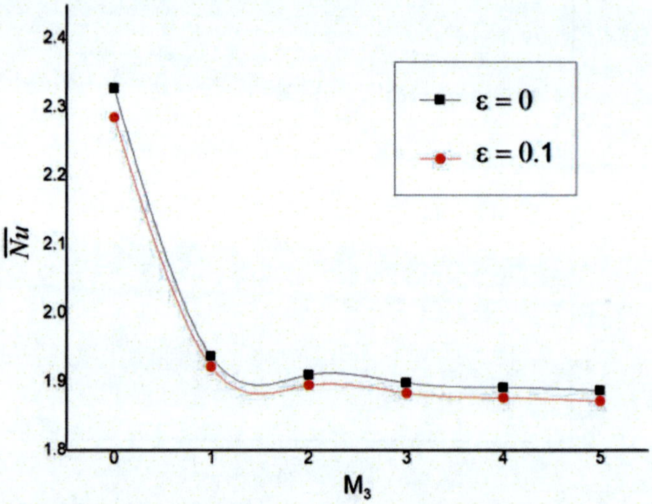


**Fig. 5** $\overline{Nu}$ against $M_3$ plot with $M_1 = 10$, $\Lambda = 0.5$, $\omega^* = 5$, $\Lambda_1 = 0.2$, $r = 5$, $Pr = 10$, $\phi = \pi/6$ and for different modulation parameter values

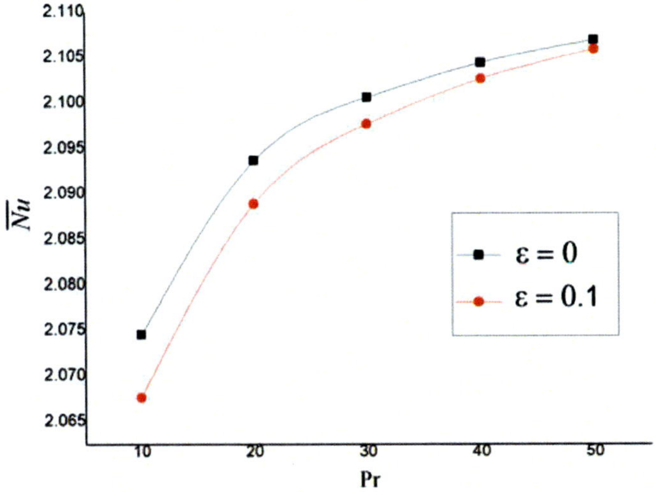


**Fig. 6** $\overline{Nu}$ against $Pr$ plot for $M_1 = 10$, $\omega^* = 5$, $\Lambda_1 = 0.2$, $\phi = \pi/6$, $M_3 = 1.1$, $\Lambda = 0.5$, $r = 5$ and for different modulation parameter values

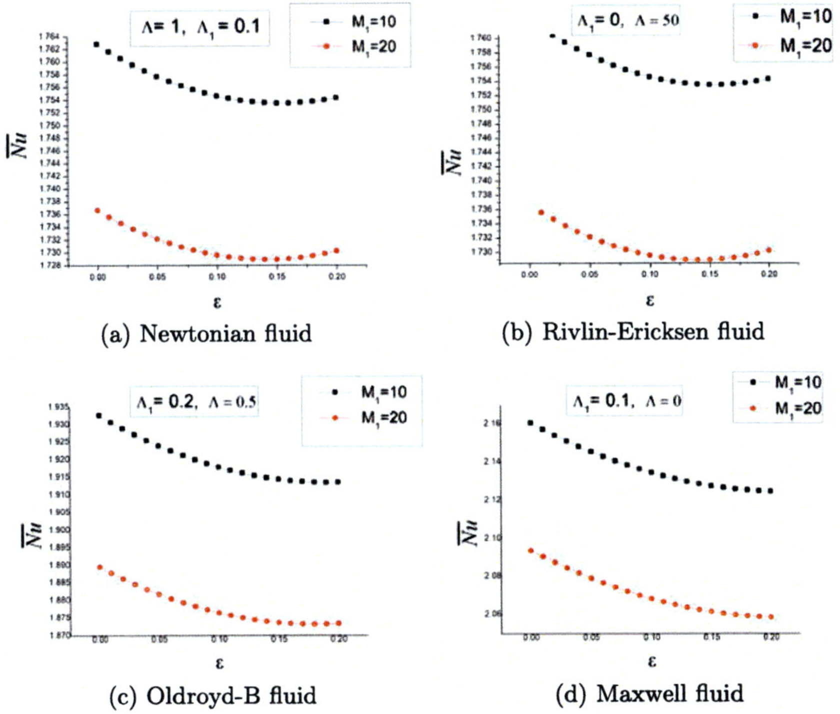


**Fig. 7** Plot of $\overline{Nu}$ against $\epsilon$ for $\phi = \pi/6$, $\omega^* = 5$, $Pr = 10$, $r = 5$, and $M_3 = 1.1$

## 7 Conclusion

- Irrespective of whether modulation is present or absent, the two magnetization parameters diminish the heat transport.
- Fluid with two relaxation times (Oldroyd fluid) has heat transfer between those of the two single-relaxation fluids.
- Newtonian fluid's heat transfer is not modified due to the presence of the strain retardation parameter and in the absence of the stress relaxation parameter.
- In the case of all four fluids considered the effect of temperature modulation is to enhance the heat transport.

**Acknowledgements** The authors express thanks to their institution & management for the encouragement, also to anonymous reviewers for beneficial comments & to Dr. P. G. Siddeshwar for valuable suggestions.

## References

1. Papell SS (1965) Low viscosity magnetic fluid obtained by the colloidal suspension of magnetic particles.U.S. Patent 3,215-572
2. Rosensweig RE (1985) Ferrohydrodynamics. Cambridge University Press, Cambridge
3. Siddheshwar PG, Abraham A (2003) Effect of time-periodic boundary temperatures/body force on Rayleigh-Bénard convection in a ferromagnetic fluid. Acta Mech 161:131–150
4. Mahajan A (2010) Stability of ferrofluids, linear and nonlinear. Lambert Academic Publishing, Germany
5. Sekhar GN, Jayalatha G, Prakash R (2013) Thermorheological and Magnetorheological effects on Rayleigh-Bénard Convection in ferromagnetic liquids with non-uniform basic temperature gradient. Fluids Eng Syst Technol IMECE2013-64522, V07AT08A021. ISBN: 978-0-7918-5631-4
6. Sekhar GN, Jayalatha G, Prakash R (2017) Thermal convection in variable viscosity ferromagnetic liquids with heat source. Int J Appl Comput Math 3:3539–3559
7. Sekhar GN, Jayalatha G, Prakash R (2017) Throughflow effects on thermal convection in variable viscosity ferromagnetic liquids. Mech Mechatron Eng 11:1269–1277
8. Mahajan A, Parashar H (2020) Linear and weakly nonlinear stability analysis on a rotating anisotropic ferrofluid layer. Phys Fluids 32:024101
9. Joseph DD (1990) Fluid dynamics of viscoelastic liquids. Springer, New York
10. Sekhar GN, Jayalatha G (2009) Elastic effects on Rayleigh-Bènard - Marangoni convection in fluids with temperature-dependent viscosity. In: Proceedings of the ASME international mechanical engineering congress & exposition, IMECE09, ISBN 978-07918-3863-1, order No.1830DV
11. Sekhar GN, Jayalatha G (2010) Elastic effects on Rayleigh-Bènard convection in fluids with temperature - dependent viscosity. Int J Thermal Sci 49:67–79, 024101
12. Siddheshwar PG, Sekhar GN, Jayalatha G (2011) Surface tension driven convection in viscoelastic liquids with thermorheological effect. Int Commun Ht Mass Transf 38:468–473
13. Yadav D, Mohamed RA, Cho HH, Lee J (2016) Effect of hall current on the onset of MHD convection in a porous medium layer Saturated by a Nanofluid. J Appl Fluid Mech 9(5)
14. Yadav D, Donglim N, Jinho L (2016) The onset of transient Soret-driven MHD convection confined within a Hele-Shaw cell with nanoparticles suspension. J Taiwan Inst Chem Eng 58:235–244

15. Yadav D (2018) Throughflow and magnetic field effects on the onset of convection in a hele-shaw cell. Rev Cubana Fis 35:108
16. Yadav D (2019) The effect of pulsating throughflow on the onset of magneto convection in a layer of nanofluid confined within a Hele-Shaw cell. In: Proceedings of the institution of mechanical engineers, Part E: journal of process mechanical engineering, vol 233, pp 1074–1085
17. Yadav D (2020) The effect of viscosity and Darcy number on the start of convective motion in a rotating porous medium layer saturated by a couple-stress fluid. In: Proceedings of the institution of mechanical engineers, part C: journal of mechanical engineering science, 0954406220942551
18. Yadav D (2020) The onset of Darcy-Brinkman convection in a porous medium layer with vertical throughflow and variable gravity field effects. Heat Transf 1–13
19. Laroze D, Martinez-Mardones J, Pèrez LM, Rojas RG (2010) Stationary thermal convection in a viscoelastic ferrofluid. J Magn Magn Mater 322:3576–3583
20. Pèrez LM, Bragardb J, Laroze D, Martinez-Mardonese J, Pleiner H (2011) Thermal convection thresholds in a Oldroyd magnetic fluid. J Magn Magn Mater 323:691–698
21. Pèrez LM, Laroze D, Díaz P, Martinez-Mardonese J, Mancini HL (2014) Rotating convection in a viscoelastic magnetic fluid. J Magn Magn Mater 364:98–105
22. Laroze D, Siddheshwar PG, Pleiner H (2013) Chaotic convection in a ferrofluid. Commun Nonlinear Sci Numer Simulat 18:2436–2447
23. Laroze D, Pleiner H (2015) Thermal convection in a nonlinear non-Newtonian magnetic fluid. Commun Nonlinear Sci Numer Simulat 26:167–183
24. Pèrez LM, Bragardb J, Dìaz P, Mancini HL, Laroze D, Pleiner H (2017) Magneto-viscous effect on thermal convection thresholds in an Oldroyd magnetic fluid. J Magn Magn Mater 444:432–438
25. Narayana M, Sibanda P, Siddheshwar PG, Jayalatha G (2013) Linear and nonlinear stability analysis of binary viscoelastic fluid convection. Appl Math Modell 37:8162–8178
26. Jayalatha G, Siddheshwar PG, Suma N (2018) Non-Linear convection in rotating viscoelastic liquids with gravity modulation. In: Proceedings of 7th international and 45th national conference on fluid mechanics and fluid power, IIT Bombay, Mumbai, India D
27. Siddheshwar PG, Pranesh S (1999) Effect of temperature/gravity modulation on the onset of magnetoconvection in weak electrically conducting uids with internal angular momentum. J Magn Magn Mater 192:159–176
28. Siddheshwar PG, Sekhar GN, Jayalatha G (2010) Effect of time - periodic vertical oscillations of the Rayleigh-Bènard system on nonlinear convection in, viscoelastic fluids. J Non - Newtonian Fluid Mech 165:1412–1418
29. Anjana K, Vanishree RK (2018) Study of effects of temperature modulation on double diffusive convection in Oldroyd-B liquids. Int J Eng Res Technol 5:335–343
30. Jayalatha G, Suma N (2020) Non linear rotating Viscoelastic liquid convection with temperature modulation. Heat Transf Wiley 49:4565–4579
31. Jayalatha G, Nivya M (2021) Modulated gravity effects on nonlinear convection in viscoelastic ferromagnetic fluids between two horizontal parallel plates. Heat Transf Wiley. https://doi.org/10.1002/htj.22077
32. Kanchana C, Siddheshwar PG, Zhao Y (2020) Regulation of heat transfer in Rayleigh-Bènard convection in Newtonian liquids and Newtonian nanoliquids using gravity, boundary temperature and rotational modulations. J Thermal Anal Calorim 142:1579–1600
33. Siddheshwar PG, Kanchana C (2018) A study of unsteady, unicellular Rayleigh-Bénard convection of nanoliquids in enclosures using additional modes. J Nanofluids 7:791–800
34. Melson AC, Sekhar GN, Siddheshwar PG (2020) Nonlinear analysis of effect of Rigid body rotation on ferroconvection. J Heat Transf 142:061802; (7 pages)Paper No: HT-19-1546
35. Melson AC, Siddheshwar PG, Sekhar GN (2021) Nonlinear analysis of the effect of viscoelasticity on ferroconvection. J Heat Transf. https://doi.org/10.1002/htj.22055

# Oscillatory Flow of Dusty Fluid Through a Narrowed Channel in the Presence of Magnetic Field

Indrajeet G. Varahadpande, V. R. K. Murthy, and P. Sudam Sekhar

## 1 Introduction

The pulsating blood flow has got consideration over the most recent couple of years because of its importance in a few physiological systems. Many researchers [1–10] worked in this field. "The portrayal of blood as a bipartite system by considering a dusty liquid for blood is given by" Saffman [1]. Nayfeh [2] "considered the Pulsating flow of a liquid in an incredibly lengthy and narrow pipe of round cross area with little solid particles". Gupta and Gupta [3] "considered the surge of a dusty gas through a waterway with convenient changing pressure gradient". Radhakrishnamacharya [4] "thought the pulsatile stream of a liquid comprising of small rigid particles through a narrow channel". Wagh and Tapi [5] "researched the pulsatile stream of a suspension in the region of mild stenosis". Afifi and Gad [6] "considered the interface of MHD Pulsatile stream and peristaltic move through a porous medium".

Nevertheless, the impact of a magnetic field on the pulsating stream of liquid with dust particles has not been given any consideration. Blood is realized to show magnetic properties. Subsequently, according to Saffaman's idea, a pulsatile stream of a liquid with dust particles in a choked channel with the impact of magnetic effect is examined. Functions of streams have been inferred for both dust and liquid particles. Assuming the length of the wave is large, the equations for gradient pressure and

I. G. Varahadpande
Department of Basic Sciences, St. Vincent Pallotti College of Engineering &Technology, Nagpur, India

Department of Basic Sciences, Vignan's Foundation for Science Technology & Research, Guntur, India

V. R. K. Murthy (✉) · P. S. Sekhar
Department of Science and Humanities, Vignan's Foundations for Science Technology & Research, Guntur, India
e-mail: vadapalli.radhakrishna@gmail.com

 
S. Srinivas et al. (eds.), *Recent Advances in Applied Mathematics and Applications to the Dynamics of Fluid Flows*, Lecture Notes in Mechanical Engineering,
https://doi.org/10.1007/978-981-19-1929-9_6

wall shear stress are obtained. Impact of parameters such as magnetic, concentration of solid particles, profiles of velocity, pressure dissemination and shear stress are examined.

## 2 Formulation of the Problem

Using [4], the flow of an electrically conducting, Newtonian, dusty fluid with constriction on the channel walls is considered. So, a transverse magnetic field (uniform) is applied to the fluid. The channel wall is given by

$$\eta = d + a \sin \tfrac{2\pi}{\lambda}\, x \tag{1}$$

where x is along the length of the channel and "a" is the height of constriction on the wall. The equations governing the fluid flow, containing solid particles, are

$$\frac{\partial \boldsymbol{V}}{\partial t} + (\boldsymbol{V}\cdot\nabla)\boldsymbol{V} = \frac{1}{\rho}\nabla p + \upsilon\nabla^2 \boldsymbol{V} + \frac{kN_o}{\rho}(\boldsymbol{V}_s - \boldsymbol{V}) = \frac{\boldsymbol{J}x\boldsymbol{B}}{\rho} \tag{2}$$

$$\frac{\partial \boldsymbol{V}_s}{\partial t} + (\boldsymbol{V}_s\cdot\nabla)\boldsymbol{V}_s = \frac{k}{m}(\boldsymbol{V} - \boldsymbol{V}_s) \tag{3}$$

$$\nabla\cdot\boldsymbol{V} = 0 \tag{4}$$

$$\nabla\cdot\boldsymbol{V}_s = 0 \tag{5}$$

where $\boldsymbol{V} = \boldsymbol{V}(u, v)$ and $\boldsymbol{V}_s = \boldsymbol{V}_s(u_s, v_s)$ represent the velocities of fluid and dust particles, respectively. $\rho$, $p$, and$\nu$ are, respectively, density, pressure and kinematic coefficient of viscosity of the fluid. m is the mass of the solid particles, k is resistance-coefficient for dust particles, **J** is the current–density, **B (= B0 + B1)** is the total magnetic field, and **B1** is the induced magnetic field.

$$\nabla^2 = \frac{\partial^2}{\partial x^2} + \frac{\partial^2}{\partial y^2}$$

conditions on channel-boundary

$$\left.\begin{aligned} u = v = 0 \\ u_s = v_s = 0 \end{aligned}\right\} \text{at } y = \pm\eta \tag{6}$$

in association with constant volume flux. The Maxwell equations and the generalized Ohm's law are (after neglecting displacement currents)

$$\nabla.B = 0$$

$$\left.\begin{aligned} \nabla \times B &= \mu_m J \\ \nabla \times E &= -\frac{\partial B}{\partial t} \\ J &= \alpha(E + V\,X\,B) \end{aligned}\right\} \tag{7}$$

where $\mu_m$ is the "magnetic permeability" which is assumed to be constant throughout the flow field, E is the "electric field" and α is the electrical conductivity of the fluid. Following Srinivasacharya [7] and Ravi Kiran [8], "It is assumed that the induced magnetic field is negligible compared to the applied magnetic field so that the magnetic Reynolds number is small. Also, the electric field is assumed to be zero".

Choosing the functions $\psi$ and $\phi$ such that (eliminating the pressure)

$$\left.\begin{aligned} u &= \frac{\partial \psi}{\partial y};\ v = -\frac{\partial \psi}{\partial x} \\ u_s &= \frac{\partial \phi}{\partial y};\ v_s = -\frac{\partial \varphi}{\partial x} \end{aligned}\right\} \tag{8}$$

and make use of the following dimensionless variables.

$x' = x/\lambda$, $y' = y/\mathrm{d}$, $\eta' = \eta/\mathrm{d}$, $\mathrm{t}' = \upsilon\, \upsilon\, \mathrm{t}/\lambda \mathrm{d}$, $\psi' = \psi/\upsilon$

$$\phi' = \frac{\phi}{kd^2/m} \tag{9}$$

in (2) and (3) the equations (after dropping primes) are changed to

$$\delta\left[\frac{\partial}{\partial t}(\nabla^2\psi) + \frac{\partial \psi}{\partial y}\nabla^2\frac{\partial \psi}{\partial x} - \frac{\partial \psi}{\partial x}\nabla^2\frac{\partial \psi}{\partial y}\right] = \nabla^4\psi + \alpha(\frac{1}{R}\nabla^2\phi - \nabla^2\psi) - M^2\frac{\partial^2\psi}{\partial y^2} \tag{10}$$

$$\delta\left[R\frac{\partial}{\partial t}(\nabla^2\phi) + \frac{\partial \phi}{\partial y}\nabla^2\frac{\partial \phi}{\partial x} - \frac{\partial \phi}{\partial x}\nabla^2\frac{\partial \phi}{\partial y}\right] = R\nabla^2\psi - \nabla^2\phi \tag{11}$$

The conditions on the boundary are transformed to

$$\left.\begin{array}{l}\frac{\partial \psi}{\partial y}=\frac{\partial \psi}{\partial x}=0\\ \frac{\partial \phi}{\partial y}=\frac{\partial \phi}{\partial x}=0\end{array}\right\} \text{at } \ y=\pm\eta \tag{12}$$

$$(\psi)_{y=\eta}=\text{Const};\ (\psi)_{y=0}=0 \tag{13}$$

where $\nabla^2 = \delta^2\frac{\partial^2}{\partial x^2} + \frac{\partial^2}{\partial y^2}$, $\varepsilon = (\frac{a}{d})$ is the ratio of amplitudes, $R(=\frac{\upsilon m}{kd^2})$ and $\alpha(=\frac{kN_0 d^2}{\rho\upsilon})$ are parameters which are dimensionless and $M(=\frac{\alpha\beta_o^2 d^2}{\rho\upsilon})$ is the parameter for the magnetic field.

## 3 Solution Method

Equations (10) and (11) are not linear. It is difficult to get exact solutions for various parametric consents. Taking channel slope δ is very small, functions of stream ($\psi$) and potential ($\phi$) are written as

$$\left.\begin{array}{l}\psi=\psi_0+\delta\psi_1+\delta^2\psi_2+..\\ \phi=\phi_0+\delta\phi_1+\delta^2\phi_2+..\end{array}\right\} \tag{14}$$

Using Eq. (14) in Eqs. (10)–(13) collecting factors of equal supremacies of δ, zeroth and first order, perturbed equations, obtained as

Zeroth order:

$$\left.\begin{array}{l}\frac{\partial^4\psi_0}{\partial y^4}+\alpha\left(\frac{1}{R}\frac{\partial^2\phi_0}{\partial y^2}-\frac{\partial^2\psi_0}{\partial y^2}\right)-M^2_{\frac{\partial^2\psi_0}{\partial y^2}}=0\\ R\frac{\partial^2\psi_0}{\partial y^2}-\frac{\partial^2\varphi_0}{\partial y^2}=0\end{array}\right\} \tag{15}$$

$$\left.\begin{array}{l}\frac{\partial \psi_0}{\partial y}=\frac{\partial \psi_0}{\partial x}=0\\ \frac{\partial \phi_0}{\partial y}=\frac{\partial \phi_0}{\partial x}=0\end{array}\right\} \text{at } \ \mathrm{y}=\pm\eta \tag{16}$$

$$(\psi_0)_{\mathrm{y}}=\eta=\text{constant, } \ \mathrm{Q}\ (\text{say});\ (\psi_0)_{\mathrm{y}=0}=0 \tag{17}$$

First order:

$$\left.\begin{array}{l}\dfrac{\partial^4\psi_1}{\partial y^4}+\alpha(\dfrac{1}{R}\dfrac{\partial^2\phi_1}{\partial y^2}-\dfrac{\partial^2\psi_1}{\partial y^2})-M^2\dfrac{\partial^2\psi_1}{\partial y^2}=\dfrac{\partial^3\psi_0}{\partial t\partial y^2}+\dfrac{\partial\psi_0}{\partial y}\dfrac{\partial^3\psi_0}{\partial x\partial y^2}-\dfrac{\partial\psi_0}{\partial x}\dfrac{\partial^3\psi_0}{\partial y^3}\\ R\dfrac{\partial^2\psi_1}{\partial y^2}-\dfrac{\partial^2\phi_1}{\partial y^2}=R\dfrac{\partial^3\phi_0}{\partial t\partial y^2}+\dfrac{\partial\phi_0}{\partial y}\dfrac{\partial^3\phi_0}{\partial x\partial y^2}-\dfrac{\partial\phi_0}{\partial x}\dfrac{\partial^3\phi_0}{\partial y^3}\end{array}\right\} \tag{18}$$

$$\left.\begin{array}{l}\dfrac{\partial\psi_1}{\partial y}=\dfrac{\partial\psi_1}{\partial x}=0\\ \dfrac{\partial\varphi_1}{\partial y}=\dfrac{\partial\varphi_1}{\partial x}=0\end{array}\right\}\text{at}\quad \mathrm{y}=\pm\eta \tag{19}$$

$$(\psi_1)_y=\eta=0,\ (\psi_1)_{y=0}=0 \tag{20}$$

Further, we assume

$$\left.\begin{array}{l}\psi_0=\psi_{00}(x,y)e^{i\,w\,t}\\ \phi_0=\phi_{00}(x,y)e^{i\,w\,t}\\ \psi_1=\psi_{10}(x,y)+\psi_{11}(x,y)e^{i\,w\,t}+\psi_{11}2(x,y)e^{2i\,w\,t}\end{array}\right\} \tag{21}$$

$$\phi_1=\phi_{10}(x,y)+\phi_{11}(x,y)\ e^{i\,w\,t}+\phi_{11}(x,y)e^{2\,i\,w\,t}.$$

Making use of identities (21) for Eqs. (15) to (20), we obtain the solutions

$$\begin{aligned}\psi&=\mathrm{QA3}(\mathrm{Sinh}(My)-My\mathrm{Cosh}(M\eta)e^{iwt}\\ &+\delta\Bigg\{\Bigg[E_1A_3A_1\Bigg(\frac{ySinh^2(M\eta)}{\eta}-\mathrm{Sinh}(My)\mathrm{Sinh}(M\eta))\\ &+E_1\mathrm{Sinh}(M\eta)\Bigg(\frac{\mathrm{y}^2\mathrm{Sinh}(My)}{4M^3}-\frac{5}{4M^4}\mathrm{y}\mathrm{Cosh}(My)\\ &+\frac{Sinh(My)}{8M^5}\Bigg)-\frac{y}{\eta}A_2+\frac{E_1\eta yCosh^2(M\eta)}{2M^3}-\frac{E_1\eta yCosh(My)Cosh(M\eta)}{2M^3}\Bigg]\\ &\Bigg[1+\frac{e^{2iwt}}{2}\Bigg]+\mathrm{A}_5e^{iwt}\Bigg[\frac{A_3A_4}{2}\Bigg(y-\frac{\eta Sinh(My)}{Sinh(M\eta)}\Bigg)\\ &-\frac{\eta Cosh(M\eta)Sinh(My)}{2M^3Sinh(M\eta)}+\frac{y}{2M^3}Cosh(My)\Bigg]\Bigg\}\end{aligned} \tag{22}$$

$$\phi = R\psi - RQe^{iwt} + R^2\delta\left(1 + \frac{e^{2iwt}}{2}\right)\frac{D_1}{2}\left\{\left(\frac{\eta^2}{M} + \frac{\eta \text{Cosh}(M\eta)\text{Sinh}(M\eta)}{M^2}\right)\right.$$
$$-\left(\frac{y\text{Cosh}(My)\text{Sinh}(M\eta)}{M^2} - \frac{2\text{Sinh}(My)\text{Sinh}(M\eta)}{M^3} - \frac{\eta\text{Cosh}(M\eta)\text{Sinh}(My)}{M^2}\right)$$
$$\left.-\left(\frac{\eta y}{M} + \frac{y\text{Cosh}(M\eta)\text{Sinh}(M\eta)}{M^2}\right) - \frac{2\text{Sin}^2 h(M\eta)}{M^3}\right\}$$
$$+R^2\delta i w(Q - \psi_{00})e^{iwt} \tag{23}$$

where

$$A_1 = \left[\frac{\eta^2 \sinh(M\eta)}{M^3} - \frac{1}{8M^4}\eta \text{Cosh}(M\eta) + \frac{\sinh(M\eta}{8M^5} + \frac{\eta^3 \cosh(M\eta)}{4M^2}\right]$$

$$A_2 = E_1 Sinh(M\eta)\left[\frac{\eta^2 Sinh(M\eta)}{4M^3} - \frac{5}{4M^4}\eta Cosh(M\eta) + \frac{Sinh(M\eta)}{8M^5}\right]$$

$$A_3 = \left[\frac{1}{(\sinh(M\eta) - M\eta\cosh(M\eta))}\right]$$

$$A_4 = \frac{\eta}{M^2} - \frac{1}{2M^3}Sinh(2M\eta)$$

$$A_5 = (1 + S)iM^2 Q A_3$$

$$E_1 = \left(\frac{1+S}{2}\right)D_1$$

$D_1 = Q^2 M^5 Sinh(M\eta)\frac{d\eta}{dx}A_3^3$ and
$S\left(\alpha R = \frac{N_0 m}{\rho}\right)$ "Concentrated mass" of dust particles.

The dimensionless gradient of pressure

$$P = \frac{\partial p/\partial x}{\rho \upsilon^2/d^3} = \nabla^2\frac{\partial\psi}{\partial y} - \delta\left\{\frac{\partial^2\psi}{\partial t\partial y} + \frac{\partial\psi}{\partial y}\frac{\partial^2\psi}{\partial x\partial y} - \frac{\partial\psi}{\partial x}\frac{\partial^2\psi}{\partial y^2}\right\} + \alpha\left(\frac{1}{R}\frac{\partial\phi}{\partial y} - \frac{\partial\psi}{\partial y}\right) \tag{24}$$

In light of Eq. (21), take

$$P = (P_{00} + \delta^1 + P_{01} + \delta^2 P_{02} + \ldots)e^{iwt} \tag{25}$$

Making use of Eqs. (14), (21) and (25) for Eq. (24), collecting like terms and rearranging, we get

$$P = \{QM^3 A_3 Cosh(My)$$
$$+\left\{\frac{(1+S)iwQA_3}{2}\left[M^2\eta A_3\, Cosh(M\eta)Cosh(My) - \frac{\eta^2 M^3 A_3 Cosh(My)}{Sinh(M\eta)}\right.\right.$$
$$+3M\text{Cosh}(My) + M^2 y\text{Sinh}(My) - \frac{\eta M^2\text{Cosh}(My)\text{Cosh}(M\eta)}{\text{Sinh}(M\eta)} \tag{26}$$
$$-2M\text{Cosh}(My) + 2M\text{Cosh}(M\eta)]\}\}e^{iwt}$$

The wall shear stress, dimensionless, is obtained as

$$\tau_{xy}\big|_{y=h} = \frac{\tau_{xy}}{\upsilon^2\rho/d^2} = A_3 Q M^2 Sinh(Mh)\, e^{iwt}$$
$$+\delta\left\{\left[A_3 E_1 Sin^2h(M\eta)\left(\frac{\eta Cosh(M\eta)}{8M^3} - \frac{\eta^2 Sinh(M\eta)}{M} - \frac{Sinh(M\eta)}{8M} - \frac{\eta^3 Cosh(M\eta)}{4}\right)\right.\right.$$
$$+E_1 Sinh(M\eta)\left(\frac{\eta^2 Sinh(M\eta)}{4M} - \frac{15 Sinh(M\eta)}{8M^3} - \frac{\eta Cosh(M\eta)}{4}\right)$$
$$\left.-\frac{E_1\eta Sinh(M\eta)Cosh(M\eta)}{M^2} - \frac{E_1\eta^2\text{Cos}^2h(M\eta)}{2M}\right]\left[1+\frac{e^{2iwt}}{2}\right]$$
$$\left.+\frac{A_5}{2M}\left[\left(\eta\text{Cosh}(M\eta)\text{Sinh}(M\eta) - M\eta^2\right) + \frac{2\text{Sinh}(M\eta)}{M}\right]e^{iwt}\right\} \tag{27}$$

## 4 Outcomes and Analysis

Streamlines of particles of fluid are depicted in Figs. 1, 2, 3, 4 and 5. It can be observed that the constricted boundary has very little influence on the streamlines closer to the centerline. This result agrees with that of Radhakrishnamacharya [4].

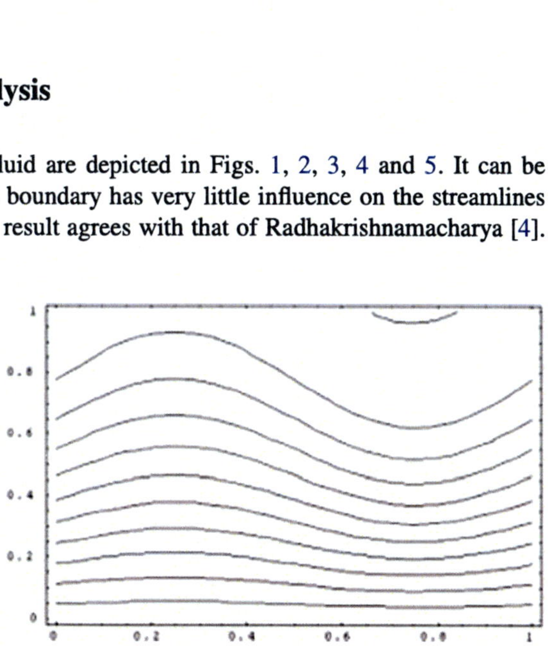


**Fig. 1** Streamlines pertaining to fluid (M = 2, S = 0.3, Q = 0.4, ε = 0.2, δ = 0.2, wt = π/4)

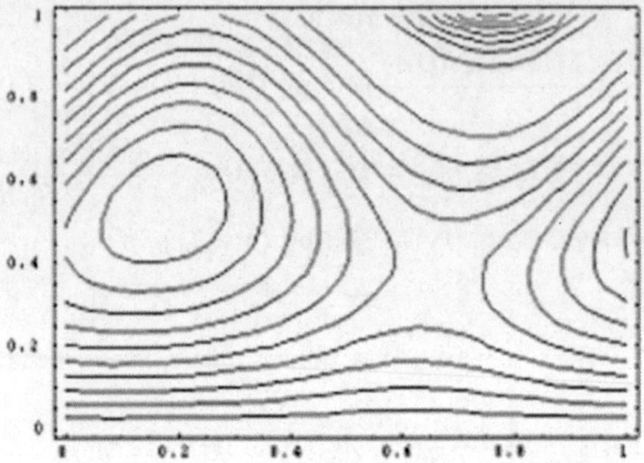

**Fig. 2** Streamlines pertaining to fluid (M = 2, S = 0.3, Q = 0.4, ε = 0.2, δ = 0.2, wt = π/2)

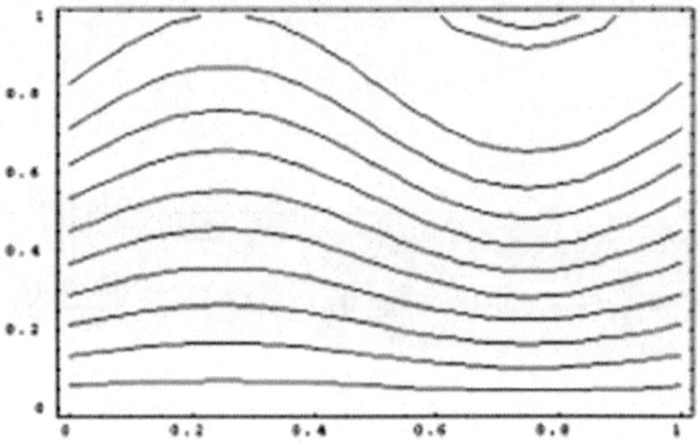

**Fig. 3** Streamlines pertaining to fluid (M = 6, S = 0.3, Q = 0.4, ε = 0.2, δ = 0.2, wt = π/4)

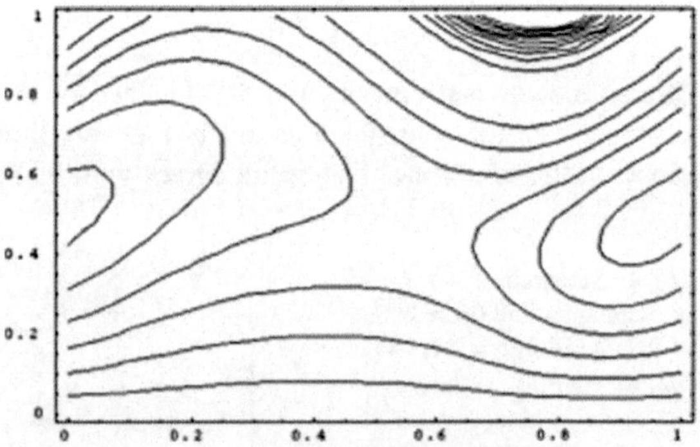

**Fig. 4** Streamlines pertaining to fluid (M = 6, S = 0.3, Q = 0.4, ε = 0.2, δ = 0.2, wt = π/2)

It is of interest to observe that closed loops formed for streamlines when wt = π/2. The effect of the magnetic parameter (M) on streamlines is not significant for wt = π/4, but for wt = π/2, the closed loops get significantly altered. Further, the effect of dust concentration (S) on streamline pattern is not very significant.

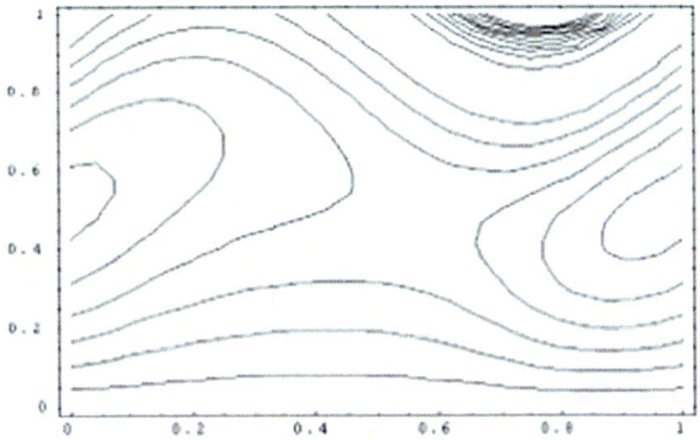

**Fig. 5** Streamlines pertaining to fluid (M = 6, S = 0.5, Q = 0.4, ε = 0.2, δ = 0.2, wt = π/2)

Profiles for the velocity of the fluid particles are shown in Figs. 6, 7 and 8. We can observe that maximum velocity is along the line of center and decreases as y increases up to some value and then increase happens. The nature of the velocity is alike to different cross-sections. It is observed that the velocity value is opposite after a certain y and the marked y changes along x. A decrease in velocity is observed with the parameter of magnetic effects but an increase in velocity is noted with the concentration of dust particles.

Profiles of the velocity of rigid particles are depicted in Figs. 9, 10 and 11. The values of velocity are positive for every value of magnetic constant and concentration value of dust particles. The velocity exhibits oscillatory behavior downstream. Also, oscillatory behavior of velocity is observed with respect to the magnetic parameter. Velocity variation with dust concentration is not significant. Velocities of dust particles close to the centerline are high and also the particles of dust exhibit reverse flow nature near the walls of the channel.

Figures 12, 13 and 14 show that shear stress on the boundary increases with magnetic constant and concentration of dust particles. Shear stress exhibits oscillatory behavior with *x*.

The fluid pressure increases with time wt (Fig. 15) and concentration of dust particles (Fig. 16) but decreases with magnetic constant (Fig. 17). Pressure rise with mass concentration is significant. Pressure change w.r.t. magnetic constant and concentration of mass is similar but opposite (Fig. 18).

Pressure distribution is not uniform in the channel. It coincides with the outcome of [4].

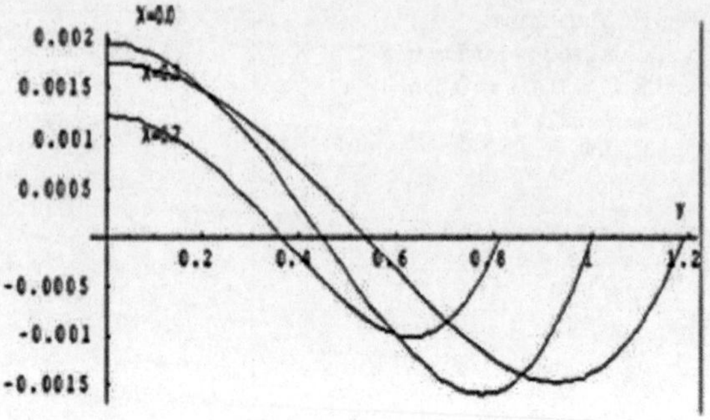


**Fig. 6** Velocity distribution for fluid (M = 2, S = 0.3, Q = 0.4, ε = 0.2, δ = 0.2, wt = π/2)

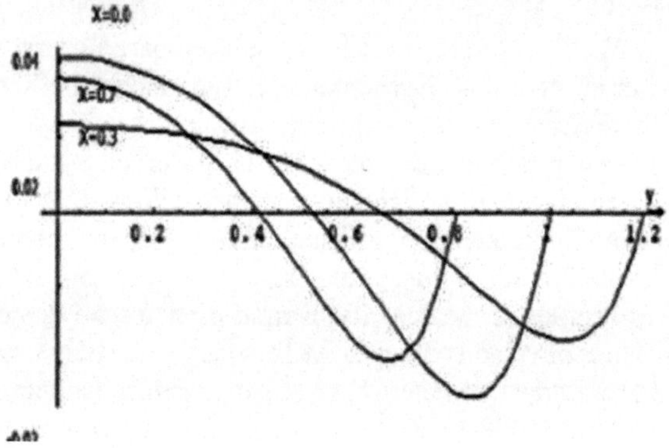


**Fig. 7** Velocity distribution for fluid (M = 6, S = 0.3, Q = 0.4, ε = 0.2, δ = 0.2, wt = π/2

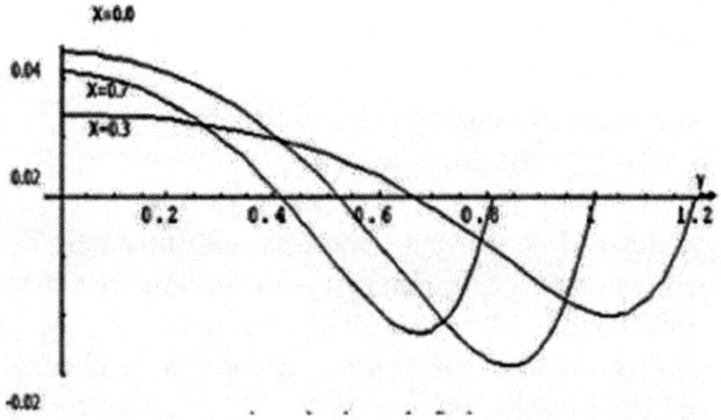


**Fig. 8** Velocity distribution for fluid (M = 6, S = 0.5, Q = 0.4, ε = 0.2, δ = 0.2, wt = π/2

## 5 Conclusion

The flow of dusty fluid in a constricted channel with a transverse magnetic field is analyzed. The perturbation method is adopted for transforming equations into ODES.

It is observed that constricted boundary has little effect on the streamlines closer to the centerline of the channel. In some cases, closed loops formed for stream-

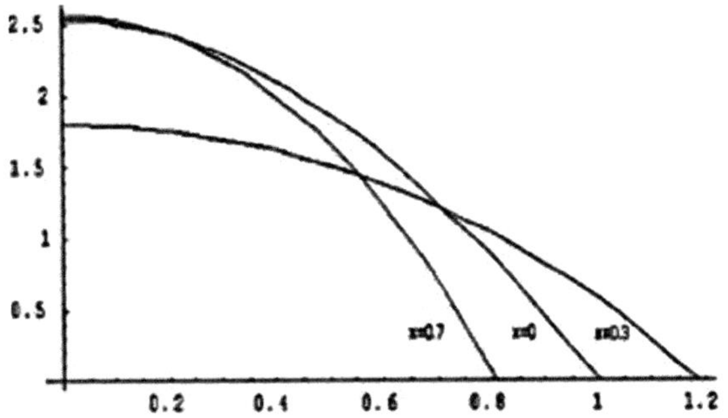


**Fig. 9** Velocity distribution-solid particles (M = 2, R = 6, S = 0.3, Q = 0.4, ε = 0.2, δ = 0.2, wt = π/2

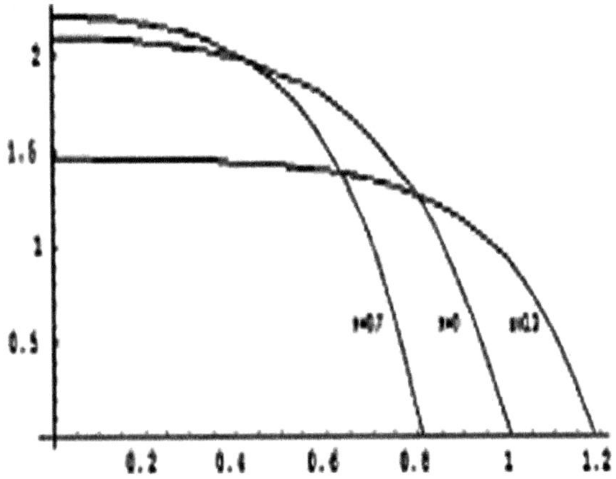


**Fig. 10** Velocity distribution-solid particles (M = 6, R = 6, S = 0.3, Q = 0.4, ε = 0.2, δ = 0.2, wt = π/2)

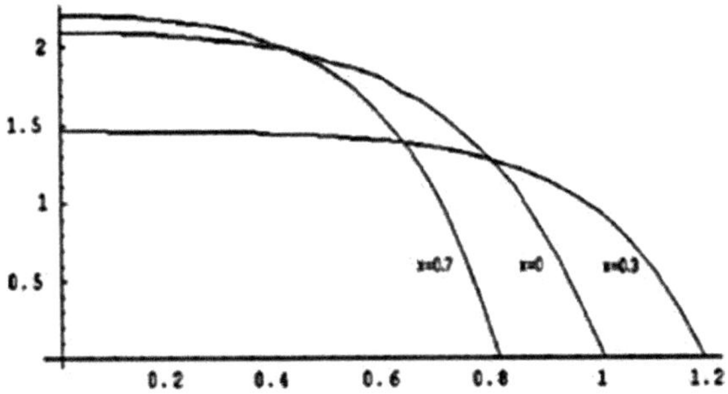


**Fig. 11** Velocity distribution-solid particles (M = 6, R = 6, S = 0.4, Q = 0.4, ε = 0.2, δ = 0.2, wt = π/2)

lines. The velocity of dust particles exhibits oscillatory behavior with respect to a magnetic parameter. Also, dust particles exhibit reverse flow nature near the walls of the channel.

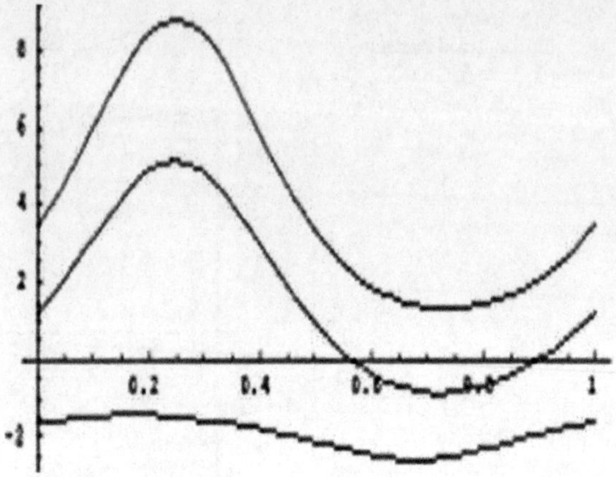


**Fig. 12** Variation of Wallshear stress (M = 5, S = 0.3, Q = 0.4, ε = 0.2, δ = 0.2)

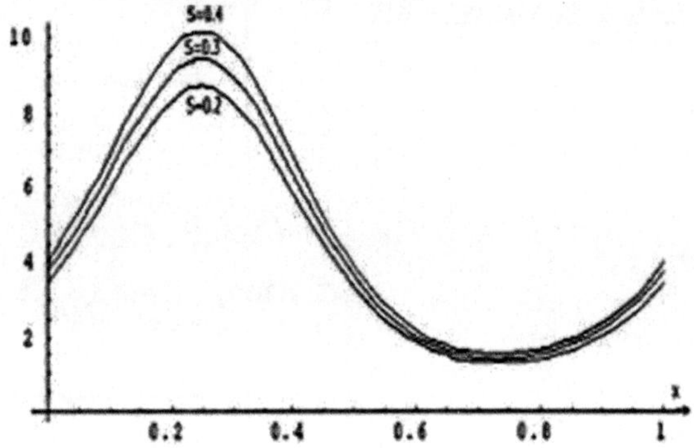


**Fig. 13** Variation of Wallshear stress (M = 6, Q = 0.4, ε = 0.2, δ = 0.2, wt = π/2)

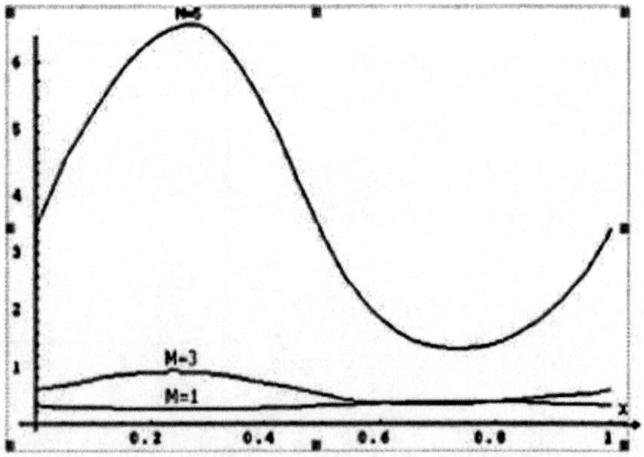


**Fig. 14** Variation of Wallshear stress (S = 0.3, Q = 0.4, ε = 0.2, δ = 0.2, wt = π/2)

**Fig. 15** Pressure variation of the fluid (M = 6, S = 0.3, Q = 0.4, ε = 0.2, δ = 0.2)

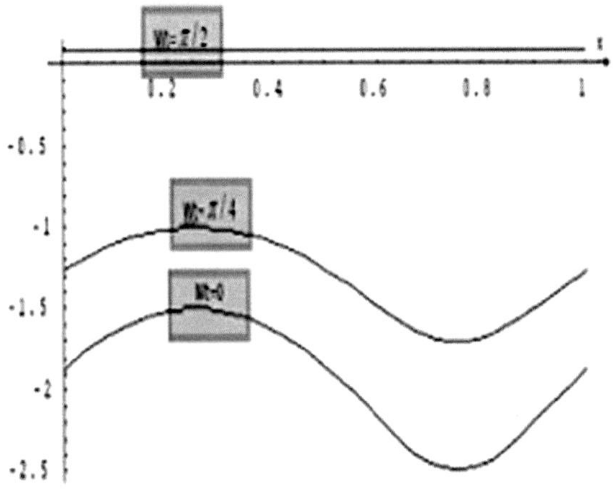

**Fig. 16** Fluid pressure variation (M = 6, Q = 0.2, ε = 0.2, δ = 0.2, wt= π/2)

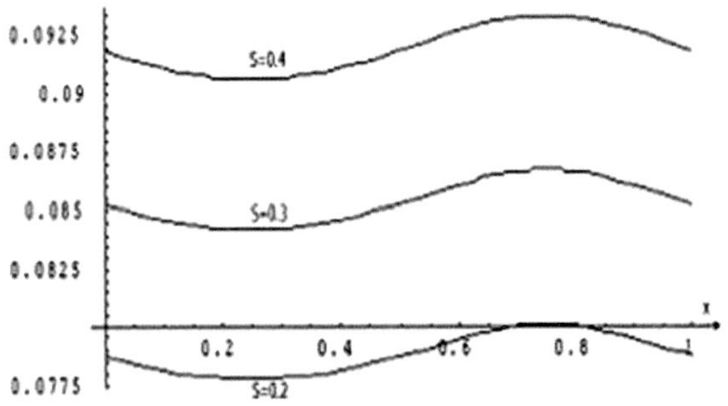


**Fig. 17** Fluid pressure variation (S = 0.3, Q = 0.4, ε = 0.2, δ = 0.2, wt = π/2)

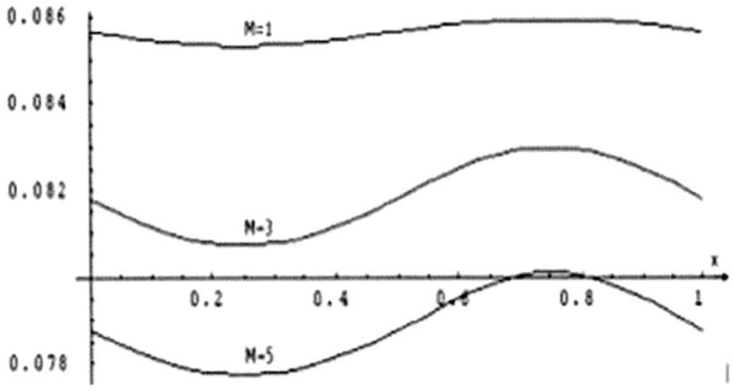

Fig. 18 Flow diagram

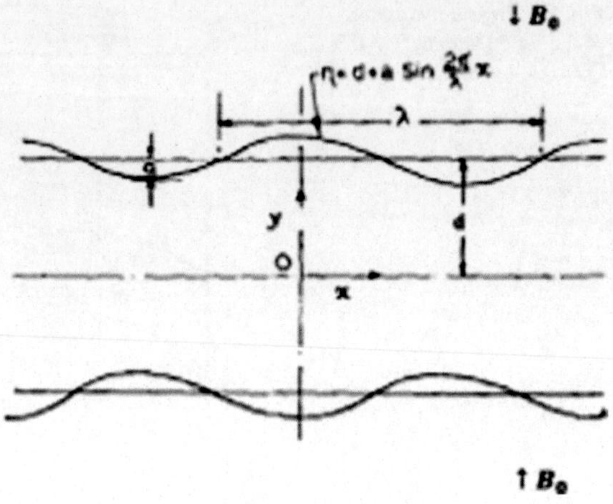


## References

1. Saffman PG (1962) On the stability of a laminar flow of a dusty gas. J Fluid Mech 13:120–128
2. Nayfeh AH (1966) Oscillating two-phase flow through a rigid pipe. AIAA J 4:1868–1870
3. Gupta RK, Gupta SC (1976) Flow of a dusty gas through a channel with arbitrarytime varying pressure gradient. ZAMP 27:119–125
4. Radhakrishnamacharya G (1978) Pulsatile flow of a dusty fluid through a constricted channel. ZAMP 29:217–225
5. Wagh DK, Tapi UD (1996) Investigation of the pulsatile flow of a suspension in the region of mild stenosis. Ind J Theor Phy 44:1–4
6. Afifi NAS, Gad NS (2001) Interaction of peristaltic flow with pulsatile magneto-fluid through a porous medium. Acta Mech 149:229–237
7. Muthuraj R, Srinivas S (2010) A note on heat transfer to MHD oscillatory flow in an asymmetric wavy channel. Int Commun Heat Mass Transf 37(9):1255–1260
8. Muthuraj R, Nirmala K, Srinivas S (2016) Influences of chemical reaction and wall properties on MHD peristaltic transport of a dusty fluid with heat and mass transfer. Alex Eng J 55(1):597–611
9. Lourdu Immaculate D, Muthuraj R, Anant Kant Shukla S (2016) Srinivas "mhd unsteady flow of a williamson nanofluid in a vertical porous space with oscillating wall temperature. Front Heat Mass Transf (FHMT) 7–12
10. Selvi RK, Muthuraj R (2018) MHD oscillatory flow of a Jeffrey fluid in a vertical porous channel with viscous dissipation. Ain Shams Eng J 9(4):2503–2516

# Influence of Radiation Absorption and Thermal Diffusion on Laminar Flow of a Chemically Reacting Fluid Past a Porous Plate in Conducting Field with Rotation

**K. V. B. Rajakumar, K. S. Balamurugan, T. Govinda Rao, and B. Satyanarayana Reddy**

## 1 Introduction

At the contemporary circumstances, there is well-known propulsion for resources edifice to ameliorate the abnormality and apotheosize the eminence in inclusive excipients of study. The last two decennaries have experienced phenomenal amendments in the exploration concerning the boundary layer flow of Newtonian fluids with reference to heterogeneous conditions across the intercontinental. It is eminence of conviction that the noteworthy researcher augmented the Newtonian fluid on Magneto Hydrodynamics (MHD) free convective rotating fluid flow past a non-parallel permeable plate in occurrence of thermal radiation as well as chemical reaction over the appropriate concentrate of base fluid beneath suitable conditions. Numerous examiners scrutinized the applications in several industrial, ion propulsion, astrophysical, geophysical, technological and engineering, design of turbo mechanics and turbines etc. Ananda Reddy et al. [1] reported impact of chemical reaction as well as thermal diffusion on MHD rotating fluid flow past a permeable plate in conducting field along with radiation absorption. Seth at al. [2, 3] addressed consequence of rotation on MHD flow of a fluid past an impetuously fluctuating non-parallel plate with ther-

K. V. B. Rajakumar (✉)
Department of Mathematics, Kallam Haranadhareddy Institute of Technology, Guntur 522019, India
e-mail: kvbrajakumar@khitguntur.ac.in

K. S. Balamurugan
Department of Mathematics, R V R & J C College of Engineering, Guntur 522019, India

T. Govinda Rao
Department of Mathematics, Gudlavalleru Engineering College, Gudlavalleru 521356, India

B. Satyanarayana Reddy
Department of Computer Science and Engineering, Kallam Haranadhareddy Institute of Technology, Guntur 522019, India



S. Srinivas et al. (eds.), *Recent Advances in Applied Mathematics and Applications to the Dynamics of Fluid Flows*, Lecture Notes in Mechanical Engineering,
https://doi.org/10.1007/978-981-19-1929-9_7

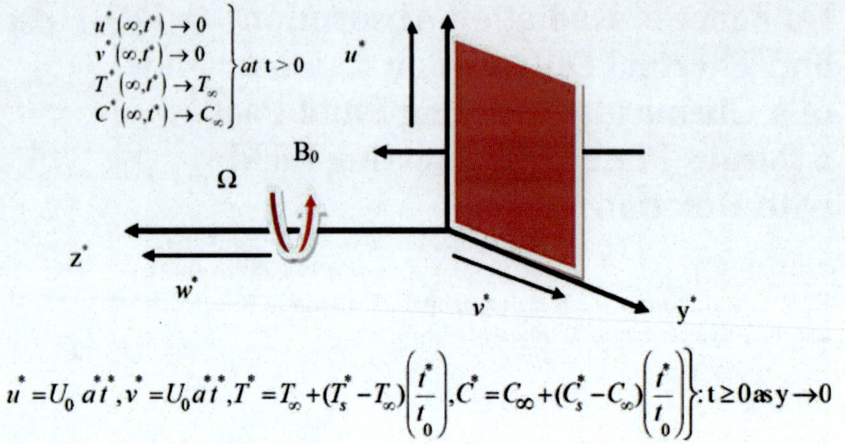


**Fig. 1** Physical model of the problem

mal diffusion along with ramped temperature. Jithender Reddy et al. [4] disclosed significance of dufour effect along with rotational on unsteady MHD natural convection fluid flow past an impetuously affecting vertical permeability plate encapsulated in propinquity of ramped temperature in presence of thermal radiation. Rudra Kanta et al. [5] contemplate two heterogeneous solutions proportional to ramped as we; as constant wall temperature in aspect of thermal ration on MHD free convective flow past an infinite vertical plate enclosing a ramped type temperature concerning time. Nandkeolyar et al. [6] explored transverse magnetic field on unsteady MHD natural convection dusty fluid flow past an impulsively moving vertical plate by means of ramped temperature as well as thermal radiation. Rajakumar et al. [7–9] and Ganapathy [10] addresseda communication on oscillatory couette flow in a rotating system (Fig. 1).

Stimulated by all these explorations along with applications of Newtonian fluid in our contemporary discipline existence at the moment, this effort is determined to scrutinize the interface of thermal diffusion and chemical reaction on MHD rotating fluid flow past a permeable plate along with radiation absorption, having scrutinized un addressed in the novelby means of the best of our acquaintance. The current exploration is that the governing PDE's are resolute employed single perturbation method.

## 2 Mathematical Formulation

Contemplate radiation absorption as well as thermal diffusion MHD rotating fluid flow past a vertical porous plate by means of variable temperature as well as variable concentration along with chemical reaction.

- Here $x^*$-axis is taken along the plate which is in vertical direction and $y^*$-axis is taken normal to the surface of the plate.
- The plate as well as the fluid in a state of rigid body rotates with a uniform angular velocity $\Omega$ about $y^*$-axis.
- At time $t^* > 0$, the plate starts moving in $x^*$-direction with uniform velocity $U_0 a^* t^*$.
- The temperature and concentration raises to $T_\infty + (T_s^* - T_\infty)\left(\frac{t^*}{t_0}\right)$ and $C_\infty + (C_s^* - C_\infty)\left(\frac{t^*}{t_0}\right)$ respectively.

Under these considerations the equations that govern the flow are as follows.

$$\left[\frac{\partial u^*}{\partial t^*}\right] + 2\,\Omega\,[v^*] = \vartheta\left[\frac{\partial^2 u^*}{\partial y^{*2}}\right] + g\beta_T\,[T^* - T_\infty^*] + g\beta_C^*\,[C^* - C_\infty^*] - \frac{\sigma B_0^2}{\rho}\,[u^*] - \frac{v}{k}[u^*] \tag{1}$$

$$\left[\frac{\partial v^*}{\partial t^*}\right] - 2\Omega\,[u^*] = \vartheta\left[\frac{\partial^2 v^*}{\partial y^{*2}}\right] - \frac{\sigma B_0^2}{\rho}\,[v^*] - \frac{v}{k}[v^*] \tag{2}$$

$$\left[\frac{\partial T^*}{\partial t^*}\right] = \frac{K_T}{\rho C_p}\left[\frac{\partial^2 T^*}{\partial y^{*2}}\right] - \frac{1}{k\rho C_p}\left[\frac{\partial q_r^*}{\partial y^*}\right] + \frac{D_m K_T}{C_S C_P}\left[\frac{\partial^2 C^*}{\partial y^{*2}}\right] + \frac{Q_0}{\rho C_p}\,[T^* - T_\infty] + \frac{R^*}{\rho C_p}\,[C^* - C_\infty] \tag{3}$$

$$\left[\frac{\partial C^*}{\partial t^*}\right] = D\left[\frac{\partial^2 C^*}{\partial y^{*2}}\right] - K_1^*\,[C^* - C_\infty^*] \tag{4}$$

The corresponding boundary conditions are

$$if\;\; t > 0 \Rightarrow \begin{cases} u^* = 0,\, v^* = 0,\;\; T^* = T_\infty^*,\, C^* = C_\infty^* \;\; for\;\; y^*, t^* \le 0 \\ \begin{cases} u^* = U_0\;\, a^* t^*,\;\; v^* = U_0\, a^* t^*, \\ T^* = T_\infty + (T_s^* - T_\infty)\left(\frac{t^*}{t_0}\right), \;\; at\;\; y^* = 0 \\ C^* = C_\infty + (C_s^* - C_\infty)\left(\frac{t^*}{t_0}\right) \end{cases} \\ u^* \to 0,\;\; v^* = 0\;\; T^* \to T_\infty,\;\; C^* \to C_\infty \quad as\;\; y^* \to \infty \end{cases} \tag{5}$$

Characteristic time $t_0$ is defined according to the non-dimensional process mentioned above as

$$t_0 = \frac{v}{U_0^2} \tag{6}$$

Let us introduce the following dimensionless quantities:

$$u = \left[\frac{u^*}{U_0}\right],\ v = \left[\frac{v^*}{U_0}\right],\ y = \left[\frac{y^* U_0}{\upsilon}\right],\ t = \left[\frac{U_0^2 t^*}{\upsilon}\right],\ a = \left[\frac{a^* \upsilon}{U_0^2}\right],\ T^* = T_\infty + \theta\left[T_s^* - T_\infty^*\right],$$
$$C^* = C_\infty + C\left[C_s^* - C_\infty^*\right] \quad (7)$$

term by using the Rosseland approximation is given by

$$q_r^* = -\frac{4\sigma_s^*}{3k_e^*}\frac{\partial T^{*^4}}{\partial y^*} \quad (8)$$

$$T^{*^4} \cong 4T_\infty^{*^3} T^* - 3T_\infty^{*^4} \quad (9)$$

Substituting Eqs. (6)–(9) in (1)–(5) then the resultant non-dimensional equations are

$$\left[\frac{\partial^2 u}{\partial y^2}\right] - \left[\frac{\partial u}{\partial t}\right] - 2k_1^2\,[v] + Gr\ [\theta] + Gm\ [C] - [u]\ N = 0 \quad (10)$$

$$\left[\frac{\partial^2 v}{\partial y^2}\right] - \left[\frac{\partial v}{\partial t}\right] + 2k_1^2\,[u] - [v]\ N = 0 \quad (11)$$

$$\left[\frac{\partial^2 \theta}{\partial y^2}\right] - \left[\frac{\partial \theta}{\partial t}\right] + \eta\,[\theta] + Dr\,\Pr\left[\frac{\partial^2 C}{\partial y^2}\right] + Ra\,\Pr\,[\phi] = 0 \quad (12)$$

$$\left[\frac{\partial^2 C}{\partial y^2}\right] - Sc\left[\frac{\partial C}{\partial t}\right] - Sc\ Kr\ [C] = 0 \quad (13)$$

The analogous conditions are

$$\begin{cases} u = 0, v = 0, \theta = 0, C = 0 & \forall\ y,\ t \le 0 \\ t > 0: \begin{cases} u = a\,t,\ \ v = a\,t,\ \theta = t,\ C = t\ \ at\ \ y = 0 \\ u = 0,\ \ v = 0,\ \ \theta = 0,\ C = 0\ \ as\ \ y \to \infty \end{cases} \end{cases} \quad (14)$$

Combined Eqs. (10) and (11) then we obtained

$$\frac{\partial^2 \phi}{\partial y^2} - \frac{\partial \phi}{\partial t} + \lambda\phi + Gr\theta + GmC = 0 \quad (15)$$

Here

$$\phi = u + i\ v\ \ \Rightarrow\ \ \overline{\phi} = u - i\ v \quad (16)$$

The boundary conditions are

$$\begin{cases} \phi = 0, \theta = 0, C = 0 & \forall\ y,\ t \le 0 \\ t > 0: \begin{cases} \phi = a\,t,\ \ \ \theta = t,\ C = t\ \ at\ \ y = 0 \\ \phi = 0,\ \ \ \theta = 0,\ C = 0\ \ as\ \ y \to \infty \end{cases} \end{cases} \quad (17)$$

$$\left\{\begin{array}{r} Ra = \frac{R^{*}\upsilon(C_s^{*}-C_{\infty})}{\rho C_p U_0^2 (T_s^{*}-T_{\infty})}, \lambda = (2k_1^2 i - N), N = \left(M + \frac{1}{K}\right), \eta = \Pr(Q - F), \\ \Gamma = \frac{1}{\Pr}, Gr = \frac{\upsilon g \beta_T (T_s^{*}-T_{\infty})}{U_0^3}, Gm = \frac{\upsilon g \beta_C (C_s^{*}-C_{\infty})}{U_0^3}, M = \frac{\sigma B_0^2 \upsilon}{\rho U_0^2}, \\ k_1^2 = \frac{\Omega \upsilon}{U_0^2}, Sc = \frac{\upsilon}{D}, K = \frac{kU_0^2}{\upsilon^2}, \Pr = \frac{\rho \upsilon C_p}{k_T}, Kr = \frac{K_1 \upsilon}{U_0^2}, Q = \frac{Q^{*}\upsilon^2}{k_T U_0^2}, F = \frac{4\sigma T_{\infty}^{*3}}{k_e k}, \\ Dr = \frac{D_m k_T (C_s^{*}-C_{\infty})}{\upsilon C_s C_p (T_s^{*}-T_{\infty})} \end{array}\right. \tag{18}$$

Solve the above equations utilizing Single perturbation method by use the following relation

$$\phi = \phi_0 e^{i\,nt},\ \theta = \theta_0\, e^{i\,nt},\ C = C_0 e^{i\,nt} \tag{19}$$

Substitute Eq. (19) in the Eqs. (12), (13) and (15)

$$\phi_0'' + (\lambda - in)\,\phi_0 = -Gr\,\theta_0 - GmC_0 \tag{20}$$

$$\Gamma\,\theta_0'' - (-\eta + in)\,\theta_0 = -Ra\,C_0 - Dr\,C_0'' \tag{21}$$

$$C_0'' - (Kr + in)ScC_0 = 0 \tag{22}$$

The interrelated boundary conditions are

$$t > 0: \begin{cases} \phi_0 = a\,te^{-\text{int}}, & \theta_0 = te^{-\text{int}},\ C_0 = te^{-\text{int}}\ \ at\ \ y = 0 \\ \phi_0 = 0, & \theta_0 = 0,\ C_0 = 0\ \ as\ \ y \to \infty \end{cases} \tag{23}$$

Solve Eqs. (20)–(22) by using (23) then we get Velocity, Temperature and concentration

$$\begin{aligned} \phi = {} & \left(at\,e^{-\text{int}} - N_2 - N_3\right) e^{-(\sqrt{\lambda - i\,n})y} e^{\text{int}} + N_2 e^{-\sqrt{\frac{(i\,n-\eta)}{\Gamma}}y} e^{\text{int}} \\ & + N_3 e^{-(\sqrt{Sc\,(Kr+in)})y} e^{\text{int}} \end{aligned} \tag{24}$$

$$\theta = \left(\left(te^{-\text{int}} - N_1\right) e^{-\sqrt{\frac{(in-\eta)}{\Gamma}}y} + N_1 e^{-(\sqrt{Sc\,(Kr+in)})y}\right) e^{\text{int}} \tag{25}$$

$$C = \left(t\,e^{-\text{int}} e^{-(\sqrt{Sc\,(Kr+in)})y}\right) e^{\text{int}} \tag{26}$$

### 2.1 Skin-Friction

$$\tau = \left(\frac{d\phi}{dy}\right)_{y=0} \tag{27}$$

### 2.2 Rate of Heat Transfer

$$N_w = -\left(\frac{d\theta}{dy}\right)_{y=0} \tag{28}$$

### 2.3 Rate of Mass Transfer

$$S_h = -\left(\frac{dC}{dy}\right)_{y=0} \tag{29}$$

## 3 Results and Discussion

The repercussions of $Pr$ on the velocity and temperature are illustrated on Figs. 2 and 3 from these figures it was ascertained that velocity as well as temperature declined with the progressive values of $Pr$. Due to the assuredness that an elevated $Pr$ fluid has partially low thermal conductivity, which derogates the conduction as a consequence temperature declined. From Figs. 4 and 5 it was establish that the accumulative values of thermal diffusion results in accelerated in the velocity as

**Fig. 2** Effect of $Pr$ on velocity

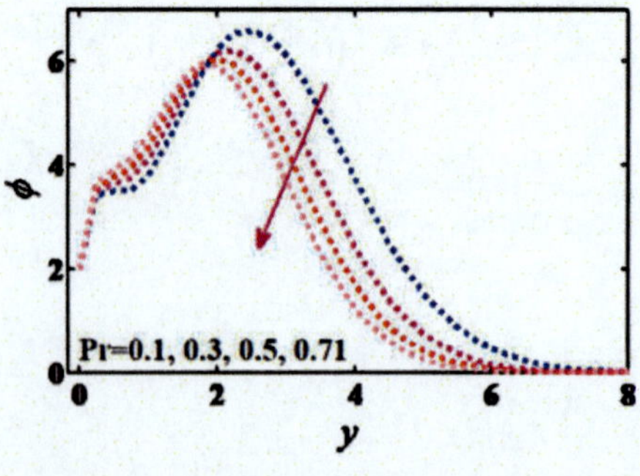

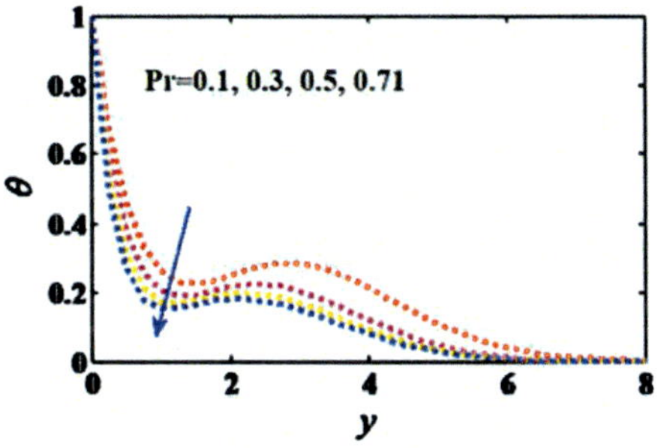


**Fig. 3** Effect of $Pr$ on temperature

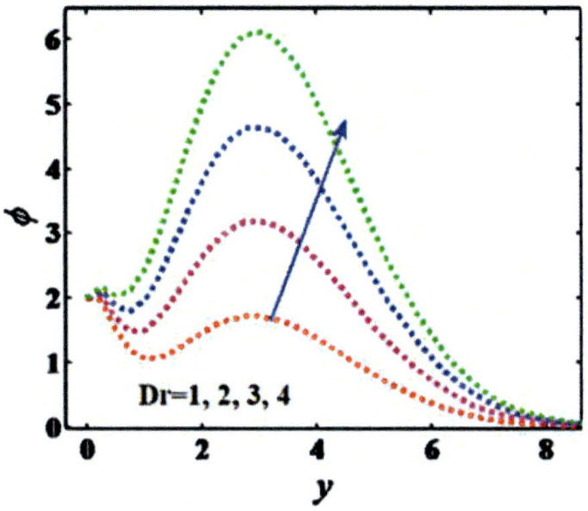


**Fig. 4** Effect of $Dr$ on velocity

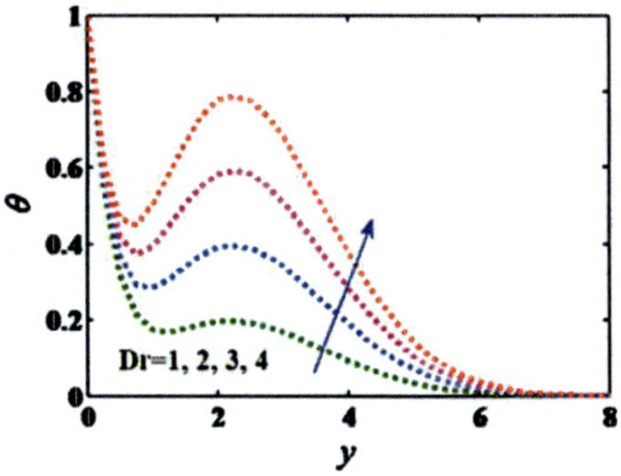


**Fig. 5** Effect of $Dr$ on velocity

well as temperature. Due to the fact that thermal diffusion mechanisms to monitor the importance of chemical concentration on fluid temperature. Impact of radiation absorption on velocity along with temperature was portrayed in Figs. 6 and 7 these figures addressed that velocity as well as temperature rises under the influence of Ra. This is somatic veracious since escalate in $Ra$ resulting heat accretion which augment the thermal boundary layer as well as temperature. Concomitant ahead this, convection current is reinforce hence, the momentum boundary layer elevates and fluid velocity emerges also. The effect of angular rotation $k1$ is illustrated in

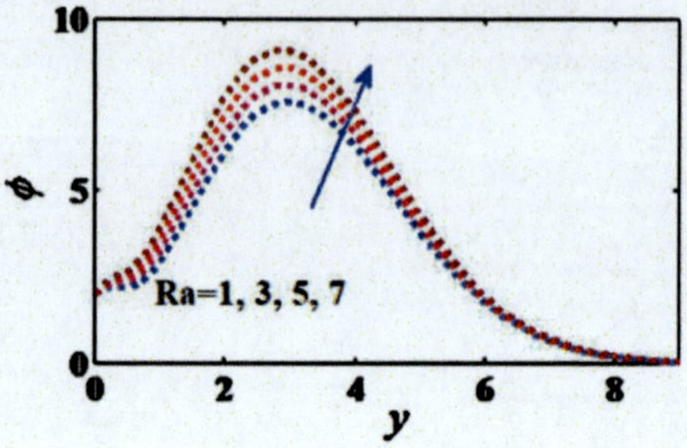


**Fig. 6** Effect of *Ra* on velocity

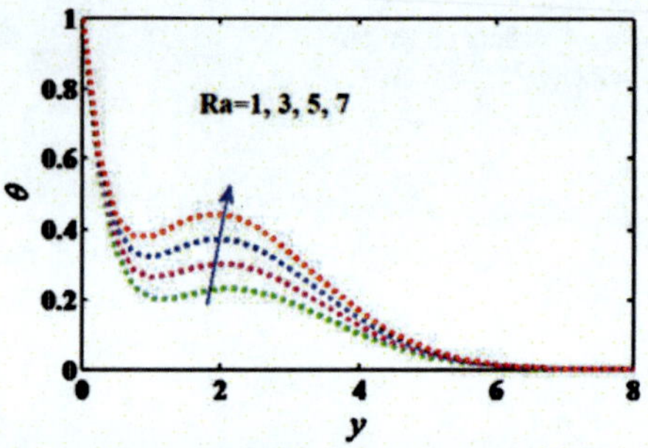


**Fig. 7** Effect of *Ra* on temperature

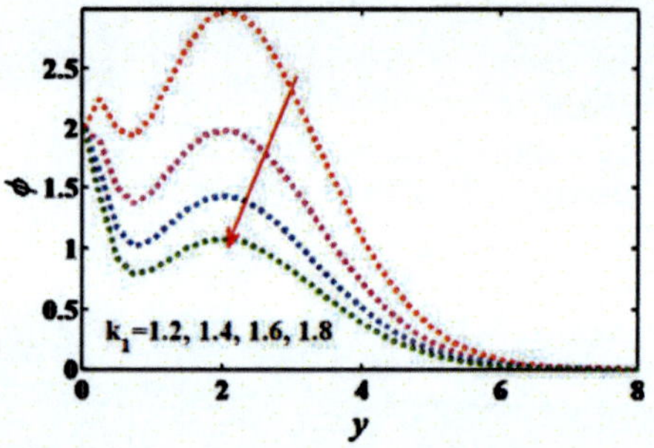


**Fig. 8** Effect of $k_1$ on velocity

Fig. 8; this figure conspicuous that velocity is lessen with accumulated values of $k_1$. The impact of chemical reaction $Kr$ was demonstrated in Fig. 9 as of this figure the results show that concentration reduced by means of the augmentation values of chemical reaction. Nevertheless, Fig. 10 expressed the results concentration declined by means of progressive values of $Sc$. The influence of Skin-friction was publicized in Fig. 11 it addressed that the improvement of $Gr,\ M,\ Pr,\ Kr,\ Sc,\ Ra,\ Dr$ and $Q$ leads to enchased in skin friction nevertheless contrary impact was take place in case of $K,\ k_1$. Figure 12 demonstrated that enhancement of $Ra,\ Pr,\ Dr$ and Qresults Nusselt number amplified. Effect of $Sc$ , $Kr$ was illustrated in Fig. 13. Here it was perceived that the accelerated estimators of $Kr$ , $Sc$ lead to reinforce in Sherwood number (Fig. 14).

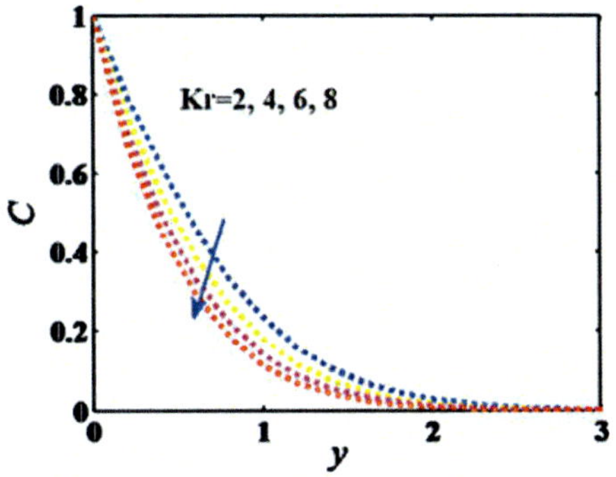


**Fig. 9** Effect of $Kr$ on concentration

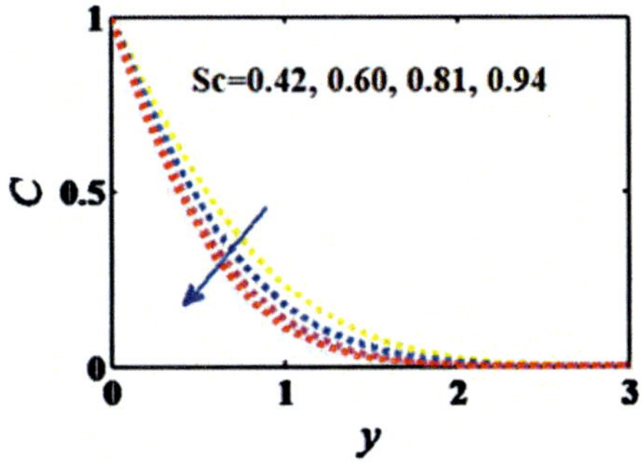


**Fig. 10** Effect of $Sc$ on concentration

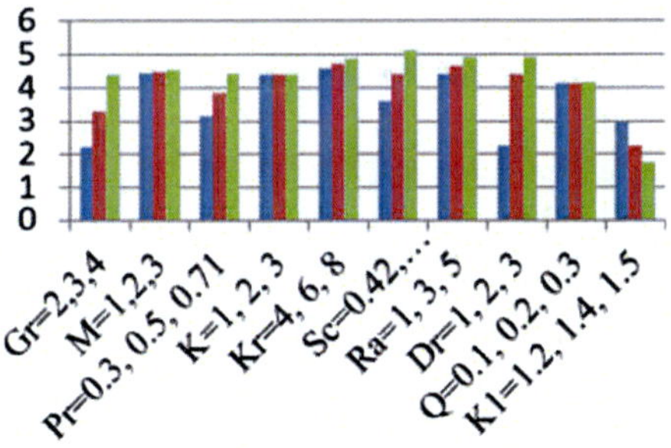


**Fig. 11** Effect of various parameters on skin friction

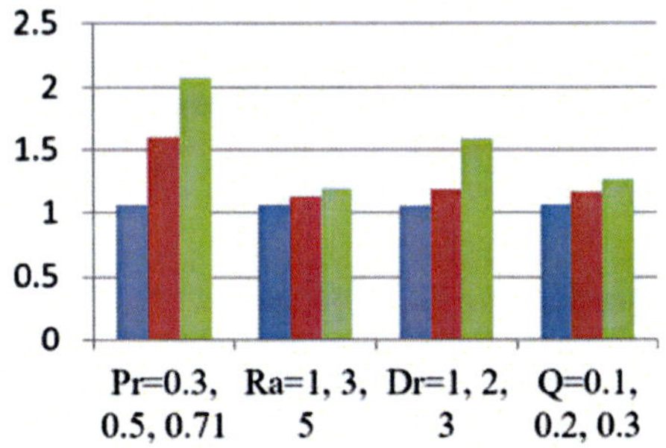


**Fig. 12** Effect of $Pr, Ra, Dr, Q$ on Nusselt number

Fig. 13 Effect of $Sc$ and $Kr$ on Sherwood number

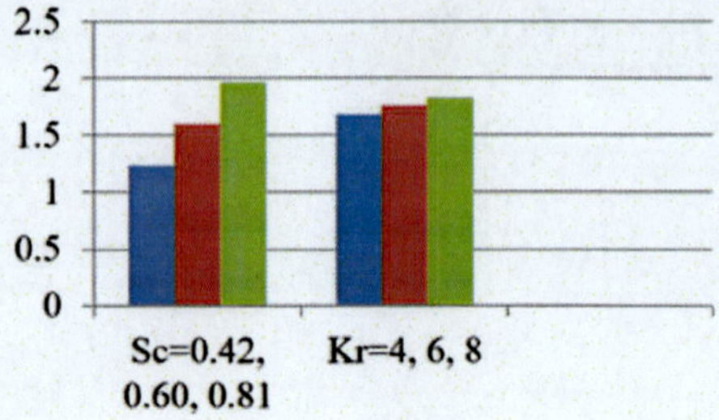


Fig. 14 Effect of $Pr$ on temperature

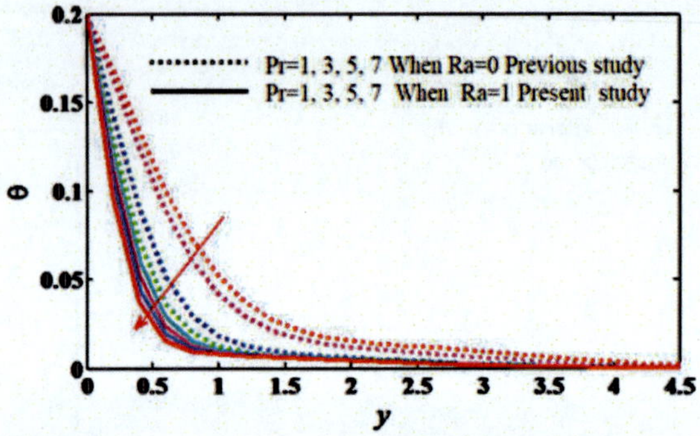


## 4 Validation of the Results

The current reconnaissance is an extended work of Ananda Reddy et al. [1] with Radiation absorption and dufour effect. In this exploration it was contemplate $Ra = 0$ when $Dr = 1, Q = 0.5, F = 0.5, a = 0.2, Kr = 0.2, K = 0.2, Gr = 5, Gm = 5, k1 = 0.5$; our outcomes are excellent agreement with the repercussions of Ananda Reddyet al. [1]. The following Fig. 13 represents current examination compared with previous analysis.

## 5 Conclusions

- Velocity and temperature are accelerated with the incremental values of thermal diffusion as well as radiation absorption.
- Velocity and temperature are declined with the enhancement values of Prandtl number.
- Velocity diminished with the incremental values of angular rotation.
- Concentration declined with the incremental values of chemical reaction and Schmidt number.
- Nusselt number enhanced by means of the incremental values of Pr, Dr, Ra & Q.

## References


1. Reddy, A.N., Reddy, C.P., Raju, M.C., Varma, S.V.K.: Radiation and dufour effects on laminar flow of a rotating fluid past a porous plate in conducting field. Front. Heat Mass Transf. **10**(4) (2018)
2. Seth, G.S., Nandkeolyar, R., Ansari, M.S.: Effect of rotation on unsteady hydro magnetic natural convection flow past an impulsively moving vertical plate with ramped temperature in a porous medium with thermal diffusion and heat absorption. Int. J. Appl. Math. Mech. **7**(21), 52-69 (2011)
3. Seth, G.S, Nandkeolyar, R., Ansari, M.S.: Effects of thermal radiation and rotation on unsteady hydro magnetic free convection flow past an impulsively moving vertical plate with ramped temperature in a porous medium. J. Appl. Fluid Mech. **6**(1), 27-38 (2013)
4. Reddy, G.J., Raju, R.S., Manideep, P., Rao, J.A.: Thermal diffusion and diffusion thermo effects on unsteady MHD fluid flow past a moving vertical plate embedded in porous medium in the presence of Hall current and rotating system. Trans. A. Razmadze Math. Inst. **170**, 243-265 (2016)
5. Rudra Kanta Deka, Sankar Kumar Das., Radiation effects on free convection flow near a vertical plate with ramped wall temperature. Engineering, 3, pp.1197-1206(2011)
6. Nandkeolyar, R., Seth, G.S., Makinde, O.D., Sibanda, P., Ansari, M.S.: Convection flow of a dusty fluid past an impulsively moving vertical plate with ramped temperature in the presence of thermal radiation. J. Appl. Mech. **80**, 1-13 (2013)
7. Rajakumar, K.V.B., Govinda Rao, T., Umasankara Reddy, M., Balamurugan, K.S.: Influence of Dufour and thermal radiation on unsteady MHD Walter's liquid modelB flow past an impulsively started infinite vertical plate embedded in a porous medium with chemical reaction, Hall and ion slip current. SN Appl. Sci. **2**, 1-71 (2020)
8. Rajakumar, K.V.B., Pavan Kumar, R.V.S.R., Balamurugan, K.S.: Unsteady MHD Casson dissipative fluid flow past a semi-infinite vertical porous plate with radiation absorption and chemical reaction in presence of heat generation. Math. Model. Eng. Prob. **7**, 160–172 (2020)
9. Rajakumar, K.V.B., Umasankara Reddy, M., Balamurugan, K.S.: Steady MHD Casson Ohmic heating and viscous dissipative fluid flow past an infinite vertical porous plate in the presence of Soret, Hall, and ionslip current. Heat Transfer. **49**, 1583–1612 (2020)
10. Ganapathy, R.: A note on oscillatory Couette flow in a rotating system. Trans. ASME J. Appl. Mech. **6**, 208–209 (1994)

# Non-linear Magnetoconvection with Modulated Rotational Speed in Viscoelastic Liquid

**G. Jayalatha and N. Suma**

## 1 Introduction

Rayleigh–Bènard convection (RBC) and heat transfer in Newtonian liquids are studied by many authors (see [1, 2]).

The convection due to surface tension and buoyancy in viscoelastic liquid of Maxwellian and Oldroyd has been studied by many authors (see [3–12]).

The stability of the liquid layer as a consequence of Coriolis force is of much interest as the rotation imparts an elastic-like behaviour. The instability in viscoelastic liquid with the Coriolis force effect is examined by these authors (see [13–18]).

The chaotic convection and aperiodic behaviour of the Oldroyd liquid B have been studied by [19]. Park and Lee [20] have studied non-linear hydrodynamic convection in viscoelastic liquids. Siddheshwar and Srikrishna [21] and Abu-Ramadan et al. [22] have considered viscoelastic liquid chaotic convection. Non-linear convection in viscoelastic liquid of Oldroyd B type has been studied by [23]. Narayana [24] examined chaotic and linear convection in the case of viscoelastic liquid of binary type.

Convection with temperature/gravity modulation has been studied by many authors (see [25–33]).

Bhadauria and Kiran [34] have carried out a theoretical investigation to understand the joint effects of rotational modulation as well as internal heating thermal instability in a horizontal liquid layer. Non-linear double-diffusive convection with rotation modulation using a Fourier series (truncated) is studied by [35]. The Fehlberg45 method is employed for measuring mass and heat transport. Consequences of rotation modulation on liquids of Oldroyd B type for double-diffusive convection is analysed by [36]. Kanchana et al. [37] examined the effect of rotational modulation in Newtonian and Newtonian nano liquids. A study of non-linear and linear con-

G. Jayalatha · N. Suma (✉)
Rashtreeya Vidyalaya College of Engineering, Bengaluru 560059, India
e-mail: sumanm@rvce.edu.in

G. Jayalatha
e-mail: jayalathag@rvce.edu.in



S. Srinivas et al. (eds.), *Recent Advances in Applied Mathematics and Applications to the Dynamics of Fluid Flows*, Lecture Notes in Mechanical Engineering,
https://doi.org/10.1007/978-981-19-1929-9_8

vection of RBC in Oldroyd B liquid under rotational modulation has been done by [38]. Few authors have considered magnetoconvection in viscoelastic liquids. Bhatia and Steiner [39–41], Takashima [42], Sharma [43], Eltayeb [44] and Sharma and Aggarwal [45] have studied magnetoconvection in constant viscosity viscoelastic liquids.

Liquids that are viscoelastic in nature are non-Newtonian liquids which are generally made use of that have vital industrial applications. Paints which are composed of a polymer in combination with a solvent are the most real-time examples of viscoelastic liquids. The corrugated surfaces' fabrication and duplication of Deoxyribonucleic acid (DNA) are a few important applications which involve viscoelastic liquids. These applications require natural convection mechanism. In many practically important problems, the control of convection is important and one of the ways to achieve this is by modulating one of the parameters of the problem. The modulated centrifugal force has applications in many fields of engineering like the chemical industry, food industry, solidification of alloys and rotating turbo machinery. The viscoelastic liquids become electrically conducting when suitable additives are introduced which makes the liquid respond to electromagnetic fields.

It is obvious that, from the literature survey, the effect of rotational modulation in the non-linear RBC pertaining to weakly electrically conducting viscoelastic liquid has not been addressed. Anjana and Vanishree [28] have worked on the linear theory of the present problem without magnetic field effect. In the present work, non-linear magnetoconvection with the modulated rotational speed in viscoelastic liquid is investigated.

## 2 Governing Equations

An infinite horizontal liquid layer of Oldroyd B viscoelastic type and with weak electrical conductivity having thickness $d$ is considered. The boundary layers of liquid are kept at different constant temperatures $T_0 + \Delta T$ & $T_0$ at $z = 0$, lower plate & $z = d$, upper plate, respectively, where $T_0$ denotes the reference temperature, and $\Delta T$ denotes the difference in temperature between the two surfaces. In the vertical direction $H_0$, a magnetic field is applied. In the rotation modulation case, it is assumed that viscoelastic liquid layer is rotating about the $z$-axis, with $\overrightarrow{\Omega_1} = \Omega_1(1 + \epsilon \cos \omega t)\hat{k}$, a time-periodic angular velocity (see Fig. 1).

The governing equations of the physical problem are as follows:

$$q_{i,i} = 0, \tag{1}$$

$$\rho_0 \left[ \frac{\partial q_i}{\partial t} + q_j \, q_{i,j} \right] = \left[ \frac{\rho_0 | \overrightarrow{\Omega_1} \times \overrightarrow{r} |^2}{2} - p \right]_{,i} + \tau'_{ij,j} + \rho g_i + 2 \, \rho_0 \, \varepsilon_{ijk} \, q_j \, \Omega_{1k} - \sigma_1 \mu_m^2 H_0^2 q_i, \tag{2}$$

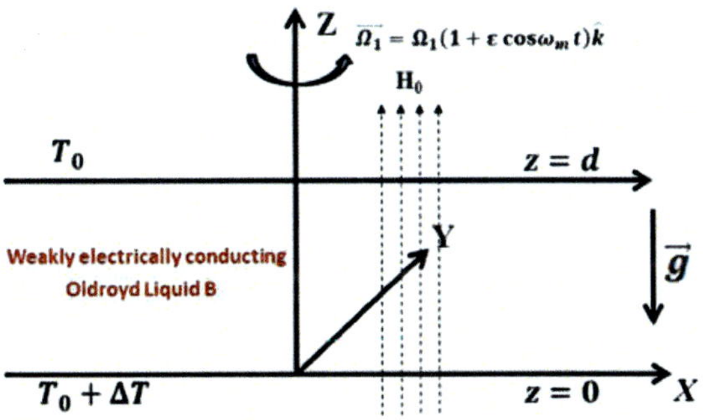


**Fig. 1** A Sketch of the physical problem

$$\left[\tau'_{ij} + \lambda_1 \frac{\partial \tau'_{ij}}{\partial t}\right] = \mu \left[1 + \lambda_2 \frac{\partial}{\partial t}\right] [\,(q_{i,j} + q_{j,i})], \tag{3}$$

$$\frac{\partial T}{\partial t} + q_j T_{,j} = [\kappa T_{,jj}], \tag{4}$$

$$\rho = [\rho_0 - \rho_0 \alpha (T - T_0)]. \tag{5}$$

Table 1 gives the nomenclature for the variables and constants used in the above equations.

The liquid is at rest in basic state and the heat transfer is due to conduction. The expression for basic state pressure and temperature are given in the following form:

$$\left.\begin{aligned} q_{ib} = (0,0,0),\ T_b(z) &= \Delta T\left(1 - \frac{z}{d}\right) + T_0, \\ \rho_b(z) &= \left[1 + \alpha \Delta T\left(\frac{z}{d} - 1\right)\right]\rho_0, \\ p_b(z) &= \left[\left(z - \frac{z^2}{2d}\right)\alpha \Delta T - z\right]\rho_0\, g + \text{constant}. \end{aligned}\right\} \tag{6}$$

Perturbations of finite amplitude are superimposed on the basic state which is given in the form:

$$q_i = q_{ib} + q'_i, \quad p = p_b + p', \quad \rho = \rho' + \rho_b \text{ and } T = T' + T_b, \tag{7}$$

where perturbed quantities are denoted by primes. Taking $(x, y, z) = (dx^*, dy^*, dz^*)$, $t = \frac{d^2}{\kappa} t^*$, $q = \frac{\kappa}{d} q^*$, $T = \Delta T T^*$ and $\omega = \frac{\kappa}{d^2}\omega^*$, eliminating the pressure, dimensionless equations which are linearized that govern small perturbations, after dropping the asterisks, are given by

**Table 1** Nomenclature

| | |
|---|---|
| $Pr$ Prandtl number | **Greek symbols**<br>$\Omega_1$ angular velocity for rotation |
| $g_i$ gravitational acceleration | $\varepsilon_{ijk}$ permutation tensor |
| $q_i$ velocity components | $\kappa$ thermal diffusivity |
| $Nu$ Nusselt number | $\mu$ viscosity |
| $a$ non-dimensional wave number | $\mu_m$ magnetic permeability |
| $Q$ Chandrasekhar number | $\omega$ frequency |
| $t$ time | $\lambda_1$ coefficient of stress relaxation |
| $d$ liquid layer thickness | $\alpha$ coefficient of thermal expansion |
| $R$ Rayleigh number | $\lambda_2$ coefficient of strain retardation |
| $T$ temperature | $\Lambda_1$ Deborah number or scaled stress relaxation parameter |
| $T_0$ upper boundary temperature | $\Lambda$ elastic ratio |
| $H_0$ uniform magnetic field | $\rho$ density |
| $Ta$ Taylor number | $\rho_0$ reference density |
| | $\tau'_{ij}$ stress components |
| | $\sigma_1$ electrical conductivity |
| | $\Lambda_2$ scaled strain retardation parameter |
| **Subscripts** | **Superscripts** |
| $b$ basic state | $'$ perturbed quantities |

$$\left[\left(1+\Lambda\Lambda_1\frac{\partial}{\partial t}\right)\nabla^2-\frac{1}{Pr}\left(\frac{\partial}{\partial t}+\Lambda_1\frac{\partial^2}{\partial t^2}\right)-Q\right]\left(\frac{\partial u}{\partial z}-\frac{\partial w}{\partial x}\right)$$
$$-\sqrt{Ta}\left(1+\Lambda_1\frac{\partial}{\partial t}\right)(1+\epsilon\cos\omega t)\frac{\partial v}{\partial z}\;+R\left(1+\Lambda_1\frac{\partial}{\partial t}\right)\frac{\partial T}{\partial x}=0, \tag{8}$$

$$\frac{1}{Pr}\left(\frac{\partial}{\partial t}+\Lambda_1\frac{\partial^2}{\partial t^2}\right)v=\left(1+\Lambda\Lambda_1\frac{\partial}{\partial t}\right)\nabla^2 v-\sqrt{Ta}\left(1+\Lambda_1\frac{\partial}{\partial t}\right)(1+\epsilon\cos\omega t)u, \tag{9}$$

$$\frac{\partial T}{\partial t}=w\frac{\partial T_b}{\partial z}-u\frac{\partial T}{\partial x}-w\frac{\partial T}{\partial z}+\nabla^2 T, \tag{10}$$

where $\nabla^2=\frac{\partial^2}{\partial x^2}+\frac{\partial^2}{\partial z^2}$.

The dimensionless parameters are

$Ta=\frac{4\rho_0^2\Omega_1^2 d^4}{\mu^2}$ (Taylor number),
$R=\frac{\alpha\rho_0 g d^3\Delta T}{\mu_0\kappa}$ (Rayleigh number),
$Pr=\frac{\mu_0}{\rho\kappa}$ (Prandtl number),
$Q=\frac{\sigma_1\mu_m^2 H_0^2 d^2}{\mu_0}$ (Chandrasekhar number),

$\Lambda_1 = \frac{\lambda_1 \kappa}{d^2}$ (Deborah number),
$\Lambda_2 = \frac{\lambda_2 \kappa}{d^2}$ (Strain retardation parameter (non-dimensional)) and
$\Lambda = \frac{\lambda_2}{\lambda_1} = \frac{\Lambda_2}{\Lambda_1}$ (Elastic ratio).

# 3 Model of Khayat–Lorenz

Introducing stream function $u = -\left(\frac{\partial \psi}{\partial z}\right)$, $w = \left(\frac{\partial \psi}{\partial x}\right)$, the quantity of heat transfer is determined by applying non-linear analysis.

$$\left(1 + \Lambda\Lambda_1 \frac{\partial}{\partial t}\right)\nabla^4\psi - \left(1 + \Lambda_1 \frac{\partial}{\partial t}\right)\left[\frac{1}{Pr}\frac{\partial}{\partial t}(\nabla^2\psi) + \sqrt{Ta}(1 + \epsilon\cos\omega t)\frac{\partial v}{\partial z} - R\frac{\partial T}{\partial x}\right] - Q\nabla^2\psi = 0, \tag{11}$$

$$\left(1 + \Lambda_1 \frac{\partial}{\partial t}\right)\left[\frac{1}{Pr}\frac{\partial v}{\partial t} - \sqrt{Ta}(1 + \epsilon\cos\omega t)\frac{\partial \psi}{\partial z}\right] = \left(1 + \Lambda\Lambda_1 \frac{\partial}{\partial t}\right)\nabla^2 v, \tag{12}$$

$$\frac{\partial T}{\partial t} - \nabla^2 T - \frac{\partial \psi}{\partial x} + \frac{\partial(\psi, T)}{\partial(x, z)} = 0. \tag{13}$$

We may rearrange Eqs. (11) and (12) to get four differential equations of first order consisting of time derivatives which reduces it to the model of Khayat–Lorenz in the form [31]:

$$\Lambda_1 \frac{\partial M}{\partial t} = -M + (1 - \Lambda)\nabla^4\psi - Q\nabla^2\psi, \tag{14}$$

$$\frac{1}{Pr}\frac{\partial}{\partial t}\nabla^2\psi = \Lambda\nabla^4\psi - \sqrt{Ta}(1 + \epsilon\cos\omega t)\frac{\partial v}{\partial z} + R\frac{\partial T}{\partial x} + M, \tag{15}$$

$$\Lambda_1 \frac{\partial N}{\partial t} = (1 - \Lambda)\nabla^2 v - N, \tag{16}$$

$$\frac{1}{Pr}\frac{\partial v}{\partial t} = \Lambda\nabla^2 v + \sqrt{Ta}(1 + \epsilon\cos\omega t)\frac{\partial \psi}{\partial z} + N, \tag{17}$$

$$\frac{\partial T}{\partial t} = -\frac{\partial(\psi, T)}{\partial(x, z)} + \frac{\partial \psi}{\partial x} + \nabla^2 T. \tag{18}$$

We consider stress free isothermal boundary conditions (BC's) which are given by

$$\psi = \frac{\partial^2 \psi}{\partial z^2} = M = T = \frac{\partial v}{\partial z} = \frac{\partial N}{\partial z} = 0 \text{ at } z = 0, 1. \tag{19}$$

From Eqs. (14) and (15), BCs on M and from Eqs. (16) and (17) that on N are derived. Also, the BCs for $T$, $\psi$, $DN$ and $Dv$ are taken as in Eq. (19). Subject to boundary conditions (Eq. (19)), the solution for Eqs. (14)–(18) is obtained in the form of a time-dependent Fourier series. Following the formulation of Lorenz–Saltzman, using Fourier representation expressing temperature and stream function with double and single terms, respectively, the non-linearity is maintained as follows.

Stream functions $\psi$, $T$, $v$, $M$ & $N$ are given by Eqs. (20)–(24).

$$\psi = A(t)\,\sin(\pi z)\,\sin(ax), \tag{20}$$

$$T = B(t)\,\sin(\pi z)\,\cos(ax) + C(t)\,\sin(2\pi z), \tag{21}$$

$$v = D(t)\,\cos(\pi z)\,\sin(ax), \tag{22}$$

$$M = E(t)\,\sin(\pi z)\,\sin(ax), \tag{23}$$

$$N = F(t)\,\cos(\pi z)\,\sin(ax). \tag{24}$$

For $M$, we take the same trial function as that of stream function $\psi$ as observed in Eq. (14). In the current work, the convective motions are assumed to be of small scale and are represented by $C(t)\,\sin 2\pi z$ which is the modification term in the temperature field.

Equations (14)–(18) are projected onto modes Eqs. (20)–(24) and the six-dimensional equations are obtained as follows:

$$\frac{dX}{d\tau} = Pr\left[\left(1 + \frac{Q}{\delta^2} + \frac{\pi^2 Ta}{\delta^6}\right)Y - \Lambda X - \left[(1-\Lambda) + \frac{Q}{\delta^2}\right] H - (1 + \epsilon\cos\omega_1\tau)\frac{\pi^2 Ta}{\delta^6}G\right], \tag{25}$$

$$\frac{dY}{d\tau} = rX - XZ - Y, \tag{26}$$

$$\frac{dZ}{d\tau} = XY - \frac{4\pi^2}{\delta^2}Z, \tag{27}$$

$$\frac{dG}{d\tau} = Pr[(1 + \epsilon\cos\omega_1\tau)X - \Lambda G - (1-\Lambda)J], \tag{28}$$

$$\frac{dH}{d\tau} = \frac{1}{\Lambda_1\delta^2}(X - H), \tag{29}$$

$$\frac{dJ}{d\tau} = \frac{1}{\Lambda_1\delta^2}(G - J), \tag{30}$$

where

$$X = \frac{\pi a A}{\sqrt{2}\delta^2}, \quad Y = \frac{\pi r B}{\sqrt{2}}, \quad Z = -\pi r C, \quad G = \frac{aD}{\sqrt{Ta}}, \quad H = \frac{a\pi E}{\sqrt{2}(1-\Lambda)\delta^6},$$

$$J = \frac{-aF}{\sqrt{2}(1-\Lambda)\delta^2}, \quad \tau = t\delta^2, \quad r = \frac{R}{R_s}, \quad R_s = \frac{\delta^6 + Q\delta^4 + \pi^2 Ta}{a^2},$$

$$\delta^2 = a^2 + \pi^2, \quad \omega_1 = \frac{\omega}{\delta^2}.$$

The above Eqs. (25)–(30) are solved using Runge–Kutta–Fehlberg45 (RKF45) which is the built-in function in Wolfram Mathematica 9.

## 4 Case of No Modulation and No Magnetic Field

The stationary Rayleigh number in the no modulation case is given by [1]

$$R_{sc} = \frac{\delta^6 + \pi^2 Ta}{a^2}, \tag{31}$$

where $\delta^2 = a^2 + \pi^2$.

Considering the equation $\frac{\partial R_{sc}}{\partial p} = 0$, where $p = a^2$, we get $(p+\pi^2)^3 - 3p(p+\pi^2)^2 + \pi^2 Ta = 0$.

We obtain $a_c^2$ by solving the cubic equation in $p$, as a function of $Ta$. In further computation, we make use of this obtained value of $a$.

## 5 Heat Transport

Here, we emphasize the outcome of the rotational modulation with magnetic field effect on the heat transport which is measurable by the Nusselt number ($Nu$):

$$Nu(\tau) = 1 + \frac{\text{Heat transport by convection}}{\text{Heat transport by conduction}},$$

$$= 1 + \frac{\left[\frac{a_c}{2\pi}\int_0^{\frac{2\pi}{a_c}} \frac{\partial T}{\partial z} dx\right]_{z=0}}{\left[\frac{a_c}{2\pi}\int_0^{\frac{2\pi}{a_c}} \frac{\partial T_b}{\partial z} dx\right]_{z=0}}. \tag{32}$$

Equation (21) is substituted in Eq. (32) and on simplification, we obtain

$$Nu(\tau) = \frac{2}{r} Z(\tau) + 1. \tag{33}$$

Different parameters' effect on the heat transfer is discussed by finding mean Nusselt number, $\overline{Nu(\tau)}$, as given by [46],
$\overline{Nu(\tau)} = \frac{\omega_1}{2\pi} \int_0^{\frac{2\pi}{\omega_1}} Nu(\tau) d\tau$.

## 6 Discussion of Results

Using a generalized model of Khayat–Lorenz, the effect of rotational speed, magnetic field and effect of elasticity of the liquid on RBC in viscoelastic liquids is investigated. Here, first the generalized Khayat–Lorenz model is resolved by Runge–Kutta–Fehlberg45 method with an 'adaptive step size' and then the numerical results are discussed. The following initial conditions for solving equations (25)–(30) are considered:

$$X(\tau) = Y(\tau) = Z(\tau) = G(\tau) = H(\tau) = J(\tau) = 1 \text{at} \tau = 0.$$

Khayat [19] and Li and Khayat [23] have shown that, for a viscoelastic liquid, the value that the Prandtl number takes is much greater than 10 and therefore we have taken $Pr = 10$ in all our calculations. For this value of $Pr$, it is found that as $Ta$ increases Nusselt number increases with or without the magnetic field effect. The Oldroyd liquid B, in the limiting case, leads to Maxwell, Newtonian and Rivlin–Ericksen as given in Table 2.

In four different kinds of liquids, viz., Newtonian, Maxwell, Oldroyd liquid B and Rivlin–Ericksen, from Fig. 2a–d, the inequalities obtained are in validation with that of Siddheshwar et al. [31] and are as follows:

$$Nu^{Rivlin-Ericksen} < Nu^{Newtonian} < Nu^{Oldroyd-B} < Nu^{Maxwell}.$$

**Table 2** Values of $\Lambda$ and $\Lambda_1$ for different kinds of liquid

| Liquid type | $\Lambda$ | $\Lambda_1$ |
|---|---|---|
| Newtonian | 1 | 0.1 |
| Maxwell | 0 | 0.1 |
| Rivlin–Ericksen | 50 | 0 |
| Oldroyd liquid B | 0.5 | 0.2 |

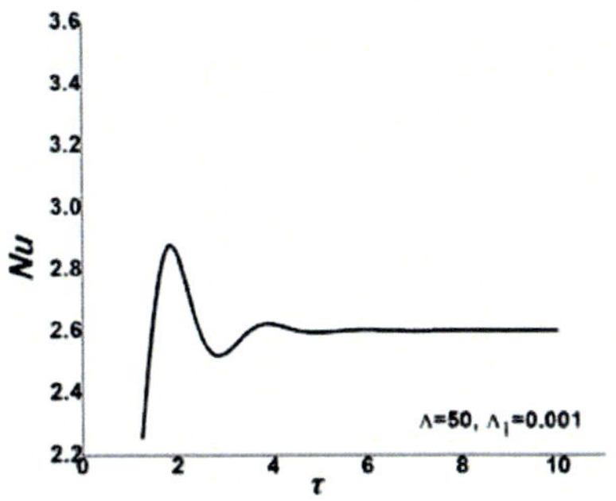


(a) Rivlin-Ericksen liquid

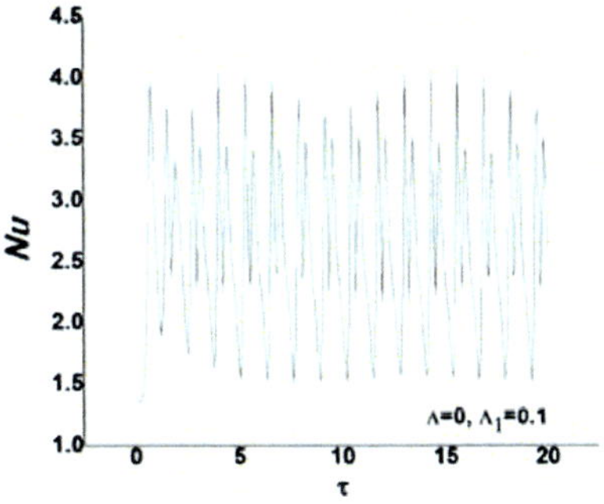


(b) Maxwell liquid

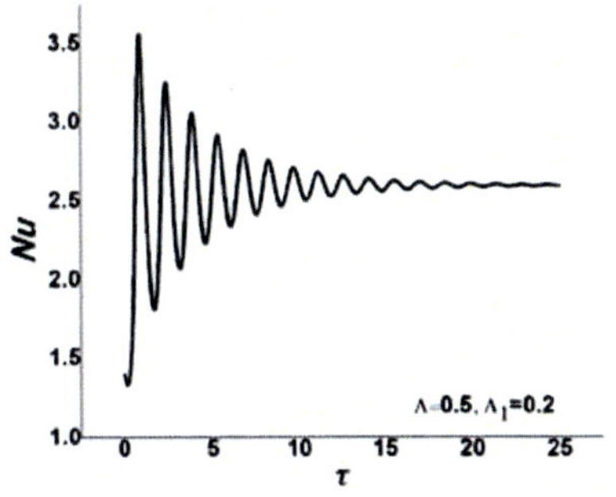


(c) Oldroyd liquid B

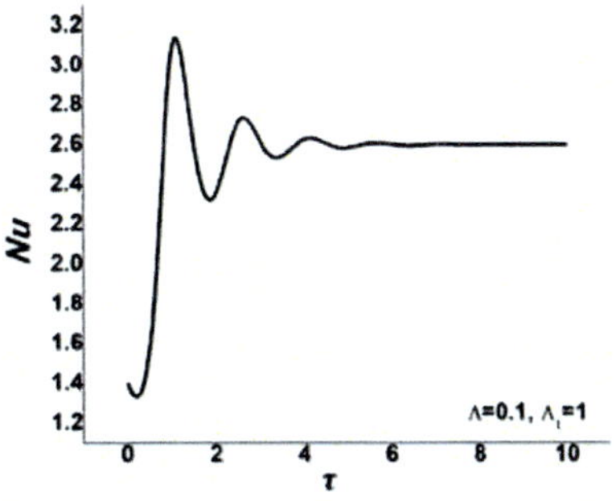


(d) Newtonian liquid

**Fig. 2** $Nu$ versus $\tau$ for $\omega_1 = 5$, $r = 5$, $Pr = 10$, $Ta = 0$, $\epsilon = 0$ and $Q = 0$

**Table 3** $\overline{Nu(\tau)}$ for $\Lambda = 0.5$, $\Lambda_1 = 0.1$, $Pr = 10$, $\omega_1 = 5$ and $r = 5$

| | $Q = 0, Ta = 0$ | $Q = 0, Ta = 10$ | $Q = 100, Ta = 0$ | $Q = 100, Ta = 10$ |
|---|---|---|---|---|
| $\epsilon = 0$ | 2.21202 | 2.2151 | 2.76662 | 2.76886 |
| $\epsilon = 0.1$ | 2.21202 | 2.21537 | 2.76662 | 2.76829 |

The graphs of $\overline{Nu(\tau)}$ versus $\Lambda_1$ (see Fig. 3a) and $\overline{Nu(\tau)}$ versus $\Lambda$ (see Fig. 3b) show that increase in both $\Lambda_1$ and $\Lambda$ increases $\overline{Nu(\tau)}$, both with/without a magnetic field effect. Increment in $Ta$ enhances $\overline{Nu(\tau)}$ (see Fig. 3c) and increase in $Pr$ also increases $\overline{Nu(\tau)}$ (see Fig. 3d) with/without the effect of a magnetic field. From Fig. 3b–d, it is clear that the joint effect of Chandrasekhar number $Q$ and modulation in rotation is to decrease the heat transport. From Table 3, it is evident that the effect of rotational modulation is to enhance the heat transport in the absence of a magnetic field, whereas in the presence of a magnetic field and $Ta$, rotational modulation gives heat transport reduction.

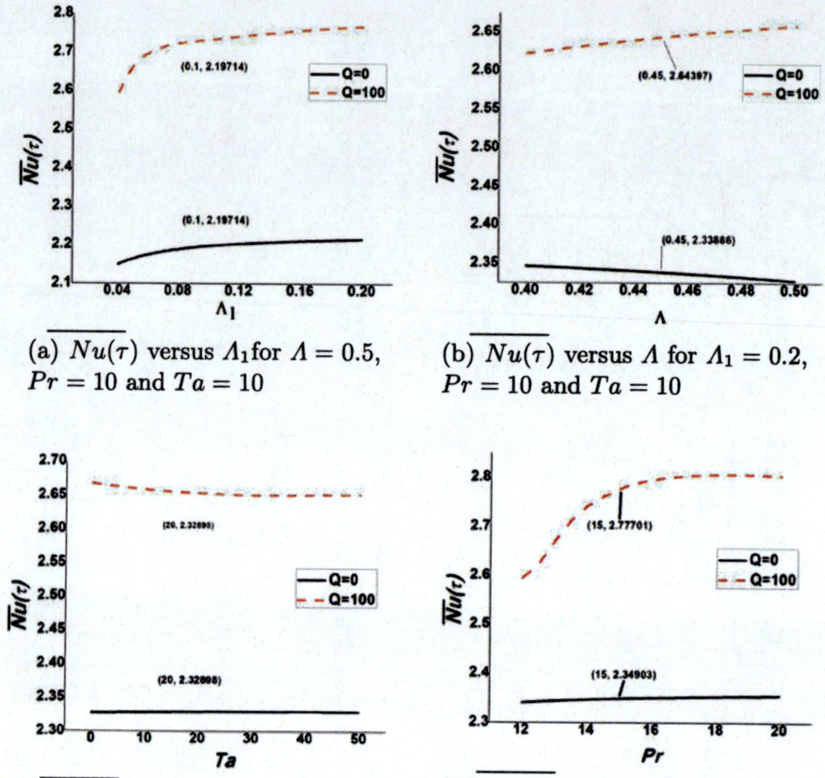


(a) $\overline{Nu(\tau)}$ versus $\Lambda_1$for $\Lambda = 0.5$, $Pr = 10$ and $Ta = 10$

(b) $\overline{Nu(\tau)}$ versus $\Lambda$ for $\Lambda_1 = 0.2$, $Pr = 10$ and $Ta = 10$

(c) $\overline{Nu(\tau)}$ versus $Ta$ for $\Lambda_1 = 0.2$, $\Lambda = 0.5$ and $Pr = 10$

(d) $\overline{Nu(\tau)}$ versus $Pr$ for $\Lambda_1 = 0.2$, $\Lambda = 0.5$ and $Ta = 10$

**Fig. 3** Plot of $\overline{Nu(\tau)}$ against different parameters for $\omega_1 = 5$ and $r = 5$

## 7 Conclusion

- The consequences of modulation of rotation on magnetoconvection in a rotating horizontal layer of viscoelastic liquid has been investigated by making non-linear stability analysis of the extended model of Khayat–Lorenz.
- The effect of both magnetic field and rotational modulation diminishes heat transport by delaying the onset of convection.
- Stress relaxation and elasticity parameters' effects are to advance convection and thereby enhance the heat transfer.
- The effect of the Taylor number is to enhance the heat transfer by advancing the onset of convection in the presence of magnetic field and modulation.
- The effect of the Prandtl number increases the convective transport of heat in the presence of magnetic field and modulation.

The authors express thanks to their institution and management for the encouragement, also to anonymous reviewers for beneficial comments and to Dr. P. G. Siddheshwar for valuable suggestions.

## References


1. Chandrasekhar S (1961) Hydrodynamic and hydromagnetic stability. Oxford University Press, Oxford
2. Platten JK, Legros JC (1984) Convection in liquids. Springer, Berlin
3. Vest CM, Arpaci VS (1969) Overstability of a viscoelastic fluid layer heated from below. J Fluid Mech 36:613–623
4. Sokolov M, Tanner RI (1972) Convective stability of a general viscoelastic fluid heated from below. Phys Fluids 15:534–539
5. Siginer DA, Valenzuela-Rendón A (1994) Natural convection of viscoelastic liquids. In: Proceedings of ASME fluids engineering division summer meeting, symposium, ASME FED, vol 179, pp 31–41
6. Dávalos-Orozco LA, Vázquez Luis E (1999) Natural convection of a viscoelastic fluid with deformable free surface. J Non-Newtonian Fluid Mech 85:257–271
7. Martinez-Mardones J, Tienmann R, Walgraef D, Zeller W (2000) Thermal convection thresholds in viscoelastic solutions. J Non-Newtonian Fluid Mech 93:1–15
8. Demir H (2003) Rayleigh-Bénard convection of viscoelastic fluid. Appl Math Comput 136:251–267
9. Dauby PC, Parmentier P, Lebon G, Grmela M (1993) Coupled buoyancy and thermocapillary convection in a vicoelastic Maxwell fluid. J Phys Cond Matter 5:4343–4352
10. Sekhar GN, Jayalatha G (2009) Elastic effects on Rayleigh-Bènard-Marangoni convection in liquids with temperature-dependent viscosity. In: Proceedings of ASME 2009 IMECE09. 1830, 978-07918-3863-1(2009)
11. Sekhar GN, Jayalatha G (2010) Elastic effects on Rayleigh-Bènard convection in liquids with temperature-dependent viscosity. Int J Therm Sci 49:67–75
12. Siddheshwar PG, Sekhar GN, Jayalatha G (2011) Surface tension driven convection in viscoelastic liquids with thermorheological effect. Int Commnu Heat Mass Transf 38:468–473
13. Takashima M (1970) The effect of rotation on thermal instability in a viscoelastic fluid layer. Phys Lett 31:379–380
14. Takashima M (1972) Thermal instability in a viscoelastic fluid layer. III. Effect of rotation. J Phys Soc Jpn 33:797–804
15. Bhatia PK, Steiner JM (1972) Convective instability in a rotating viscoelastic fluid layer. ZAMM 52:321–327
16. Laroze D,Mart'inez-Mardones J, Bragard J (2007) Thermal convection in a rotating binary viscoelastic liquid mixture. Eur Phys J Special Topics 146:291–300
17. Jayalatha G (2011) Free convection problems in variable viscosity viscoelastic liquids under terrestrial or microgravity condition. Ph.D. Thesis, Visweswaraya Technological University, India
18. Jayalatha G, Siddheshwar PG, Prakash R, Sekhar GN (2018) Effect of coriolis force on Rayleigh Bènard convection in ferromagnetic liquids with variable viscosity. In: Proceedings of 7th international and 45th national conference on fluid mechanics and fluid power. IIT Bombay, Mumbai, India
19. Khayat RE (1994) Chaos and overstability in the thermal convection of viscoelastic fluids. J Non-Newtonian Fluid Mech 53:227–255
20. Park HM, Lee HS (1995) Nonlinear hydrodynamic stability of viscoelastic fluids heated from below. J Non-Newtonian Fluid Mech 60:1–26

21. Siddheshwar PG, Srikrishna CV (2002) Unsteady nonlinear convection in second order fluid. Int J Non-Lin Mech 37:321–330
22. Abu-Ramadan E, Hay JM, Khayat RE (2003) Characterization of chaotic thermal convection of viscoelastic fluids. J Non-Newtonian Fluid Mech 115:79–113
23. Li Z, Khayat RE (2005) Finite-amplitude Rayleigh-Bénard convection and pattern selection for viscoelastic fluids. J Fluid Mech 529:221–251
24. Narayana M, Sibanda P, Siddheshwar PG, Jayalatha G (2013) Linear and nonlinear stability analysis of binary viscoelastic fluid convection. Appl Math Model 37:8162–8178
25. Venezian GI (1969) Effect of modulation on the onset of thermal convection. J Fluid Mech 35:243–254
26. Yih CS, Li CH (1972) Instability of unsteady flows or configurations-2 convective instability. J Fluid Mech 54:143–152
27. Siddheshwar PG, Pranesh S (1999) Effect of temperature/gravity modulation on the onset of magneto-convection in weak electrically conducting fluids with internal angular momentum. JMMM 192:159–176
28. Anjana K, Vanishree RK (2016) Study of effects of temperature modulation on double diffusive convection in Oldroyd-B liquids. Int J Eng Res Technol 5:335–343
29. Malashetty MS, Mahantesh S (2008) Effect of thermal modulation on the onset of convection in a rotating fluid layer. Int J Heat Mass Transf 51:2814–2823
30. Siddheshwar PG, Uma D, Bhavya S (2019) Effects of variable viscosity and temperature modulation on linear Rayleigh-Bènard convection in Newtonian dielectric liquid. Appl Math Mech 40:1601–1614
31. Siddheshwar PG, Sekhar GN, Jayalatha G (2010) Effect of time-periodic vertical oscillations of the Rayleigh Bènard system on nonlinear convection in viscoelastic liquids. J Non- Newtonian Fluid Mech 165:1412–1418
32. Jayalatha G, Siddheshwar PG, Suma N (2018) Non-Linear convection in rotating viscoelastic liquids with gravity modulation. In: Proceedings of 7th international and 45th national conference on fluid mechanics and fluid power. IIT Bombay, Mumbai, India
33. Jayalatha G, Suma N (2020) Nonlinear rotating viscoelastic liquid convection with temperature modulation, heat transfer. http://doi.org/10.1002/htj.21841
34. Bhadauria BS, Kiran P (2014) Effect of rotational speed modulation on heat transport in a fluid layer with temperature dependent viscosity and internal heat source. Ain Shams Eng J 5:1287–1297
35. Premila K, Sudhaamsh GM (2017) Weak non linear double diffusive convection with rotational modulation. Int J Math Arch 8:97–101
36. Vanishree RK, Anjana K (2018) The study of effects of rotation modulation on double diffusive convection in Oldroyd-B liquids. IOSR J Eng 8:27–40
37. Kanchana C, Siddheshwar PG, Yi Z (2020) Regulation of heat transfer in Rayleigh-Bènard convection in Newtonian liquids and Newtonian nanoliquids using gravity, boundary temperature and rotational modulations. J Therm Anal Calorim. 10.1007/s10973-020-09325-3
38. Kenath A, Soibam D, Khanum S (2020) Linear and nonlinear study of Rayleigh Bénard convection in Oldroyd-B fluids under rotational modulation. Int J Fluid Mech Res 47:387–397
39. Bhatia PK, Steiner JM (1971) Convection in a viscoelastic fluid layer in hydromagnetics. Phys Lett A 37:419–420
40. Bhatia PK, Steiner JM (1972) Oscillatory convection in a viscoelastic fluid layer in hydromagnetics. Aust J Phys 25:695–702
41. Bhatia PK, Steiner JM (1973) Thermal instability in a viscoelastic fluid layer in hydromagnetics. J Math Anal Appl 41:271–283
42. Takashima M (1972) Thermal instability in a viscoelastic fluid layer. III. Effect of a magnetic field. J Phys Soc Jpn 33:1142–1148
43. Sharma RC (1975) Thermal instability in a viscoelastic fluid in hydromagnetics. Acta Phys Acad Sci Hungaricae Tomus 38:293–298
44. Eltayeb IA (1976) On thermal instability in a viscoelastic fluid layer in hydromagnetics. J Math Anal Appl 54:846–848

45. Sharma RC, Aggarwal AK (2006) Effect of compressibility and suspended particles on thermal convection in a Walters' B elastico-viscous fluid in hydromagnetics. Int J Appl Mech Eng 11:391–399
46. Siddheshwar PG, Meenakshi N (2019) Comparison of the effects of three types of time-periodic body force on linear and non-linear stability of convection in nanoliquids. Eur J Mech/B Fluids 77:221–229

# MHD Flow of Nanofluids over a Stretching Sheet with the Effect of Heat and Mass Transfer

R. Jayakar and B. Rushi Kumar

## 1 Introduction

In consideration of their higher thermal properties, investigators have gained much interest from the nanofluids at present. These substances consist of solid particles of a nanometer size, and the basic fluid is called nanoparticles such as oxides, metals, carbon nanotubes, or carbides. The basic fluids recognized include water, oil, ethylene glycol as well as bioliquids. Nanofluids are, in other terms, two-phase nanometer-sized fluids in which the base fluid reaches up to 100 nm. For the very first time, [1] utilized nanofluids. Because of its broad range of uses in solar cells, transformer oil, motor cooling, solar water heating, refrigerators, electricity and fuel cells, etc. [2, 3], it is interesting. Nanotechnology has reached all industries by direct efforts. Due to its diversity, several research reports in the literature are being carried out on an analysis of the electrically conductive fluid flow or MHD (magnetohydrodynamics). MHD has several applications, such as the MHD accelerator, generator power as well as the science of fusion. Studies [4–6] can cite several recent studies with the impact of magnetic fields on fluid flow. Due to gross moments, the fluid bulk causes convection term by mixing a portion of one fluid into another portion. Hayat et al. [7] implemented thixotropic nanofluid joule heating as well as solar radiation in the MHD flow. In fluid flow issues associated with heat transfer, nonlinear thermal radiation effects are also significant. Kataria and Patel [8] observed MHD flow through a convection area with thermal radiation non-linear impacts. The issues of flow, mass, as well as heat transfer through a stretching surface through a porous media are one of the most interesting fields of study in few years since this is a crucial method of flow in several engineering as well as industrial domains like polymer technology.

However, no studies on the MHD flow of nanofluids through a stretching sheet along with the impact of mass as well as heat transfer have been reported according to the reports. The momentum boundary layer thickness is raised, whereas the thickness

R. Jayakar · B. Rushi Kumar (✉)
Department of Mathematics, School of Advanced Sciences, VIT, Vellore, TN, India
e-mail: rushikumar@vit.ac.in



S. Srinivas et al. (eds.), *Recent Advances in Applied Mathematics and Applications to the Dynamics of Fluid Flows*, Lecture Notes in Mechanical Engineering,
https://doi.org/10.1007/978-981-19-1929-9_9

of thermal boundary layer is decreased by the parameter of the fraction amount of the higher nanoparticle. It should be noted that the heat transfer levels rise with the increases in the nanoparticle's fraction value.

In the present work, the model equations are transformed into ODE using similarity transformation. The numerical shooting approach along with 4th order Runge-Kutta approach would be used to provide numerical model solutions. When the magnetic nanofluids in the presence of outer magnetic fields as coolant fluids, the results of non-dimensional governing parameters like permeability parameter, Schmidt No., magnetic parameter, nanoparticle volume fraction, the thermal conductivity of nanoparticle, chemical reaction parameter, velocity slip, and radiation parameter are discussed.

## 2 Mathematical Analysis

As seen in Fig. 1, 3-D, steady, laminar, as well as electrical conducting MHD flow of nanofluid boundary layer by the permeable stretched surface saturated with a porous media. The stretching sheet surface is applied normally with a uniform magnetic field. The magnet field strength in the z-direction is often taken into account. The stretching surface should be kept at uniform concentration $C_w$ and temperature $T_w$ and values are deemed larger as compared to ambient temperature $T_\infty$ as well as

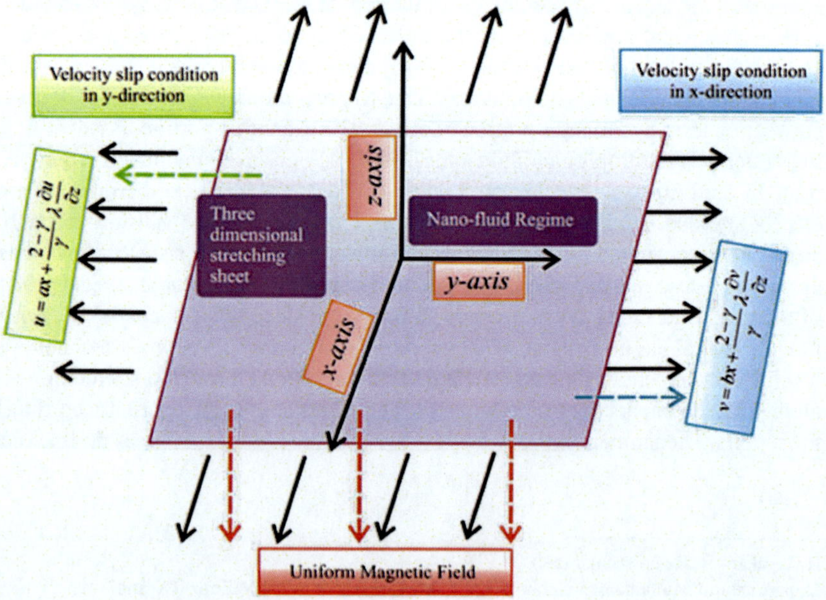


**Fig. 1** Geometry of the problem

concentration $C_\infty$. The nanofluid is provided with standardized heat through the stretching sheet as well as chemical species concentration. The influence of density variance on concentration and temperature is assumed to exist only on the duration of the body strength and hence variations in temperature and concentration cause buoyancy force.

The flow, heat transfer, and concentration are provided as simplified three-dimensional boundary layer equations as follows:

$$\frac{\partial u}{\partial x}+\frac{\partial v}{\partial y}+\frac{\partial w}{\partial z}=0 \tag{1}$$

$$u\frac{\partial u}{\partial x}+v\frac{\partial u}{\partial y}+w\frac{\partial u}{\partial z}=\upsilon_{nf}\frac{\partial^2 u}{\partial z^2}-\frac{\sigma B_0^2 u}{\rho_{nf}}-\frac{\upsilon_{nf}u}{K} \tag{2}$$

$$u\frac{\partial v}{\partial x}+v\frac{\partial v}{\partial y}+w\frac{\partial v}{\partial z}=\upsilon_{nf}\frac{\partial^2 v}{\partial z^2}-\frac{\sigma B_0^2 v}{\rho_{nf}}-\frac{\upsilon_{nf}v}{K} \tag{3}$$

$$u\frac{\partial T}{\partial x}+v\frac{\partial T}{\partial y}+w\frac{\partial T}{\partial z}=\frac{k_{nf}}{\rho(Cp)_{nf}}\frac{\partial^2 T}{\partial z^2}-\frac{1}{(\rho Cp)_{nf}}\frac{\partial q_r}{\partial z} \tag{4}$$

$$u\frac{\partial C}{\partial x}+v\frac{\partial C}{\partial y}+w\frac{\partial C}{\partial z}=D_m\frac{\partial^2 C}{\partial z^2}-k_0\left(C-C_\infty\right) \tag{5}$$

given boundary limits,

$$u=\frac{ax\gamma+(2-\gamma)\lambda(\partial u/\partial z)}{\gamma},\quad v=\frac{bx\gamma+(2-\gamma)\lambda(\partial v/\partial z)}{\gamma}$$

$$W=-U_0,\, T=T_W,\ C=C_W\ \ at\ \ z=0 \tag{6}$$

$$u\to 0,\ \ v\to 0,\ \ T\to T_\infty,\ \ C\to C_\infty\ \ \ at\ z\to\infty$$

the electric conductivity of the nanofluid is represented by $\sigma_{nf}$ and kinematic viscosity of the base fluid $\upsilon_{nf}$ as given below:

$$\rho_{nf}=(1-\varphi)\,\rho_f+\varphi\rho_s,\ \ \mu_{nf}=\mu_f/(1-\varphi)^{2.5},\ \ \alpha_{nf}=\frac{k_{nf}}{(\rho Cp)_{nf}}$$

$$(\rho Cp)_{nf}=(1-\varphi)\,(\rho Cp)_f+\varphi(\rho Cp)_s, \tag{7}$$

$$k_{nf}/k_f=\frac{\left(k_s+2k_f-2\varphi\left(k_f-k_s\right)\right)}{\left(k_s+2k_f+2\varphi\left(k_f-k_s\right)\right)},\ \ \upsilon_{nf}=\frac{\mu_{nf}}{\rho_{nf}},$$

We also converted the governing equations into ODE, with the following similarity transformations;

$$f'(\eta) = \tfrac{u}{ax},\; g'(\eta) = \tfrac{v}{ay},\; (f+g) = -\tfrac{w}{\sqrt{a v_f}},$$
$$\eta = \sqrt{\tfrac{a}{v_f}}z,\; \theta(\eta) = \tfrac{T-T_\infty}{T_w-T_\infty},\; S(\eta) = \tfrac{C-C_\infty}{C_w-C_\infty} \tag{8}$$

Here, $u$, $v$ and $w$ velocity components are presented in directions $x$, $y$ and $z$, kinematic viscosity signified by $\upsilon$ and $\sigma$ represents electrical conductivity. The radiative heat flux with the approximation of Rosseland is represented by

$$q_r = -\frac{4\sigma^*}{3k}\frac{\partial T^4}{\partial z}, \tag{9}$$

Suppose that temperature variations in flow may be represented as a linear temperature combination $T^4$. Therefore, we extend $T^4$ into Taylor series can be rewritten as $T_\infty$, also terms with higher-order will be dismissed, we will get,

$$T^4 = 4T_\infty^3 T - 3T_\infty^4 \tag{10}$$

So,

$$\frac{\partial q_r}{\partial z} = -\frac{16\sigma^* T_\infty^3}{3k}\frac{\partial^2 T}{\partial z^2} \tag{11}$$

Hence $\theta(\eta) = \frac{T-T_\infty}{T_w-T_\infty}$ temperature Equation may deduce as;

$$T = T_\infty\left(1 + (\theta_w - 1)\theta\right) \tag{12}$$

By taking Eqs. (8) to (12), and Eqs. (1) to (5) can be rewritten as,

$$\left[\begin{array}{l}\frac{1}{\left((1-\varphi)+\varphi\frac{\rho_s}{\rho_f}\right)(1-\varphi)^{2.5}}f''' + \left[(f+g)f'' - (f')^2\right] \\ -\frac{k_1 f'}{\left((1-\varphi)+\varphi\frac{\rho_s}{\rho_f}\right)(1-\varphi)^{2.5}} - \frac{M f'}{(1-\varphi)+\varphi\frac{\rho_s}{\rho_f}}\end{array}\right] = 0 \tag{13}$$

$$\left[\begin{array}{l}\frac{1}{\left((1-\varphi)+\varphi\frac{\rho_s}{\rho_f}\right)(1-\varphi)^{2.5}}g''' + \left[(f+g)g'' - (g')^2\right] \\ -\frac{k_1 g'}{\left((1-\varphi)+\varphi\frac{\rho_s}{\rho_f}\right)(1-\varphi)^{2.5}} - \frac{M g'}{(1-\varphi)+\varphi\frac{\rho_s}{\rho_f}}\end{array}\right] = 0 \tag{14}$$

$$\left[\begin{array}{l}\left(1 + \frac{R k_f}{k_{nf}}\right)\theta'' + \Pr\frac{\left((1-\varphi)+\varphi\frac{(\rho Cp)_s}{(\rho Cp)_f}\right)k_f}{k_{nf}}(f+g)\theta' \\ +\frac{R k_f}{k_{nf}}(\theta_w - 1)^3\left(3\theta^2\theta'^2 + \theta^3\theta''\right) \\ +\frac{3R k_f}{k_{nf}}(\theta_w - 1)^2\left(2\theta\theta'^2 + \theta^2\theta''\right) + \frac{3R k_f}{k_{nf}}(\theta_w - 1)\left(\theta'^2 + \theta\theta'\right)\end{array}\right] = 0 \tag{15}$$

$$S'' + Sc\,(f + g)\,S' - Sc\,Cr\,S = 0 \tag{16}$$

Here boundary terms represented as;

$$f'(0) = \left[1 + \alpha f''(0)\right],\; g'(0) = \left[\lambda + \alpha g''(0)\right],$$
$$f(0) + g(0) = V_0,\; \theta(0) = 1,\; S(0) = 1, \tag{17}$$
$$f'(\infty) \to 0,\; g'(\infty) \to 0,\; \theta(\infty) \to 0,\; S(\infty) \to 0$$

Here $M$ (magnetic parameter) = $\frac{\sigma B_0^2}{\rho_f a}$, Pr (Prandtl number) = $\frac{\upsilon_f}{\alpha_f}$, $K$ (Porosity parameter)= $\frac{\upsilon_f}{Ka}$, $R$ (Radiation parameter) = $\frac{4T_\infty^3\sigma^*}{3K^*k}$, $Sc$(Schmidt number) = $\frac{\upsilon_f}{D_m}$, $\lambda$(ratio of stretching rates) = $\frac{b}{a}$, $\alpha$ (velocity slip) = $\frac{2-\gamma}{\gamma}\sqrt{\frac{a}{\upsilon_f}}\beta$, $V_0$ (suction velocity) = $\frac{U_0}{\sqrt{a\upsilon_f}}$, $Cr$ (chemical reaction) = $\frac{k_0}{a}$, The Nusselt Number, the skin friction coefficient, the Sherwood Number are expressed by,

$$C_{fx} = (\mathrm{Re}_x)^{-1/2}\frac{f''(0)}{(1-\varphi)^{2.5}},\; Nu_x = -\left(1 + R\theta_w^3\right)(\mathrm{Re}_x)^{1/2}\theta'(0)$$
$$C_{fy} = \left(\mathrm{Re}_y\right)^{-1/2}\frac{g''(0)}{(1-\varphi)^{2.5}},\; Sh_x = -S'(0)\,(\mathrm{Re}_x)^{1/2}$$

## 3 Results and Discussions

Equations (13) to (16) were numerically resolved using the Runge-Kutta approach using the shooting method with the given initial as well as boundary limits (17). The impact on temperature, velocity, as well as concentration fields of the non-dimensional governing parameters. The findings achieved are analyzed and depicted in graphs.

The influence of $M$ magnetic numbers on velocity is seen in Figs. 2 and 6. Then we have found that as magnetic numbers increase the velocity is decreased. The use of applied magnetic fields tends to reduce the fluid displacement, which causes the reduction in thickness of the boundary layer and velocity.

Figures 3 and 7 show the effect of the porosity parameter $k_1$ on velocity profiles. Here we can see that the parameter's greater porosity values $k_1$ are the same as the lower velocity. Figure 4 illustrates the impact of the slip-parameter velocity $\lambda$ effect on the velocity. The variation in the slip-parameter velocity $\lambda$ contributes to a reduction in the stretching sheet velocity as well as the fluid. The nanofluid velocity has been identified as a declining function $\lambda$.

Figure 5 illustrates a decelerated velocity in the fluid area as the suction parameter values rise. It happens as suction eliminates the warm fluid from the surface plate as well as thereby increases the thickness of the boundary layer of the velocity. In

Fig. 2 Velocity $f'$ for different $M$

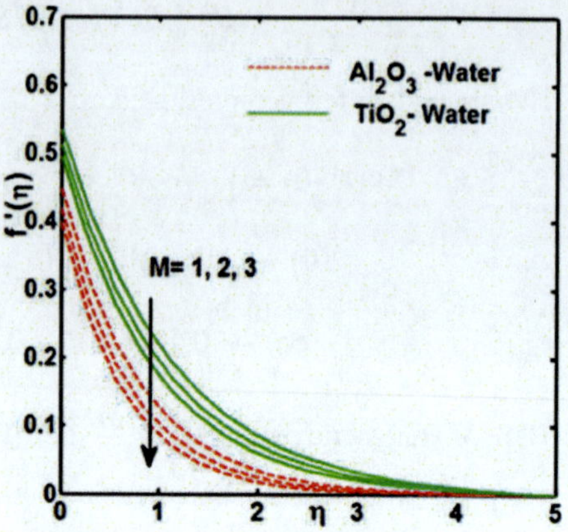


Fig. 3 Velocity $f'$ for different $k_1$

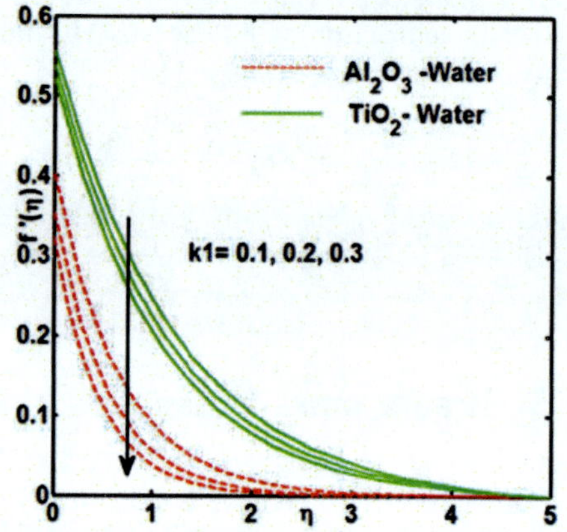


the temperature profile, Pr the Prandtl number is decreased as seen in Fig. 8. This finding justifies that the fluid's thermal conductivity decays the thermal boundary layer thickness thus decreases.

Moreover, the rising value Pr reduces the conduction, as well as therefore the variation in thermal properties increases. Figure 9 shows that the value of the temperature parameter rises as the heat transfer mechanism improves. The reason is increasing the thermal boundary layer thickness. The influence of $R$ on temperature profile is illustrated in Fig. 10. The rising values of $R$ thermal nanofluid radiation-induced an exponential rise in the temperature profile. For increased thermal radiation values, the nanofluid produces large amounts of heating that increase the nanofluid's temperature and the thermal boundary layer thickness rises. The concentration on different Schmidt number values is depicted in Fig. 11. The ratio between mass and momentum diffusivity can be termed as the Schmidt number. Thus, the higher Schmidt value limits the boundary layer thickness concentration.

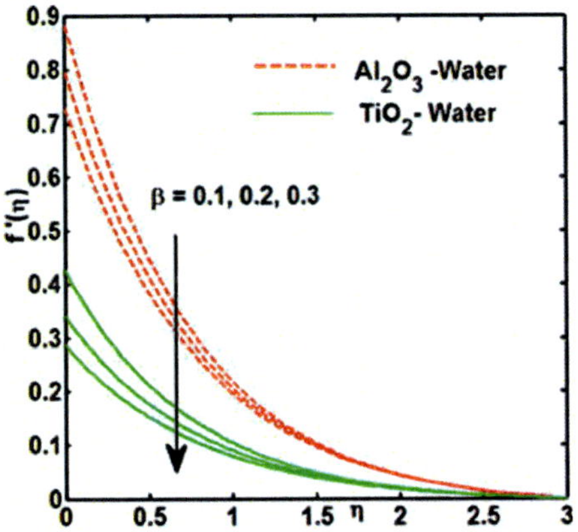


**Fig. 4** Velocity $f'$ for different $\beta$

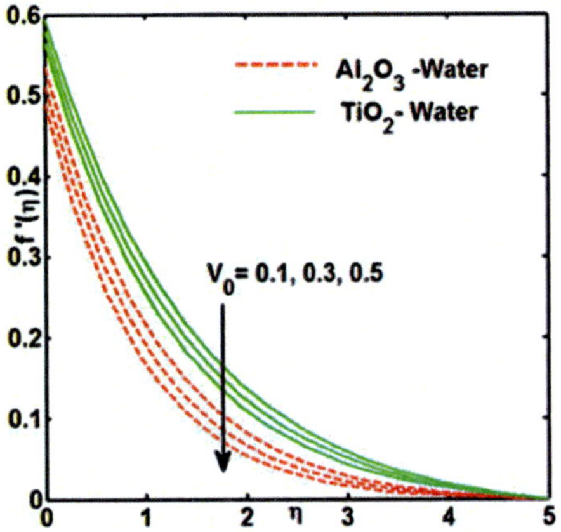


**Fig. 5** Velocity $f'$ for different $V_0$

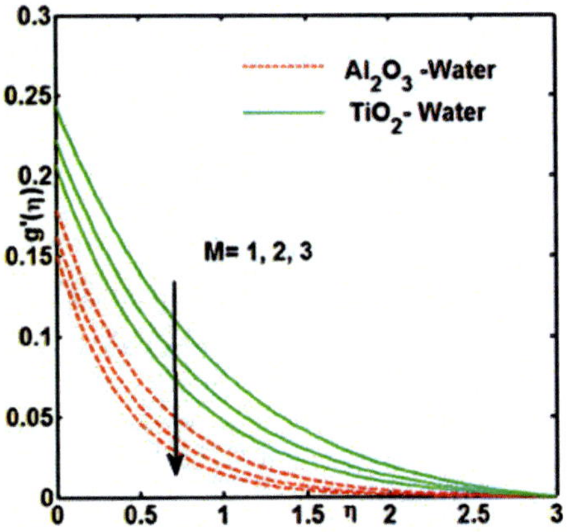


**Fig. 6** Velocity $g'$ for various $M$

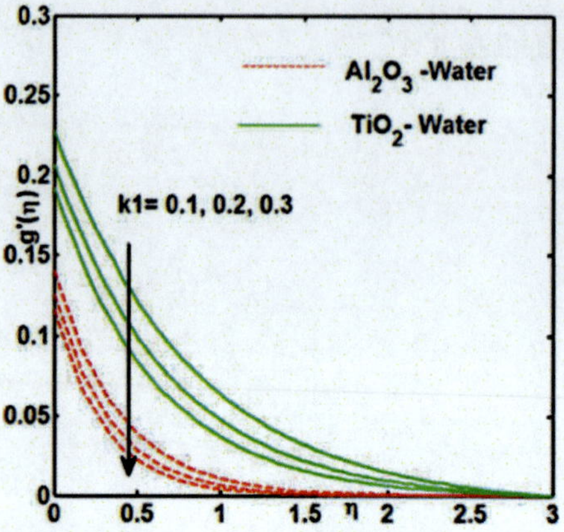


**Fig. 7** Temperature for different $k_1$

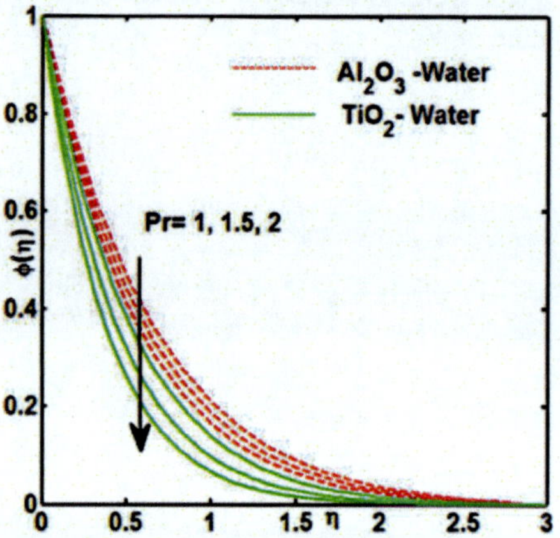


**Fig. 8** Temperature for different $Pr$

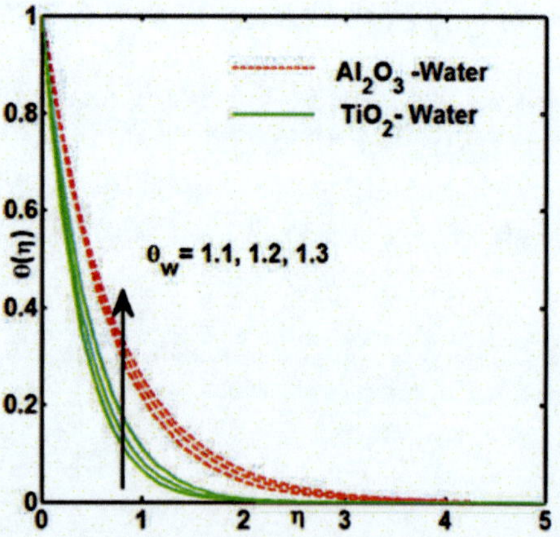


**Fig. 9** Temperature for different $\theta_w$

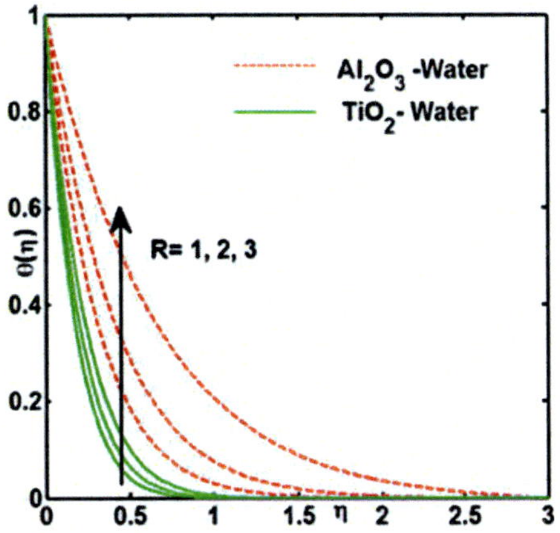


**Fig. 10** Temperature for different $R$

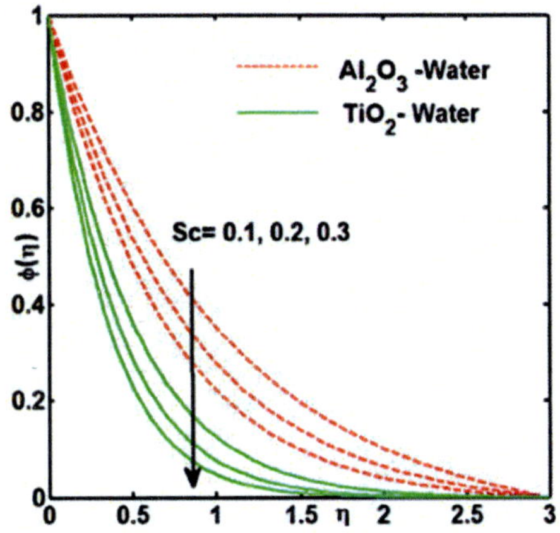


**Fig. 11** Concentrationfor different $Sc$

# 4 Concluding Remarks

- The magnetic field is likely to delay the motion of the fluid, resulting in a reduction in the momentum as well as the velocity of the boundary layer thickness.
- The decreasing $Pr$ values decrease the conduction as well as therefore the changes in thermal characteristics increased.
- The higher Schmidt value limits the thickness of the concentration boundary layer.
- Nanofluid velocity has been identified as a declining function of $\lambda$.

## References

1. Choi SU, Eastman JA (1955) Enhancing thermal conductivity of fluids with nanoparticles, Technical report, Argonne National Lab., IL (United States)
2. Irfan M, Khan W, Khan M, Gulzar MM (2019) Influence of arrhenius activation energy in chemically reactive radiative flow of 3d carreau nanofluid with nonlinear mixed convection. J Phys Chem Solids 125:141–152
3. Sheikholeslami M, Zareei A, Jafaryar M, Shafee A, Li Z, Smida A, Tlili I (2019) Heat transfer simulation during charging of nanoparticle enhanced pcm within a channel. Physica A: Stat Mech Appl 525:557–565
4. Hayat T, Waqas M, Shehzad S, Alsaedi A (2016) A model of solar radiation and joule heating in magnetohydrodynamic (mhd) convective flow of thixotropic nanofluid. J Molecular Liquids 215:704–710
5. Kumar A, Sugunamma V, Sandeep N (2018) Impact of non-linear radiation on mhd non-aligned stagnation point flow of micropolar fluid over a convective surface. J Non-Equilib Thermodyn 43(4):327–345
6. Kumar KA, Sugunamma V, Sandeep N, Reddy JR (2019) Numerical examination of mhd nonlinear radiative slip motion of non-newtonian fluid across a stretching sheet in the presence of a porous medium. Heat Trans Res. 50(12)
7. Hayat T, Ullah S, Khan MI, Alsaedi A, Zia QZ (2018) Non-darcy flow of water-based carbon nanotubes with nonlinear radiation and heat generation/absorption. Results Phys 8:473–480
8. Kataria HR, Patel HR (2019) Effects of chemical reaction and heat generation/absorption on magnetohydrodynamic (mhd) casson fluid flow over an exponentially accelerated vertical plate embedded in porous medium with ramped wall temperature and ramped surface concentration. Propuls Power Res 8(1):35–46

# Non-Darcian Combined Convection of Water Near Its Maximum Density in a Porous Lid-Driven Box with Linearly Heating

S. Sivasankaran and K. Janagi

## 1 Introduction

Convective study in porous cavities is of great attentiveness in plentiful geothermal, biological, agriculture, and engineering disciplines due to its vast applications in various fields [1–10]. Numerical model on joint convection with enclosures of lid-driven is interest of researcher for several cases [11, 12]. Khanafer and Chamkha [13] mathematically deliberated the combined convective flow in a porous wall-driven closed area. A porous lid-driven box considered by Al-Amiri [14] to perform a work on convection. Mixed convective stream in a wall-driven box occupied with aluminum bubbles was explored by Jeng and Tzeng [15]. Sivasankaran and Pan [16] numerically examined the combined convection in a square lid-driven porous box with non-uniform thermal condition on barriers. They observed that energy efficiency in transport is raised with the increase in porosity and Da.

Convective stream and thermal transport in a square porous box treating with position of sinusoidal heating was explored by Sivasankaran and Bhuvaneswari [17]. They identified the rise in averaged Nu with rise in Darcy and Grashof numbers. Wu et al. [18] explored the stream and thermal energy diffusion in a porous box with linear thermal border condition. Gorla et al. [19] explored numerically the MHD free convection in a square porous box with energy source and sink on four sidewalls filled with nanoliquid. Thermal sink location, thermal source, and length influenced the MHD combined convection in a porous box nanoliquid with partial slip effect is identified by Chamkha et al. [20]. Baytas and Baytas [21] explored the convection of

S. Sivasankaran (✉)
Department of Mathematics, King Abdulaziz University, Jeddah 21589, Saudi Arabia
e-mail: sd.siva@yahoo.com

K. Janagi
Department of Mathematics, KPR Institute of Engineering and Technology, Coimbatore, India

 
S. Srinivas et al. (eds.), *Recent Advances in Applied Mathematics and Applications to the Dynamics of Fluid Flows*, Lecture Notes in Mechanical Engineering,
https://doi.org/10.1007/978-981-19-1929-9_10

non-Darcy stream in a box contains heat-generating porous layer and thermal non-equilibrium models. Combined convective stream of alumina-water nanoliquid in a lid-driven enclosed area with 2 porous layers was numerically explored by Astanina et al. [22]. The impact on partial slide effect and heat generation on MHD combined convective stream of nanoliquid is examined in the area enclosed by porous materials by Chamkha et al. [23]. Sheikholeslami et al. [24] considered a porous cubical box containing centrally placed hot sphere by LBM to investigate the magneto-convective flow.

Convective stream of water about its maximum density region is a thorny phenomenon in our environment and in many engineering applications [25–30]. Tong [31] explored the influence of the geometry on free convective flow of water due to density inversion. Sivasankaran and Ho [32] considered the influence of temperature-dependent water properties on convective stream around maximum density. In a medium with porous materials, a convection study of water near 4 °C is made by Zheng et al. [33]. Free convection of water around maximum density in right-angled triangular porous box is done numerically by Oztop et al. [34]. They used non-Boussinesq and Darcy models to formulate the governing equations. It reveals that the energy transference is augmented on diminishing of the L/d ratio. Varol et al. [35] searched the buoyant convection and energy transport of cold-water in a trapezoidal porous box.

It is found from the review of literatures that the non-uniform thermal border has a major effect on the convection stream in enclosed areas. It provides multi-cellular stream structures and complicated energy transport features. This will affects the global enactment of temperature dependent devices. By motivating the use of non-uniform thermal conditions on the thermal systems, the current study is to discover the convective stream and thermal energy transference in a porous box with water nearby its maximum density.

## 2 Mathematical Modelling

Consider incompressible, laminar, unsteady, combined convective stream, and energy transport in the two dimensional porous enclosed space of size $L$ filled with cold water around its density maximum as in Fig. 1. The left vertical barrier is heated linear fashion, and the right side border is cooled. The flat barriers of the box are insulated. The top barrier is permitted to move toward a constant velocity $U_0$. The gravity ($g$) acts directed toward vertically downward. The Darcy-Brinkman-Forchheimer model is utilized to model the medium consists of porous materials. The isotropic and homogeneous porous medium is taken, which is in thermodynamic equilibrium with the liquid. The thermal and physical characteristics of the liquid are uniform excluding the density in buoyancy term. The non-linear variation of water density $\rho = \rho_m\left[1 - \beta|\theta - \theta_m|^b\right]$, where $\rho_m$ (= 999.972) is maintained and is considered as the extreme density of water, $\theta_m$ = 4.029325 °C, $\beta = 9.297173 \times 10^{-6}$, $b =$

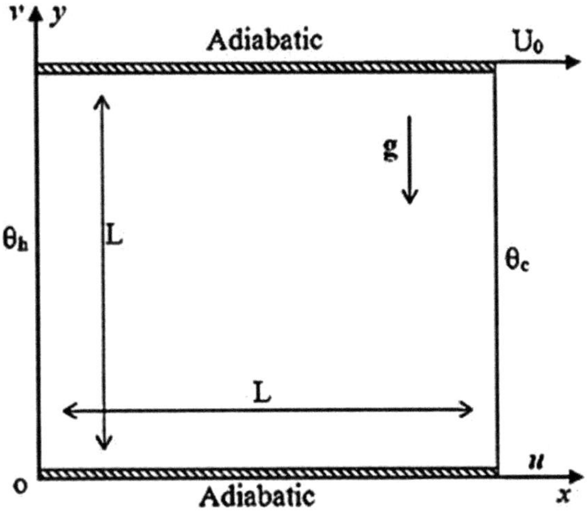


**Fig. 1** Physical configuration

1.894816, and $\beta$ is the coefficient of thermal expansion. The Boussinesq estimate is adopted, and viscous dissolution is insignificant.

Agreeing to named earlier norms, the ruling model can be inscribed as

$$\frac{\partial u}{\partial x} + \frac{\partial v}{\partial y} = 0 \tag{1}$$

$$\frac{1}{\varepsilon}\frac{\partial u}{\partial t} + \frac{1}{\varepsilon^2}\left[u\frac{\partial u}{\partial x} + v\frac{\partial u}{\partial y}\right] = -\frac{1}{\rho_0}\frac{\partial p}{\partial x} + \frac{\nu}{\varepsilon}\left[\frac{\partial^2 u}{\partial x^2} + \frac{\partial^2 u}{\partial y^2}\right] - \frac{\nu}{K}u - \frac{Fc}{\sqrt{K}}u\sqrt{u^2 + v^2} \tag{2}$$

$$\frac{1}{\varepsilon}\frac{\partial v}{\partial t} + \frac{1}{\varepsilon^2}\left[u\frac{\partial v}{\partial x} + v\frac{\partial v}{\partial y}\right] = -\frac{1}{\rho_0}\frac{\partial p}{\partial y} + \frac{\nu}{\varepsilon}\left[\frac{\partial^2 v}{\partial x^2} + \frac{\partial^2 v}{\partial y^2}\right] - \frac{\nu}{K}v - \frac{Fc}{\sqrt{K}}v\sqrt{u^2 + v^2} + g\beta|\theta - \theta_m|^b \tag{3}$$

$$\sigma\frac{\partial \theta}{\partial t} + u\frac{\partial \theta}{\partial x} + v\frac{\partial \theta}{\partial y} = \alpha\left[\frac{\partial^2 \theta}{\partial x^2} + \frac{\partial^2 \theta}{\partial y^2}\right] \tag{4}$$

where $Fc = \frac{1.75}{\sqrt{150}\,\varepsilon^{3/2}}$, and $p$, $t$, $c_p$, $K$, $\alpha$, ε, $v$, and $\sigma$ are the pressure, time, specific heat, permeability, thermal diffusivity, porosity, specific heat ratio, and kinematic viscosity respectively. The following are the constraints at the walls.

$$t = 0:\ u = v = 0\ \ \theta = \theta_c\ \ 0 \le (x, y) \le L$$

$$t > 0: u = v = 0\ \ \frac{\partial \theta}{\partial y} = 0\ \ y = 0$$

$$u = U_0, \quad v = 0, \quad \frac{\partial \theta}{\partial y} = 0, \quad y = L \tag{5}$$

$$u = v = 0 \quad \theta = \theta_h(y) = (\theta_h - \theta_c)\left(\frac{y}{L}\right) + \theta_c \quad x = 0$$

$$u = v = 0 \quad \theta = \theta_c \quad x = L$$

Using the non-dimensional quantities $X = \frac{x}{L}$, $Y = \frac{y}{L}$, $U = \frac{u}{U_0}$, $V = \frac{v}{U_0}$, $T = \frac{\theta - \theta_c}{\theta_h - \theta_c}$, $\tau = \frac{tU_0}{L}$, and $P = \frac{p}{\rho_0 U_0^2}$, the equations could be modeled non-dimensionally as follows:

$$\frac{\partial U}{\partial X} + \frac{\partial V}{\partial Y} = 0 \tag{6}$$

$$\frac{1}{\varepsilon}\frac{\partial U}{\partial \tau} + \frac{1}{\varepsilon^2}\left[U\frac{\partial U}{\partial X} + V\frac{\partial U}{\partial Y}\right] = -\frac{\partial P}{\partial X} + \frac{1}{\varepsilon \mathrm{Re}}\nabla^2 U - \frac{U}{Da\mathrm{Re}} - \frac{Fc}{\sqrt{Da}}U\sqrt{U^2 + V^2} \tag{7}$$

$$\frac{1}{\varepsilon}\frac{\partial V}{\partial \tau} + \frac{1}{\varepsilon^2}\left[U\frac{\partial V}{\partial X} + V\frac{\partial V}{\partial Y}\right] = -\frac{\partial P}{\partial Y} + \frac{1}{\varepsilon \mathrm{Re}}\nabla^2 V - \frac{V}{Da\mathrm{Re}} - \frac{Fc}{\sqrt{Da}}V\sqrt{U^2 + V^2} + Ri|T - T_m|^b \tag{8}$$

$$\sigma\frac{\partial T}{\partial \tau} + U\frac{\partial T}{\partial X} + V\frac{\partial T}{\partial Y} = \frac{1}{\Pr \mathrm{Re}}\nabla^2 T \tag{9}$$

The non-dimensional constraints presented in the equations mentioned above are the Darcy number $Da = \frac{K}{L^2}$, the density inversion parameter, $T_m = \frac{\theta_m - \theta_c}{\theta_h - \theta_c}$, the Grashof number, $Gr = \frac{g\beta|\theta_h - \theta_c|^b L^3}{\nu^2}$, the Prandtl number, $\Pr = \frac{\nu}{\alpha}$, the Reynolds number, $\mathrm{Re} = \frac{U_0 L}{\nu}$, and the Richardson number, $Ri = \frac{Gr}{\mathrm{Re}^2}$.

$$\tau = 0\text{: } U = \mathrm{V} = 0; \quad T = 0; \quad 0 \le (X, Y) \le 1,$$

$$\tau > 0\text{: } U = \mathrm{V} = 0; \quad Y = 0; \quad X = 0 \;\&\; 1,$$

$$\mathrm{U} = 1, \ \mathrm{V} = 0; \ Y = 1, \tag{10}$$

$$\frac{\partial T}{\partial Y} = 0; \ Y = 0 \;\&\; 1$$

$$T = Y; \ X = 0,$$

**Table 1** Comparison of averaged Nusselt number for a porous box with $\varepsilon = 0.4$ and $Pr = 1.0$

| *Da* | *Ra* | $\overline{Nu}$ | |
|---|---|---|---|
| | | Nithiarasu et al. [36] | Present |
| $10^{-4}$ | $10^5$ | 1.067 | 1.074 |
| | $10^6$ | 2.550 | 2.688 |
| $10^{-2}$ | $10^3$ | 1.010 | 1.007 |
| | $10^4$ | 1.408 | 1.343 |
| | $10^5$ | 2.983 | 2.993 |

$$T = 0;\ X = 1,$$

Magnitude of transfer of energy through the enclosure is considered to be the significant factor for thermal applications and other systems. The identity $Nu = -\left(\frac{\partial T}{\partial X}\right)\Big|_{X=0}$ represents the local Nu with the barrier in the left side of the box. The averaged Nu is computed as $\overline{Nu} = \int_0^1 Nu dY$.

## 3 Process of Solution

The control volume approach is used to discretize the modeled Eqs. (6)–(9) having border conditions (10). The detailed solution procedure is found in the paper [1, 16, 17]. To obtain the accurateness and authority of the code, some mathematical reproductions acquired from the program are compared with the outcomes in the literature [36]. The comparison is presented in the Table 1. It is witnessed from the tables that the results of current outcomes agreed adequately with the previous outcomes.

## 4 Results and Discussion

Numerical solutions have been completed to examine the influence of density inversion on stream and energy through convective mode as combined form and analyzed the characteristics of water in a closed porous wall-driven enclosure. The non-dimensional constraints involved are density reversal parameter from 0 to 1 and the *Da* from $10^{-5}$ to $10^{-1}$. The Richardson number (*Ri*) is varying from 0.01 to 100, while Grashof numbers (*Gr*) are ranging from $10^2$ to $10^6$ for fixed Reynolds number, Re = 100. The porosity $\in$ is selected as 0.1, 0.3, 0.5, and 0.8. The Prandtl number is Pr = 11.6 which relates to the cold water around 4 °C. The density reversal parameter aids us to trace the density extreme plane within the box due to the existence of two dissimilar flat layers of water with dissimilar changes in density divided by this extreme level density.

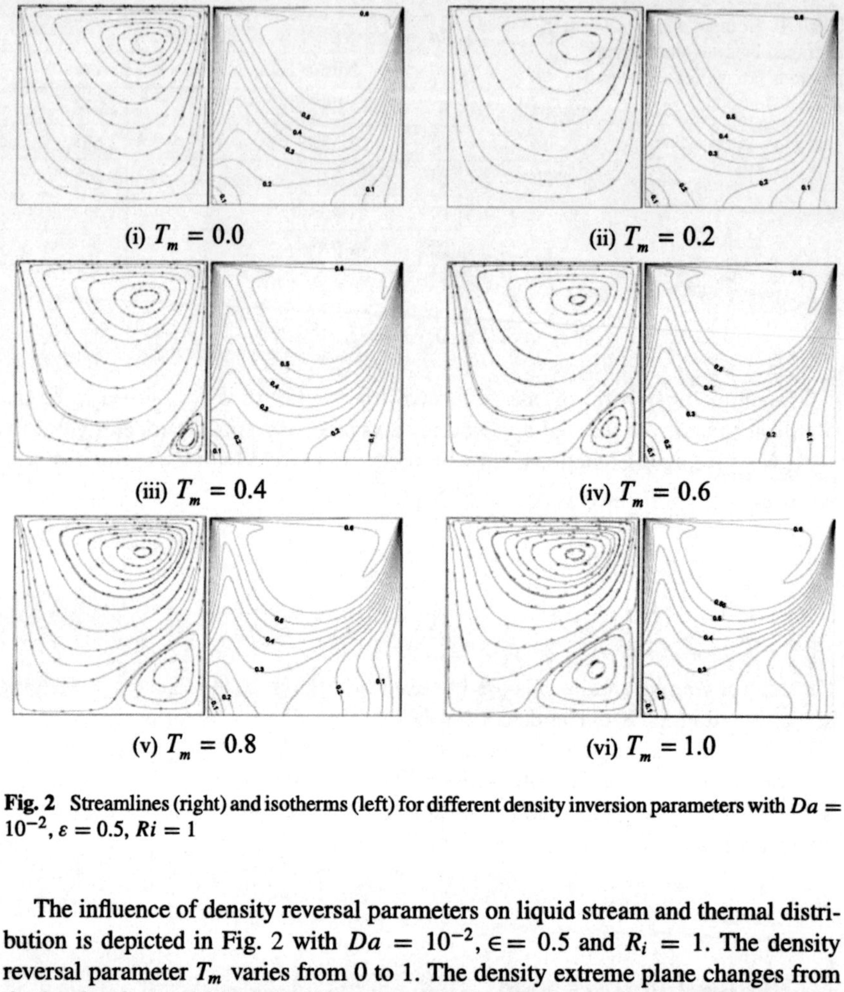

**Fig. 2** Streamlines (right) and isotherms (left) for different density inversion parameters with $Da = 10^{-2}$, $\varepsilon = 0.5$, $Ri = 1$

The influence of density reversal parameters on liquid stream and thermal distribution is depicted in Fig. 2 with $Da = 10^{-2}$, $\in = 0.5$ and $R_i = 1$. The density reversal parameter $T_m$ varies from 0 to 1. The density extreme plane changes from the hot barrier to cold barrier when growing the density reversal values from 0 to 1. When $T_m = 0(1)$, the density extreme plane is at the cold (hot) wall of the box. The stream consists of a single right-handed rotating vortex and occupies the full enclosed space when $T_m = 0$. The central region of eddy is near to the top (moving) barrier because of the shear force produced by stirring barrier. The major area of the main eddy is placed in a same place for all values of the density reversal parameter. However the secondary stream exists along the cold barrier due to density reversal effect on growing $T_m$. When $T_m = 0.4$, a small re-circulating eddy exists in the down to the right of the box. The size of the re-circulating eddy rises on raising $T_m$. The streamlines are to be dependent with the values of the density reversal parameter. The stream structure is changed from single cell to dual cell pattern while shifting the density reversal parameter from 0 to 1.

It shows the density reversal effect clearly inside the box, that is, both vertical walls produce an equal rising buoyancy force, which yields a 2-cell stream arrangement inside the enclosed space. The liquid stream rising through the hot/cold walls and descends in the density extreme plane. The hot left cell stretched out in the upper area of box due to shear force, which enhances the circulation of the hot cell, so that the right cold cell shrinks in the lower part of the box. The relevant isotherms indicate the convection model of transfer of energy across the box. The thermal layers are shaped along the upper part of the isothermal walls. The impact of Darcy number is in the stream through porous media is significant. To find the influence of $Da$ on stream thermal fields, the streamlines and isotherms are portrayed for diverse values of $Da$ in Fig. 3 with $T_m = 0.5$, $\in = 0.3$ and $R_i = 1$.

When $Da = 10^{-5}$ the stream is weaken and the core region of the eddy is stretched out elliptically along the top wall because of shear force by moving wall. The isotherms are almost equally distributed inside the box and indicate the conduction mode of energy transport. When raising the values of $Da$, the power of the eddy rises and energy transport is improved. When $Da \leq 10^{-2}$, the twin eddy is formed within the box. The temperature boundary layers form beside the vertical

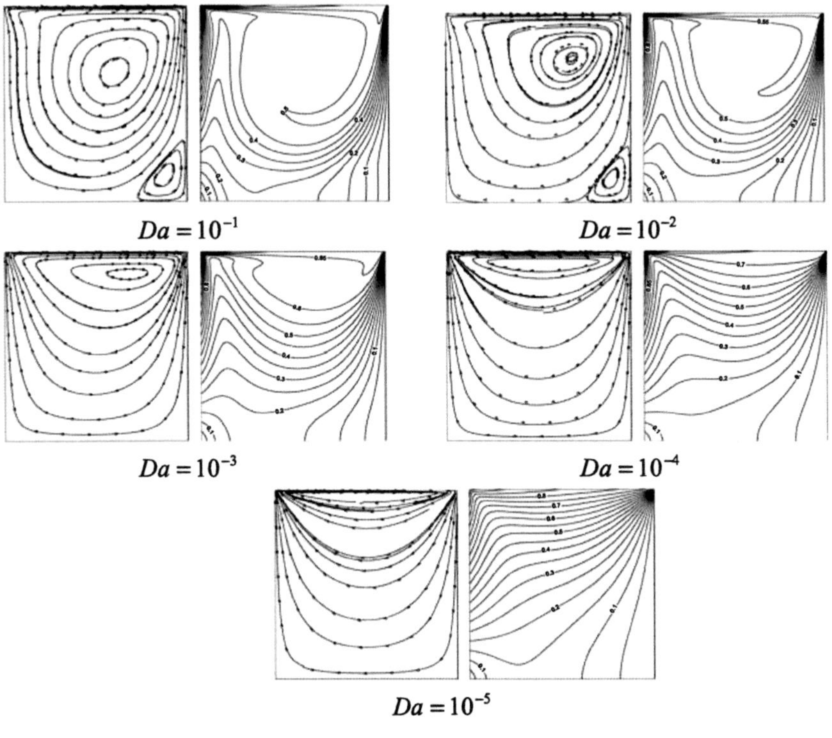


**Fig. 3** Streamlines (right) and isotherms (left) for different Darcy numbers with $T_m = 0.5$, $\varepsilon = 0.3$, $Ri = 1$

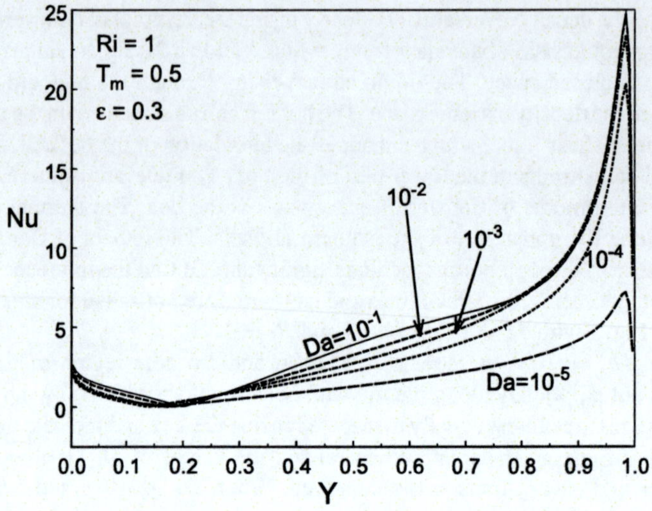


**Fig. 4** Local Nusselt number for different Darcy numbers with $T_m = 0.5$, $\varepsilon = 0.3$, $Ri = 1$

walls. Hence, convection mode of energy transport is dominated for lower values of $Da\left(Da \leq 10^{-2}\right)$.

Since the rate of energy transport across the box is most important for thermal system applications, the local and averaged energy transport rates are calculated along the hot wall for diverse combinations of pertinent constraints involved here. Figure 4 represents the plot of local Nu versus diverse values of $Da$ with $T_m = 0.5$, $\in = 0.3$ and $R_i = 1$. The local energy transport is increased with the hot wall height. The peak value of Nu (that is, the highest energy transport rate) attains the top wall. The movement in the top wall causes the reason of highest transfer of energy. That is, the particles of fluid close to the upper portion of the hot wall move rapidly by shear force produced by moving wall. The increment in Darcy number raises the local energy transfer.

The averaged Nu for various combinations of physical parameters driven the stream and energy transport inside the box is displayed in Figs. 5, 6, and 7. Figure 5 indicates the averaged Nusselt number for numerous $Ri$ and $Da$ with $T_m = 0.5$, and $\in = 0.3$. The averaged energy transport is almost constant when growing Richardson number up to 10 and then rises on growing Richardson number for a given $Da$. The energy transport rate rises on raising the Darcy number. The stream velocity becomes significant due to the increase in $Da$ from $10^{-5}$ to $10^{-1}$. The permeability rises with Darcy number and henceforth the solid frictional resistance gradually reduced, which improves the rotation of the vortex considerably inside the enclosure. Henceforth, the rate of energy transport is enhanced. Energy transfer phenomenon to convection from conduction can be correlated with this behavior on growing the values of $Da$.

Cumulated Nu for various density reversal parameter with $R_i = 1$ is plotted in Fig. 6. Energy transport rate behaves constantly for all values of $T_m$ when Darcy

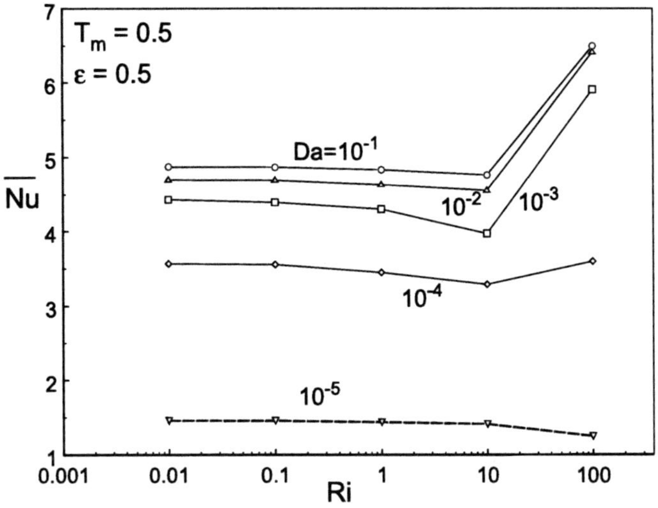


**Fig. 5** Average Nusselt number versus Richardson number for different Darcy numbers with $T_m = 0.5$, $\varepsilon = 0.3$

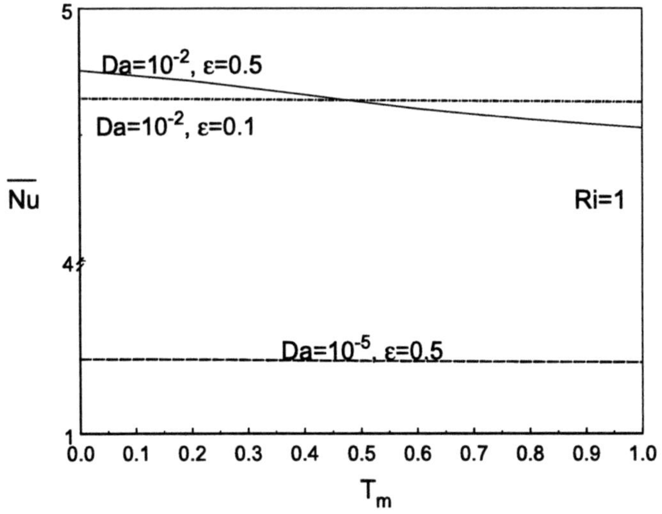


**Fig. 6** Average Nusselt number versus density inversion parameter for different Darcy numbers and porosity with $Ri = 1$

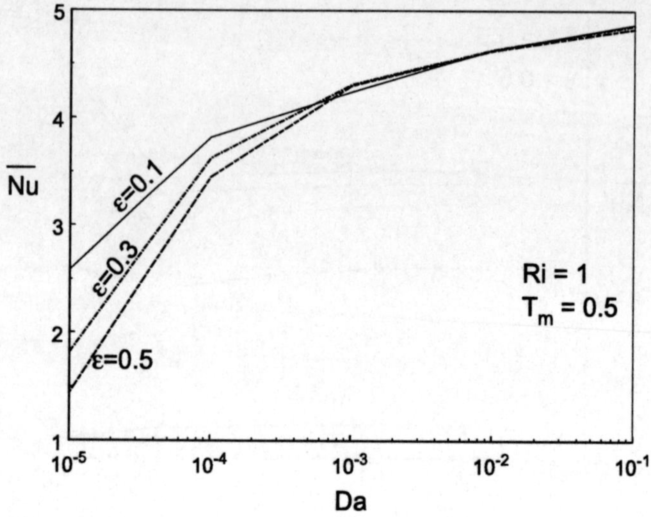


**Fig. 7** Average Nusselt number versus Darcy number for different porosity with $T_m = 0.5$, $Ri = 1$

number is small. Averaged Nusselt number decreases slightly when increasing $T_m$ from 0 to 1 for $Da = 10^{-2}$. The decreasing rate of the averaged Nu against $T_m$ is slower for $\in= 0.1$, and it is high for $\in= 0.5$. When the density reversal occurs inside the enclosure, the stream involves of 2 counteracting eddies. Since thermal energy is conveyed from one vortex to other vortex by conduction, the energy transferal rate is diminished. The thermal energy between these 2 vortices is exchanged by conduction. The dual-cell arrangement interdicts convection type of energy transport across the enclosure. However, the lid-driven force makes the hot cell always dominant one inside the box and occupies the majority of the enclosure. So, most of the energy transport through the hot vortex provides enhancement of global energy transport. Suppose, if the two vortices are in nearly equal size (symmetry with horizontal), the rate of energy transport touches the least value. It is exciting to mark that the averaged Nusselt number approaches almost constant value on changing density inversion parameter $T_m$ with $Da = 10^{-5}$. It is concluded that the density inversion effect is negligible for low values of $Da$. Figure 7 demonstrates the effect of porosity for different Darcy numbers with $R_i = 1$ and $T_m = 0.5$. The impact of porosity is highly influenced for least values of the Darcy numbers (that is, $Da < 10^{-3}$ ). The porosity effect on average energy transport is negligible for $Da > 10^{-2}$. It is also witnessed that averaged Nusselt number declines on growing the porosity values.

# 5 Conclusions

A numerical work on mixed convection stream of water around $4°C$ in a wall-driven porous box of square size is carried out. The investigation is primarily focused on the impact of density inversion of cold water on mixed convective stream and energy transport in a porous box. The following inferences are obtained in the existing analysis.

- The density inversion parameter affects the stream field and gives dual cell structure. The 2-cell structure of density extreme prohibits energy transport across the box. Since the exchange of thermal energy in the middle of these 2 cells by conduction, the energy transport rate reduces at density extreme region.
- The energy transport rate is almost constant for distinct values of $T_m$ and $Ri$ with small $Da$. It is perceived that energy efficiency rate rises on rising the Darcy number.
- The averaged Nu drops on raising the porosity of the medium.
- The averaged energy transport is almost constant when growing the Richardson number up to 1 and then increases on growing the Richardson number for a given Darcy number.

# References


1. Bhuvaneswari M, Sivasankaran S, Kim YJ (2011) Effect of aspect ratio on convection in a porous enclosure with partially active thermal walls. Comput Math Appl 62:3844–3856
2. Chamkha AJ, Selimefendigil F, Ismael MA (2016) Mixed convection in a partially layered porous cavity with an inner rotating cylinder. Numer Heat Transf Part A 69(6):659–675
3. Alsabery AI, Tayebi T, Chamkha AJ, Hashim I (2018) Effect of rotating solid cylinder on entropy generation and convective heat transfer in a wavy porous cavity heated from below. Int Commun Heat Mass Transf 95:197–209
4. Biswas N, Manna NK (2017) Enhanced convective heat transfer in lid-driven porous cavity with aspiration. Int J Heat Mass Transf 114:430–452
5. Sheikholeslami M, Shamlooei M (2017) Convective flow of nanofluid inside a lid driven porous cavity using CVFEM. Phys B 521(15):239–250
6. Sivasankaran S, Bhuvaneswari M, Alzahrani AK (2020) Numerical simulation on convection of non-Newtonian fluid in a porous enclosure with non-uniform heating and thermal radiation. Alexandria Eng J 59:3315–3323
7. Sivasankaran S, Cheong HT, Bhuvaneswari M (2019) Natural convection in an inclined porous triangular enclosure with various thermal boundary conditions. Therm Sci 23(2A):537–548
8. Cheong HT, Sivasankaran S, Bhuvaneswari M (2018) Effect of aspect ratio on natural convection in a porous wavy cavity. Arab J Sci Eng 43:1409–1421
9. Cheong HT, Sivasankaran S, Bhuvaneswari M (2017) Natural convection in a wavy porous cavity with sinusoidal heating and internal heat generation. Int J Numer Meth Heat Fluid Flow 27(2):287–309
10. Sivasankaran S, Pan KL (2014) Natural convection of nanofluids in a cavity with non-uniform temperature distributions on side walls. Numer Heat Transf A 65:247–268

11. Sivasankaran S, Malleswaran A, Bhuvaneswari M, Poo Balan G (2017) Hydro-magnetic mixed convection in a lid-driven cavity with partially thermally active walls. Sci Iran Trans B Mech Eng 24(1):153–163
12. Sivasankaran S, Ananthan SS, Abdul Hakeem AK (2016) Mixed convection in a lid-driven cavity with sinusoidal boundary temperature at the bottom wall in the presence of magnetic field. Sci Iran Trans B Mech Eng 23(3):1027–1036
13. Khanafer KM, Chamkha AJ (1999) Mixed Convection flow in a lid-driven enclosure filled with a fluid-saturated porous medium. Int J Heat Mass Transf 42:2465–2481
14. Al-Amiri AM (2000) Analysis of momentum and energy transfer in a lid-driven cavity filled with a porous medium. Int J Heat Mass Transf 43:3513–3527
15. Jeng TM, Tzeng SC (2008) Heat transfer in a lid-driven enclosure filled with water-saturated aluminum foams. Numer Heat Transf A 54:178–196
16. Sivasankaran S, Pan KL (2012) Numerical simulation on mixed convection in a porous lid-driven cavity with non-uniform heating on both sidewalls. Numer Heat Transf A 61:101–121
17. Sivasankaran S, Bhuvaneswari M (2013) Natural convection in a porous cavity with sinusoidal heating on both sidewalls. Numer Heat Transf A 63:14–30
18. Wu F, Zhou W, Wang G, Ma X, Wang Y (2015) Numerical simulation of natural convection in a porous cavity with linearly temperature distributions under the local thermal non equilibrium condition. Numer Heat Transf A 68:1394–1415
19. Gorla RSR, Siddiqa S, Mansour MA, Rashad AM, Salah T (2017) Heat source/sink effects on a hybrid nano fluid-filled porous cavity. J Thermo Phys Heat Transf 31(4):847–857
20. Chamkha AJ, Rashad AM, Mansour MA, Armaghani T, Ghalambaz M (2017) Effects of heat sink and source and entropy generation on MHD mixed convection of a Cu-water nanofluid in a lid-driven square porous enclosure with partial slip. Phys Fluids 29:052001
21. Baytaş AF, Baytaş AC (2017) Thermal non-equilibrium natural convection in a square enclosure with heat-generating porous layer on inner walls. Trans Por Media 120(1):167–182
22. Astanina MS, Sheremet MA, Oztop HF, Abu-Hamdeh N (2018) Mixed convection of Al2O3-water nanofluid in a lid-driven cavity having two porous layers. Int J Heat Mass Transf 118:527–537
23. Chamkha AJ, Rashad AM, Armaghani T, Mansour MA (2018) Effects of partial slip on entropy generation and MHD combined convection in a lid-driven porous enclosure saturated with a Cu–water nanofluid. J Therm Anal Calori 132(2):1291–1306
24. Sheikholeslami M, Shehzad SA, Li Z (2018) Water based nanofluid free convection heat transfer in a three dimensional porous cavity with hot sphere obstacle in existence of Lorenz forces. Int J Heat Mass Transf 125:375–386
25. Sivasankaran S (2018) Natural convection of cold water near its density maximum in a porous wavy cavity. Flow Transp Subsurf Environ (Part 3, Chap 4) 305–324
26. Bindhu R, Sai Sundara Krishnan G, Sivasankaran S, Bhuvaneswari M (2019) Magneto-convection of water near its maximum density in a cavity with partially thermally active walls. Energy Environ 30(5):833–853
27. Janagi K, Sivasankaran S, Bhuvaneswari M, Eswaramurthi M (2017) Numerical study on free convection of cold water in a square porous cavity with sinusoidal wall temperature. Int J Numer Meth Heat Fluid Flow 27(4):1000–1014
28. Sivasankaran S, Ho CJ (2010) Buoyancy and thermo-capillary induced convection of cold water in an open enclosure with variable fluid properties. Numer Heat Transf A 58(6):457–474
29. Sivasankaran S, Ho CJ (2008) Effect of temperature dependent properties on MHD convection of water near its density maximum in a cavity. Int J Thermal Sci 47:1184–1194
30. Lin DS, Nansteel MW (1987) Natural convection heat transfer in a square enclosure containing water near its density maximum. Int J Heat Mass Transf 30:2319–2329
31. Tong W (1999) Aspect ratio effect on natural convection in water near its density maximum temperature. Int J Heat Fluid Flow 20:624–633
32. Sivasankaran S, Ho CJ (2008) Effect of temperature dependent properties on convection of water near its density maximum in enclosures. Numer Heat Transf A 53:507–523

33. Zheng W, Robillard L, Vasseur P (2001) Convection in a square cavity filled with an anisotropic porous medium saturated with water near 4°C. Int J Heat Mass Transf 44:3463–3470
34. Oztop HF, Varol Y, Pop I (2009) Investigation of natural convection in triangular enclosure filled with porous media saturated with water near 4°C. Energy Convers Manage 50:1473–1480
35. Varol Y, Oztop HF, Pop I (2010) Maximum density effects on buoyancy-driven convection in a porous trapezoidal cavity. Int Commun Heat Mass Transf 37:401–409
36. Nithiarasu P, Seetharamu KN, Sundararajan T (1997) Natural convective heat transfer in a fluid saturated variable porosity medium. Int J Heat Mass Transf 40:3955–3967

# Influence of Tilting Angle and Heater Locations on MHD Mixed Convection in a Tilted Box

**S. Sivasankaran, M. Bhuvaneswari, and R. Bindhu**

## 1 Introduction

Convection situation governed by the joint action of free and forced convection such that each dominates the other is commonly referred to as mixed or combined convection. The problems involving such mixed convection in a closed box have been given a significant attention by researchers [1–3]. Some general applications in industries are glass production, crystal growth, nuclear reactors, and food processing [4–8]. While magnetic field is exposed to an electrically conducting fluid, Lorentz force is resulted and this leads to magneto-hydrodynamics. Interactions of these forces with the convective force to govern the heat exchange and flow characteristics are studied in [9–12]. The Lorentz force holds back the convection currents thereby reducing the velocities. This force employs an external magnetic field and this aspect takes an important role in industrial applications whereof as controlling mechanisms are much necessary [13–17]. Sivasankaran and Ho [18] explained the effects numerically of temperature-dependent properties of water near its maximum density while magnetic field is applied uniformly. Sivasankaran et al. [19] analyzed numerically the study on temperature-dependent properties and its effect on cold water near its maximum density on Marangoni MHD convection. Sivasankaran et al. [20] analyzed on combined convection within a square box with sinusoidal boundary conditions and magnetic influence.

S. Sivasankaran (✉)
Department of Mathematics, King Abdulaziz University, Jeddah, Saudi Arabia
e-mail: sd.siva@yahoo.com

M. Bhuvaneswari
Department of Mathematics, Kongunadu Polytechnic College, D.Gudalore, Tamilnadu, India

R. Bindhu
Department of Mathematics, PSG College of Technology, Coimbatore 641004, India



S. Srinivas et al. (eds.), *Recent Advances in Applied Mathematics and Applications to the Dynamics of Fluid Flows*, Lecture Notes in Mechanical Engineering,
https://doi.org/10.1007/978-981-19-1929-9_11

The convection under the condition of isothermal vertical walls maintained fully or partially at different temperatures in cavities has been considerably focused and studied in recent times. Kuhn and Oosthuizen [21] proposed results on free convective, unsteady flow inside a partially heated rectangular box. Numerical study on natural convection in a cubical box with a cold vertical wall and a hot square sector on the opposite wall was made by Frederick and Quiroz [22]. Oztop [23] examined numerically the combined convective flow within a partially heated porous lid-driven box. Kuznetsov and Sheremet [24, 25] investigated on conjugate convective flow in a rectangular-shaped region with localized heat resource. Bhuvaneswari et al. [26] investigated the effects of partially active thermal zones and aspect ratio of rectangular-shaped porous box on convective heat transfer. Sivasankaran et al. [27] analyzed the effect of discrete heating on natural convection in a rectangular porous box. Recently, the studies of convection in inclined boxes have been considered in literature to discuss the power of inclination on thermally driven flows as they originate in many engineering applications [28, 29]. Al-Najem et al. [30] examined on free convective flow's effect inside an inclined box in the incident of magnetic force. Numerical results proved that the heat transfer is by pure conduction mode for the inclination angle –90°. Cianfrini et al. [31] studied on free convective flow of heat in a tilted square box filled with air, which has differentially heated opposite walls. Sharif [32] analyzed the combined convective flow pattern within a shallow two-dimensional rectangular box for varying values of Richardson number. Their results showed that the increase rate in mean Nusselt number value is mild for enforced convection and much steeper for natural convection with box inclination.

It can be understood that most of the previous studies have been committed on learning the effects of different thermal boundary conditions such as fully or partially active walls and for the MHD mixed convection effects in lid-driven cavities. In many industrial applications, it is required to keep the box inclined since the inclination angle of the box may be helpful to assist or oppose the buoyancy. This motivated the present study, and it aims to provide numerical results that give the unexplained effects of location changes of thermal source kept on the left wall of an inclined square lid-driven box when uniform magnetic field is applied.

## 2 Mathematical Formulation

The configuration of an inclined square box of length L filled with an electrically conducting fluid taken for this study is illustrated in Fig. 1a. The flow assumed is two-dimensional and it is unsteady incompressible and laminar. The left sidewall is partially heated by a heater of constant length $L_{H}$. Higher temperature $\theta_h$ is prevailed there. The right wall of the inclined box prevails a lesser temperature $\theta_c$, such that $\theta_h > \theta_c$. The left out portions on the left sidewall and horizontal walls are thermally insulated. The locations of the heater are varied for three positions along the left wall namely top, middle, and bottom. The upper wall of the box is permitted to slide with a constant speed $U_0$ in its own surface of the plane. The velocity components $u$ and $v$

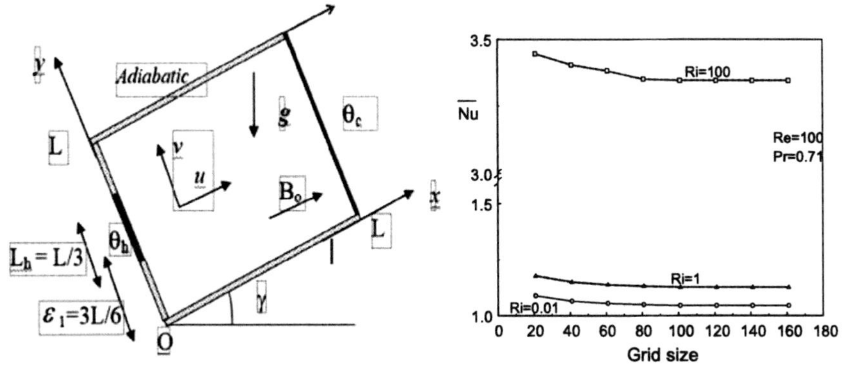


**Fig. 1** **a** Schematic diagram, **b** grid independence test

are taken in $x$- and $y$-directions, respectively. The inclination angle ($\gamma$) of the box is measured from horizontal in the anti-clockwise direction. Acceleration due to gravity acts downward. Externally supplied magnetic field is allowed to along $x$-direction, and it is applied uniformly with a constant magnitude $B_0$. The induced magnetic field produced within the cavity by the movement of electrically conducting fluid is negligible when compared to the applied magnetic field since the magnetic Reynolds number is assumed to be too small. In the modeled equations for momentum, except the body force term, the other properties are considered as a constant. The properties such as Joule heating or resistance heating and viscous dissipation are ignored. On employing Boussinesq approximation, the governing equations become

$$\frac{\partial U}{\partial X} + \frac{\partial V}{\partial Y} = 0 \tag{1}$$

$$\frac{\partial U}{\partial \tau} + U\frac{\partial U}{\partial X} + V\frac{\partial V}{\partial Y} = -\frac{\partial P}{\partial X} + \frac{\partial^2 U}{\partial X^2} + \frac{\partial^2 U}{\partial Y^2} + \frac{Gr}{Re^2}\mathrm{Tsin}(\gamma) \tag{2}$$

$$\frac{\partial V}{\partial \tau} + U\frac{\partial V}{\partial X} + V\frac{\partial V}{\partial Y} = -\frac{\partial P}{\partial Y} + \frac{\partial^2 V}{\partial X^2} + \frac{\partial^2 V}{\partial Y^2} + +\frac{Gr}{Re^2}T cos(\gamma) - Ha^2 V \tag{3}$$

$$\frac{\partial T}{\partial \tau} + U\frac{\partial T}{\partial X} + V\frac{\partial T}{\partial Y} = \frac{1}{PrRe}\left[\frac{\partial^2 T}{\partial X^2} + \frac{\partial^2 T}{\partial Y^2}\right] \tag{4}$$

The following dimensionless quantities are used to non-dimensionalize the equations: $(X, Y) = \frac{(x,y)}{L}$, $(U, V) = \frac{(u,v)}{U_0}$, $T = \frac{\theta-\theta_c}{\theta_h-\theta_c}$, $\tau = \frac{tU_0}{L}$, $\varepsilon_2 = \frac{\varepsilon_1}{L}$, $P = \frac{p}{\rho U_o^2}$.

The non-dimensional parameters are defined as $Gr = g\beta\Delta T L^3/\nu^2$, the Grashof number, $\mathrm{H}a = B_0 L\sqrt{\sigma_e/\mu}$, the Hartmann number, $\mathrm{Pr} = \nu/\alpha$, the Prandtl number, $Re = U_0 L/\nu$, the Reynolds number and $Ri = Gr/\mathrm{Re}^2$, the Richardson number. The appropriate dimensionless initial conditions and boundary conditions are

$$\text{For } \tau = 0 : U = 0,\ V = 0,\ T = 0,\ 0 \leq (X, Y) \leq 1$$

$$\text{For } \tau > 0 : U = 0,\ V = 0,\ \partial T/\partial Y = 0,\ Y = 0$$

$$U = 1,\ V = 0,\ \partial T/\partial Y = 0,\ Y = 1 \tag{5}$$

$$U = 0,\ V = 0\ T = 0,\ X = 1\ U = 0,\ V = 0,$$
$$T = 1, X = 0,\ \varepsilon_2 - 1/6 \leq Y \leq \varepsilon_2 + 1/6$$
$$\frac{\partial T}{\partial X} = 0,\ X = 0,\ 0 \leq Y \leq \varepsilon_2 - 1/6,\ \varepsilon_2 + 1/6 \leq Y \leq 1$$

where $\varepsilon_2 = 1/6$, 3/6, and 5/6. When $\varepsilon_2 = 1/6$, the heater is held at the bottom, for $\varepsilon_2 = 3/6$ it is held at middle and when $\varepsilon_2 = 5/6$, it is kept at top of the wall. To provide the relative strengths between convective to conductive heat transfers, the Nusselt number values are measured. The local values of Nusselt number of the heater are attained by means of the expression $Nu = -\left(\frac{\partial T}{\partial X}\right)_{X=0}$. The averaged Nusselt number is calculated for three distinct places based on the value of $\varepsilon_2$. For $L_h$ denoting the length of the heater, it is calculated by $\overline{Nu} = \frac{1}{L_h} \int\limits_{\varepsilon_2 - 1/6}^{\varepsilon_2 + 1/6} Nu dY$.

## 3 Computational Procedure

The dimensionless Eqs. (1)–(4) together with the boundary conditions (5) are meshed to produce a set of simultaneous equations by adopting control volume formulation. Using the Gauss-Seidel point by point iteration procedure, the numerical solutions for the discretized sets of algebraic equations are calculated with uniform mesh system carried along both axes. The convergence criterion to achieve is set to satisfy $10^{-6}$. Grid independence tests are performed to adopt an appropriate grid system by experimenting with various uniform grid sizes ranging from 41 × 41 to 161 × 161 for $Ri = 0.01,\ 1, 100$, $Pr = .054$ and $Ha = 0$ and this is exhibited in Fig. 1b. It is observed that a uniform grid with size 121 × 121 is sufficient to calculate the desired accuracy of solutions. Accuracy of the numerical results is established by validating the present computational code against the solution obtained from convective flow in a lid-driven box [32], see Table 1. It can be verified that the present solutions show a rational conformity with the available solutions.

**Table 1** Comparison of $\overline{Nu}$ with available results in the literature for lid-driven box

| *Gr* | *Re* = 400 | | *Re* = 1000 | |
|---|---|---|---|---|
| | Present work | Sharif [32] | Present work | Sharif [32] |
| $10^2$ | 4.08 | 4.05 | 6.48 | 6.55 |
| $10^4$ | 3.84 | 3.82 | 6.47 | 6.50 |
| $10^6$ | 1.10 | 1.17 | 1.66 | 1.81 |

## 4 Discussion and Results

The parameters that govern the flow speed and heat characteristics are the locations of heater ($\varepsilon_2 = 1/6, 3/6\&5/6$), the inclination angle of the box ($\gamma = 0°, 30°, 45°, 60°$, and $90°$), the Hartmann number ($Ha = 0, 25, 100$), and the Richardson number ($\mathrm{Ri} = \mathrm{Gr}/\mathrm{Re}^2$). Reynolds number and Prandtl number are taken to be Re = 100 & Pr = 0.054. The range in which Richardson number varies is from $0.01 \leq Ri \leq 100$ and this is set by varying the Grashof numbers from $10^2$ to $10^6$.

The isotherms and streamlines for three locations of heater and inclination angles 0°, 45°, and 90° of the box are depicted in Fig. 2a–f at $Ha = 25$, Ri = 0.01, and Ri = 100. The centers of the heater at different locations, namely, bottom, middle, and top are $\varepsilon_2 = 1/6$, $\varepsilon_2 = 3/6$, and $\varepsilon_2 = 5/6$, respectively. In the forced convection mode (Ri = 0.01), Fig. 2a–c displays the isotherms that are almost straightened for all the inclination angles. Even in the vicinity of the active region, the isotherms are nearly parallel to the surface with no formations of boundary layers. As far as the heater positions are considered, almost all the isotherms are parallel to active left wall for all positions of heater. These indicate that the heat distribution occurs by conduction mode of heat transfer for all the varied inclination angles and for all the three locations of heater. Obviously, the temperature gradients in the box are not significant due to the dominancy of conduction mode. But, heat transport mechanism is changed to pure convection in the buoyancy dominated (Ri = 100) regime. In fact, the convection becomes more dominant since the value of the Grashof number increases. The temperature gradients near the heater are appreciable, and this results in the advance of thermal boundary layer. In addition to it, isotherms are crowded close to the right-top of the cold wall and hence thin boundary layers are formed for the non-inclined ($\gamma = 0°$) box and for lower inclination angle $\gamma = 45°$. Similar fashion of heat distribution occurs for all the locations of heater. In fact, the temperature distribution slightly tends to increase when inclination angle is $\gamma = 45°$ at all the locations. At $\gamma = 90°$, the isotherms are clustered and thin thermal boundary layers appear in the vicinity of left heater wall and also about the corresponding locations of right wall. This indicates a good evidence for the event of better heat transfer. In general, the heat transport phenomena occur in the region of heating and rising heat transfer reaches next to the leading edge of the heater close to the adiabatic portion on the left side heater wall. Moreover, at this inclination angle, isotherms are distributed almost in the whole box representing that the heat movement is enhanced for the middle position of the heater. Generally, the heat transport is boosted up for the

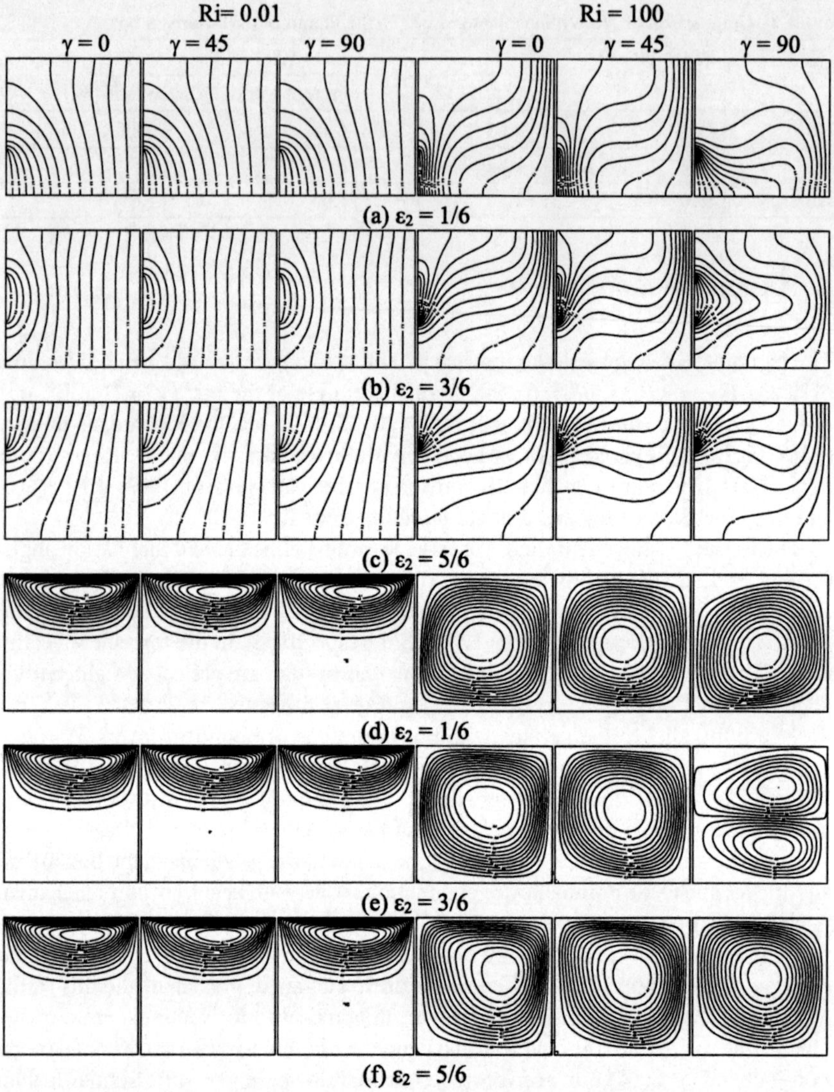


**Fig. 2** **(a–f)** Isotherms **(a–c)** and streamlines **(d–f)** for different inclinations, location of heater and Ri with Ha = 25 and $L_H = 1/3$

inclination angle $\gamma = 90°$ for the middle location of heater. An enhanced convection heat transfer is noted for high value of Richardson number ($Ri = 100$).

Figure 2d–f exemplifies the flow scenarios for various locality of the heater and box inclination angles while $Ha = 25$, Ri = 0.01, and Ri = 100. The flow pattern shows that although the box is skewed with various inclinations, when lid-driven (forced) convection takes control, flow structures are almost similar for all inclinations and locations considered. Its flow pattern consists of a clockwise turning eddy in elliptical like shape and it is sighted only in the top-portion of the box near the upper wall. Since the inertia force generated by the lid velocity dominates over the strength of buoyancy, the movement of the fluid flow is controlled and the core or the nucleus region of the vortex is close to the upper part of the box. Moreover, the flow near to the bottom of the box is almost sluggish. At the same time, a major change in the flow pattern can be viewed when buoyant-driven convection takes control. Flow consists of a circulating eddy which appears primary fills almost the entire box. The core area of the eddy is witnessed in the center of the box when it is inclined or non-inclined. On increasing the inclination angle to $\gamma = 45°$, the rotation in the clockwise direction becomes stronger. When the box is tilted to $\gamma = 45°$, the core region of the single vortex slightly directs the circulation of the flow towards the bottom sidewall. At $\gamma = 90°$, a dual circulation pattern is observed and the flow is approximately symmetric about the mid line of the box for the middle location of heater. It is noticeable that the alteration in the locations of the heater does not influence the flow pattern to any considerable extent. Fluid element heated near the active region(s) is moved upward along the left vertical wall and the fluid elements that are cold starts descending to replace the heated elements of fluid. Thereby, the circulation inside the box becomes stronger in the clockwise direction for all the locations of heater since the box is tilted.

The local Nusselt number values calculated for various box inclinations and locations of heater have been plotted in Fig. 3 for Ha = 25, Ri = 0.01, and Ri = 100. In the domination of forced convection, no substantial variation is experienced in the local values of Nusselt number although the box is inclined for all the angles. But, the modifications in the local transfer rate of heat are signified clearly in the case of free convective flow domination. For the bottom and middle locations of the partial heater, the local heat transport rate seems high for the inclination angle $\gamma = 90°$ and it is in the top edge of the heater. The local heat transport rate is low for the non-inclined box and the deviations in the Nusselt number values calculated locally

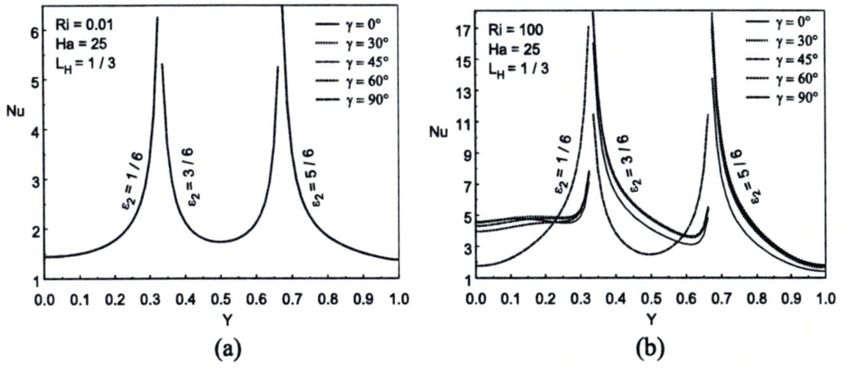


**Fig. 3** **a, b** Local Nusselt numbers for various inclinations and Ri with $Ha = 25$

is insignificant for every angle varying from $\gamma = 30°$ and $\gamma = 90°$ at the top location. Generally, the local transfer rate of heat increases for $Ri > 1$. When heater is placed about bottom and top locations, the local transfer rate attains its high values, respectively, at the top edge and rear edge of the partial heaters in forced convection mode. The U-shaped curve obtained from the local values of Nusselt number for middle location confirms the heat transport enhancement at both the heater edges placed near the left wall along the insulated portions. But it is reversed in the scenario of free convection. In this case, the heat transport rate becomes larger at the middle position when compared to the remaining two locations. For low inclinations such as $\gamma = 0°$ and $\gamma = 45°$ and in the free convection regime, the top and middle heater locations result in high rate of heat transport. There the transfer rate rises for the top and bottom locations when $\gamma = 90°$. The local values of Nusselt number attain its highest value at an inclination angle $\gamma = 45°$ and this is prevailed for the middle location when Ri = 100.

Total rate of heat exchange at the heating surface is measured using average Nusselt number. Figure 4a–c illustrate the whole heat exchange rate as a function of the box inclinations for different locations of heater at various Richardson numbers.

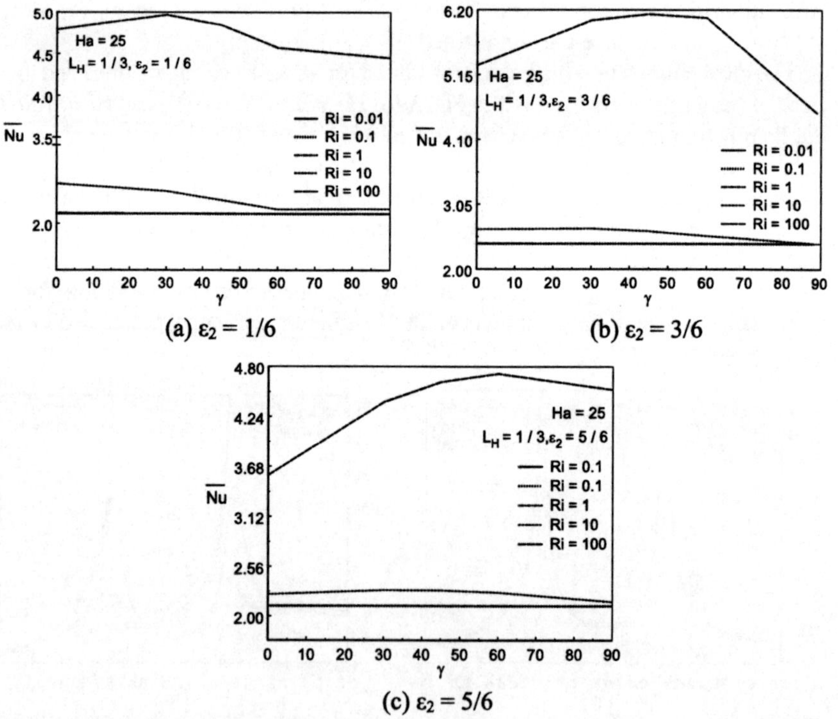


**Fig. 4** **a–c** Averaged Nusselt number vs. inclination for different Ri with $Ha = 25$ and $L_H = 1/3$

The average heat transport rate shows no remarkable variations for all the box inclinations when $Ri \leq 1$. It is noted that the inclination angle is an insignificant factor in the forced flow situation. An interesting fact noticed is that the average Nusselt number value rises along with box inclinations up to $\gamma = 30°$ at bottom location, up to $\gamma = 45°$ at middle location, and up to $\gamma = 60°$ at top location for $Ri = 100$. Thereafter, it decreases for the increase in angle of inclinations. Consequently, the average rate of heat transport is larger for the angles $\gamma = 30°$, $\gamma = 45°$, and $\gamma = 60°$, respectively, at bottom, middle, and top heater locations. In all the locations of heater, the average heat exchange rate reaches high values for $Ri > 1$. On comparing the heat exchange rate attained for all the three locations of heater, the average rate in which the heat gets transferred seems to be high for the middle location of heater in the regime of natural convection. In general, an improved heat transport rate is observed for the buoyant-driven free convection than that for the lid-driven forced convection.

Figure 5 describes the influence of varying Richardson number values from 0.01 to 100 with inclinational angles $\gamma = 0°, 45°, 90°$ of the box that imbibes on average heat transport rate. The Nusselt number is approximately constant while rising the Richardson number values from 0.01 to 1 and it increases after $Ri > 1$ for Ha = 0. When Ha = 100, the Nusselt number value obtained on average is almost constant for $0.01 \leq Ri \leq 10$ and it elevates high after $Ri > 10$. Rate of change of heat enhances when the box is inclined at $\gamma = 45°$ in most of the cases. Also, for a particular inclination angle and a Richardson number, middle heating of the wall provides enhanced transmission of heat.

## 5 Conclusions

A numerical investigation on discrete heating effects on the combined convection flow in a square shaped inclined box when magnetic force exists is explored. Three different heater locations such as bottom, middle, and top that are kept along the left wall of the box are examined. Numerical results obtained lead to the following conclusions. The overall heat exchange rate inside the box shows no remarkable variations for inclinations when $Ri \leq 1$. The average Nusselt number value elevates along with inclinations up to $\gamma = 30°$ at bottom heating location, up to $\gamma = 45°$ at middle heating location, and up to $\gamma = 60°$ at top heating location. The value decreases for further increase of inclination angles. The average rate of heat movement is found superior when box is heated from middle regardless of the existence of magnetic field. For all the locations, on increasing $Ri$ the average rate of heat exchange increases. In the non-existence of magnetic field and when the heater is located at the center of the box, the overall rate of heat transfer reached to its highest value when $\gamma = 45°$. The inclination of the geometry is an insignificant factor in the scenario of forced convection but the same gives a major impact when natural convection prevails. These results are very useful to the thermal engineering applications, particularly, in the cooling of electronics equipment.

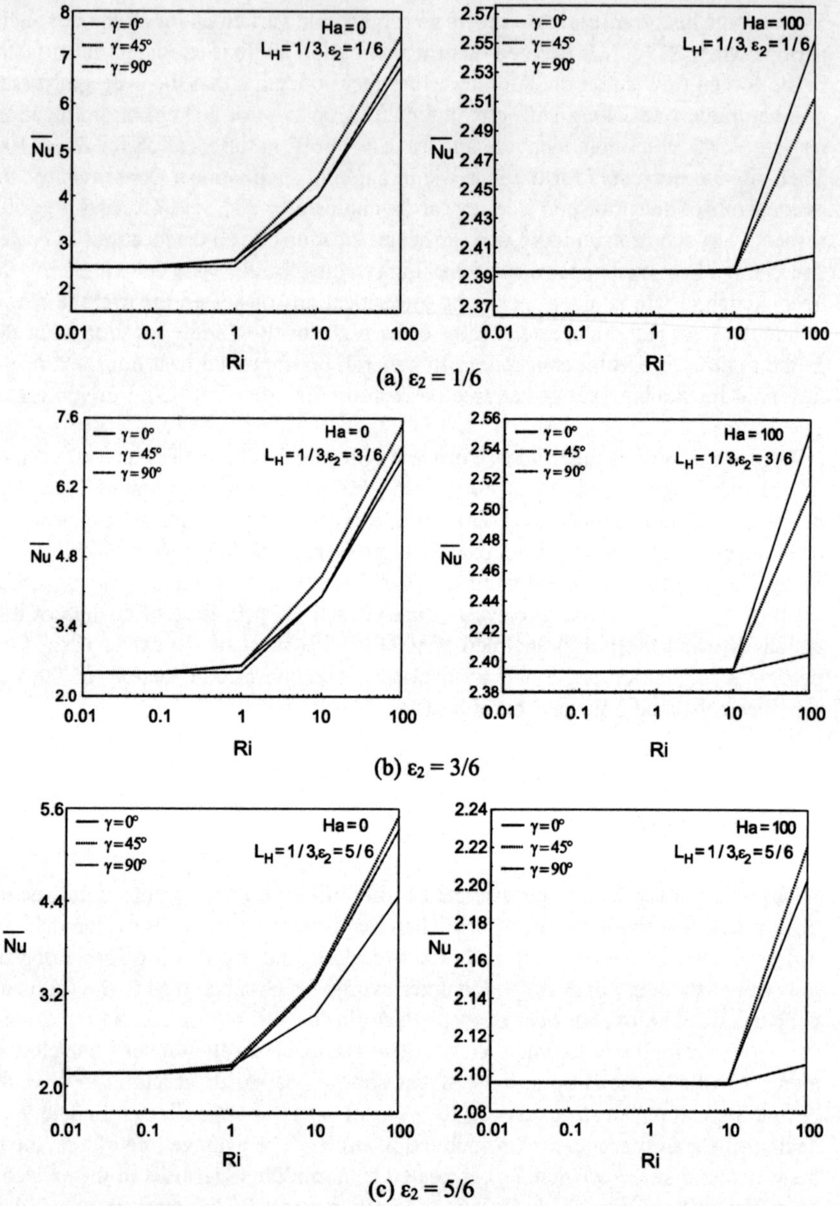


**Fig. 5** **a–c** Averaged Nusselt number for $L_H = 1/3$, Ha = 0, and Ha = 100

## References


1. Oztop HF, Dagtekin I (2004) Mixed convection in two-sided lid-driven differentially heated square cavity. Int J Heat Mass Transf (47):1761–1769
2. Sivasankaran S, Chamkha AJ, Mallawi FOM, Alghamdi MS, Alqahtani AM (2020) Effects of entropy generation, thermal radiation and moving-wall direction on mixed convective flow of nanofluid in an enclosure. Mathematics (8):1471
3. Sivasankaran S, Ananthan SS, Bhuvaneswari M, AbdulHakeem AK (2017) Double-diffusive mixed convection in a lid-driven cavity with non-uniform heating on sidewalls. Sadhana (42):1929–1941
4. Sivasankaran S Pan KL (2016) Lattice Boltzmann simulation for a lid-driven cavity with discrete heating/cooling sources. AIAA J Thermophys Heat Transf (30):573–586
5. Sivasankaran S, Ananthan SS, Abdul Hakeem AK (2016) Mixed convection in a lid-driven cavity with sinusoidal boundary temperature at the bottom wall in the presence of magnetic field. Iran J Sci Technolo Trans B Mech Eng (23):1027–1036
6. Sivasankaran S, Cheong HT, Bhuvaneswari M, Ganesan P (2016) Effect of moving wall direction on mixed convection in an inclined lid-driven square cavity with sinusoidal heating. Numer Heat Transf Part A Appl (69):630–642
7. Sivakumar V, Sivasankaran S, Prakash P, Lee J (2010) Effect of heating location and size on mixed convection in lid-driven cavities. Comput Math Appl (59):3053–3065
8. Sivasankaran S, Sivakumar V, Prakash P (2010) Numerical study on mixed convection in a lid-driven cavity with non-uniform heating on both sidewalls. Int J Heat Mass Transf (53):4304–4315
9. Chamkha AJ (2002) Hydromagnetic combined convection flow in a vertical lid-driven cavity with internal heat generation or absorption. Numer Heat Transf Part A Appl 41:529–546
10. Sivasankaran S, Narrein K (2020) Influence of geometry and magnetic field on convective flow of nanofluids in trapezoidal microchannel heat sink. Iran J Sci Technol Trans Mech Eng (44):373–382
11. Bindhu R, Sai SundaraKrishnan G, Sivasankaran S, Bhuvaneswari M (2019) Magneto-convection of water near its maximum density in a cavity with partially thermally active walls. Energy Environ (30):833–853
12. Sarris IE, Kakarantzas SC, Grecos AP, Vlachos NS (2005) MHD natural convection in a laterally and volumetrically heated square cavity. Int J Heat Mass Transf (48):3443–3453
13. Rashad AM, Sivasankaran S, Mansour MA, Bhuvaneswari M (2017) Magneto-convection of nanofluids in a lid-driven trapezoidal cavity with internal heat generation and discrete heating. Numer Heat Transf Part A (71):1223–1234
14. Xu B, Li BQ, Stock DE (2006) An experimental study of thermally induced convection of molten gallium in magnetic field. Int J Heat Mass Transf (49):2009–2019
15. Sivasankaran S, Malleswaran A, Bhuvaneswari M, Poo Balan G (2017) Hydro-magnetic mixed convection in a lid-driven cavity with partially thermally active walls. Iran J Sci Technol Trans B Mech Eng (24):153–163
16. Narrein K, Sivasankaran S, Ganesan P (2016) Influence of transverse magnetic field on microchannel heat sink performance. J Appl Fluid Mech 9:3159–3166
17. Jalil JM, Al-Tae'y KA (2007) The effect of non uniform magnetic field on natural convection in an enclosure. Numer Heat Transf A (51):899–917
18. Sivasankaran S, Ho CJ (2008) Effect of temperature dependent properties on MHD convection of water near its density maximum in a square cavity. Int J Therm Sci (47):1184–1194
19. Sivasankaran S, Bhuvaneswari M, Kim YJ, Ho CJ, Pan KL (2011) Numerical study on magneto-convection of cold water in an open cavity with variable fluid properties. Int J Heat Fluid Flow 32:932–942
20. Sivasankaran S, Malleswaran A, Lee J, Sundar P (2011) Hydro-magnetic combined convection in a lid-driven cavity with sinusoidal boundary conditions on both sidewalls. Int J Heat Mass Transf (54):512–525

21. Kuhn D Oosthuizen PH (1987) Unsteady natural convection in a partially heated rectangular cavity. J Heat Transf (109):798–801
22. Frederick RL, Quiroz F (2001) On the transition from conduction to convection regime in a cubical enclosure with partially heated wall. Int J Heat Mass Transf (44):1699–1709
23. Oztop HF (2006) Combined convection heat transfer in a porous lid-driven enclosure due to heater with finite length. Int J Heat Mass Transf (33):72–779
24. Kuznetsov GV, Sheremet MA (2006) Modelling of thermo gravitation convection in closed volume with local sources of heat release. Thermophys Aeromech (13):565–574
25. Kuznetsov GV, Sheremet MA (2008) Mathematical modeling of heat and mass transfer under conditions of mixed convection in rectangular region with heat source and heat-conducting walls. Thermophys Aeromech (15):99–112
26. Bhuvaneswari M, Sivasankaran S, Kim YJ (2011) Effect of aspect ratio on convection in a porous enclosure with partially active thermal walls. Comput Math Appl (62):3844–3856
27. Sivasankaran S, Do Y, Sankar M (2011) Effect of discrete heating on natural convection in a rectangular porous enclosure. Transp Porous Media (86):291–311
28. Sivasankaran S, Cheong HT, Bhuvaneswari M (2019) Natural convection in an inclined porous triangular enclosure with various thermal boundary conditions. Therm Sci (23):537–548
29. Sivasankaran S, Alsabery A, Hashim I (2018) Internal heat generation effect on transient natural convection in a nanofluid-saturated local thermal non-equilibrium porous inclined cavity. Phys A 509:275–293
30. Al-Najem NM, Khanafer KM, El-Refaee MM (1998) Numerical study of laminar natural convection in tilted enclosure with transverse magnetic field. Int J Numer Meth Heat Fluid Flow (8):651–672
31. Cianfrini C, Corcione M, Paolo Dell'Omo P (2005) Natural convection in tilted square cavities with differentially heated opposite walls. Int J Therm Sci (44):441–451
32. Sharif MAR (2007) Laminar mixed convection in shallow inclined driven cavities with hot moving lid on top and cooled from bottom. Appl Therm Eng (27):1036–1042

# Impact of Radiation on Flow of Copper-Water Nanofluid Squeezed Between Parallel Plates Filled with Darcy Porous Medium

T. Chandrapushpam, M. Bhuvaneswari, S. Sivasankaran, and S. Karthikeyan

## 1 Introduction

Flow of fluid between two parallel objects approaching each other-commonly referred to as squeezing flow-deserves more detailed research. In many of the engineering models, existence of squeezing flow is an undeniable fact in both non-Newtonian and Newtonian fluids and these applications are apparent in food processing, chocolate fillers, hydraulic lifts, polymers processing, moving pistons, electric motors, compression, power transmission squeezed film, etc. In 1874, Stefan initiated the pioneering work on squeeze flow and this was followed by Reynolds in 1886. Owing to its extensive industrial applications, in the past two decades logical and systematic investigations have been steadily increasing on squeeze flow. Engmann et al. [1] established that squeeze flow is an adaptable and consistent process of rheological description. Considerable insights in fluid flow behavior under different conditions in the field of MHD were provided through many researches [2–5]. Saadatmandi et al. [6] and Ran et al. [7] explored the flow of a Newtonian incompressible fluid which is axisymmetric and squeezed between parallel plates. Rashidi et al. [8] solved the problem for circular plates.

T. Chandrapushpam
Nehru Arts and Science College, Coimbatore 641105, India

M. Bhuvaneswari
Department of Mathematics, Kongunadu Polytechnic College, D. Gudalur, Dindigul, Tamilnadu, India

S. Sivasankaran (✉)
Department of Mathematics, King Abdulaziz University, Jeddah, Saudi Arabia
e-mail: sd.siva@yahoo.com

S. Karthikeyan
Erode Arts and Science College, Erode 638009, India

 
S. Srinivas et al. (eds.), *Recent Advances in Applied Mathematics and Applications to the Dynamics of Fluid Flows*, Lecture Notes in Mechanical Engineering,
https://doi.org/10.1007/978-981-19-1929-9_12

Heat transfer and thermal radiation play a major character in illustrating engineering design. Heat transmission happens in various systems, viz., electronic devices, heat exchange machines, automobiles, refrigerators, and so on. Size reduction in heat exchangers helps to increase the heat conductive property and hence miniaturization techniques of recent days are very much constructive. Technology provided us a newer material with desirable features along with high heat conductivity property called nanofluid, and this helps in improving the design of heat exchangers. A significant number of researchers investigated the flow behavior of nanofluids in different fields under different conditions which offer a significant insight as in [9–11].

Fluid flows through porous medium take a significant role in many fields of mechanical applications, earth, and environmental sciences. Khan et al. [12] analyzed the flow of nanofluids over a stretching surface through a porous medium. Impacts of chemical reaction and other fluid flow parameters on MHD joint convection stagnation-point stream through a porous medium were discussed by many authors [13–17]. Studies on squeeze flow of fluid between parallel objects filled with porous medium under different conditions were carried out as on [18–20].

The foremost intend of the current study is to explore the influence of radiative heat transfer in the flow behavior of a copper-water nanofluid squeezed between two parallel plates filled with Darcy porous medium along with externally applied magnetic field. To predict the flow behavior in converting the primary equations—Partial Differential Equations (PDE) to Ordinary Differential Equations (ODE), similarity transformation is applied. In the course of current study, an analytical method called Differential Transformation Method (DTM) which enables the convergent solution while resolving the resultant ODEs. Impacts of physical parameters are demonstrated, and the outcomes are analyzed with the results of earlier investigators [11].

## 2 Formulation of the Problem

The squeezed flow of 2D copper-water nanofluid between two parallel plates is considered in this paper. Both plates are set in a Darcy porous medium. The horizontal axis X is taken along the side of the plate and Y-axis is perpendicular to the plates as given in Fig. 1. The gap among the plates at any time is $h_1(t) = H_1(1 - at)^{0.5}$ where $H_1$ the position of the plates at $t = 0$ and $a$ is named as characteristic parameter with dimension $t^{-1}$. Here $a > 0$ denotes both plates are approaching one another through the velocity $v(t) = h_1'(t)$ until the plates meet one another at $t = a^{-1}$ while negative sign of $a$ indicates the outside motion of the plates. The intensity of the magnetic field given by $B(t) = B_o(1-at)^{-0.5}$ is exerted in the direction normal to the flow of fluid and $B_o$ is the initial strength of the magnetic field. Table 1 shows the material properties of the components of the nanofluid considered in this problem.

The mathematical model of the problem is constructed with the assumptions that (i) the porous medium is both thermally and hydrodynamically isotropic, (ii) no

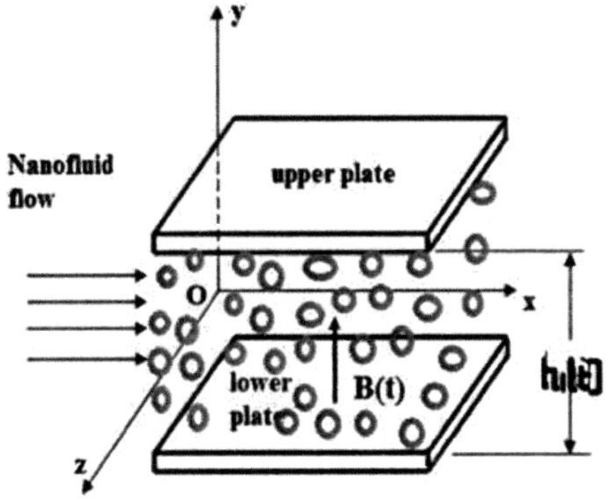


**Fig. 1** Fluid flow system-model

**Table 1** Physical characteristics of the components

| | Thermal conductivity (W/mK) | Specific heat capacity (J/KgK) | Density (Kg/m$^3$) |
|---|---|---|---|
| Pure water | 997.1 | 0.613 | 4179 |
| Copper | 8933 | 401 | 385 |

slip among the plates, (iii) thermal equilibrium occurs between the base liquid and the nanoparticle, (iv) radiative heat transfer takes place, (v) there should not be any chemical reaction among the nanofluid components, and (vi) physical characteristics of nanoparticle and base fluid are constants. The associated governing equations are given below.

$$\rho_{nf}\left(\frac{\partial u}{\partial t}+u\frac{\partial u}{\partial x}+v\frac{\partial u}{\partial y}\right)=-\frac{\partial p}{\partial x}+\mu_{nf}\left(\frac{\partial^2 u}{\partial x^2}+\frac{\partial^2 u}{\partial y^2}\right)-\sigma_{nf}B^2(t)u-\frac{\mu_{nf}u}{k_p} \tag{1}$$

$$\rho_{nf}\left(\frac{\partial v}{\partial t}+u\frac{\partial v}{\partial x}+v\frac{\partial v}{\partial y}\right)=-\frac{\partial p}{\partial y}+\mu_{nf}\left(\frac{\partial^2 v}{\partial x^2}+\frac{\partial^2 v}{\partial y^2}\right)-\frac{\mu_{nf}v}{k_p} \tag{2}$$

$$\begin{aligned}\frac{\partial T}{\partial t}+u\frac{\partial T}{\partial x}+v\frac{\partial T}{\partial y}&=\frac{K_{nf}}{(\rho C_p)_{nf}}\left(\frac{\partial^2 T}{\partial x^2}+\frac{\partial^2 T}{\partial y^2}\right)\\&+\frac{\mu_{nf}}{(\rho C_p)_{nf}}\left[4\left(\frac{\partial u}{\partial x}\right)^2+\left(\frac{\partial u}{\partial y}+\frac{\partial v}{\partial x}\right)^2\right]\\&+\frac{1}{(\rho C_p)_{nf}}\frac{16\sigma^* T_0^3}{3k'}\left[\frac{\partial^2 T}{\partial x^2}+\frac{\partial^2 T}{\partial y^2}\right]\end{aligned} \tag{3}$$

Here, the velocity components along the axes x and y are denoted by $u$ and $v$ respectively. P, T, $k'$, and $\sigma^*$ represent the pressure, the temperature, the coefficient of mean absorption and the constant due to Stefan-Boltzmann. $\rho_{nf}$, $K_{nf}$, $\mu_{nf}$, and $(\rho C_p)_{nf}$ represent the effective—density, thermal conductivity, dynamic viscosity, and specific heat capacity, respectively.

The base liquid and the nanoparticles are connected by the relations as follows:

$$\rho_{nf} = (1-\varphi)\rho_f + \varphi\rho_s \; (\rho C_p)_{nf} = \varphi(\rho C_p)_s + (1-\varphi)(\rho C_p)_f \tag{4}$$

$$\mu_{nf} = \frac{\mu_f}{(1-\varphi)^{2.5}} \;, \sigma_{nf} = \sigma_f\left[1 + \frac{3\left\{\left(\sigma_s\sigma_f^{-1}\right)-1\right\}\varphi}{\left\{\left(\sigma_s\sigma_f^{-1}\right)+2\right\}\varphi - \left\{\left(\sigma_s\sigma_f^{-1}\right)-1\right\}\varphi}\right]a \tag{5}$$

$$K_{nf} = K_f\frac{2K_f + K_s - 2\varphi(K_f - K_s)}{2K_f + K_s + \varphi(K_f - K_s)}, \; (\rho C_p)_{nf} = \varphi(\rho C_p)_s + (1-\varphi)(\rho C_p)_f \tag{6}$$

The boundary settings are as follows:

$$\left.\begin{array}{llll} u = 0, \; v = \frac{dh}{dt}, & -k\frac{\partial T}{\partial y} = h_f(T_f - T) & at & y = h_1(t) \\ \frac{\partial u}{\partial y} = 0, \; v = 0, & \frac{\partial T}{\partial y} = 0 & at & y = 0 \end{array}\right\} \tag{7}$$

During conversion of the primary equations (1)–(4) as ODEs introduced the dimensionless functions $g$ and $\theta$ and the similarity variable $\eta$ as below

$$\left.\begin{array}{c} u = \frac{ax}{2(1-at)}g_1'(\eta), \; v = \frac{-aH_1}{2(1-at)^{0.5}}g_1(\eta) \\ \eta = \frac{y}{H_1\sqrt{1-at}}, \; \theta = \frac{T-T_H}{T_f-T_H}, \; B(t) = B_o(1-at)^{-0.5} \end{array}\right\} \tag{8}$$

Applying similarity transformation and dimensionless functions in Eqs. (2)–(4), after removing pressure gradients the following equations are obtained.

$$g_1^{IV} - A_1S(1-\varphi)^{2.5}[3g_1'' + \eta g_1''' + g_1'g_1'' - g_1g_1'''] - \left(M + \frac{1}{Da}\right)g_1'' = 0 \tag{9}$$

$$\left(1 + \frac{4}{3}R_d\right)\theta'' + \Pr S\left(\frac{A_2}{A_3}\right)(g_1\theta' - \eta\theta') + \frac{P_rE_e}{A_3(1-\varphi)^{2.5}}[g_1^{2\prime} + 4\delta^2 g_1^{2\prime}] = 0 \tag{10}$$

The associated limiting conditions are

$$\left.\begin{array}{c} g_1(0) = 0, \; g_1''(0) = 0, \; \theta'(0) = 0 \; at \; \eta = 0 \\ g_1(1) = 1, \; g_1'(1) = 0, \; \theta'(1) = Bi[\theta(1) - 1] \; at \; \eta = 1 \end{array}\right\} \tag{11}$$

Here $A_3 = \frac{2k_f + k_s - 2\varphi(k_f - k_s)}{2k_f + k_s + \varphi(k_f - k_s)}$, $A_1 = (1-\varphi) + \varphi\frac{\rho_s}{\rho_f}$, $A_2 = (1-\varphi) + \varphi\frac{(\rho C_p)_s}{(\rho C_p)_f}$, $\Pr = \frac{(\rho C_p)_f}{k_f}\frac{\mu_f}{\rho_f}$, $Ec = \frac{\rho_f}{(\rho C_p)_f (T_f - T_H)}\left(\frac{ax}{2(1-at)}\right)^2$, $M = \frac{\sigma_s H_1^2 B_0^2}{\mu_{nf}}$, $Da = \frac{k_p}{H_1^2}$, $R_d = \frac{4\sigma^* T_0^3}{A_3 k_f k'}$, $\delta = \frac{H_1\sqrt{1-at}}{x}$, $Bi = \frac{h_f h_1(t)}{k}$.

$S = \frac{aH_1^2}{2\upsilon_f}$ where Pr—Prandtl number, $Ec$—modified Eckert number, $M$—Hartmann number, $Da$—Darcy number, $R_d$—Radiation parameter, $\delta$—dimensionless length, $Bi$—Biot number, and S—Squeeze number, which can be positive or negative. Negative and positive signs of $S$ represent closer and far away movement of the plates, respectively, and kinematic viscosity is denoted by $\upsilon_f$.

The non-dimensional Skin friction coefficient $Cf_r$ and Nusselt number $Nu_r$ obtained are

$$Cf_r = \frac{\mathrm{Re}_x Cf H_1^2\sqrt{1-at}}{2x^2} = \frac{g''(1)}{A_1(1-\varphi)^{2.5}} \quad \text{where } \mathrm{Re}_x = \frac{xaH_1}{2\upsilon_f} \tag{12}$$

$$Nu_r = Nu\sqrt{1-at} = -A_3\left(1 + \frac{4}{3}R_d\right)\theta'(1) \tag{13}$$

## 3 Differential Transformation Method

The inverse differential transform of $F(n)$ is given by

$$f(t) = \sum_{n=0}^{\infty} F(n).(t - t_0)^n \tag{14}$$

where

$$F(n) = \frac{1}{n!}\frac{d^n f(t)}{dt^n}\bigg|_{t=t_0} \quad t_0 \in I \tag{15}$$

Applying DTM in Eqs. (9) and (10) yields

$$\left.\begin{aligned}&(\lambda+1)(\lambda+2)(\lambda+3)(\lambda+4)G_1[\lambda+4]-3A_1S(1-\varphi)^{2.5}(\lambda+1)(\lambda+2)G_1[\lambda+2]\\&-A_1S(1-\varphi)^{2.5}\sum_{q=0}^{\lambda}\delta(\lambda-q-1)(q+1)(q+2)(q+3)G_1[q+3]\\&-A_1S(1-\varphi)^{2.5}\sum_{q=0}^{\lambda}(\lambda-q+1)G_1[\lambda-q+1](q+1)(q+2)G_1[q+2]\\&+A_1S(1-\varphi)^{2.5}\sum_{q=0}^{\lambda}G_1[\lambda-q](q+1)(q+2)(q+3)G_1[q+3]\\&-\left(M+\frac{1}{Da}\right)(\lambda+1)(\lambda+2)G_1[\lambda+2]=0\end{aligned}\right\} \tag{16}$$

$$\begin{aligned}&\left(1+\frac{4}{3}R_d\right)(\lambda+1)(\lambda+2)\Theta[\lambda+2]+S\Pr\left(\frac{A_2}{A_3}\right)\sum_{q=0}^{\lambda}G[\lambda-q](q+1)\Theta[q+1]\\&-S\Pr\left(\frac{A_2}{A_3}\right)\sum_{q=0}^{\lambda}\delta[\lambda-q-1](q+1)\Theta[q+1]\\&+\frac{\Pr Ec}{A_3(1-\varphi)^{2.5}}\sum_{q=0}^{\lambda}(\lambda-q+1)(\lambda-q+2)G[\lambda-q+2](q+1)(q+2)G[q+2]\\&+\frac{4\Pr Ec\delta^2}{A_3(1-\varphi)^{2.5}}\sum_{q=0}^{\lambda}(\lambda-q-1)G[\lambda-q+1](q+1)G[q+1]=0\end{aligned} \tag{17}$$

The iterated limiting conditions $G_1[\lambda]$ and $\Theta[\lambda]$ are obtained as below

$$\left.\begin{array}{c}G_1[0]=0,\ G_1[1]=C_1,\ G_1[2]=0,\ G_1[3]=C_2,\ G_1[4]=0\\ G_1[5]=\left[\frac{3}{20}SA_1(1-\varphi)^{2.5}+\frac{1}{20}\left(M+\frac{1}{Da}\right)\right]C_2,\ G_1[6]=\frac{1}{60}SA_1(1-\varphi)^{2.5}C_2\end{array}\right\} \tag{18}$$

$$\left.\begin{array}{c}\Theta[0]=C_3,\ \Theta[1]=0,\ \Theta[2]=\frac{-2\Pr Ec\delta^2}{A_3(1-\varphi)^{2.5}}\frac{C_1^2}{\left(1+\frac{4}{3}R_d\right)},\ \Theta[3]=0\\ \Theta[4]=S\Pr\left(\frac{A_2}{A_3}\right)\left(\frac{-C_1}{6\left(1+\frac{4}{3}R_d\right)}\right)\Theta[2]-\frac{\Pr Ec}{A_3(1-\varphi)^{2.5}}\left[3C_2^2+2\delta^2C_1C_2\right]\end{array}\right\} \tag{19}$$

Using Eqs. (18) and (19), Eqs. (16) and (17) results in

$$\begin{aligned}g_1(\eta)=&C_1\eta+C_2\eta^3+\left[\frac{3}{20}SA_1(1-\varphi)^{2.5}+\frac{1}{20}\left(M+\frac{1}{Da}\right)\right]C_2\eta^5\\&+\frac{1}{60}SA_1(1-\varphi)^{2.5}C_2\eta^6+\ldots\end{aligned} \tag{20}$$

$$\theta[\eta] = C_3 + \left(\frac{-2\Pr Ec\delta^2 C_1^2}{A_3(1-\varphi)^{2.5}\left(1+\frac{4}{3}R_d\right)}\right)\eta^2 + \left[\begin{array}{c} S\Pr\left(\frac{A_2}{A_3}\right)\left(\frac{-C_1}{6\left(1+\frac{4}{3}R_d\right)}\right)\Theta[2]- \\ \frac{\Pr Ec}{A_3(1-\varphi)^{2.5}}\left[3C_2^2 + 2\delta^2 C_1 C_2\right] \end{array}\right]\eta^4 + \ldots \tag{21}$$

Taking $\eta = 1, S = 0.5, \Pr = 6.7, Ec = 0.1, \delta = 0.1, M = 1.5, \varphi = 0.01, Da = 0.1$ in Eqs. (20) and (21) and from Table 1 the constants $C_1$, $C_2$, and $C_3$ become $C_1 = 1.3568321719$, $C_2 = -0.2146057496$, $C_3 = 0.069135786$.

## 4 Results and Discussion

This work presents details pertaining to the progress of a mathematical representation of copper-water nanofluid flow squeezed between parallel plates in a Darcy porous medium as presented in Fig. 1. The significant consequences of thermal radiation together with related parameters like squeezing number, Hartmann number, Darcy number, Biot number, and Eckert number on the profiles of flow behavior and temperature of copper-water nanofluid system are discussed with the help of tables and graphs. While replicating the model the values for the parameters are fixed as follows: $\Pr = 6.7$, $\varphi = 0.03$, $M = 2$, $R_d = 2$, $Ec = 0.5$, $Da = 0.1$, $\delta = 0.1$, and $\beta = 4$. A comparison of the results of the current work with the outcomes on the influences of M, ϕ on $Cf_r$ of the cu-water nanofluid obtained in [11] is offered in Table 2 and this confirms the validity of the present mathematical model used.

Figure 2a and b describes the consequence of squeezing number $S$ on velocity $f'$

**Table 2** Effects of $M$, $\varphi$ on $Cf_r$

| $M$ | $\varphi$ | $Cf_r$ | | | |
|---|---|---|---|---|---|
| | | Acharya et al. [11] | | Present work | |
| | | S = 0.5 | S = −0.5 | S = 0.5 | S = −0.5 |
| 0.5 | 0.02 | −3.096871 | −2.388320 | −3.09686954 | −2.38831880 |
| – | 0.04 | −2.889905 | −2.180419 | −2.88990435 | −2.18041956 |
| – | 0.06 | −2.736134 | −2.025792 | −2.73613423 | −2.02579186 |
| 1.5 | 0.02 | −3.402609 | −2.773329 | −3.40260760 | −2.77332814 |
| 3 | – | −4.277192 | −3.811202 | −4.27719194 | −3.81119972 |
| 4.5 | – | −5.414585 | −5.074220 | −5.41458423 | −5.07421764 |

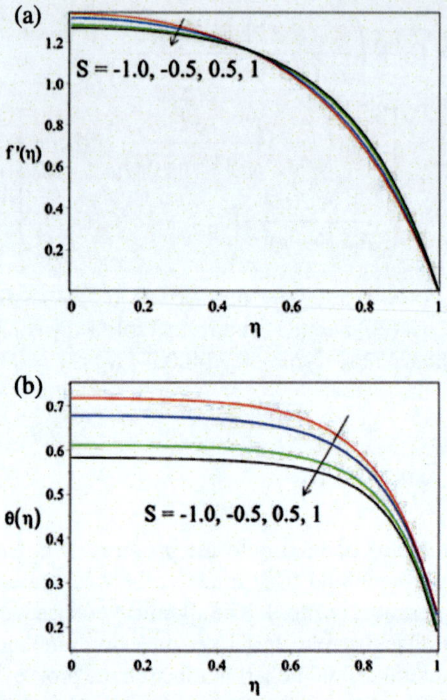


**Fig. 2** **a** Effect of $S$ on $f'$, **b** Effect of S on $\theta$

and temperature $\theta$ of the nanofluid considered. They explored that $\theta$ and $f'$ values are slightly less for every value of $S$, and the overall spread in velocity is reduced and not significant in temperature values when compared to the outcomes of [15] in which the authors investigated the case where the radiation term and porous medium were absent. Figure 2a shows that $f'$ come down for enhancing $S$ for each $\eta$ over $0 \leq \eta \leq 0.49$ (approx.) and starts increasing for the values of $\eta$ in $0.49 < \eta \leq 1$. As a consequence of reduce in fluid flow next to the boundary area the velocity gradient increases there. While retaining the conventional mass flow rate, the reduced fluid flows close to the border line of the plates are balanced by the greater fluid flows close to the middle region. As a result, a separation point occurs at $\eta = 0.49$. Figure 2b depicts the consequences of S on $\theta$ of the nanofluid system for various values of S. With enhancing squeeze number S the corresponding θ values decreases steadily for the values of $\eta$ between 0 and 1. When S is maximum the kinematic viscosity becomes minimum, which makes higher the speed rate of separation of the plates. Hence the contact pressure is reduced amid the plates that results in reduced heat of the nanofluid.

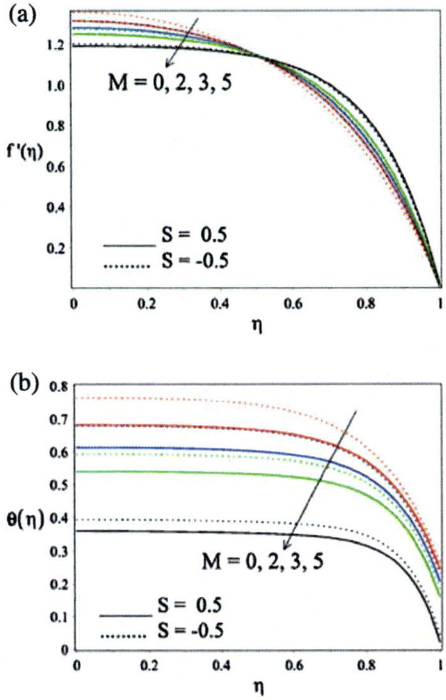


**Fig. 3** **a** Effect of M on $f'$, **b** effect of M on $\theta$

Figure 3a and b shows the influences of Hartmann number M on velocity $f'$ and temperature field $\theta$ of the nanofluid. From Fig. 3a, it is noted that for both $S = 0.5$ and $S = -0.5$, as M increases correspondingly $f'$ decreases between the area $\eta = 0 \& \eta = 0.49$ and starts increasing between the region $\eta > 0.49$ & $\eta \leq 1$. With existing thermal radiation, a raise in Hartmann number is associated with raise in magnetic field and a power called Lorentz force is generated that reduces the fluid movement in the region closure to the boundary during squeezing movement. Moreover when the plates are moving closure to each other, the consequence of Lorentz force is found to be very less. Zero velocity is attained at $\eta = 1$ due to skin friction that stops the flow of nanofluid. Also from the graph plotted showing variation in velocity, the cross over point occurs at $\eta = 0.49$ and the velocity of the nanofluid for negative 0.5 of S is greater in contrast with $S = 0.5$ in the area $0 \leq \eta \leq 0.49$ and an opposite trend in $0.49 < \eta \leq 1$.Table 3 explains that with enhancing M, the dimensionless skin friction reduces. Figure 3b portrays that the values of temperature turn down with increasing M for both positive and negative 0. 5 of S. Figure 3a and b exhibits with existing radiation the range in temperature and

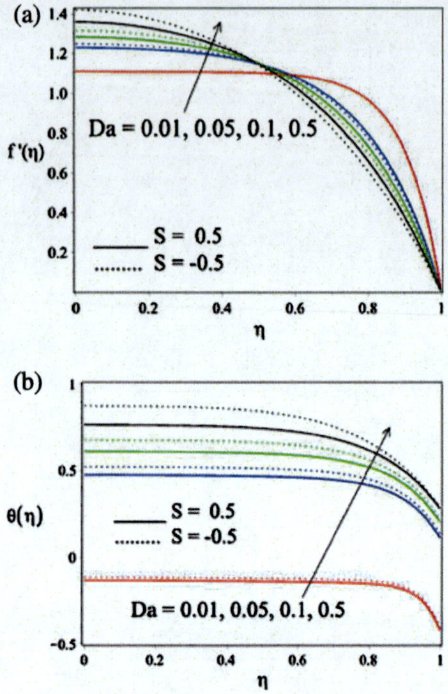


**Fig. 4** **a** Effect of $Da$ on $f'$, **b** Effect of $Da$ on $\theta$

velocity values are enhanced which are greater to some extent for all the values of S in comparison with [11].

Figure 4a and b presents the impact of Darcy number $Da$ on the nanofluid's flow behavior $f'$ and temperature $\theta$. A decrease in Darcy number implies lower permeability, which in turn increases the resistance to fluid flow. It is viewed from Fig. 4a that the fluid flow raises with enhancement in Darcy number. When the Darcy number is very low, $f'$ is constant over $0 \leq \eta \leq 0.6$ and the values beyond $\eta = 0.6$ decrease rapidly and become zero near the boundary of the plate. As $Da$ raises, the flow is more concentrated in the center and the velocity decreases steadily from the center to the end of the plate, also with the effect of skin friction velocity becomes low near the boundary. Figure 4b depicts the effect of variation in $Da$ on temperature profile. Increasing Darcy number results in higher temperature. The temperature enhances with enhancing Darcy number, and increase in temperature values analogous to negative 0.5 of S are higher than for S = 0.5.

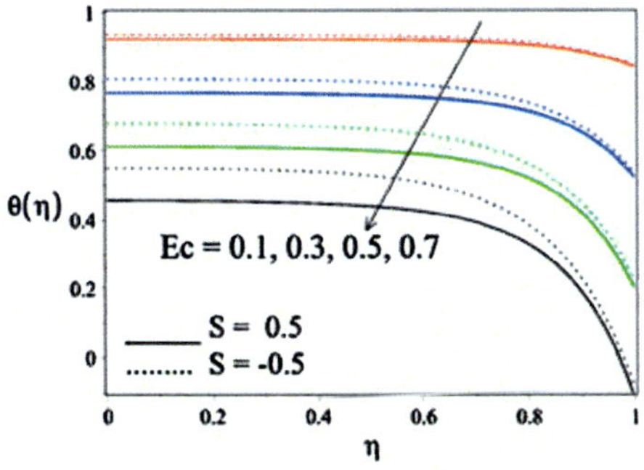


**Fig. 5** Effect of $Ec$ on $\theta$

Figure 5 depicts the consequences of Eckert number $Ec$ on temperature $\theta$. With increasing $Ec$ the temperature decreases for both 0.5 and $-0.5$ of S. But the subsequent temperature values related to $S = 0.5$ are below than values related to $S = -0.5$ for each value of Ec. While comparing the effect with [11], the stretch in the values of temperature attained in the current work for both positive and negative 0.5 of S are less due to the existence of radiation. The influence of radiation parameter on temperature is shown in Fig. 6 through which it follows that as Rd increases, $\theta$ also increases for positive and negative 0.5 of S. It is owing to the existence of radiation, the temperature of the entire fluid flow area enhances. Generally, it is true that increase in the Rosseland diffusion approximation for radiation enhances the fluid temperature, and hence the distribution of temperature shows greater values of $\theta$ for greater values of Rd. The temperatures are seen to decrease closer to the plate ($\eta = 1$), and this can be due to the conduction of heat from the plates. Figures 7 and 8 depict the effect of Biot number on temperature field and Nusselt number. Figure 7

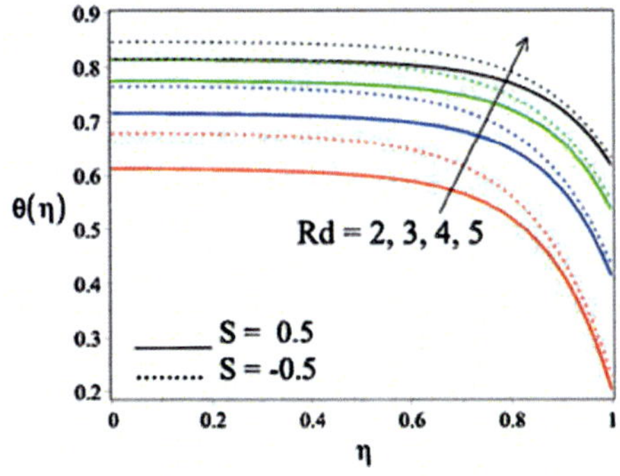


**Fig. 6** Effect of $R_d$ on $\theta$

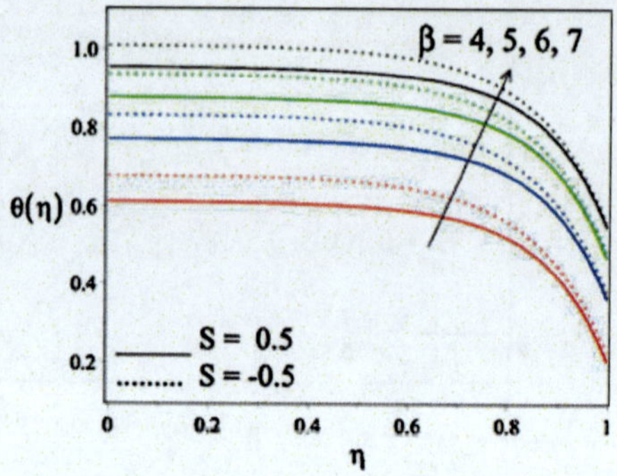


Fig. 7 Effect of $\beta$ on $\theta$

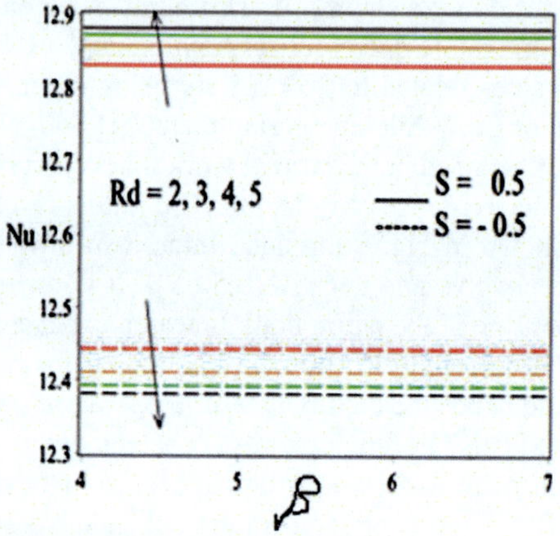


Fig. 8 Effects of $R_d$ & $\beta$ on $Nu$

explores when enhancing Biot number, the thermal profile rises for the positive and negative 0.5 of S but the values analogous to −0.5 are larger than those for 0.5. As the Biot number is involving the heat transfer coefficient, increasing values of Biot number implies reduced thermal conductivity giving rise to higher temperature. Figure 8 depicts the effect of radiation (Rd) and Biot number (β) on Nusselt number. Augmenting values of β and Rd the heat transfer values corresponding to S = 0.5 are increasing whereas the values corresponding to S = −0.5 are decreasing. Moreover, for both positive and negative 0.5 of S, the heat transfer values do not vary for the given values of Rd.

Figure 9a and b presents the influences of variation in volume fraction $\varphi$ and M on skin friction and Nusselt number, respectively. It reveals from Fig. 9a that with increasing $\phi$ the value of $Cf$ is seen to increase, whereas with increasing M, represents the magnetic field strength causing slow movement of the nanofluid, the $Cf$ values come lower for both positive and negative 0.5 of S. The $Cf$ values for S = 0.5 are lower compared to the values analogous to negative 0.5 of S and a less significant differences between the values of S. Figure 9b reveals that with enhancing $\phi$ and M, the local heat transfer values becomes high for both −0.5 and 0.5 of S.

Figure 10a and b depicts the impacts of $\phi$ and Rd on $Cf$ and $Nu$ respectively. Both figures reveal that the enhancing values of $\phi$ and Rd associated with enhancing values of $Nu$ and $Cf$. Also noted that the values of $Cf$ and $Nu$ do not vary significantly for different values of Rd but for the specific values of S and $\phi$. Figure 10a reveals that the values related to S for positive 0.5 are smaller than for S = - 0.5, and a reverse trend for Nusselt number as in Fig. 10b.

Figure 11a and b presents the effects of $Da$ and S on skin friction $Cf$ and Nusselt number $Nu$ respectively. In enhancing $Da$ and S, the $Cf$ increases, whereas $Nu$ shows a decreasing trend. The cause for this trend is attributed to the fact that higher porosity results in higher Darcy number which increases the resistance to fluid flow

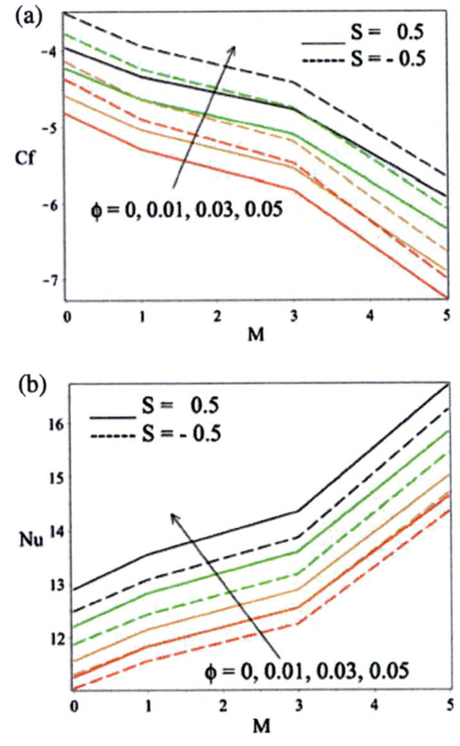


**Fig. 9** **a** Effects of $\phi$ and M on $Cf$, **b** effects of $\phi$ and M on $Nu$

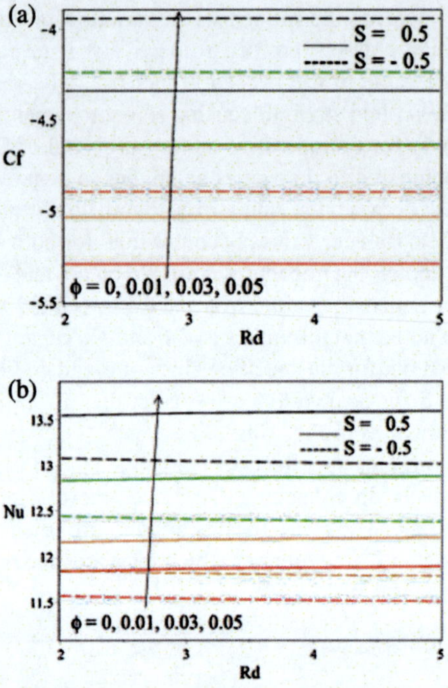


**Fig. 10** **a** Effects of $\phi$ and $Rd$ on $Cf$, **b** effects of $\phi$ and $Rd$ on $Nu$

and hence higher skin friction. Figure 12 depicts the effects of Eckert and squeeze numbers on $Nu$. Higher values of Ec and S result in higher Nusselt number and also the increment rate appears to be constant.

## 5 Conclusions

In the current study, the consequence of radiation parameter on the temperature profiles and flow behavior of the copper-water nanofluid squeezed between parallel plates in a Darcy porous medium is analyzed. By means of similarity transformation, the primary PDEs are transformed into nonlinear ODEs which are resolved analytically using DTM and the outcomes are conferred with the help of graphs and tables are compared with results of [11]. The following conclusions are observed based on the resultant solutions through our investigation.

The velocity and temperature values reduced while enhancing both squeezing and Hartmann numbers and as a result, steady reduction in the dimensionless skin friction coefficient occurred for S = 0.5 and −0.5. With enhanced Darcy numbers, both the temperature and the velocity profiles enhances for given S. With enhanced thermal radiation, the values of temperature raised and while comparing these values,

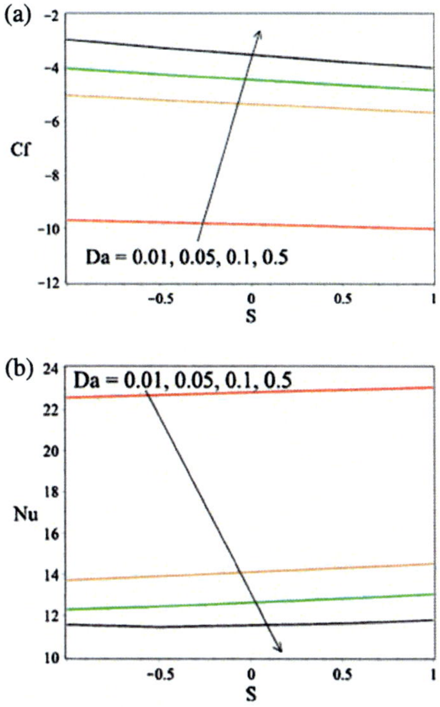


**Fig. 11** **a** Effects of *Da* & *S* on *Cf*, **b** effects of *Da* & *S* on *Nu*

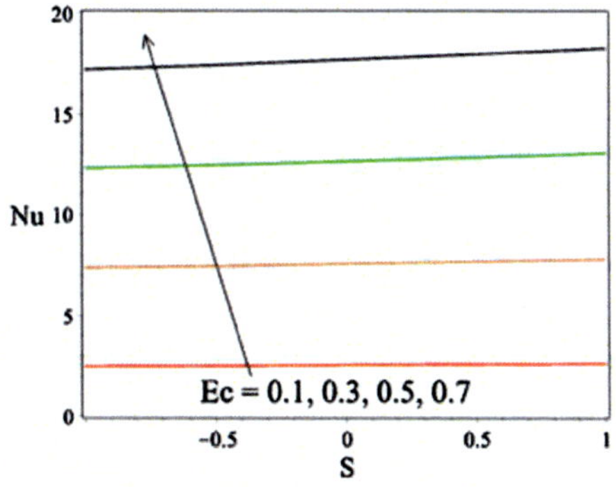


**Fig. 12** Effects of *Ec* & *S* on *Nu*

the values corresponding to positive S are lesser than the values corresponding to negative S. In enhancing Eckert number the temperature values decreases but a reverse trend when an increment in Biot number occurs.

## References


1. Engmann J, Servais C, Burbidge AS (2005) Squeeze flow theory and applications to rheometry: a review. J Non-Newton Fluid Mech 132:1–27
2. Sivasankaran S, Ho CJ (2008) Effect of temperature dependent properties on MHD convection of water near its density maximum in a cavity. Int J Therm Sci 47:1184–1194
3. Sivasankaran S, Bhuvaneswari M, Kim YJ, Ho CJ, Pan KL (2011) Magneto-convection of cold water near its density maximum in an open cavity with variable fluid properties. Int J Heat Fluid Flow 32:932–942
4. Sivasankaran S, Malleswaran A, Lee J, Sundar P (2011) Hydro-magnetic combined convection in a lid-driven cavity with sinusoidal boundary conditions on both sidewalls. Int J Heat Mass Trans 54:512–525
5. Malleswaran A, Sivasankaran S, Bhuvaneswari M (2013) Effect of heating location and size on MHD convection in a lid-driven cavity. Int J Numer Meth Heat Fluid Flow 23:867–884
6. Sivasankaran S, Narrein K (2020) Influence of geometry and magnetic field on convective flow of nanofluids in trapezoidal microchannel heat sink. Iran J Sci Tech Trans Mech Eng 44:373–382
7. Saadatmandi A, Asadi A, Eftekhari A (2016) Collocation method using quintic B-spline and sinc functions for solving a model of squeezing flow between two infinite plates. Int J Comput Mat 93:1921–1936
8. Ran XJ, Zhu QY, Li Y (2009) An explicit series solution of the squeezing flow between two infinite plates by means of the homotopy analysis method. Commu Non-linear Sci Numer Simul 14:119–132
9. Rashidi MM, Siddiqui AM, Rastegari MT (2012) Analytical solution of squeezing flow between two circular plates. Int J Comput Meth Eng Sci Mech 13:342–349
10. Dib A, Haiahem A, Bou-said B (2015) Approximate analytical solution of squeezing unsteady nanofluid flow. Powder Tech 269:193–199
11. Acharya N, Das K, Kundu PK (2016) The squeezing flow of Cu-water and Cu-kerosene nanofluids between two parallel plates. Alexandria Eng J. 55:1177–1786
12. Khan U, Ahmed N, Asadullah M, Mahmud-din ST (2015) Effects of viscous dissipation and slip velocity on two-dimensional and axisymmetric squeezing flow of Cu-water and Cu-kerosene nanofluids. Prop Power Res 4:40–49
13. Noghrehabadi A, Saffarian MR, Pourrajab R, Ghalambaz M (2013) Entropy analysis for nanofluid flow over a stretching sheet in the presence of heat generation/absorption and partial slip. J Mech Sci Tech 27:927–937
14. Mkwizu MH, Makinde OD (2015) Entropy generation in a variable viscosity channel flow of nanofluids with convective cooling. CR Mec 343(1):38–56
15. Niranjan H, Sivasankaran S, Bhuvaneswari M (2016) Analytical and numerical study on magnetoconvection stagnation-point flow in a porous medium with chemical reaction, radiation and slip effects. Math Prob Eng 2016:1–12
16. Niranjan H, Sivasankaran S, Bhuvaneswari M (2017) Chemical reaction, soret and dufour effects on MHD mixed convection stagnation point flow with radiation and slip condition. Sci Iran B Mech Eng 24:698–706
17. Sivasankaran S, Niranjan H, Bhuvaneswari M (2017) Chemical reaction, radiation and slip effects on MHD mixed convection stagnation-point flow in a porous medium with convective boundary condition. Int J Numer Meth Heat Fluid Flow 27:454–470
18. Mohyud-din S, Khan SI, Khan U, Ahmed N, Xiao-jun Y (2018) Squeezing flow of MHD fluid between parallel disks. Int J Comp Meth Eng Sci Mech 19:42–47
19. Pandey AK, Kumar M (2018) Squeezing unsteady MHD Cu-water nanofluid flow between two parallel plates in porous medium with suction/injection. Comp App Math J 4:31–42
20. Sivasankaran S, Bhuvaneswari M, Chandrapushpam T, Karthikeyan S (2021) Influence of Thermal radiation on squeezing flow of copper-water nanofluid between parallel plates. Mater Today Proc 42(P2):457–464

# MHD Flow of Casson Nanofluid Over An Inclined Porous Stretching Surface

**Suripeddi Srinivas, Challa Kalyan Kumar, Satyanarayana Badeti, and Anala Subramanyam Reddy**

## 1 Introduction

In 1995, Choi [1] first introduced the nanofluid to study the suspension of nanoparticles in a base fluid, for example, ethylene glycol, water, and oil. Buongiorno [2] endeavored to describe the rise in thermal conductivity of nanofluids and created a method that considered the particle Brownian motion and thermophoresis parameter. To predict the flow behavior of pigment-oil suspensions, Casson developed the Casson fluid model in 1959. Some popular Casson fluid examples include concentrated fruit juices, soup, honey, tomato sauce, jelly, and human blood [3–6]. Srinivas et al. [7] reported the MHD pulsatile flow of Casson fluid in a porous channel. The pulsating Casson nanofluid flow in a vertical channel was recently delineated by Kumar et al. [8].

Study on the boundary layer flow over a stretching sheet has attracted many researchers because of its various applications in crystal growing, manufacture of foods, drawing plastic films, metal spinning, glass blowing, the aerodynamic extrusion of plastic sheets, paper production, and polymer industries [9–15]. Ishak et al. [16] worked on the stagnation-point mixed convection flow against a vertical permeable stretching surface. Heat transfer and stagnation point flow of Casson fluid towards a stretching sheet was analyzed by Mustafa et al. [17]. Stagnation-point flow of a visco-elastic nanofluid over a stretching sheet was delineated by Nadeem et al. [18]. Nandy [19] examined the hydromagnetic boundary layer stagnation point flow of Casson fluid towards a stretching sheet. The impact of magnetic field and

S. Srinivas · S. Badeti
Department of Mathematics, School of Advanced Sciences, VIT-AP University, Inavolu 522237, Vijayawada, India
e-mail: srinivas.s@vitap.ac.in

C. Kalyan Kumar (✉)
Department of Mathematics, Narayana Engineering College, Gudur 524101, SPSR Nellore, India
e-mail: kalyankumar.challa@gmail.com

A. S. Reddy
Department of Mathematics, School of Advanced Sciences, Vellore Institute of Technology, Vellore 632014, India



S. Srinivas et al. (eds.), *Recent Advances in Applied Mathematics and Applications to the Dynamics of Fluid Flows*, Lecture Notes in Mechanical Engineering,
https://doi.org/10.1007/978-981-19-1929-9_13

heat transfer on stagnation point flow towards stretching sheet in nanofluid was discussed by Ibrahim et al. [20]. Unsteady Casson nanofluid flow towards a stretching surface was analyzed by Oyelakin et al. [21]. Study on magnetohydrodynamic Casson nanofluid flow with nonlinear thermal radiation impact over an inclined porous stretching surface was done by Ghadikolaei et al. [22]. Kumar and Srinivas [23] reported the influences of thermal radiation and Joule heating on unsteady MHD Eyring-Powell nanofluid flow over an inclined permeable stretching surface. Kumar and Srinivas [24] examined the pulsating hydromagnetic flow of Casson fluid in a vertical channel. Hydromagnetic Maxwell fluid flow over a stretching surface embedded in a porous medium was recently delineated by Megahed [25].

The fundamental objective of the current examination is to describe the cross diffusion impacts on MHD flow of Casson nanofluid over an inclined permeable stretching surface. Runge-Kutta 4th order method along with shooting technique has been utilized to solve the system of nonlinear ordinary differential equations.

## 2 Mathematical Formulation

Consider the unsteady stagnation point flow of laminar and incompressible electrically conducting Casson nanofluid over a permeable inclined stretching surface with an acute angle $\alpha$ to the vertical. The $B(t) = B_0(1 - a_3 t)^{-\frac{1}{2}}$ magnetic field is applied orthogonal to the surface, where $B_0$ is a constant defining the strength of the magnetic field at $t = 0$. The physical schematic diagram of problem is appeared in Fig. 1. Assume that the sheet begins to stretching from time $t = 0$ along the $x$-axis with the velocity $u_w(x, t)$, where the $y$-axis is perpendicular to the surface. Rheological

Fig. 1 Schematic view and physical geometry

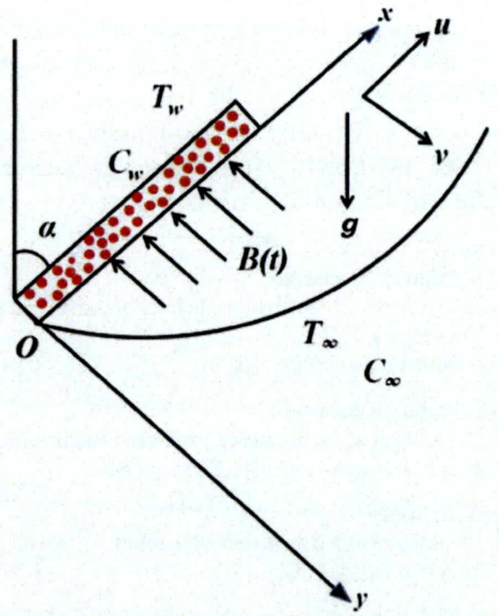

model for Casson nanofluid [8, 22] is given by

$$\tau_{ij} = \begin{cases} 2\left(\mu_B + \frac{P_y}{\sqrt{2\pi_c}}\right) e_{ij}, & \pi < \pi_c \\ 2\left(\mu_B + \frac{P_y}{\sqrt{2\pi}}\right) e_{ij}, & \pi > \pi_c \end{cases} \tag{1}$$

The $(i, j)$th component of the deformation rate and shear stress tensor are $e_{ij}$ and $\tau_{ij}$, respectively. The product of the component of deformation rate with itself is $\pi(= e_{ij}^2)$, the critical value of $\pi$ is $\pi_c$, $\mu_B$ is the plastic dynamic viscosity of the non-Newtonian fluid, and $P_y$ is the yield stress of the fluid. The stretching velocity $u_w(x, t)$, surface temperature $T_w(x, t)$, nanoparticle concentration $C_w(x, t)$, and free stream velocity $U_\infty(x, t)$ are assumed as

$$u_w = \frac{a_0 x}{1 - a_3 t}, \; T_w = T_\infty + \frac{a_1 x}{1 - a_3 t}, \; C_w = C_\infty + \frac{a_2 x}{1 - a_3 t}, \; U_\infty = \frac{a_4 x}{1 - a_3 t} \tag{2}$$

where $a_0, a_1, a_2, a_3$, and $a_4$ are the constants such that $a_4 > 0$, $a_3 \geq 0$, $a_2 \geq 0$, $a_1 \geq 0$, $a_0 > 0$ and $a_3 t < 1$. $C_\infty$, $T_\infty$ are the nanoparticle concentration and temperature far from the surface. The governing equations are described by [15, 19, 21]:

$$\frac{\partial u}{\partial x} + \frac{\partial v}{\partial y} = 0 \tag{3}$$

$$\begin{aligned} \frac{\partial u}{\partial t} + u\frac{\partial u}{\partial x} + v\frac{\partial u}{\partial y} &= U_\infty \frac{dU_\infty}{dx} + \nu\left(1 + \frac{1}{\beta}\right)\frac{\partial^2 u}{\partial y^2} \\ &+ \left[g\beta_T^*(T - T_\infty) + g\beta_C^*(C - C_\infty)\right] cos\alpha + \frac{\sigma B^2(t)}{\rho}(U_\infty - u) \\ &+ \frac{\nu}{k_0(t)}(U_\infty - u) \end{aligned} \tag{4}$$

$$\begin{aligned} \frac{\partial T}{\partial t} + u\frac{\partial T}{\partial x} + v\frac{\partial T}{\partial y} &= \frac{\kappa}{\rho C_p}\frac{\partial^2 T}{\partial y^2} + \tau\left[D_B\left(\frac{\partial C}{\partial y}\frac{\partial T}{\partial y}\right) + \frac{D_T}{T_\infty}\left(\frac{\partial T}{\partial y}\right)^2\right] \\ &+ \frac{\mu}{\rho C_p}\left(1 + \frac{1}{\beta}\right)\left(\frac{\partial u}{\partial y}\right)^2 + \frac{\sigma B^2(t)}{\rho C_p}(U_\infty - u)^2 \\ &- \frac{1}{\rho C_p}\frac{\partial q_r}{\partial y} + \frac{q'''}{\rho C_p} \end{aligned} \tag{5}$$

$$\frac{\partial C}{\partial t} + u\frac{\partial C}{\partial x} + v\frac{\partial C}{\partial y} = D_B\frac{\partial^2 C}{\partial y^2} + \frac{D_T}{T_\infty}\frac{\partial^2 T}{\partial y^2} + \frac{D_m k_0}{Tm}\frac{\partial^2 T}{\partial y^2} - K_r(t)(C - C_\infty). \tag{6}$$

where $q_r$ is the radiative heat flux, $D_B$ is Brownian diffusion coefficient, $\kappa$ is the thermal conductivity, $k_0$ is the thermal diffusivity, $\beta_T^*$ is the coefficient of thermal expansion, $g$ is the acceleration due to gravity, $\rho$ is density of the fluid, $Tm$ is the mean temperature, $D_m$ is the mass diffusivity, $D_T$ is thermophoretic diffusion coefficient, $C_p$ is the specific heat at constant pressure, the components of velocity along the $x$-axis and $y$-axis are $u$ and $v$, $\tau = (\rho C_p)_p/(\rho C_p)_f$ is the ratio of the heat capacity of the nanoparticle material and heat capacity of the fluid, $T$ is the temperature, $\sigma$ is the electrical conductivity, $\mu$ is the dynamic viscosity, $C$ is the concentration, $\nu$ is the kinematic viscosity, $k_0(t) = k_1(1 - a_3 t)$ is the time-dependent permeability parameter, $\beta_C^*$ is the coefficient of concentration expansion, $K_1$ is a constant, $K_r(t) = K_1(1 - a_3 t)^{-1}$ represents a dimensional chemical reaction parameter, and $\beta = \frac{\mu_B\sqrt{2\pi_c}}{P_y}$ is the Casson nanofluid parameter. Assuming the Rosseland approximation to $q_r$, is known as [5, 7]

$$q_r = -\left(\frac{4\sigma^*}{3\chi}\right)\frac{\partial T^4}{\partial y} \tag{7}$$

Here $\chi, \sigma^*$ are Rosseland mean absorption coefficient and Stefan-Boltzmann constant respectively. Assuming a sufficiently small temperature difference in the flow and indicating $T^4$ by Taylor series at $T_\infty$ we acquire, $T^4 \cong 4T_\infty^3 T - 3T_\infty^4$ (Higher order terms are omitted). The non-uniform heat source/sink, $q'''$ is considered as follows (Srinivas et al. [12]):

$$q''' = \frac{\kappa u_w(x,t)}{x\nu}[A^*(T_w - T_\infty)f' + B^*(T - T_\infty)] \tag{8}$$

The negative values of $A^*$, $B^*$ correspond to internal heat absorption while positive values correspond to internal heat generation.

The dimensional boundary conditions are

$$u = u_w + N\mu\left(1 + \frac{1}{\beta}\right)\left(\frac{\partial u}{\partial y}\right),\ v = v_w,\ T = T_w + K\frac{\partial T}{\partial y},$$
$$D_B\frac{\partial C}{\partial y} + \frac{D_T}{T_\infty}\frac{\partial T}{\partial y} = 0 \ \text{ at } \ y = 0 \tag{9}$$

$$u \to 0,\ T \to T_\infty,\ C \to C_\infty \ \text{ as } \ y \to \infty, \tag{10}$$

Here $v_w$ indicates the injection/suction velocity provided by

$$v_w = -\sqrt{\frac{\nu u_w}{x}}S \tag{11}$$

Equation (11) implies that the mass transfer at the surface with negative values of $v_w$ for suction and positive values of $v_w$ for injection. Here $K = K_0\sqrt{1 - a_3 t}$ is the thermal slip factor, $N = N_0\sqrt{1-a_3t}$ is the velocity slip factor. We are introducing self-similar transformations [18, 19]:

$$\eta = \left(\frac{u_w}{\nu x}\right)^{\frac{1}{2}} y,\ \psi = \sqrt{\nu x u_w} f(\eta),\ \theta(\eta) = \frac{T - T_\infty}{T_w - T_\infty},\ \phi(\eta) = \frac{C - C_\infty}{C_w - C_\infty} \quad (12)$$

where $\psi$ is the stream function and $\eta$ is the similarity variable. The components of the velocity are $u = \frac{\partial\psi}{\partial y}$, $v = -\frac{\partial\psi}{\partial x}$, which identically satisfies Eq. (3).
Now substituting Eqs. (2), (7), (8) and (12) into Eqs. (4)–(6), gives the dimensionless equations

$$\left(1+\frac{1}{\beta}\right) f''' + ff'' - (f')^2 - A\left(f' + \frac{1}{2}\eta f''\right) + \left(\frac{Gr_x}{Re_x^2}\theta + \frac{Gc_x}{Re_x^2}\phi\right)\cos\alpha$$
$$- M^2(\varepsilon - f') - \frac{1}{k_2}(\varepsilon - f') + \frac{a_4^2}{a_0^2} = 0 \quad (13)$$

$$\left(1+\frac{4}{3}Rd\right)\frac{1}{Pr}\theta'' + f\theta' - f'\theta - A\left(\theta + \frac{1}{2}\eta\theta'\right) + Nb\theta'\phi' + Nt(\theta')^2$$
$$+ (A^*f' + B^*\theta)\frac{1}{Pr} + \left(1+\frac{1}{\beta}\right) Ec\left(f''\right)^2 + M^2Ec(\varepsilon - f')^2 = 0 \quad (14)$$

$$\phi'' - LePr\left[A\left(\phi + \frac{1}{2}\eta\phi'\right) + f'\phi - f\phi' - Sr\theta'' + \gamma\phi\right] + \frac{Nt}{Nb}\theta'' = 0. \quad (15)$$

subject to the boundary conditions

$$f' = 1 + S_f\left(1+\frac{1}{\beta}\right) f''(0),\ f = S,\ \theta = 1 + S_t\theta'(0),\ Nb\phi' + Nt\theta' = 0$$
$$\text{at } \eta = 0 \quad (16)$$

$$f' \to 0,\ \theta \to 0,\ \phi \to 0 \ \text{ as } \ \eta \to \infty \quad (17)$$

where $Pr = \frac{\nu}{\alpha_0}$ is Prandtl number, $\alpha_0 = \frac{\kappa}{\rho C_p}$ is thermal diffusivity, $Le = \frac{\alpha_0}{D_B}$ is Lewis number, $A = \frac{c}{a}$ is unsteadiness parameter, $Sr = \frac{D_m k_0(T_w - T_\infty)}{Tm\nu(C_w - C_\infty)}$ is Soret number, $\gamma = \frac{K_1}{a}$ is chemical reaction parameter, $Nt = \frac{\tau D_T(T_w - T_\infty)}{\nu T_\infty}$ is thermophoresis parameter, $Nb = \frac{\tau D_B(C_w - C_\infty)}{\nu}$ is Brownian motion parameter, $Rd = \frac{4\sigma^* T_\infty^3}{\kappa\chi}$ radiation parameter, $Ec = \frac{u_w^2}{C_p(T_w - T_\infty)}$ is Eckert number, $M = B_0\sqrt{\frac{\sigma}{\rho a}}$ is Hartmann number, $Gc_x = \frac{g\beta_C^* x^3(C_w - C_\infty)}{\nu^2}$ is local modified Grashof number, $Re_x = \frac{u_w x}{\nu}$ is the local Reynolds number, $k_2 = \frac{ak_1}{\nu}$ is the permeability parameter, $Gr_x = \frac{g\beta_T^* x^3(T_w - T_\infty)}{\nu^2}$ is local Grashof number.

In Eq. (16), $S > 0$ and $S < 0$ correspond to suction and injection, respectively. The dimensionless velocity slip $S_f$ and thermal slip $S_t$ are defined by

$$S_f = N_0 \rho \sqrt{a\nu}, \ S_t = K_0 \sqrt{\frac{a}{\nu}} \tag{18}$$

where prime denotes differentiation with respect to $\eta$. Also, skin friction coefficient ($C_f$), local Nusselt number ($Nu_x$), and local Sherwood number ($Sh_x$) are defined as

$$C_f = \frac{\tau_w}{\rho u_w^2}, \ Nu_x = \frac{xq_w}{\kappa(T_w - T_\infty)}, \ Sh_x = \frac{xm_w}{D(C_w - C_\infty)} \tag{19}$$

The shear stress, surface heat, and mass fluxes near the wall are
$\tau_w = \mu\left(1+\frac{1}{\beta}\right)\left(\frac{\partial u}{\partial y}\right)_{y=0}, q_w = -\kappa\left(\frac{\partial T}{\partial y}\right)_{y=0} + (q_r)_{y=0}, m_w = -D\left(\frac{\partial C}{\partial y}\right)_{y=0}$
On simplifying Eq. (19), we obtain

$$\begin{aligned} &Re_x^{\frac{1}{2}} C_f = \left(1+\frac{1}{\beta}\right) f''(0), \ Re_x^{-\frac{1}{2}} Nu_x = -\left(1+\frac{4}{3}Rd\right)\theta'(0), \\ &Re_x^{-\frac{1}{2}} Sh_x = -\phi'(0). \end{aligned} \tag{20}$$

where $Re_x^{\frac{1}{2}} C_f$, $Re_x^{-\frac{1}{2}} Nu_x$ and $Re_x^{-\frac{1}{2}} Sh_x$ are the surface drag force, heat and mass transfer rates, respectively.

## 3 Results and Discussion

The system of dimensionless nonlinear coupled ordinary differential equations (13)–(15) subject to the boundary conditions Eqs. (16)–(17) are solved numerically by shooting method along with 4th order Runge-Kutta method (as explained in Ibrahim and Shankar [20]). Throughout the calculations, the parametric values are fixed to be $\beta = 2$, $\alpha = \frac{\pi}{3}$, $S = 0.5$, $A = 0.3$, $S_f = 2.0$, $k_2 = 0.4$, $M = 1.0$, $Ec = 0.5$, $\gamma = 0.5$, $Nb = 0.3$, $A^* = 1.0$, $Gc_x = 2.0$, $Gr_x = 5.0$, $S_t = 1.0$, $Rd = 0.3$, $Pr = 21$, $B^* = 1.0$, $\varepsilon = 0.2$, $Nt = 0.3$, $Le = 1$, $Sr = 1$, $Re_x = 1$, unless otherwise stated. For theoretical/experimental studies (Misra and Sinha [11], Srinivas et al. [12], Eswaramoorthi et al. [13]), the dimensionless distinct parameters are considered.

$\beta = 2, 3, 4.$ | $Le = 1, 2, 3, 4$
$Nb, Nt = 0.1, 0.2, 0.3, 0.4, 0.5.$ | $\alpha = 0, \frac{\pi}{4}, \frac{\pi}{3}, \frac{\pi}{2}.$
$Rd = 0, 0.5, 1, 1.5.$ | $M = 0, 0.5, 1, 2.$
$Gc_x, Gr_x = 1, 3, 5.$ | $A = 0.1, 0.2, 0.3, 0.4.$
$k_2 = 0, 0.3, 0.6, 1.$ | $Ec = 0.1, 0.3, 0.5, 0.7.$

The influences of various physical parameters on velocity, temperature, nanoparticle concentration profiles are elucidated graphically in Figs. 2, 3, and 4. Figure 2a, b describes the impact of the Casson nanofluid parameter ($\beta$) and suction parameter ($S$) on the velocity profile. Figure 2a shows that the rise in Casson nanofluid parameter results in rise of velocity profile. Figure 2b illustrates the effect of $S$ on the velocity profile. It is seen that rise in $S$ decrease the velocity distributions. This is because the heated fluid is pushed towards the wall where the buoyancy forces can act to retard the fluid due to high viscosity effect.

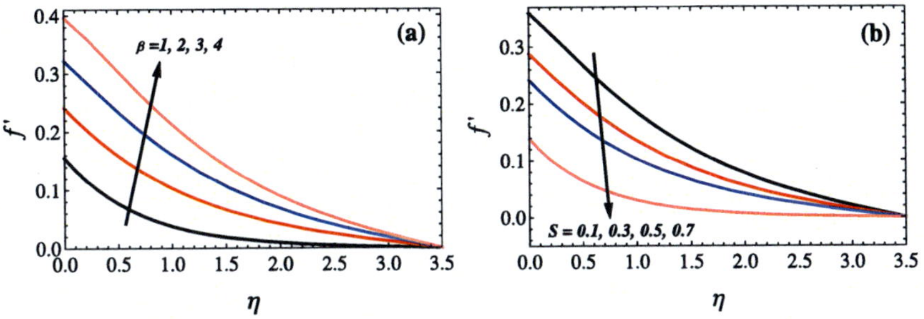


**Fig. 2** Influences of $\beta$, $S$ on velocity distribution

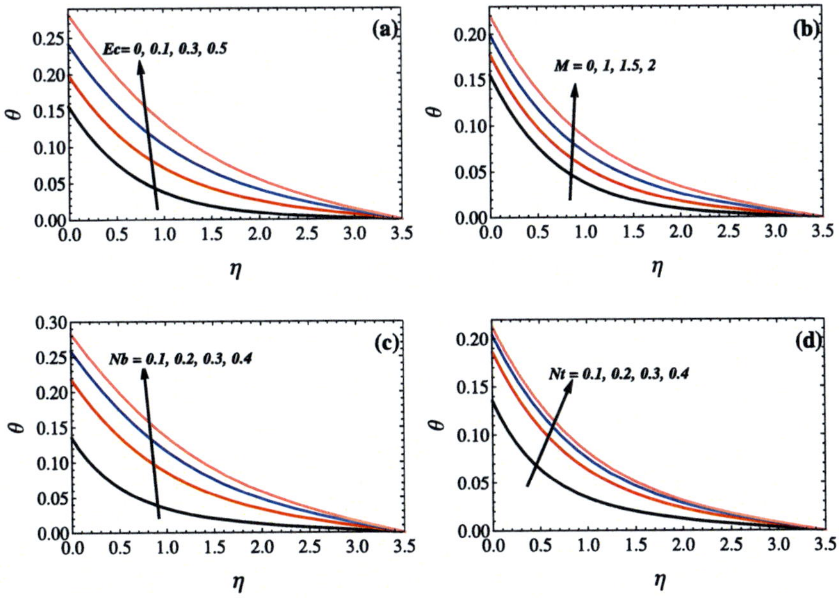


**Fig. 3** Influences of $Ec$, $M$, $Nb$, and $Nt$ on temperature distribution

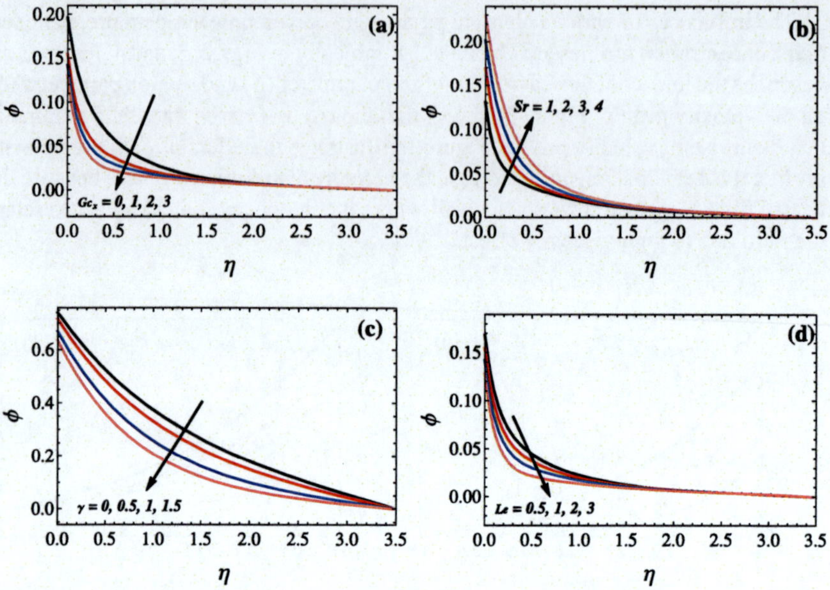


**Fig. 4** Influences of $Gc_x$, $Sr$, $\gamma$, $Le$ on nanoparticles concentration distribution

Figure 3a–d displays the change of the temperature profile ($\theta$) for the different values of $Ec$, $M$, $Nb$, $Nt$. Figure 3a illustrates that the $\theta$ rises with an enhancing in Eckert number. It is noticed that the heat generated by viscous dissipation leads to rise in temperature. Figure 3b demonstrates that the growth in $M$ results in an increase in temperature profile. From Fig. 3c, d, it is noticed that there is an increase temperature of the fluid as well as the boundary layer thickness with rise in $Nb$ and $Nt$, respectively.

Figure 4 illustrate the influences of $Gc_x$, $Sr$, $\gamma$ and $Le$ on nanoparticle concentration distribution. Figure 4a depicts $Gc_x$ effect on $\phi$. It is seen that the concentration falls with a rise in $Gc_x$. Figure 4b illustrates that an enhancement in $Sr$ results in rise of nanoparticle concentration distribution. Figure 4c shows the $\gamma$ effect on $\phi$. It is noticed that chemical reaction parameter enhancement leads to decreasing $\phi$. Figure 4d reveals that an increase in $Le$ results in decrease of nanoparticle concentration distribution because the lesser molecular diffusivity.

It can be seen from Table 1 that the present study agrees well with the numerical results previously published. In addition, from Table 2. the local skin friction coefficient decreases with the rise in the unsteadiness parameter, the angle of inclination and chemical reaction parameter, while it rises with the rise of Brownian motion parameter and thermophoresis parameter. Further, with an enhancement in the thermophoresis parameter and Brownian motion parameter, the value of the local Nusselt number decreases, while it rises with the rise of the unsteadiness

**Table 1** Comparison of the $f'(0)$ values with the current literature values for the different $\varepsilon$ values when $\beta \to \infty$

| $\varepsilon$ | Ishak et al. [16] | Mustafa et al. [17] | Oyelakin et al. [21] | Present |
|---|---|---|---|---|
| 0.1 | −0.9694 | −0.96939 | −0.96937 | −0.9694 |
| 0.2 | 0.9181 | 0.918107 | 0.918111 | 0.9181 |
| 0.5 | −0.6673 | −0.66735 | −0.66726 | −0.6673 |
| 2.0 | 2.0175 | 2.01757 | 2.01750 | 2.0175 |
| 3.0 | 4.7294 | 4.72964 | 4.72928 | 4.7294 |

**Table 2** Variations of $Re_x^{\frac{1}{2}} C_f$, $Re_x^{-\frac{1}{2}} Nu_x$ and $Re_x^{-\frac{1}{2}} Sh_x$ for different values of $A$, $\alpha$, $\gamma$, $Nb$, $Nt$

| Parameter | Values | $Re_x^{\frac{1}{2}} C_f$ | $Re_x^{-\frac{1}{2}} Nu_x$ | $Re_x^{-\frac{1}{2}} Sh_x$ |
|---|---|---|---|---|
| $A$ | 0.0 | −0.273972 | 0.877451 | −0.626751 |
| | 0.1 | −0.283422 | 0.901433 | −0.643881 |
| | 0.3 | −0.300217 | 0.943027 | −0.673590 |
| | 0.5 | −0.314640 | 0.977907 | −0.698505 |
| $\alpha$ | 0 | −0.179281 | 0.814910 | −0.582079 |
| | $\frac{\pi}{6}$ | −0.211753 | 0.852954 | −0.609253 |
| | $\frac{\pi}{4}$ | −0.249692 | 0.894651 | −0.639037 |
| | $\frac{\pi}{3}$ | −0.300217 | 0.943027 | −0.673590 |
| $\gamma$ | 0.0 | -0.292449 | 0.925544 | −0.661103 |
| | 0.5 | −0.300217 | 0.943027 | −0.673590 |
| | 1.0 | −0.307543 | 0.958724 | −0.684803 |
| | 1.5 | −0.314365 | 0.972696 | −0.694783 |
| $Nb$ | 0.1 | −0.320040 | 1.036230 | -2.220490 |
| | 0.2 | −0.307664 | 0.981582 | -1.051700 |
| | 0.3 | −0.300217 | 0.943027 | −0.673590 |
| | 0.4 | −0.295074 | 0.914043 | −0.489666 |
| $Nt$ | 0.1 | −0.301248 | 0.962249 | −0.229107 |
| | 0.2 | −0.300719 | 0.952953 | −0.453787 |
| | 0.3 | −0.300217 | 0.943027 | −0.673590 |
| | 0.4 | −0.299767 | 0.932419 | −0.888018 |

parameter, inclination angle, and chemical reaction parameter. The local Sherwood number decreases with the rise in the unsteadiness parameter, angle of inclination, chemical reaction parameter, and thermophoresis parameter, while it rises with the rise of the Brownian motion parameter.

## 4 Conclusion

An analysis is made with Brownian diffusion and thermophoresis mechanisms on unstable MHD Casson nanofluid flow over an inclined permeable stretching surface. The governing flow equations are numerically solved along with the shooting technique by the Runge-Kutta 4th order method. The following is a summary of the results:

- The velocity rises with the rising Casson nanofluid parameter as it decreases with an increasing suction parameter.
- Increasing the Brownian motion and thermophoresis parameters cause an increase in the temperature profile.
- The nanoparticle concentration decreases by rising chemical reaction parameter and rises by rising the Soret number.
- Local Nusselt number increases with rising unsteadiness parameter while it decreases with rising $Nb$, $Nt$.
- Further, the local Sherwood number decreases with rising $\alpha$ and $\gamma$, whereas it rises with an enhancing Brownian motion parameter.
- The results of Kumar and Srinivas [15] for case the Casson fluid can be recovered by taking $\varepsilon = Nb = Nt = Le = Sr = 0$.

This investigation is appropriate to a circumstance where the lumen of the blood vessel has turned into a porous medium because of a deposition of fatty substances, cholesterol, etc. (Srinivas et al. [12]). It is expected that the current examination will be helpful in evaluating the precision of the future investigations in this direction, accounting greater number of physical parameters.

## References

1. Choi SUS (1995) Enhancing thermal conductivity of fluids with nanoparticles. In: Proceeding of the 1995 ASME international mechanical engineering congress and exposition, vol 66. San Francisco, USA. ASME, FED 231/MD, pp 99–105
2. Buongiorno J (2006) Convective transport nanofluids. J Heat Trans 128:240–250
3. Dash RK, Jayaraman G, Mehta KN (2000) Shear augmented dispersion of a solute in a Casson fluid flowing in a conduit. Ann Biomed Eng 28:373–385
4. Nagarani P, Sarojamma G, Jayaraman G (2009) Effect of boundary absorption on dispersion in Casson fluid flow in an annulus: application to catheterized artery. Acta Mech 202:47–63
5. Kumar CK, Srinivas S (2017) Simultaneous effects of thermal radiation and chemical reaction on hydromagnetic pulsatile flow of a Casson fluid in a porous space. Eng Trans 65:461–481
6. Alzahrani AK, Sivasankaran S, Bhuvaneswari M (2020) Numerical simulation on convection and thermal radiation of Casson fluid in an enclosure with entropy generation. Entropy 22:229
7. Srinivas S, Kumar CK, Reddy AS (2018) Pulsating flow of Casson fluid in a porous channel with thermal radiation, chemical reaction and applied magnetic field. Nonlinear Anal Model Control 23:213–233
8. Kumar CK, Srinivas S, Reddy AS (2020) MHD pulsating flow of Casson nanofluid in a vertical porous space with thermal radiation and Joule heating. J Mech 36:535–549

9. Crane LJ (1970) Flow past a stretching plate. Zeitschrift für angewandte Mathematik und Physik 21:645–647
10. Pantokratoras A (2008) Study of MHD boundary layer flow over a heated stretching sheet with variable viscosity: A numerical reinvestigation. Int J Heat Mass Trans 51:104–110
11. Misra JC, Sinha A (2013) Effect of thermal radiation on MHD flow of blood and heat transfer in a permeable capillary in stretching motion. Heat Mass Trans 49:617–628
12. Srinivas S, Reddy PBA, Prasad BSRV (2014) Effects of chemical reaction and thermal radiation on MHD flow over an inclined permeable stretching surface with non-uniform heat source/sink: an application to the dynamics of blood flow. J Mech Med Biol 14:1450067–1450091
13. Eswaramoorthi S, Sivasankaran S, Alshomrani AS (2018) Effect of thermal radiation and heat absorption of MHD Casson nanofluid over a stretching surface in a porous medium with convective heat and mass conditions. J Phys Conf Ser 1139:012017
14. Bhuvaneswari M, Eswaramoorthi S, Sivasankaran S, Hussein AK (2019) Cross-diffusion effects on MHD mixed convection over a stretching surface in a porous medium with chemical reaction and convective condition. Eng Trans 67(1):3–19
15. Kumar CK, Srinivas S (2019) Influence of Joule heating and thermal radiation on unsteady hydromagnetic flow of chemically reacting Casson fluid over an inclined porous stretching sheet. Special Top Rev Porous Media Int J 10:385–400
16. Ishak A, Nazar R, Arifin NM, Pop I (2007) Mixed convection of the stagnation-point flow towards a stretching vertical permeable sheet. Malay J Math Sci 1:217–226
17. Mustafa M, Hayat T, Pop I, Hendi A (2011) Stagnation-point flow and heat transfer of a Casson fluid towards a stretching sheet. Z Naturforsch 67a:70–76
18. Nadeem S, Mehmood R, Akbar NS (2013) Non-orthogonal stagnation point flow of a nano non-Newtonian fluid towards a stretching surface with heat transfer. Int J Heat Mass Trans 57:679–689
19. Nandy SK (2013) Analytical solution of MHD stagnation-point flow and heat transfer of Casson fluid over a stretching sheet with partial slip. ISRN Therm 2013/108264. Article ID 108264
20. Ibrahim W, Shankar B, Nandeppanavar MM (2013) MHD stagnation point flow and heat transfer due to nanofluid towards a stretching sheet. Int J Heat Mass Trans 56:1–9
21. Oyelakin IS, Mondal S, Sibanda P (2016) Unsteady Casson nanofluid flow over a stretching sheet with thermal radiation, convective and slip boundary conditions. Alex Eng J 55:1025–1035
22. Ghadikolaei SS, Hosseinzadeh Kh, Ganji DD, Jafari B (2018) Nonlinear thermal radiation effect on magneto Casson nanofluid flow with Joule heating effect over an inclined porous stretching sheet. Case Stud Therm Eng 12:176–187
23. Kumar B, Srinivas S (2020) Unsteady hydromagnetic flow of Eyring-Powell nanofluid over an inclined permeable stretching sheet with Joule heating and thermal radiation. J Appl Comput Mech 6:259–270
24. Kumar CK, Srinivas S (2021) Pulsating hydromagnetic flow of Casson fluid in a vertical channel filled with non-Darcian porous medium. Heat Trans 50:5225–5239
25. Megahed AM (2021) Improvement of heat transfer mechanism through a Maxwell fluid flow over a stretching sheet embedded in a porous medium and convectively heated. Math Comput Simul 187:97–109

# Dispersion of Rayleigh Wave in a Shielded Anisotropic Generalized Thermoelastic Layer

S. Kumar, D. Prakash, Narsu Sivakumar, and B. Rushi Kumar

## 1 Introduction

Rayleigh wave is a surface wave that includes both longitudinal and transverse motion that travels near the surface of solids. It was Rayleigh [1], who predicted the existence of this surface wave in a homogeneous elastic half-space. Later on, many authors like Wilson [2], Newlands [3], Stonely [4], etc. put a vital contribution in the analysis of Rayleigh wave propagation in heterogeneous medium. Gupta et al. [5] and Vishwakarma et al. [6] considered Rayleigh wave propagation with a rigid boundary to get more real scenario of the earth surface. The anisotropic and thermal behaviour of Earth made the researcher to study the seismic wave propagation in anisotropic thermoelastic medium. In this context, study started with classical dynamical theory of thermo-elasticty. The theories come to remove the paradox of infinite speed of heat propagation in the classical theory of thermo-elasticity that was due to Lord and Shulman [7] and Fox [8], but this study was limited to a homogeneous isotropic medium. The next is due to Green and Lindsay [9], who conducted a study without violating law of Fourier for heat conduction in both isotropic and anisotropic media. Ivanov [10] studied the Rayleigh wave propagation in a thermoelastic half-space. The papers that can be cited for remarking works in the field of thermo-elasticity as Sharma and Pal [11], Kumar and Pal [12], Pal et al. [13], Biswas et al. [14], Yadav [15–19] etc.

S. Kumar (✉) · D. Prakash · N. Sivakumar
Department of Mathematics, Faculty of Engineering and Technology, College of Engineering and Technology, SRM Institute of Science and Technology, SRM Nagar, Kattankulathur, Kanchipuram, Chennai 603203, TN, India
e-mail: santoshh@srmist.edu.in

B. R. Kumar
Department of Mathematics, School of Advanced Sciences, Vellore Institute of Technology, Vellore, India



S. Srinivas et al. (eds.), *Recent Advances in Applied Mathematics and Applications to the Dynamics of Fluid Flows*, Lecture Notes in Mechanical Engineering,
https://doi.org/10.1007/978-981-19-1929-9_14

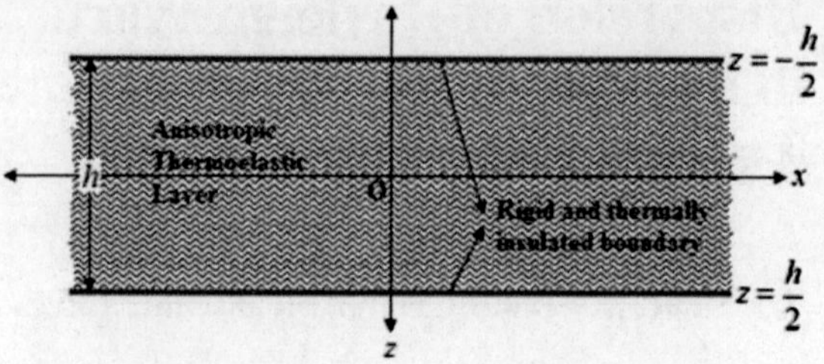


**Fig. 1** Physical geometry of the problem

In this paper, we have incorporated the propagation of the Rayleigh wave in a layer with rigid and adiabatic interfaces which makes it unique from the other papers. The studies were made jointly with single heat conduction equation for three different theories. The anisotropy of the layer is taken as tri-clinic. The theories of generalized thermo-elasticity which are taken into account are CD theory, LS theory and GL theory with two thermal relaxation times. An analytical solution obtained and procured a dispersion relation subjected to boundaries as rigid and insulated. In order to deal the numerical and physical scenario of the problem, a particular model is taken into account. The results were interpreted in terms of graphs. The graphs are drawn between non-dimensional real and imaginary phase velocities and wave number for different values of thickness.

## 2 Formulation of the Problem

Considering the homogeneous anisotropic-thermally conducting uniform layer at consistent temperature $\theta_0$, the rigid walls of layer have been assumed as thermally insulated. Also, the upper and lower walls of layer are positioned at $x_3 = -\frac{h}{2}$ and $x_3 = \frac{h}{2}$, whereas the $x_3$-axis is considered with direction of vertically downwards. The geometry of the problem depicted in Fig. 1. Based on the assumptions of the physical geometry, the governing basic equations for generalized thermo-elasticity are given as follows:

$$\frac{\partial \sigma_{ij}}{\partial x_j} = \rho \frac{\partial^2 u_i}{\partial t^2} \tag{1}$$

$$\frac{\partial}{\partial x_i}\left(K_{ij}\frac{\partial \theta}{\partial x_j}\right) - \rho C_e\left(\frac{\partial \theta}{\partial t} + \tau_0 \frac{\partial^2 \theta}{\partial t^2}\right) = \theta_0 \beta_{ij} \frac{\partial}{\partial x_j}\left\{\left(\frac{\partial u_i}{\partial t} + \tau_0 \Omega \frac{\partial^2 u_i}{\partial t^2}\right)\right\} \tag{2}$$

$$\sigma_{ij} = C_{ijkl} e_{kl} - \beta_{ij}\left(\theta + \tau_1 \frac{\partial \theta}{\partial t}\right) \tag{3}$$

where $\sigma_{ij}$-strain tensor, $e_{ij}$-stress tensor, $\beta_{ij}\left(= C_{ijkl}\alpha_{kl},\quad i, j, k, l = 1, 2, 3\right)$-thermal module, $\alpha_{ij}$-thermal expansion tensor, $\rho$-density, $t$-time, $u_i$-distortion in the direction of $x_i$, $K_{ij}$-thermal conductivity, $C_e$, $\tau_0$ and $\tau_1$-specific heat at constant stress and thermal relaxation times, respectively, $\theta$-the absolute temperature. The fourth-order tensor of the elasticity $C_{ijkl}$ satisfies the (Green) symmetry conditions:

$$C_{ijkl} = C_{klij} = C_{ijlk} = C_{jikl},\ \alpha_{ij} = \alpha_{ji},\ \beta_{ij} = \beta_{ji},\ K_{ij} = K_{ji}. \tag{4}$$

The parameters in Eq. (2) and the relaxation time parameters $\tau_0$ and $\tau_1$ will lead these rudimentary equations feasible for the theories described below:

1. For CD theory: $\tau_0 = \tau_1 = 0$ and $\Omega = 0$
   which lead Eqs. (2) and (3) has the following form:

$$\frac{\partial}{\partial x_i}\left(K_{ij}\frac{\partial\theta}{\partial x_j}\right) - \rho C_e\frac{\partial\theta}{\partial t} = T_0\beta_{ij}\frac{\partial^2 u_i}{\partial t\partial x_j} \tag{5}$$

$$\sigma_{ij} = C_{ijkl}e_{kl} - \beta_{ij}\theta \tag{6}$$

2. For LS theory: $\tau_1 = 0,\ \Omega = 1,\ \tau_0 > 0$
   which lead Eq. (3) has the same form as Eq. (5). But Eq. (2) has the following form:

$$\frac{\partial}{\partial x_i}\left(K_{ij}\frac{\partial\theta}{\partial x_j}\right) - \rho C_e\left(\frac{\partial\theta}{\partial t} + \tau_0\frac{\partial^2\theta}{\partial t^2}\right) = \theta_0\beta_{ij}\frac{\partial}{\partial x_j}\left\{\left(\frac{\partial u_i}{\partial t} + \tau_0\frac{\partial^2 u_i}{\partial t^2}\right)\right\} \tag{7}$$

3. For GL theory: $\Omega = 0,\ \ \tau_1 \geq \tau_0 \geq 0$. Then Eq. (2) has form

$$\frac{\partial}{\partial x_i}\left(K_{ij}\frac{\partial\theta}{\partial x_j}\right) - \rho C_e\left(\frac{\partial\theta}{\partial t} + \tau_0\frac{\partial^2\theta}{\partial t^2}\right) = \theta_0\beta_{ij}\frac{\partial^2 u_i}{\partial t\partial x_j} \tag{8}$$

The respective condition for the propagation of Rayleigh wave in $x_1 - x_3$ plane assumed as

$$u_1 \equiv u_1(x_1, x_3, t),\ u_3 \equiv u_3(x_1, x_3, t),\ u_2 \equiv 0,\ \frac{\partial}{\partial x_3} \equiv 0 \tag{9}$$

Posing the above conditions (9) in $x_1 - x_3$ plane for generalized thermo-elasticity, Rayleigh wave propagation will have the following equations of motion:

$$C_{11}\frac{\partial^2 u_1}{\partial x_1^2} + C_{15}\frac{\partial^2 u_3}{\partial x_1^2} + C_{55}\frac{\partial^2 u_1}{\partial x_3^2} + C_{35}\frac{\partial^2 u_3}{\partial x_3^2} + 2C_{15}\frac{\partial^2 u_1}{\partial x_1\partial x_3} + (C_{13} + C_{55})\frac{\partial^2 u_3}{\partial x_1\partial x_3}$$
$$-\beta_{11}\left(\frac{\partial\theta}{\partial x_1} + \tau_1\frac{\partial^2\theta}{\partial x_1\partial t}\right) = \rho\frac{\partial^2 u_1}{\partial t^2} \tag{10}$$

$$C_{15}\frac{\partial^2 u_1}{\partial x_1^2}+C_{55}\frac{\partial^2 u_3}{\partial x_1^2}+C_{35}\frac{\partial^2 u_1}{\partial x_3^2}+C_{33}\frac{\partial^2 u_3}{\partial x_3^2}+(C_{13}+C_{55})\frac{\partial^2 u_1}{\partial x_1\partial x_3}+2C_{35}\frac{\partial^2 u_3}{\partial x_1\partial x_3}$$
$$-\beta_{33}\left(\frac{\partial\theta}{\partial x_3}+\tau_1\frac{\partial^2\theta}{\partial x_3\partial t}\right)=\rho\frac{\partial^2 u_3}{\partial t^2} \quad (11)$$

$$K_{11}\frac{\partial^2\theta}{\partial x_1^2}+K_{33}\frac{\partial^2\theta}{\partial x_3^2}-\rho C_e\left(\frac{\partial\theta}{\partial t}+\tau_0\frac{\partial^2\theta}{\partial t^2}\right)=\theta_0\frac{\partial}{\partial t}\left(1+\tau_0\Omega\frac{\partial}{\partial t}\right)\left\{\beta_{11}\frac{\partial u_1}{\partial x_1}+\beta_{33}\frac{\partial u_3}{\partial x_3}\right\} \quad (12)$$

where,

$$\beta_{11}=C_{11}\alpha_{11}+C_{12}\alpha_{22}+C_{13}\alpha_{33}+C_{14}\alpha_{23}+C_{15}\alpha_{13}+C_{16}\alpha_{12}$$
$$\beta_{33}=C_{13}\alpha_{11}+C_{23}\alpha_{22}+C_{33}\alpha_{33}+C_{34}\alpha_{23}+C_{35}\alpha_{13}+C_{36}\alpha_{12}$$

## 3 Boundary Conditions

The boundaries of layer are assumed as rigid and adiabatic and are mathematically expressed as

$$u_1=u_3=0,\ \frac{\partial T}{\partial x_3}=0 \text{ at } x_3=-\frac{h}{2} \text{ and } x_3=\frac{h}{2} \quad (13)$$

## 4 Solution of the Problem

The respective solution may be taken in the form

$$\{u_1,u_3,\theta\}(x_1,x_3,t)=(A,B,C)\exp\left[\{ik(-sx_3+x_1-ct)\}\right] \quad (14)$$

Now, placing the above solution in (10), (11) and (12), the equations transformed as

$$\left[s^2-2is\xi_1-\xi_2\eta_1^2\right]A+\left[\xi_3 s^2-i\xi_4 s-\xi_1\right]B+\left[-\xi_6\left(\frac{i}{k}+\tau_1 c\right)\right]C=0 \quad (15)$$

$$\left[\xi_3 s^2-i\xi_4 s-\xi_1\right]A+\left[\xi_5 s^2-2i\xi_3 s-\eta_2^2\right]B+\left[s\xi_7\left(\frac{1}{k}+i\tau_1 c\right)\right]C=0 \quad (16)$$

$$\left[\xi_8 c\left(1-ikc\tau_0\Omega\right)\right]A+\left[\xi_9 c\left(i+kc\tau_0\Omega\right)s\right]B+\left[-s^2+\xi_{10}-\frac{c}{\nu}\left(\frac{i}{k}+\tau_0 c\right)\right]C=0 \quad (17)$$

where, $\xi_1=\frac{C_{15}}{C_{55}}$; $\xi_2=\frac{C_{11}}{C_{55}}$; $\xi_3=\frac{C_{35}}{C_{55}}$; $\xi_4=\frac{(C_{13}+C_{55})}{C_{55}}$; $\xi_5=\frac{C_{33}}{C_{55}}$; $\xi_6=\frac{\beta_{11}}{C_{55}}$; $\xi_7=\frac{\beta_{33}}{C_{55}}$; $\xi_8=\frac{\theta_0\beta_{11}}{K_{33}}$; $\xi_9=\frac{\theta_0\beta_{33}}{K_{33}}$; $\xi_{10}=\frac{K_{11}}{K_{33}}$; $\nu=\frac{\rho C_e}{K_{33}}$; $\eta_1^2=1-\frac{c^2}{c_1^2}$; $\eta_2^2=1-\frac{c^2}{c_2^2}$; $c_1^2=\frac{C_{11}}{\rho}$; $c_2^2=\frac{C_{55}}{\rho}$.

The trivial solution of simultaneous homogeneous equations (15), (16) and (17), we should have $|a_{ij}|=0$, $(i,j=1,2,3)$ where, $a_{ij}$ $(i,j=1,2,3)$ is mentioned in Appendix 1, from which we obtain an equation in s of order 6 as

$$a_0 s^6 + a_1 s^5 + a_2 s^4 + a_3 s^3 + a_4 s^2 + a_5 s + a_6 = 0 \tag{18}$$

where, $a_i$ ($\{i = 0, 1, \ldots, 6\}$) mentioned in Appendix 2. Let us take roots of (18) $s_j$ ($\{j = 1, \ldots, 6\}$) subject to to $s = s_j$, the ratio of the component $U_{1j}$, $U_{3j} U_{1j}$ and $\Theta_j$ is

$$\frac{U_{3j}}{U_{1j}} = \frac{B_j}{A_j} = \frac{a_{13,j} a_{21,j} - a_{11,j} a_{23,j}}{a_{12,j} a_{23,j} - a_{13,j} a_{22,j}} = \delta_j \tag{19}$$

$$\frac{\Theta_j}{U_{1j}} = \frac{C_j}{A_j} = \frac{a_{11,j} a_{22,j} - a_{12,j} a_{21,j}}{a_{12,j} a_{23,j} - a_{13,j} a_{22,j}} = \gamma_j \tag{20}$$

where, $a_{ij,k}$ $(i, j = 1, 2, 3; k = 1, \ldots, 6)$ mentioned in Appendix 3.
Therefore, the final solution for displacement component and temperature function of (10), (11) and (12) is

$$u_1 (x_1, x_3, t) = \sum_{j=1}^{6} A_j \exp\left[-k s_j x_3 + ik (x_1 - ct)\right] \tag{21}$$

$$u_3 (x_1, x_3, t) = \sum_{j=1}^{6} \delta_j A_j \exp\left[-k s_j x_3 + ik (x_1 - ct)\right] \tag{22}$$

$$\theta (x_1, x_3, t) = \sum_{j=1}^{6} \gamma_j A_j \exp\left[-k s_j x_3 + ik (x_1 - ct)\right] \tag{23}$$

Using boundary condition (13) and (21), (22) and (23), we get six simultaneous homogeneous equations in $A_j$ $(j = 1, \ldots, 6)$.

$$\sum_{j=1}^{6} \exp\left[\frac{k s_j h}{2}\right] A_j = 0; \qquad \sum_{j=1}^{6} \exp\left[\frac{k s_j h}{2}\right] \delta_j A_j = 0$$
$$\sum_{j=1}^{6} \gamma_j s_j \exp\left[\frac{k s_j h}{2}\right] A_j = 0; \qquad \sum_{j=1}^{6} \exp\left[-\frac{k s_j h}{2}\right] A_j = 0$$
$$\sum_{j=1}^{6} \exp\left[-\frac{k s_j h}{2}\right] \delta_j A_j = 0; \qquad \sum_{j=1}^{6} \gamma_j s_j \exp\left[-\frac{k s_j h}{2}\right] A_j = 0$$

Form the above equation, the non-trivial solution exists if the determinant of the coefficient is zero, i.e.

$$\left|M_{ij}\right| = 0, (i, j = 1, \ldots, 6) \tag{24}$$

where, $M_{ij}$ $(i, j = 1, \ldots, 6)$ is the elements of coefficient-matrix of the above equations and mentioned in the Appendix 4. The dispersion relation for the propagation of Rayleigh wave obtained from the real part of (24).

## 5 Numerical Results and Discussion

The following data is taken in consideration for obtaining the numerical results (Refs. [12, 20]) $K_{11} = K_{33} = 1.24\times 10^2\,\mathrm{Wm^{-1}deg^{-1}}\,T_0 = 296^o\,\mathrm{K}$, $\tau_0 = 0.04$, $\tau_1 = 0.05$., $C_{11} = 106.8\times 10^9\mathrm{Pa}$, $C_{22} = 99.0\times 10^9\mathrm{Pa}$, $C_{33} = 54.57\times 10^9\mathrm{Pa}$, $C_{12} = 27.1\times 10^9\mathrm{Pa}$, $C_{13} = 9.68\times 10^9\mathrm{Pa}$, $C_{14} = -0.03\times 10^9\mathrm{Pa}$, $C_{15} = 0.28\times 10^9\mathrm{Pa}$, $C_{16} = 0.12\times 10^9\mathrm{Pa}$, $C_{23} = 18.22\times 10^9\mathrm{Pa}$, $C_{24} = 1.49\times 10^9\mathrm{Pa}$, $C_{25} = 0.13\times 10^9\mathrm{Pa}$, $C_{26} = -0.58\times 10^9\mathrm{Pa}C_{34} = 2.44\times 10^9\mathrm{Pa}$, $C_{35} = -1.69\times 10^9\mathrm{Pa}$, $C_{36} = -0.75\times 10^9\mathrm{Pa}$, $C_{44} = 25.97\times 10^9\mathrm{Pa}$, $C_{45} = 1.98\times 10^9\mathrm{Pa}$, $C_{46} = 0.43\times 10^9\mathrm{Pa}$, $C_{55} = 25.05\times 10^9\mathrm{Pa}$, $C_{66} = 37.82\times 10^9\mathrm{Pa}$, $C_{56} = 1.44\times 10^9\mathrm{Pa}$, $\rho = 2.727\times 10^3\,\mathrm{kg/m^3}$, $\beta_{11} = 5.75\times 10^6\,\mathrm{Nm^{-2}deg^{-2}}$, $\beta_{33} = 5.17\times 10^6\,\mathrm{Nm^{-2}deg^{-2}}$, $C_e = 3.9\times 10^2\mathrm{J\,kg^{-1}deg^{-1}}$.

The graphs are drawn between non-dimensional real and imaginary phase velocities and wave number for various thickness values of layer for all three theories. In Fig. 2, the graphs drawn for CD theory for dimensionless phase velocities $\frac{c}{c_1}$ with dimensionless wave number kh. It can be seen that from Fig. 2, as the wave number increases both real and imaginary phase velocities decreases. As we increase the thickness of layer $h(= 1.0\,\mathrm{Km}, 1.2\,\mathrm{Km}, 1.4\,\mathrm{Km})$, then also both real and imaginary velocities decrease consistently for all $kh$ and does not meet any curves. In Fig. 3, we noted that the nature of curves is similar to those of the Fig. 2, but the marginal changes in phase velocity have been observed with the wavenumber and layer thickness. In Fig. 3, the plots drawn for LS theory for dimensionless phase velocities $\frac{c}{c_1}$ with dimensionless wave number kh. From the figure, it is elucidated that the nature of the curve followed by both real and imaginary phase velocities are the same as in CD theory, but there is minimal change in the phase velocities for both real and imaginary that can be significantly identified. In Fig. 4, the graphs were drawn for GL theory for dimensionless phase velocities $\frac{c}{c_1}$ with $kh$. The same trends for graphs can be also viewed in GL theory as CD theory and LS theory.

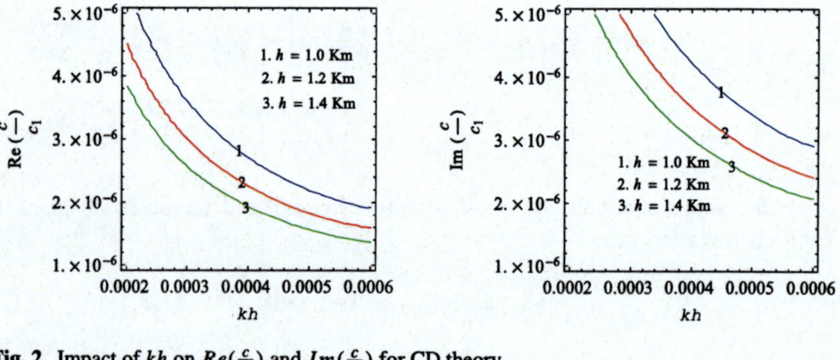


**Fig. 2** Impact of $kh$ on $Re(\frac{c}{c_1})$ and $Im(\frac{c}{c_1})$ for CD theory

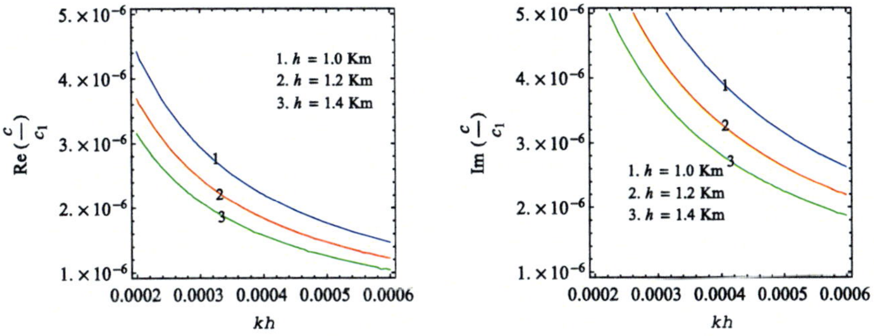


**Fig. 3** Impact of $kh$ on $Re(\frac{c}{c_1})$ and $Im(\frac{c}{c_1})$ for LS theory

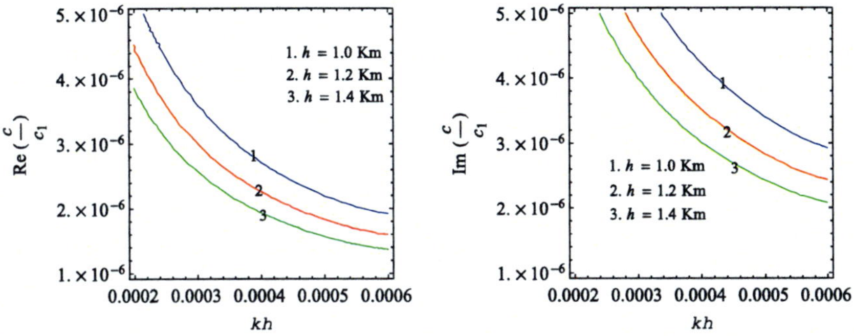


**Fig. 4** Impact of $kh$ on $Re(\frac{c}{c_1})$ and $Im(\frac{c}{c_1})$ for GL theory

# 6 Conclusion

The propagation of the Rayleigh wave in a layer with rigid and adiabatic interfaces has been incorporated. The studies were made jointly with single heat conduction equation for three theories of generalized thermo-elasticity. It is assumed that the anisotropy of the layer is tri-clinic. The theories of generalized thermo-elasticity which are taken in account are CD theory, LS theory and GL theory with two thermal relaxation times. An analytical solution obtained and procured a dispersion relation subjected to boundaries as rigid and insulated. In order to deal with the numerical and physical scenario of the problem, a particular model is taken into account. The results were interpreted in terms of graphs. From the graphs, it is noted that phase velocity decreases as wavenumber increases for all theories. Also, there is a significant effect of thickness that can be observed from the graphs. As we increase the thickness of layer, the phase velocity of wave decreases. Thus, thickness of layer plays a vital role put the impact on the velocities of waves.

## Appendix 1

$$a_{11} = s^2 - 2is\xi_1 - \xi_2\eta_1^2 \; a_{12} = \xi_3 s^2 - i\xi_4 s - \xi_1 \quad a_{13} = -\xi_6\left(\tfrac{i}{k} + \tau_1 c\right)$$
$$a_{21} = \xi_3 s^2 - i\xi_4 s - \xi_1 \quad a_{22} = \xi_5 s^2 - 2i\xi_3 s - \eta_2^2 \; a_{23} = s\xi_7\left(\tfrac{1}{k} + i\tau_1 c\right)$$
$$a_{31} = \xi_8 c\,(1 - ikc\tau_0\Omega) \; a_{32} = \xi_9 c\,(i + kc\tau_0\Omega)\,s \; a_{33} = -s^2 + \xi_{10} - \tfrac{c}{\nu}\left(\tfrac{i}{k} + \tau_0 c\right)$$

## Appendix 2

$$a_0 = \xi_3^2 - \xi_5,\; a_1 = 2i\xi_3 - \xi_3\xi_4 + 2i\xi_1\xi_5$$

$$a_2 = 2\xi_1\xi_3 + \tfrac{ic\xi_3^2}{kv} - \xi_{10}\xi_3^2 - \tfrac{ic\xi_5}{kv} + \xi_{10}\xi_5 - \tfrac{ic\xi_7\xi_9}{k} + \xi_2\xi_5\eta_1^2 + \eta_2^2 + \tfrac{c^2\xi_3^2\tau_0}{v} - \tfrac{c^2\xi_5\tau_0}{v}$$
$$- \tfrac{c^2k\omega\xi_7\xi_9\tau_0}{k} + c^2\xi_7\xi_9\tau_1 - ic^3k\omega\xi_7\xi_9\tau_0\tau_1$$

$$a_3 = -\tfrac{2c\xi_3}{kv} - 2i\xi_{10}\xi_3 + 2i\xi_1\xi_4 + \tfrac{2c\xi_3\xi_4}{kv} + 2i\xi_{10}\xi_3\xi_4 - \tfrac{2c\xi_1\xi_5}{kv} - 2i\xi_1\xi_{10}\xi_5 + \tfrac{c\xi_3\xi_7\xi_8}{k} - \tfrac{c\xi_3\xi_6\xi_9}{k}$$
$$- \tfrac{2c\xi_1\xi_7\xi_9}{k} - 2i\xi_2\xi_3\eta_1^2 - 2i\xi_1\eta_2^2 + \tfrac{2ic^2\xi_3\tau_0}{v} - \tfrac{2ic^2\xi_3\xi_4\tau_0}{v} + \tfrac{2ic^2\xi_1\xi_5\tau_0}{v} - \tfrac{ickc\omega\xi_3\xi_7\xi_8\tau_0}{k}$$
$$+ \tfrac{ickm\omega\xi_3\xi_6\xi_9\tau_0}{k} + \tfrac{2ickc\omega\xi_1\xi_7\xi_9\tau_0}{k} + ic^2\xi_3\xi_7\xi_8\tau_1 + ic^2\xi_3\xi_6\xi_9\tau_1 - 2ic^2\xi_1\xi_7\xi_9\tau_1$$
$$+ c^2kc\omega\xi_3\xi_7\xi_8\tau_0\tau_1 + c^2kc\omega\xi_3\xi_6\xi_9\tau_0\tau_1 - 2c^2kc\omega\xi_1\xi_7\xi_9\tau_0\tau_1$$

$$a_4 = \xi_1^2 + \tfrac{2ic\xi_1\xi_3}{kv} - 2i\xi_1\xi_{10}\xi_3 - \tfrac{ic\xi_4^2}{kv} + \xi_{10}\xi_4^2 - \tfrac{ic\xi_5\xi_6\xi_8}{k} - \tfrac{ic\xi_4\xi_7\xi_8}{k} + \tfrac{ic\xi_4\xi_6\xi_9}{k} + \tfrac{ic\xi_2\xi_5\eta_1^2}{kv} - \xi_{10}\xi_2\xi_5\eta_1^2$$
$$+ \tfrac{ic\xi_2\xi_7\xi_9\eta_1^2}{k} - \tfrac{ic\eta_2^2}{kv} - \xi_{10}\eta_2^2 - \xi_2\eta_1^2\eta_2^2 + \tfrac{2c^2\xi_1\xi_3\tau_0}{v} - \tfrac{c^2\xi_4^2\tau_0}{v} - \tfrac{ckc\omega\xi_5\xi_6\xi_8\tau_0}{k} - \tfrac{ckc\omega\xi_4\xi_7\xi_8\tau_0}{k}$$
$$+ \tfrac{ckc\omega\xi_4\xi_6\xi_9\tau_0}{k} + \tfrac{c^2\xi_2\xi_5\eta_1^2\tau_0}{v} + \tfrac{ckc\omega\xi_2\xi_7\xi_9\eta_1^2\tau_0}{k} + \tfrac{c^2\eta_2^2\tau_0}{v} - c^2\xi_5\xi_6\xi_8\tau_1 + c^2\xi_4\xi_7\xi_8\tau_1 + c^2\xi_4\xi_6\xi_9\tau_1$$
$$- c^2\xi_2\xi_7\xi_9\eta_1^2\tau_1 + ic^2kc\omega\xi_5\xi_6\xi_8\tau_0\tau_1 - ic^2kc\omega\xi_4\xi_7\xi_8\tau_0\tau_1 - ic^2kc\omega\xi_4\xi_6\xi_9\tau_0\tau_1 + ic^2kc\omega\xi_2\xi_7\xi_9\eta_1^2\tau_0\tau_1$$

$$a_5 = -\tfrac{2ic\xi_1\xi_4}{kv} - 2i\xi_1\xi_{10}\xi_4 - \tfrac{2c\xi_3\xi_6\xi_8}{k} - \tfrac{c\xi_1\xi_7\xi_8}{k} + \tfrac{c\xi_1\xi_6\xi_9}{k} + \tfrac{2c\xi_2\xi_3\eta_1^2}{k} + 2i\xi_{10}\xi_2\xi_3\eta_1^2 + \tfrac{2c\xi_1\eta_2^2}{kv}$$
$$+ 2i\xi_1\xi_{10}\eta_2^2 + \tfrac{2ic^2\xi_1\xi_4\tau_0}{v} + \tfrac{2ic^2k\omega\xi_1\xi_7\xi_8\tau_0}{k} + \tfrac{ic^2k\omega\xi_1\xi_7\xi_8\tau_0}{k} - \tfrac{2ic^2k\omega\xi_1\xi_7\xi_8\tau_0}{k} - \tfrac{ic^2k\omega\xi_1\xi_6\xi_9\tau_0}{k}$$
$$- \tfrac{2ic^2k\omega\xi_2\xi_3\eta_1^2\tau_0}{k} - \tfrac{2ic^2\xi_1\eta_2^2\tau_0}{k} + 2ic^2\xi_3\xi_6\xi_8\tau_1 - ic^2\xi_1\xi_7\xi_8\tau_1 - ic^2\xi_1\xi_6\xi_9\tau_1 + 2c^3k\omega\xi_3\xi_6\xi_8\tau_0\tau_1$$
$$- c^3k\omega\xi_1\xi_7\xi_8\tau_0\tau_1 - c^3k\omega\xi_1\xi_6\xi_9\tau_0\tau_1$$

$$a_6 = \tfrac{ic\xi_1^2}{kv} - \xi_1^2\xi_{10} + \tfrac{ic\xi_6\xi_8\eta_2^2}{k} + \xi_{10}\xi_2\eta_1^2\eta_2^2 + \tfrac{c^2\xi_1^2\tau_0}{v} + \tfrac{c^2k\omega\xi_6\xi_8\eta_2^2\tau_0}{k} - \tfrac{c^2\xi_2\eta_1^2\eta_2^2\tau_0}{v} + c^2\xi_6\xi_8\eta_2^2\tau_1$$
$$- ic^3k\omega\xi_6\xi_8\eta_2^2\tau_1\tau_0$$

## Appendix 3

$$a_{11,j} = s_j^2 - 2is_j\xi_1 - \xi_2\eta_1^2,\; a_{12,j} = \xi_3 s_j^2 - i\xi_4 s_j - \xi_1, a_{13,j} = -\xi_6\left(\tfrac{i}{k} + \tau_1 c\right),$$
$$a_{21,j} = \xi_3 s_j^2 - i\xi_4 s_j - \xi_1 a_{22,j} = \xi_5 s_j^2 - 2i\xi_3 s_j - \eta_2^2,\; a_{23,j} = s_j\xi_7\left(\tfrac{1}{k} + i\tau_1 c\right), (j = 1, \ldots, 6)$$

# Appendix 4

$$M_{1j} = \exp\left[\frac{ks_j h}{2}\right],\ M_{2j} = \delta_j \exp\left[\frac{ks_j h}{2}\right],\ M_{3j} = \gamma_j s_j \exp\left[\frac{ks_j h}{2}\right],$$
$$M_{4j} = \exp\left[-\frac{ks_j h}{2}\right],\ M_{5j} = \delta_j \exp\left[-\frac{ks_j h}{2}\right],\ M_{6j} = \gamma_j s_j \exp\left[-\frac{ks_j h}{2}\right],\ (j = 1, \ldots, 6)$$

# References


1. Rayleigh L (1885) On waves propagated along the plane surface of an elastic solid. Proc Lond Math Soc s1-17(1):4–11
2. Wilson JT (1942) Surface waves in a heterogeneous medium. Bull Seismol Soc Am 32:297–305
3. Newlands M (1950) Rayleigh waves in a two layer heterogeneous medium. Mon Notices R Astron Soc Geophys Suppl 6:S2109
4. Stonely R (1934) The transmission of Rayleigh waves in a heterogeneous medium. Mon Notices R Astron Soc Geophys Suppl 3:S222
5. Gupta S, Chattopadhayay A, Vishwakarma SK, Majhi DK (2011) Influence of rigid boundary and initial stress on the propagation of Love wave. Appl Math 2:586–594
6. Vishwakarma SK, Gupta S, Majhi DK (2013) Influence of rigid boundary on the Love wave propagation in elastic layer with void pores. Acta Mechanica Solida Sinica 26(5):551–558
7. Lord HW, Shulman YA (1967) Generalized dynamical theory of thermo-elasticity. J Mech Phys Solids 15:299–309
8. Fox N (1969) Generalised thermoelasticity. Int J Eng Sci 7:437–445
9. Green AE, Lindsay KA (1972) Thermoelasticity. J Elast 2:1–7
10. Ivanov TP (1988) On the propagation of thermoelastic Rayleigh waves. Wave Motion 10:73–82
11. Sharma JN (2004) Pal M: Rayleigh-Lamb waves in magneto-thermoelastic homogeneous isotropic plate. Int J Eng Sci 42:137–155
12. Kumar S, Pal PC (2014) Wave propagation in an inhomogeneous anisotropic generalized thermoelastic solid under the effect of gravity. Comput Therm Sci 6:241–250
13. Pal PC, Kumar S, Mandal D (2014) Wave propagation in an inhomogeneous anisotropic generalized thermoelastic solid. J Therm Stress 37:817–831
14. Biswas S, Mukhopadhyay B, Shaw S (2017) Rayleigh surface wave propagation in orthotropic thermoelastic solids under three-phase-lag model. J Therm Stress 40:403–419
15. Yadav D, Kim MC (2016) Theoretical and numerical analyses on the onset and growth of convective instabilities in a horizontal anisotropic porous medium. J Porous Media 17(12):1061–1074
16. Yadav D (2020) Numerical examination of the thermal instability in an anisotropic porous medium layer subjected to rotation and variable gravity field. Special Top Rev Porous Media Int J 11(4):395–407
17. Yadav D (2020) The density-driven nanofluid convection in an anisotropic porous medium layer with rotation and variable gravity field: a numerical investigation. J Appl Comput Mech 6(3):699–712
18. Yadav D (2020) Numerical solution of the onset of Buoyancy?driven nanofluid convective motion in an anisotropic porous medium layer with variable gravity and internal heating. Heat Trans Asian Res 49(3):1170–1191
19. Yadav D, Mohamad AM, Rana GC (2020) Effect of throughflow on the convective instabilities in an ?anisotropic porous medium layer with inconstant gravity. J Appl Comput Mech (2020). https://doi.org/10.22055/jacm.2020.32381.2006
20. Rasolofosaon PNJ, Zinszner BE (2002) Comparison between permeability anisotropy and elasticity anisotropy of reservoir rocks. Geophysics 67:230–240

# Nonlinear Equation for Wave Motion with Undulated Bottom

**M. Muniyappan and S. K. Mohanty**

## 1 Introduction

The study of gravity wave is attracted by Mathematicians, Physicists, and Engineers for several decades. Craik [1] studied the interaction of waves and fluid flows. Current generated due to the change of wind speed and direction and Coriolis effect plays a significant role in the gravity wave motion. Peregrine [2] studied the gravity wave interaction with current, and Jonsson and Wang [3] analyzed the wave propagation in a running stream by assuming that the sea bed is undulated in nature. Mohanty et al. [4] discussed the effect of flexural gravity wave motion in the presence of current in single and two layer fluid using Green's function technique. Also, the small undulated bottom plays an important role in gravity wave motion. Bora and Martha [5] analyzed the gravity wave motion with undulated bottom using Green's function technique. Martha et al. [6] discussed the wave motion with undulated permeable bottom for obliquely incident wave. Mohanty [7] discussed the effect of current, surface tension, and porosity parameter on reflection and transmission coefficients.

All the above works are done under the assumption of small amplitude water wave theory. First-order amplitude theory for gravity wave motion is discussed by Lokenath [8]. The homotopy analysis method is an important technique for solving the nonlinear partial differential equation and is studied by Liao [9]. Liu et al. [10] studied the nonlinear interaction between unidirectional bichromatic wave train and exponentially sheared current by considering the water depth as infinite. It may be noted that the wave current interaction in the presence of undulated permeable bottom is not addressed in the literature.

In the present study, the gravity wave motion in the presence of current and undulated permeable bottom is examined using HAM, perturbation technique, and Fourier

M. Muniyappan · S. K. Mohanty (✉)
Department of Mathematics, SAS, VIT, Vellore, India
e-mail: sanjaymath@gmail.com



S. Srinivas et al. (eds.), *Recent Advances in Applied Mathematics and Applications to the Dynamics of Fluid Flows*, Lecture Notes in Mechanical Engineering,
https://doi.org/10.1007/978-981-19-1929-9_15

transform technique. The effect of current and porosity parameter on phase and group velocities are analyzed in the case of progressive wave solution. The velocity potential and surface elevations are derived using the higher order deformation equation and HAM. The effect of current and porosity parameter on reflection coefficients are analyzed in the presence of a small undulated permeable bed.

## 2 Mathematical Formulation

In the present study, the interaction between the nonlinear equation for gravity waves motion in a running stream with infinitely deep permeable bottom is discussed in the two dimensional Cartesian co-ordinate system with $x$-axis is in the horizontal direction and $z$-axis is positive in the vertically downward direction. Here, the bottom is of an infinitely deep porous layer with undulated bottom topography and a uniform current is flowing with current speed $U$ in the fluid region. The fluid domain occupies the region $-\infty < x < \infty$, $0 \leq z \leq h + \epsilon C(x)$ with $h$ is the water depth, $\epsilon$ being a small undulated parameter and $C(x)$ is of with compact support as in Fig. 1. The existence of the total velocity potential $\Phi(x, z, t) = Ux + \phi(x, z, t)$ is evident from the assumed nature of fluid which is not compressible, not viscous, and irrotational. The potential function of this type satisfies the two dimensional Laplace equation

$$\phi_{xx} + \phi_{zz} = 0, \text{ in the fluid region,} \tag{1}$$

The nonlinear kinematic and dynamic boundary conditions near the free surface are derived as [10]

$$\phi_z = \zeta_t + \phi_x \zeta_x + U\zeta_x, \text{ on } z = 0, \tag{2}$$

$$\phi_t + U\phi_x + \frac{1}{2}|\nabla\phi|^2 = g\zeta, \text{ on } z = 0, \tag{3}$$

where g is the acceleration due to gravity. Also, it is assumed that the bottom is undulated and permeable in nature. So the boundary condition for permeable bottom topography with porosity parameter $G$ is given by

$$\phi_z = G\phi, \ z = h + \epsilon C(x). \tag{4}$$

Before analyzing the nonlinear gravity wave motion, the progressive wave solution is discussed.

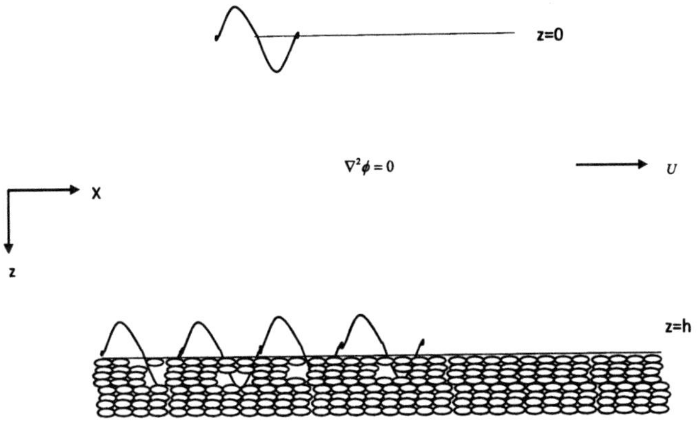


**Fig. 1** Schematic diagram for gravity wave motion with undulated permeable bed

# 3 Progressive Wave Solution

The velocity potential $\phi(x, z, t)$ is assumed to be time periodic in nature and is given by $\phi(x, z, t) = Re[\psi(x, z)e^{-i\omega t}]$, with $Re$ being the real part, $\omega$ is the circular frequency. The progressive wave solution for the velocity potential and plate deflection are derived using the method (similar as [7]) and is given by

$$\psi_n(x, z) = \alpha_n \frac{\sinh nk(h - z) - G \cosh nk(h - z)}{nk \sinh nkh - G \cosh nkh} e^{inx}, \tag{5}$$

$$\zeta_n(x) = 0. \tag{6}$$

Here, the wave number $k$ satisfies the dispersion relation

$$(\omega - Uk)^2(k - G \tanh kh) - k(k \tanh kh - G) = 0. \tag{7}$$

For the clarity of the presentation, the units and dimensions of different physical parameters are removed form legends and captions of the figure. In Fig. 2, the variation of group velocities for several values of current speed and porosity parameter are plotted. It is noted that the wave energy propagates faster for higher values of porosity parameter in a fixed range of wave number. A similar observation in the case of flexural gravity wave motion in the presence of current is discussed in [4] in the case of rigid bottom.

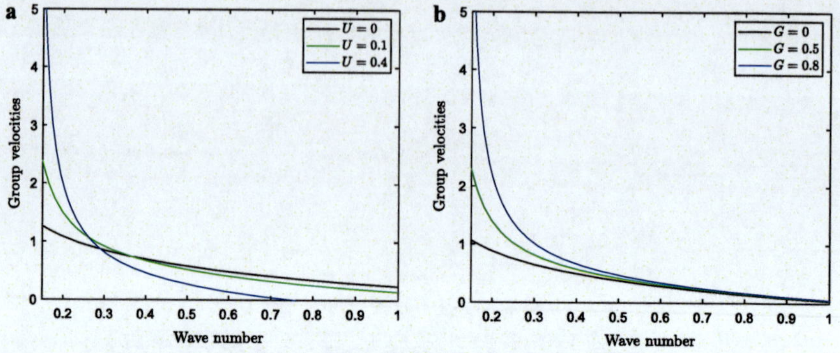


**Fig. 2** Variation of group velocity for several values of **a** current speed **b** porosity parameter

### 3.1 HAM for Wave-Current Interaction with Permeable Bottom

In this subsection, the HAM for wave current interaction is used to derive the velocity potential and surface elevation. The velocity potential for the progressive waves derived in Eq. (5) is used for the initial approximation and is chosen as

$$\psi_{0,1}(x,z) = \alpha_{0,1}[(\sinh k_0(h-z) - G\cosh k_0(h-z))/(k_0 \sinh k_0 h - G\cosh k_0 h)]e^{ik_0 x}.$$

The initial surface elevation is chosen as $\zeta_0(x) = 0$ with $k_0$ being the real root of the dispersion relation. From Eqs. (2) and (3), the nonlinear operators are constructed in the following manner:

$$\mathcal{F}_1[\phi(Y,z;p),\beta(Y;p)] = \omega(\omega - Uk)\phi_{YY}(Y,z;p) - g\phi_z(Y,z;p) - gk\beta_Y(Y;p) \times \left[k\phi_Y(Y,z;p) + U_1\right] - \omega f_Y, \quad (8)$$

$$\mathcal{F}_2[\phi(Y,z;p),\beta(Y;p)] = -\omega\phi_Y(Y,z;p) - g\beta(Y;p) + kU\phi_Y(Y,z;p) + f. \quad (9)$$

The corresponding linear operators are constructed as

$$\mathcal{G}_1[\phi(Y,z;p),\beta(Y;p)] = \omega^2\phi_{YY}(Y,z;p) - g\phi_z(Y,z;p), \quad (10)$$

$$\mathcal{G}_2[\phi(Y,z;p),\beta(Y;p)] = \frac{\partial^4\beta(Y;p)}{\partial Y^4} + \beta(Y;p), \quad (11)$$

where the circular frequency $\omega$ is given by Eq. (7). The construction of zeroth- order deformation is quite tricky and is chosen by keeping an eye on the convergence of the method. From Eqs. (10)–(11), the zeroth-order deformation equation is derived as

$$k^2\phi_{YY}(Y,z;p)+\phi_{zz}(Y,z;p)=0, \text{ in the fluid region,} \tag{12}$$

$$\phi_z(Y,z;p)=G\phi(Y,z;p),\ z=h, \tag{13}$$

$$(1-p)\mathcal{G}_1[\phi(Y,z;p)-\psi_0(Y,z)]=pc_0\mathcal{F}_1[\phi(Y,z;p),\beta(Y;p)], \text{ on } z=\beta(Y;p), \tag{14}$$

$$(1-p)\mathcal{G}_2[\beta(Y;p)-\zeta_0(Y)]=pc_0\mathcal{F}_2[\phi(Y,z;p),\beta(Y;p)], \text{ on } z=\beta(Y;p), \tag{15}$$

with $c_0$ being the convergence control parameter. It may be noted that the above defined function $\phi(Y,z;p)$ describes the variation of the solutions from the initial solutions $\psi_0(Y,z)$ to the exact solution $\phi(Y,z;p)$ when $p$ varies from 0 to 1. In a similar way, the function $\beta(Y;p)$ describes the variation of the solutions from the initial solutions $\zeta_0(Y)$ to the exact solution $\beta(Y)$ when $p$ varries from 0 to 1. Using Taylor's series expansion on $\phi(Y,z;p)$ and $\beta(Y;p)$ at $p=0$, it is derived that

$$\phi(Y,z;p)=\psi_0(Y,z)+\sum_{j=1}^{\infty}\psi_j(Y,z)p^j, \tag{16}$$

$$\beta(Y;p)=\zeta_0(Y)+\sum_{j=1}^{\infty}\zeta_j(Y)p^j, \tag{17}$$

$$\text{where } \zeta_j(Y)=\frac{1}{j!}\frac{\partial^j}{\partial p^j}\Big[\beta(Y;p)\Big]_{p=0},\quad \psi_j(Y,z)=\frac{1}{j!}\frac{\partial^j}{\partial p^j}\Big[\phi(Y,z;p)\Big]_{p=0}.$$

Here $c_0$ is choosen properly so that at $p=1$, the Eqs. (16) and (17) converge. Using the Eqs. (16) and (17) and zeroth-order derformation equation and equating the like power of $p$, the higher order deformation equations are derived as

$$\begin{aligned}\mathcal{G}_1(\psi_n)-\chi_n\mathcal{G}_1(\psi_{n-1})=c_0\Big\{&\omega(\omega-Uk)[(\psi_{n-1})_{YY}]-g(\psi_{n-1})_z\\&-gk^2\sum_{k=0}^{n-1}(\psi_k)_Y[(\zeta_{n-k-1})_Y]-2\omega\sum_{k=0}^{n-1}(\psi_k)_Y[(\psi_{n-k-1})_{YY}]\\&-2\omega\sum_{k=0}^{n-1}(\psi_k)_z[(\psi_{n-k-1})_{Yz}]+gkU(\zeta_{n-1})_Y\Big\},\end{aligned} \tag{18}$$

$$\begin{aligned}\mathcal{L}_2(\zeta_n)-\chi_n\mathcal{L}_2(\zeta_{n-1})=c_0\Big\{&-\omega(\psi_n)_Y-g\zeta_{n-1}-\frac{1}{2}\sum_{k=0}^{n-1}(\psi_k)_Y[(\psi_{n-k-1})_Y]\\&+(\psi_k)_z[(\psi_{n-k-1})_z]+k_x[U_1(\psi_{n-1})_Y]\Big\},\end{aligned} \tag{19}$$

where $\chi_n$ is 0 for $n\le 1$ and 1 for $n>1$. The velocity potential and surface elevation are obtained from the above decoupled system of differential equations by using the initial solution and the inverse of the linear operator. Next, the convergence of the above defined series are discussed. To guarantee the precision of HAM-based series solutions, the total square error is defined as follows:

$$\epsilon_n^T = \epsilon_{n,1} + \epsilon_{n,2}, \tag{20}$$

where $\epsilon_{n,1}$ and $\epsilon_{n,2}$ are denoted as residual square errors and, respectively, defined as

$$\epsilon_{n,1} = \frac{1}{(1+N)^2}\sum_{i=0}^{N}\sum_{j=0}^{N}(\mathcal{G}_1[\phi(y,\varphi;p)]|_{y=i\delta y,\varphi=j\delta\varphi})^2, \tag{21}$$

$$\epsilon_{n,2} = \frac{1}{(1+N)}\sum_{i=0}^{N}(\mathcal{G}_2[\phi(y,\varphi;p),\delta(p),\gamma(p)]|_{y=i\delta y,\varphi=0})^2, \tag{22}$$

and $\delta y = \delta\varphi = \frac{\pi}{N}$. By choosing larger values of $N$, the convergence control parameter $c_0$ obtained by the least esteem of $\epsilon_n^T$.

## 4 Wave Motion with Small Undulation Sinusoidal Bottom Topography

Here, the case of small bottom undulation is considered and the effect of porosity parameter and uniform current on reflection coefficient is discussed. The surface elevation and velocity potential are assumed to be time periodic in nature and are of the form $\phi(x,z,t) = \varphi(x,z)e^{-i\omega t}$ and $\beta(x,t) = \gamma(x)e^{-i\omega t}$. The linearized form of Eqs. (10)–(11) for the case of small undulated bottom topography is derived as

$$\varphi_{xx} + \varphi_{zz} = 0, \quad \text{in } 0 \le z \le h + \epsilon C(x), \tag{23}$$

$$U\gamma_x - \mathrm{i}\omega\beta - \varphi_z = 0, \quad \text{on } z = 0, \tag{24}$$

$$g\gamma + \mathrm{i}\omega\varphi - U\varphi_x = 0, \quad \text{on } z = 0, \tag{25}$$

$$\varphi_n - G\varphi = 0, \quad \text{on } z = h + \epsilon C(x). \tag{26}$$

Also, the incidence wave reflects and transmitted partially over the undulated bottom. The radiation condition at far field is given by $\varphi(x,z) \to \varphi_i(x,z) + R\varphi_i(-x,z)$ as $x \to -\infty$, and $\varphi(x,z) \to T\varphi_i(x,z)$ as $x \to \infty$, where $R$ is the reflection and $T$ is the transmission coefficients and are to be determined. The mathematical technique used here is the perturbation method and is similar to [7]. Proceeding in a similar way as in [7], the velocity potential for the first order bottom topography is derived as

$$\varphi_1(x,z) = \frac{1}{2\pi}\int_0^\infty \left[\frac{q\cosh qz - (\omega - Uq)^2 \sinh qz}{qB1(q)}\right] B2(q,x)dq, \tag{27}$$

$$\text{where}\quad B1(q) = (q + G\omega_u^2/q)\sinh qh - (G + \omega_u^2)\cosh qh,$$
$$B2(q) = I_s(q)e^{iqx} + I_s(-q)e^{-iqx},\quad \omega_u = \omega - Uq,$$
$$(\psi_1)_z - G\psi_1 = \left\{C(x)(\psi_0)_x\right\}' + GC(x)(\psi_0)_z = I_s, \quad \text{on } z = h.$$

Now, the above integral is simplified using contour integration. The sinusoidal bottom topography which is non-vanishing on $-m\pi < x < m\pi$ and has the value $a_1\sin(x/\lambda)$ is considered. Here $\lambda$ is the wave number and $m$ is the number of ripples. For this specific case of bottom topography, the reflection and transmission coefficients are derived as

$$R_1 = -\frac{(\tilde{\alpha}_0 A_1 + D1(\alpha)G)}{B1'(\alpha_0)}\frac{2a_1\lambda(-1)^m}{1 - 4\alpha_0^2\lambda^2}\sin(2\pi\alpha_0^2\lambda^2), \tag{28}$$

with $A_1 = -\omega_u^2\sinh\alpha_0$, $B_1 = -\omega_u^2\cosh\alpha_0$, $\omega_u = \omega - F_r\alpha_0$, $B1' = \frac{d}{dk}B1$, $D1(k) = (k + G\omega_u^2/k)\sinh k - (G + \omega_u^2)\cosh k$ and of zero transmission coefficient. It may be noted that for $\lambda = 1/2\alpha_0$, the Bragg resonance occurs. For computational purpose, the values of $G = 0.2$, and $U = 0.2$ is fixed unless it is mentioned. In Fig. 3, the wave number versus reflection coefficients are plotted for several values of current speed and porosity parameter. It has been found that there is an increase in the maximum value of the reflection coefficient whenever the values of the porosity parameter increases. Also, it is found that the current speed has a small impact on reflection coefficients. It may be noted that using $U = 0$ the above results exactly matches with [5].

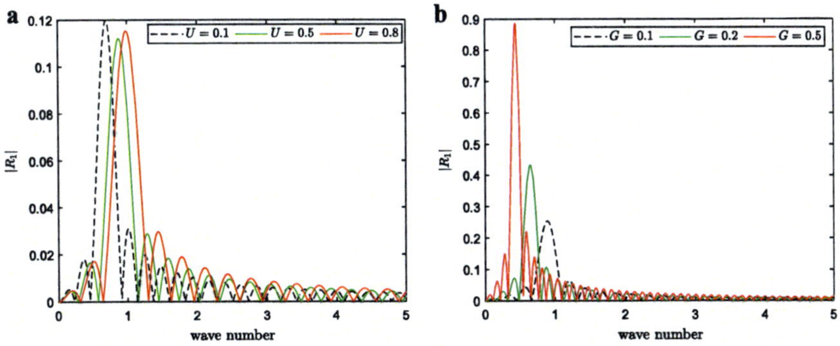


**Fig. 3** variation of reflection coefficient for several values of **a** current speed **b** porosity parameter

## 5 Conclusion

In this paper, the nonlinear gravity wave motion in the presence of current and undulated porous bed is discussed. For plane progressive wave, the group velocities are derived from dispersion relation. It is examined that the current has a significant impact on wave energy propagation. The associated surface elevation and velocity potential in case of finite-amplitude wave motion is derived using HAM. It is noted that the current and porosity parameter has a mild impact on the reflection coefficient in the case of small undulated bottom.

## References

1. Craik ADD (1985) Wave Interactions and Fluid Flows. Cambridge University Press, Cambridge
2. Peregrine DH (1976) Interaction of water waves and currents. Adv Appl Mech 16:9–117. https://doi.org/10.1016/S0065-2156(08)70087-5
3. Jonsson IV, Wang JD (1980) Current-depth refraction of water waves. Ocean Eng 7(1):153–171. https://doi.org/10.1016/0029-8018(80)90035-9
4. Mohanty SK, Mondal R, Sahoo T (2014) Time dependent flexural gravity waves in the presence of current. J Fluids Struct 45:28–49. https://doi.org/10.1016/j.jfluidstructs.2013.11.018
5. Bora SN, Martha SC (2008) Scattering of surface waves over an uneven sea-bed. Appl Math Lett 21(10):1082-1089 (2008). https://doi.org/10.1016/j.aml.2007.12.005
6. Martha SC, Bora SN, Chakrabarti A (2007) Oblique water-wave scattering by small undulation on a porous sea-bed. Appl Ocean Res 29(1):86–90. https://doi.org/10.1016/j.apor.2007.07.001
7. Mohanty SK (2021) Time-dependent wave motion with undulated bottom. Acta Mech. https://doi.org/10.1007/s00707-020-02838-w
8. Debnath L (1994) Nonlinear water waves. Elsevier Science, United Kingdom
9. Liao SJ (2003) Beyond perturbation: introduction to the homotopy analysis method. CRC Press, USA. https://doi.org/10.1201/9780203491164
10. Liu Z, Lin Z, Liao S (2014) Phase velocity effects of the wave interaction with exponentially sheared current. Wave Motion 51(6):967–985. https://doi.org/10.1016/j.wavemoti.2014.03.009

# Numerical Study on Nanofluid Flow and Heat Transfer Over a Thin Moving Needle with Arrhenius Pre-Exponential Factor Law and Slip Effect

**I. Sadham Hussain, D. Prakash, and Santosh Kumar**

## 1 Introduction

The activation energy is the energy that starts the reaction process. It plays a significant role in the application of dynamics of oil, geothermal reservoir engineering, water pigment, petroleum engineering applications, thermal reactions food processing, and many more. On the other hand, heat transfer in nanoliquids has gained considerable interest because of its extensive applications, such as biomedicine, chemical industry microelectronics, solar cells, nuclear reactors, processing of generic drugs, space technology, etc. Choi [1] accounted that by nanoparticles inclusion the thermal features of the base fluids are tremendously enhanced. Buongiorno [2] explains that Brownian movement and thermophoresis are two main characteristics that are helpful for heightening base fluid thermal conductivity. Hayat et al. [3] has investigated the effect of binary chemical reaction and activation energy. Examination of the radiative mixed convective flow of Casson nanofluid is addressed by Ijaz khan et al. [4]. Waqas et al. [5] characterized the powell-Eyring fluid in the regime of magnetic dipole. Magneto hydrodynamic flow with Arrhenius activation energy was explored by Asma et al. [6]. Hayat et al. [7] discussed Ree-Eyring nanofluid flow with Arrhenius activation energy. The viscous fluid flow around a slender needle was investigated by Lee [8]. Afridi and Qasim [9] investigate heat transport by moving a slender needle with entropy generation into a parallel stream. Mabood et al. [10] has investigated the flow of micropolar fluids in a parallel stream with Arrhenius activation energy. Salleh et al. [11] discussed the effect on the steady 2D nanofluid flow over a vertical needle with the presence of velocity slip on the surface. Heat transfer with several surface temperature using a thin needle was primarily

I. Sadham Hussain · D. Prakash (✉) · S. Kumar
Department of Mathematics, Faculty of Engineering and Technology,
College of Engineering and Technology, SRM Institute of Science and Technology,
SRM Nagar, Kattankulathur 603203, TN, India
e-mail: prakashd1@srmist.edu.in



S. Srinivas et al. (eds.), *Recent Advances in Applied Mathematics and Applications to the Dynamics of Fluid Flows*, Lecture Notes in Mechanical Engineering,
https://doi.org/10.1007/978-981-19-1929-9_16

studied by Grosan and Pop [12]. The nanofluid flux of heat flux caused by a thin needle near a stagnation point was examined by Hayat et al. [13]. Mabood et al. [14] has investigated electromagnetohydrodynamic flow of non-Newtonian nanofluids over thin needle with Arrhenius pre-exponential factor law. In this paper, the novelty of the present work is to investigate the influences of binary chemical reaction with activation energy on slender needle moving of nanofluid using Buongiorno model with slip effect. The governing equations are solved numerically by Runge–Kutta–Fehlberg method with shooting technique. A careful review of existing work reveals that no one has attempted the slip effects of activation energy on steady flow of Buongiorno's nanofluid over a thin moving needle.

## 2 Mathematical Formulation

We have taken here a boundary layer flow, heat, and mass transfer characteristics of nanofluid over a thin moving needle in the presence of slip effect with constant velocity $u_w$ and free stream velocity $u_\infty$ in steady state. The composite velocity $U$ is assumed to be the combinations of needle velocity $u_w$ and free stream velocity $u_\infty$. The flow velocities u and v are taken in the direction of x and r axes, respectively, which is shown in Fig. 1. Assuming a negligible pressure gradient along the needle, the function $r = R(x) = \sqrt{\frac{vcx}{U}}$ specifies the thin needle shape. The needle surface is maintained at a uniform temperature $T_w\ (> T_\infty)$ and concentration $C_w\ (> C_\infty)$. After posing all the necessary boundary layer approximations, we have the following equation:

$$\frac{\partial}{\partial x}(ru) + \frac{\partial}{\partial r}(rv) = 0 \tag{1}$$

$$u\frac{\partial u}{\partial x} + v\frac{\partial u}{\partial r} = \frac{\nu}{r}\frac{\partial}{\partial x}\left(r\frac{\partial u}{\partial x}\right) \tag{2}$$

$$u\frac{\partial T}{\partial x} + v\frac{\partial T}{\partial r} = \frac{\alpha}{r}\frac{\partial}{\partial r}\left(r\frac{\partial T}{\partial r}\right) + \tau\left(D_B\frac{\partial T}{\partial x}\frac{\partial C}{\partial x} + \frac{D_T}{T_\infty}\left(\frac{\partial T}{\partial r}\right)^2\right) \tag{3}$$

$$u\frac{\partial C}{\partial x} + v\frac{\partial C}{\partial r} = \frac{D_B}{r}\frac{\partial}{\partial r}\left(r\frac{\partial C}{\partial r}\right) + \frac{D_T}{T_\infty}\frac{1}{r}\frac{\partial}{\partial r}\left(r\frac{\partial T}{\partial r}\right) - k_r^2(C - C_\infty)\left(\frac{T}{T_\infty}\right)^n \exp\left(\frac{-E_a}{kT}\right) \tag{4}$$

The boundary condition associated with Eqs. (1)–(4)

$$u = u_w,\ v = 0, T = T_w, D_B\frac{\partial C}{\partial r} + \frac{D_T}{T_\infty}\frac{\partial T}{\partial r} = 0\ at\ r = R(x)$$

$$u \to u_{\infty,} T \to T_\infty, C \to C_\infty\ as\ r \to \infty \tag{5}$$

where $\nu$ stands for kinematic viscosity, $C$ stands for concentration of nanofluid, $D_T$ stands for thermophoretic diffusion coefficient, $D_B$ stands for Brownian diffusion

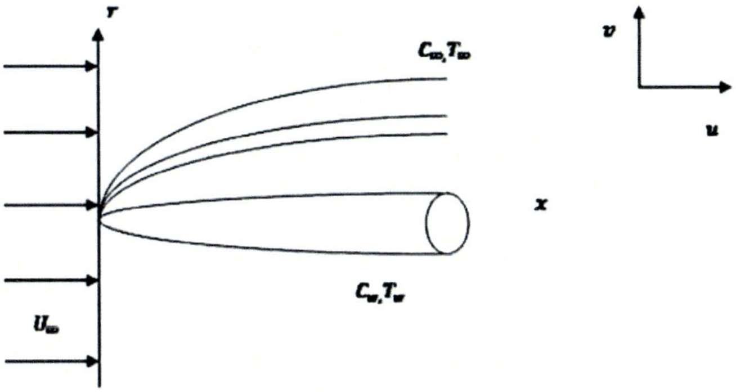


**Fig. 1** Physical geometry of the problem

coefficient, $k_r$ stands for chemical reaction rate, $\alpha$ stands for thermal diffusivity, $\tau$ stands for ratio of nanofluid effective heat capacity, $T$ stands for temperature of nanofluid.

Considering the appropriate similarity variable

$$\eta = \frac{Ur^2}{\nu x}, \psi = \nu x f', \phi(\eta) = \frac{C - C_\infty}{C_\infty}, \theta(\eta) = \frac{T - T_\infty}{T_w - T_\infty}, \chi(\eta) = \frac{n - n_\infty}{n_w - n_\infty} \tag{6}$$

where $u = \frac{1}{r}\frac{\partial\psi}{\partial r}$ and $v = -\frac{1}{r}\frac{\partial\psi}{\partial x}$. Equations (1) and (4) transform into the ordinary differential equations.

$$ff' + 2\eta f''' + 2f'' = 0 \tag{7}$$

$$\frac{2}{\Pr}\left(\eta\theta'' + \theta'\right) + 2\eta\left(Nt\theta'^2 + Nb\theta'\phi'\right) + f\theta' = 0 \tag{8}$$

$$Scf\phi' + 2\left(\phi' + \eta\phi''\right) + 2\frac{Nt}{Nb}\left(\eta\theta'' + \theta'\right) - Sc\sigma\left(\delta\theta + 1\right)^n \exp\left(\frac{-E}{\delta\theta + 1}\right) = 0 \tag{9}$$

Here $Sc$ the Schmidt number, $Nb$ the Brownian motion parameter, $Pr$ the Prandtl number and $Nt$ the thermophoresis parameter, $E$ the activation parameter, $\delta$ the temperature difference parameter and $\sigma$ the reaction rate parameter are defined as

$$Pr = \frac{\nu}{\alpha}, \; Nb = \frac{\tau D_B C_\infty}{\nu}, Nt = \frac{\tau D_T\left(T_w - T_\infty\right)}{\nu T_\infty}, E = \frac{E_a}{kT_\infty}, \delta = \frac{T_w - T_\infty}{T_\infty} \tag{10}$$

The transformed boundary conditions are

$$f = \frac{\lambda a}{2}, f' = \frac{\lambda}{2}, \theta = 1, \phi' + \frac{Nt}{Nb}\theta' = 0 \; at \; \eta = a$$
$$f' = \frac{1-\lambda}{2}, \theta \to 0, \phi \to 0 \; at \; \eta \to \infty. \tag{11}$$

where $\lambda$ is the ratio of needle velocity to composite velocity. The physical aspects of the pre-eminent interest are skin friction, local Nusselt number, and local Sherwood number which are, respectively, defined by

$$Cf = 4Re^{\|frac12}a^{\frac{1}{2}}f''(a), Nu = -2Re^{\frac{1}{2}}a^{\frac{1}{2}}\theta'(a), Sh = -2Re^{\frac{1}{2}}a^{\frac{1}{2}}\phi'(a). \quad (12)$$

## 3 Numerical Solution

The formulated problem given in Eqs. (7)–(9) has been treated by a convenient shooting approach coupled with Runge–Kutta–Fehlberg scheme. We obtain a system comprising seven first-order equations given below:

$$f = y_1, \; f' = y_2, f'' = y_3, \theta = y_4, \; \theta' = y_5, \phi = y_6, \phi' = y_7$$

$$y_1' = y_2, \; y_2' = y_3, \; y_3' = \frac{1}{2\eta}(-2y_3 - y_1y_3)$$

$$y_4' = y_5, \; y_5' = \frac{\Pr}{2\eta}\left(-\frac{2y_5}{\Pr} - 2\eta\left(Nty_5'^2 + Nby_5y_7\right) - y_1y_5\right), \; y_6' = y_7$$

$$y_7' = -\frac{1}{2\eta}(2y_7 + 2\frac{Nt}{Nb}\left(\eta y_5' + y_5\right) + Scy_1y_7 - Sc\sigma(\delta y_4 + 1)^n \exp\left(\frac{-E}{\delta y_4 + 1}\right),$$

and the transformed conditions are

$$At \; \eta = a: \; y_1 = \frac{\lambda a}{2}, y_2 = \frac{\lambda}{2}, y_3 = g_1, y_4 = 1, y_5 = g_2, y_6 = g_3,$$

$$y_7 = -\frac{Nt}{Nb}y_5$$

$$As \; \eta \to \infty: y_2(\eta) = \frac{1-\lambda}{2}, \; y_4(\eta) = 0, y_6(\eta) = 0, \; as \; \eta \to \infty$$

## 4 Results and Discussion

In this present study, the governing Eqs. (7)–(9) are solved numerically for various parameter values. Figures and tables show the effects of these parameters that cause the flow problem. Figures 2, 3, 4, 5, and 6 exemplify the important features of several variables like $a$, $E$, $\delta$, $\sigma$, $\lambda$, $Pr$, $n$, $Sc$, $Nb$, and $Nt$ versus $\theta$, $f'$, and $\phi$. we used the range of parameters all over the simulation as $0 \leq E \leq 5, 0 \leq \sigma \leq 5, 0 \leq \delta \leq 7, 0 \leq Sc \leq 3, 0.71 \leq Pr \leq 10, 0 \leq n \leq 5, 0 \leq \lambda \leq 1, 0 \leq Nb \leq 1, 0 \leq Nt \leq 1, 0 \leq \omega$

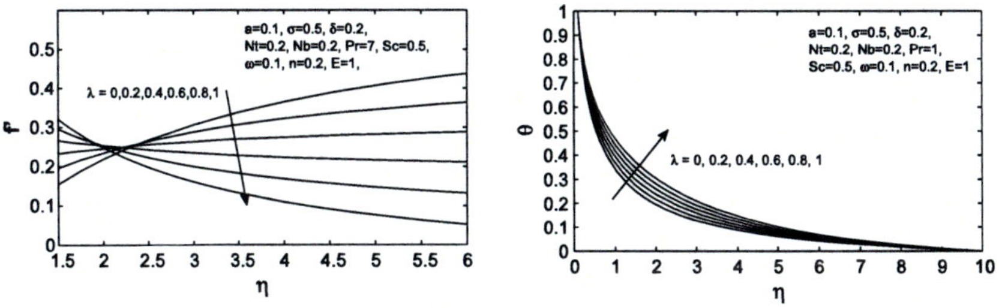


**Fig. 2** Disparity of velocity ratio parameter on $f'(\eta)$ and $\theta$

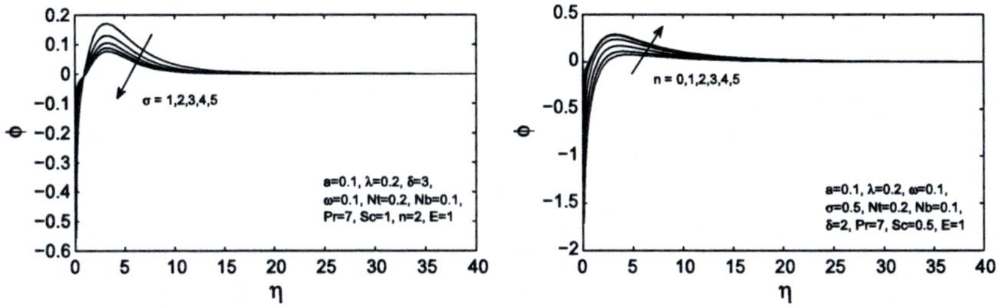


**Fig. 3** Effects of $\sigma$ and $n$ on $\phi$

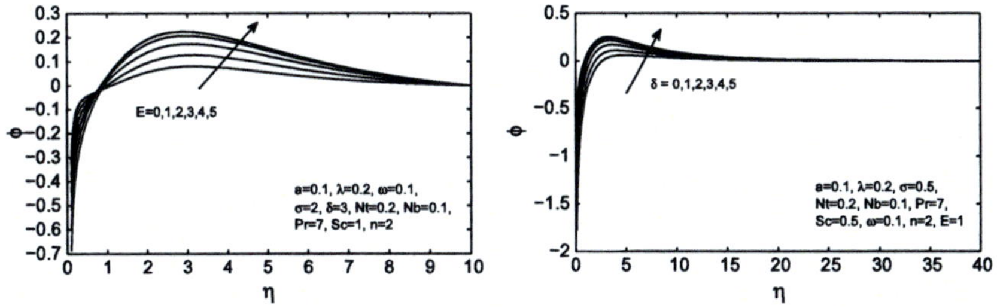


**Fig. 4** Disparity of $E$ and $\delta$ on $\phi$

$\leq 1$. Comparison is made with accessible results of $f''(a)$ with those of Afridi and Qasim [9], Ishak et al. [16], Chen and Smith [15] are described in Table 1 for different values of needle size parameter. The variations of skin friction coefficient with $\lambda$ and $n$ are indicated in Table 2. The skin friction coefficient is regularized at the wall by the slope of the velocity of the fluid. It is noticed that the skin friction coefficient increases with the decrease in both needle size and velocity ratio parameters.

Table 3 tabulates numerical values of $Nu(Re)^{-0.5}$ and $Sh(Re)^{-0.5}$ for various parameter values. A significant fall in Nusselt number and higher in Sherwood number are expected when $a$ is increased from $a = 0.1$ to $a = 0.4$. This means that the larger the needle size, the lower the rate of heat transfer at the needle surface. The next three entries in this table predict that the Nusselt number decreases and the Sherwood

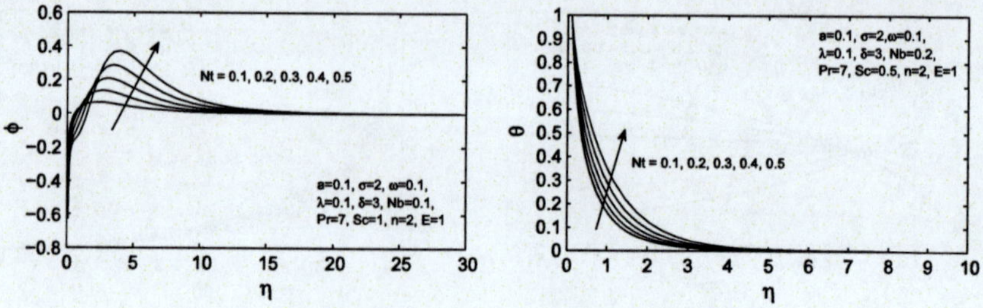


**Fig. 5** Dimensionless concentration and temperature for various values $Nt$

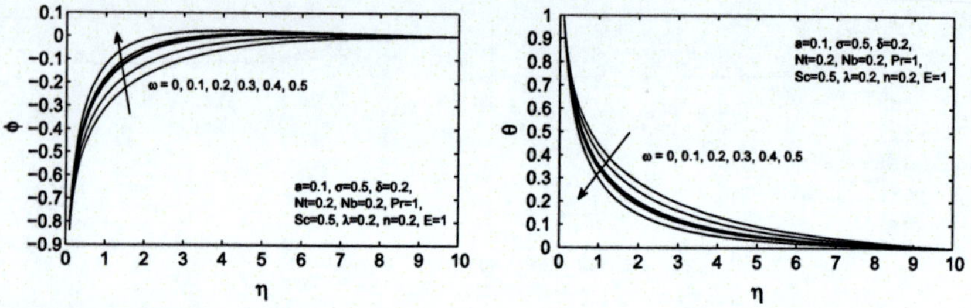


**Fig. 6** Disparity of $\omega$ on $\phi$ and $\theta$

**Table 1** Comparison of $f''(a)$ when $Pr = 1$

| $a$ | Chen and Smith [15] | Ishak et al. [16] | Afiridi and Qasim [9] | Current results |
|---|---|---|---|---|
| 0.1 | 1.2888 | 1.2888 | 1.2887 | 1.288700 |
| 0.01 | 8.4924 | 8.4924 | 8.4913 | 8.491663 |
| 0.001 | 62.1637 | 62.1637 | 62.1637 | 62.167783 |

**Table 2** Comparison values for skin friction coefficient for $a$ and $\lambda$

| $a$ | $\lambda$ | Afiridi and Qasim [9] | Current results |
|---|---|---|---|
| 0.2 | 0.3 | 1.100352 | 1.100322 |
| 0.01 | | 2.662191 | 2.662438 |
| 0.001 | | 6.139201 | 6.140168 |
| 0.1 | 0.0 | 6.792972 | 6.793330 |
| | 0.4 | 1.320171 | 1.320291 |
| | 0.7 | 1.320171 | 1.320291 |
| | 1 | −6.111587 | −6.099525 |

**Table 3** Numerical values for Nusselt number and Sherwood number of various parameters

| $a$ | $E$ | $\delta$ | $n$ | $\omega$ | $Nu(Re)^{-0.5}$ | $Sh(Re)^{-0.5}$ |
|---|---|---|---|---|---|---|
| 0.1 | 1.0 | 0.1 | 0.1 | 0.1 | 1.717846 | –1.717846 |
| 0.2 | | | | | 1.417619 | –1.417619 |
| 0.3 | | | | | 1.277221 | –1.277221 |
| 0.4 | | | | | 1.191132 | –1.191132 |
| 0.1 | 2.0 | 0.1 | 0.1 | 0.1 | 1.722766 | –1.722766 |
| | 3.0 | | | | 1.724951 | –1.724951 |
| | 4.0 | | | | 1.725857 | –1.725857 |
| | 5.0 | | | | 1.726221 | –1.726221 |
| $a$ | $E$ | $\delta$ | $n$ | $\omega$ | $Nu(Re)^{-0.5}$ | $Sh(Re)^{-0.5}$ |
| 0.1 | 1.0 | 0.2 | 0.1 | 0.1 | 1.717354 | –1.717354 |
| | | 0.4 | | | 1.716476 | –1.716476 |
| | | 0.5 | | | 1.716083 | –1.716083 |
| 0.1 | 1.0 | 0.1 | 0.2 | 0.1 | 1.717794 | –1.717794 |
| | | | 0.3 | | 1.717742 | –1.717742 |
| | | | 0.4 | | 1.717690 | –1.717690 |
| 0.1 | 1.0 | 0.1 | 0.1 | 0.2 | 1.752926 | –1.752926 |
| | | | | 0.4 | 1.795964 | –1.795964 |

number increases as the velocity ratio parameter values increase. It means that the rate of needle cooling can be changed by increasing the velocity of the needle. This table revealed that $Nu$ decreases when the Schmidt number temperature difference parameter and fitted rate constant increases. It shows a large increase in the local Sherwood number when the $Sc$ or $\sigma$ is incremented. Thus, the concentration boundary layer becomes slimmer and the mass transfer rate grows when schmidt number increases. The impact of $\lambda$ on profiles of $f'$ is presented in Fig. 2a. $f'$ increments monotonously with the $\eta$ similarity variable when $0 \leq \lambda < 0.5$. Furthermore, the velocity increases right next to the needle surface and decreases away from it. Figure 2b demonstrates the impact of temperature $\theta$ for distinct values of $\lambda$. Apparently, $\theta$ increases as $\lambda$ increase. This means that the heat perforates deeper into the nanofluid when the needle velocity is greater. Figure 3a, b are arranged to notice the impact of reaction rate $\sigma$ and fitted rate constant $n$ on $\phi$, respectively. It has noted that a growth in $n$ or $\delta$ results in a rise in the term $\sigma(\delta\theta + 1)^n exp(-E/(\delta\theta + 1))$. The decrease is followed by a greater concentration gradient to the wall. $\sigma$ decreases the nanoparticles concentration and eventually controls the concentration boundary layer. Moreover, Fig. 3b observed that the growing trend in the concentric field was achieved because of an increase in $n$. For each $n$ on overshoot seems near the surface of the needle. The presence of activation energy refers to higher nanoparticle concentrations. Figure 4a brings out that the increase in activation energy causes the concentration boundary layer to thicken. Figure 4b revealed that $\phi$ increases when temperature difference parameter $\sigma$

increases. Higher concentration occurs near the needle surface. Figure 5a represents the effectiveness of the thermophoresis parameter $Nt$ on the concentration curves. From this plot, thermophoretic enhances nanoparticle concentration since hot needle repels submicron particles from it and forming a particle free layer near the surface if $Nt > 0$. Furthermore, decreasing trend happens in the concentration profile at near the surface of needle and opposite reaction happens away from the needle. Figure 5b shows that temperature profile increases when $Nt$ increases. Figure 6a exhibits that a higher slip parameter fluid possesses lower thermal diffusivity, and therefore, has shortest size of the thermal layer. Figure 6b conveys the change in $\phi$ with the change in slip parameter. Concentration profile increases when increasing the slip parameter values.

## 5 Conclusion

Here, we examined the flow of nanofluids on a moving thin needle with a slip effect. It is negotiated that the thermophoretic force leads to enhance the concentration and axial velocity near the needle surface and decreases away from it. An improvement in the rate constant and nanofluid parameters is to stimulate the concentration, whereas it decreases with the temperature difference parameter and fitted rate constant. An improvement in the slip parameter indicates a decrease in the temperature field and the reverse pattern was found in the concentration field.

## References

1. Choi SUS (1995) Enhancing thermal conductivity of fluids with nanoparticles. ASME FED 231:99–105. https://doi.org/10.1115/1.2150834
2. Buongiorno J (2006) Convective transport in nanofluids. ASME J Heat Trans 128:240–250. https://doi.org/10.1115/1.2150834
3. Hayat T, Riaz R, Aziz A, Alsaedi A (2020) Influence of Arrhenius activation energy in MHD flow of third grade nanofluid over a nonlinear stretching surface with convective heat and mass conditions. Physica A 549:124006. https://doi.org/10.1016/j.physa.2019.124006
4. Ijaz Khan M, Waleed Ahmad Khan M (2020) Entropy generation optimization in flow of non-Newtonian nanomaterial with binary chemical reaction and Arrhenius activation energy. Stat Mech Appl 538:122806. https://doi.org/10.1016/j.physa.2019.122806
5. Waqas M, Shagufta Jabeen, Hayat T, Shehzad S, Alsaedi A (2020) Numerical simulation for nonlinear radiated Eyring-Powell nanofluid considering magnetic dipole and activation energy. Int Commun Heat Mass Trans 112:104401. https://doi.org/10.1016/j.icheatmasstransfer.2019.104401
6. Asma M, Othman WAM, Muhammad T, Mallawi F, Wong BR (2019) Numerical study for magnetohydrodynamic flow of nanofluid due to a rotating disk with binary chemical reaction and Arrhenius activation energy. Symmetry 11:1282. https://doi.org/10.3390/math7100921
7. Hayat T, Sohail Kha A, Ijaz Kha M, Alsaedi A (2019) Ree-Eyring nanofluid flow with entropy optimization and Arrhenius activation energy between two rotating disks. Comput Methods Programs Biomed 177:57–68. https://doi.org/10.1016/j.cmpb.2019.05.012

8. Lee LL (1967) Boundary layer over a thin needle. Phys Fluids 10:820–822. https://doi.org/10.1063/1.1762194
9. Afridi MI, Qasim M (2018) Entropy generation and heat transfer in boundary layer flow over a thin needle moving in a parallel stream in the presence of nonlinear Rosseland radiation. Int J Therm Sci 123:117–128. https://doi.org/10.1016/j.ijthermalsci.2017.09.014
10. Mabood F, Nayak MK, Ali J (2019) Heat transfer on the cross flow of micropolar fluids over a thin needle moving in a parallel stream influenced by binary chemical reaction and Arrhenius activation energy. Euro Phys J Plus 134–427. https://doi.org/10.1140/epjp/i2019-12716-9
11. Salleh SN, Bachok N, Arifin NM, Ali FM (2019) Slip effect on mixed convection flow past a thin needle in nanofluid using buongiorno's model. J Adv Res Fluid Mech Thermal Sci 59:243–253
12. Grosan T, Pop I (2011) Forced convection boundary layer flow past non isothermal thin needles in nanofluids. J Heat Trans 133. https://doi.org/10.1115/1.4003059
13. Hayat T, Khan MI, Farooq M, Yasmeen T, Alsaedi A (2016) Water-carbon nanofluid flow with variable heat flux by a thin needle. J Mol Liq 224:786–791. https://doi.org/10.1016/j.molliq.2016.10.069
14. Mabood F, Muhammad T, Nayak MK, Waqas H, Makinde OD (2020) EMHD flow of non-Newtonian nanofluids over thin needle with Robinson's condition and Arrhenius pre-exponential factor law. Phys Scr 95:115219. https://doi.org/10.1088/1402-4896/abc0c3
15. Chen JLS, Smith TN (1978) Forced convection heat transfer from nonisothermal thin needles. J Heat Trans 100:358–362. https://doi.org/10.1115/1.2150834
16. Ishak A, Nazar R, Pop I (2007) Boundary layer flow over a continuously moving thin needle in a parallel free stream. Chin Phys Lett 24:2895–2897. https://doi.org/10.1115/1.3450809

# Effect of Magnetic Field on Unsteady Flow of Dusty Fluid Due to Constant Pressure Gradient Through a Circular Cylinder: An Analytical Treatment

**Naresh Reddimalla, J. V. Ramana Murthy, V. Radha Krishna Murthy, and Srinivas Jangili**

## 1 Introduction

The study of dusty fluid flows received special attention recently due to their two-phase nature. This phenomenon can be noticed in liquid or gas flows containing the distributed solid particles. For example, motion of the dusty gas in fluidization problems. The solid particles appearing in a dusty fluid can exist either naturally or deliberately in the form of dust. This model was first pioneered by Saffman [1] in 1961. He obtained the stability characteristics of a dusty gas in a plane parallel flow by using the Orr–Sommerfield equations. Singh and Pathak [2] studied the laminar flow of time-dependent dusty fluid through a circular channel with the assumption of exponential pressure gradient. They noticed that the velocity of the dust particles is more than that of viscous fluid. Singh [3] reported an exact solution to study the effect of oscillatory pressure gradient on unsteady flow of viscous fluid with the assumption of uniformly distributed dust particles through a circular cylinder. The laplace transform technique was employed to study the dusty fluid flows by the authors [4–6]. Rukmangadachari [4] examined the dusty fluid flow via a circular cylindrical tube. He observed that the resistance offered by the viscous forces to the pressure forces across a section of the cylinder is greater for a dusty fluid than for a clean fluid. Mandal [5] investigated the flow of elastic-viscous liquid with the inclusion of dust particles through an oscillatory cylinder. They observed that the dust particles decelerate the velocity. Gireesha et al. [6] investigated the

N. Reddimalla (✉) · J. V. Ramana Murthy · S. Jangili
Department of Mathematics, National Institute of Technology, Warangal, Telangana 506004, India
e-mail: nareshr@student.nitw.ac.in

V. Radha Krishna Murthy
Department of Mathematics, Vignan University, Vadlamudi, Guntur, Andhra Pradesh 522213, India



S. Srinivas et al. (eds.), *Recent Advances in Applied Mathematics and Applications to the Dynamics of Fluid Flows*, Lecture Notes in Mechanical Engineering,
https://doi.org/10.1007/978-981-19-1929-9_17

time-dependent dusty fluid flow through a circular pipe in the presence of porous material. Singh et al. [7] reported that the numerical series solution of uniform slip flow along a vertical cylinder and they observed that at the surface, the heat transfer increases with increase in slip parameter. Om Prakash et al. [8] employed a variable separable method to obtain the expressions for temperature, velocity profiles, and skin friction to examine MHD flow of dusty fluid in a vertical channel filled with a porous material. Sandeep [9] reported heat transfer characteristics of dusty nanofluid under the impact of magnetic force. Attia et al. [10] applied Crank–Nicolson implicit method for unsteady flow of dusty viscous Bingham fluid through a circular pipe.

The MHD dusty fluid flow via a circular cylinder has attracted researchers recently due to its enormous applications in MHD generators, aerodynamics, MHD pumps, accelerators, purification of crude oil, etc. Dey [11] studied MHD flow of dusty stratified fluid. They pointed out that the elevation in magnetic parameter decelerates the speed of the dusty fluid. Attia and Ewis [12] utilized the finite difference method to investigate the effects of continuous unsteady flow of dust and power-law fluid in a circular channel. They highlighted that the increase in magnetic parameter causes phase change of the particles. Saqib et al. [13] discussed time-varying blood flow through a circular tube with uniformly distributed magnetite dust particles through a cylindrical tube. They noticed that both velocities decreased.

Motivated by the aforementioned works, in this paper, the authors intend to examine the effect of constant transverse magnetic force on unsteady flow of dusty fluid in a circular pipe. The impact of diverse nondimensional parameters on fluid velocity and dust velocity is investigated.

## 2 Mathematical Description of the Problem

Consider an unsteady, laminar, and incompressible MHD flow of dusty fluid passing through a circular pipe. The fluid flow is generated due to a constant pressure gradient. Both the fluid and dust particles are considered to be at rest in the beginning. It is supposed that the dust particles are spherical in shape and uniform in size having the radius $a$ and the dust particles contain the constant number density ($N$). The magnetic field of strength $B_0$ is acting in $\bar{e_\theta}$ direction with external magnetic field $\overline{B} = B_0\overline{e_\theta}$. We consider cylindrical coordinate system with $z$-axis along the axis of the cylinder and origin on the axis (see Fig. 1). The clean fluid and the dusty fluid are moving in the $z$-direction.

$$\nabla_0 \cdot \bar{q} = 0 \tag{1}$$

$$\rho\frac{\partial \bar{q}}{\partial t^*} + (\bar{q} \cdot \nabla_0)\bar{q} = -\nabla_0 P^* - \mu\nabla_0^2\bar{q} + KN(\bar{q_0} - \bar{q}) + (\bar{J} \times \bar{B}) \tag{2}$$

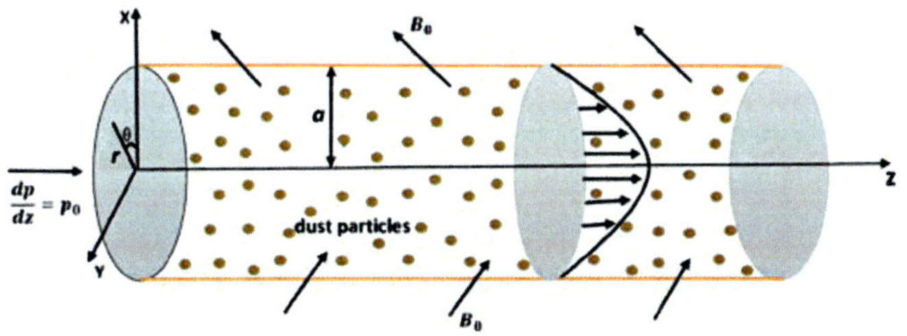


**Fig. 1** Geometry of the dusty fluid flow in a circular cylinder

$$m\left[\frac{\partial \bar{q_0}}{\partial t^*} + (\bar{q_0} \cdot \nabla_0)\bar{q_0}\right] = K(\bar{q} - \bar{q_0}) \tag{3}$$

$$\frac{\partial N}{\partial t^*} + \nabla_0 \cdot (N\bar{q_0}) = 0 \tag{4}$$

where $\nabla_0^2 = \frac{\partial^2}{\partial r^{*2}} + \frac{1}{r^*}\frac{\partial}{\partial r^*}$ a dimensional Laplacian.

In the above equations, $\rho$-density of the fluid, $\mu$-dynamic viscosity of the fluid, $p$-pressure of the fluid, $m$-mass of a dust particle, $t^*$-time, $K$-Stokes drag coefficient (=$6\pi\mu a$ [2] for a spherical dust particle of radius $a$), $\bar{q}$-the clean fluid velocity, $\bar{q_0}$-dust particle velocity, and $N$-number density of a dust particle.

Since the flow direction is along $z$-axis and is axially symmetric, we assume that the clean fluid velocity and dust particle velocity to be in the form

$$\bar{q} = u^*\bar{e_z} \; and \; \bar{q_0} = v^*\bar{e_z} \tag{5}$$

From (5), the equations of motion in cylindrical coordinates reduces to

$$\frac{\partial p^*}{\partial r^*} = 0, \frac{\partial p^*}{\partial \theta} = 0 \tag{6}$$

In Eq. (2), the Lorentz force term due to magetic force is given by $(\bar{J} \times \bar{B})$ and takes the form $-\sigma B_0^2 u^*$. Here $\bar{J}$ is the current density and $B_0$ is the intensity of the imposed transverse magnetic field.

Now the Eqs. (3) and (4) reduce to

$$\rho\frac{\partial u^*}{\partial t^*} = -\frac{\partial p^*}{\partial z^*} - \mu\nabla_0^2 u^* + KN(v^* - u^*) - \sigma B_0^2 u^* \tag{7}$$

$$\frac{\partial u^*}{\partial t^*} = \frac{K}{m}(u^* - v^*) \tag{8}$$

Before proceeding, we first introduce the following nondimensional groups

$$u = \frac{u^* a\rho}{\mu}, v = \frac{v^* a\rho}{\mu}, r = \frac{r^*}{a}, p = \frac{p^*}{\rho U_0^2}, t = \frac{\mu t^*}{\rho a^2}, z = \frac{z^*}{a}, \nabla_0 = \frac{1}{a}\nabla \quad (9)$$

In the light of above nondimensional scheme, the Eqs. (7) and (8) become

$$\frac{\partial u}{\partial t} = -Re^2\frac{dp}{dz} + \left[\frac{\partial^2 u}{\partial r^2} + \frac{1}{r}\frac{\partial u}{\partial r}\right] + \frac{f_0}{\tau}(v-u) - M^2 u \quad (10)$$

$$\frac{\partial v}{\partial t} + \frac{v}{\tau} = \frac{1}{\tau}u \quad (11)$$

where $Re = \frac{\rho U_0 a}{\mu}$ is a Reynolds number, $M = B_0 a\sqrt{\frac{\sigma}{\mu}}$ is Hartmann number (or magnetic parameter), $\tau = \frac{m\mu}{\rho K a^2}$ is relaxation time of the dust, and $f_0 = \frac{mN}{\rho}$ is mass concentration of dust. We consider the special case $\frac{dp}{dz} = p_0$ is constant. The initial and boudary conditions are:

$$\begin{aligned} &t < 0, u(r,0) = 0; v(r,0) = 0 \text{ for } 0 \le r \le 1, \\ &t \ge 0, u(1,t) = 0; v(1,t) = 0; \frac{\partial u}{\partial r} = 0 \text{ for } r = 0 \end{aligned} \quad (12)$$

and as $t \to \infty$, the fluid and dust particles will attain steady state.

Further, when $f_0 = 0$ and $M = 0$, the Eq. (10) reduces to classical Navier–Stokes equation.

## 3 Exact Solution by Method of Separation of Variables

By eliminating $v$ from Eqs. (10) and (11), we get

$$\frac{\partial^2 u}{\partial t^2} + (\eta + M^2)\frac{\partial u}{\partial t} - \frac{\partial}{\partial t}\nabla^2 u + \frac{M^2}{\tau}u - \frac{1}{\tau}\nabla^2 u = -Re^2\left(\frac{\partial}{\partial t} + \frac{1}{\tau}\right)\frac{\partial p}{\partial z} \quad (13)$$

where $\eta = \frac{1+f_0}{\tau}$ and $\nabla^2 = \frac{\partial^2}{\partial r^2} + \frac{1}{r}\frac{\partial}{\partial r}$.

Now consider $u = u_s(r) + u_t(r,t)$ where $u_s(r)$ and $u_t(r,t)$ are the steady state and transient state solutions, respectively.

Then the equations for $u_s$ and $u_t$ are given by

$$\left(\frac{d^2}{dr^2} + \frac{1}{r}\frac{d}{dr} - M^2\right) u_s = Re^2 p_0 \tag{14}$$

and

$$\frac{\partial^2 u_t}{\partial t^2} + \frac{\partial}{\partial t}(\nabla^2 - M^2 - \eta)u_t - \frac{1}{\tau}(\nabla^2 - M^2)u_t = 0 \tag{15}$$

with the boundary conditions

$$u_s(1) = 0, u_t(1,t) = 0, u_t(r,0) = 0 \text{ and } u_t(r, t \to \infty) = 0 \tag{16}$$

By solving Eqs. (14) and (15), and implementing the boundary conditions, we get the solution for $u$ as

$$u = -\frac{Re^2 p_0}{M^2}\left(1 - \frac{I_0(Mr)}{I_0(M)}\right) + \sum_{n=1}^{\infty} c_n e^{-m_1 t} J_0(\beta_n r) \tag{17}$$

By substituting Eq. (17) in (11) and solving, we get

$$v(r,t) = -\frac{Re^2 p_0}{M^2}\left(1 - \frac{I_0(Mr)}{I_0(M)}\right) + \sum_{n=1}^{\infty} \frac{c_n}{1 + m_1 t} e^{-m_1 t} J_0(\beta_n r) \tag{18}$$

where $c_n = \dfrac{2Re^2 p_0}{\beta_n (M^2 + \beta_n^2) J_1(\beta_n)}, m_1 = -\dfrac{\eta + \tau^2\lambda_n^2}{2}\left[1 + \sqrt{1 - \dfrac{4\tau^2\lambda_n^2}{(\eta + \tau^2\lambda_n^2)^2}}\right]$ and $\beta_n = \sqrt{\lambda_n^2\tau^2 - M^2}$ are the roots of the Bessel function $J_0$ and $\lambda_n$ are arbitrary constants.

The expressions of skin friction at the boundary $r = 1$ for clean fluid and dust particles, respectively, are given by

$$c_{f_1} = \frac{Re^2 p_0}{M^2}\left(\frac{I_1(M)}{I_0(M)}\right) - \sum_{n=1}^{\infty} \beta_n c_n e^{-m_1 t} J_1(\beta_n) \tag{19}$$

$$c_{f_2} = \frac{Re^2 p_0}{M^2}\left(\frac{I_1(M)}{I_0(M)}\right) - \sum_{n=1}^{\infty} \frac{\beta_n c_n}{1 + m_1 t} e^{-m_1 t} J_1(\beta_n) \tag{20}$$

## 4 Results and Discussion

The governing equations of MHD dusty flow have been derived and solved by employing the method of separation of variables. The exact solutions are obtained. The flow of dusty fluid in a circular pipe under the action uniform magnetic force for diverse values of the parameters is discussed. The following parametric values are fixed at: $M = 1$, $Re = 1$, $f_0 = 0.2$, $\tau = 0.4$, $p_0 = -1$, $t = 0.2$.

***Temporal variations of fluid and dust particles***: The temporal variations of fluid and dust particles velocities in a circular pipe channel are depicted in Fig. 2a. It is observed that at steady state (attained nearly at $t = 0.6$) the velocities of the dust particles and fluid are same though initially dust particle velocity is greater than the clean fluid velocity.

***Influence of magnetic field on velocity profiles***: Figure 2b elucidates the impact of magnetic parameter ($M$) on the fluid and dust particle velocities, it is noticed that the increase in the values of $M$ decelerates the fluid particles motion significantly. This may be due to the fact that the Lorentz force of magnetic field is opposing the fluid motion. Hence, the speed of dust particles is decreased drastically as compared to fluid particles. This indicates retarding effect of the imposed magnetic field on the velocity of fluid and dust.

***Influence of Reynolds number on velocity profiles***: The response of fluid and dust particles with respect to various values of Reynolds number ($Re$) are demonstrated in Fig. 2c. As the Reynolds number increases, the inertial force increases due to which, force on fluid particles increases, and hence velocity increases as expected. A similar trend is noticed in the case of dust particles with small variations.

***Influence of relaxation time on velocity profiles***: The effect of nondimensional relaxation time ($\tau$) on both the velocities is described through Fig. 2d. The nondimensional relaxation time represents the time required for the dust particles to adjust to changes in the fluid velocity. It is clear from the figure that the velocity profiles have shown decreasing nature as relaxation time increases.

***Influence of mass concentration of dust on velocity profiles***: Figure 2e shows the impact of the mass concentration of dust ($f_0$) on both velocity distributions. It is evident from the figure that the elevation in the values of $f_0$, enhances the velocity of clean fluid and decreases the velocity of dust particles. The higher values of $f_0$ represent the heavy mass of dust particles in the fluid medium. Due to the inertia effect of dust particles, the velocity variations in dust particles are almost negligible. Further, as we can observe that the velocity profiles are almost constant.

***Influence of various parameters on the skin friction***: The variations of skin friction due to the flow of clean fluid and dust particles for different parameters are depicted in

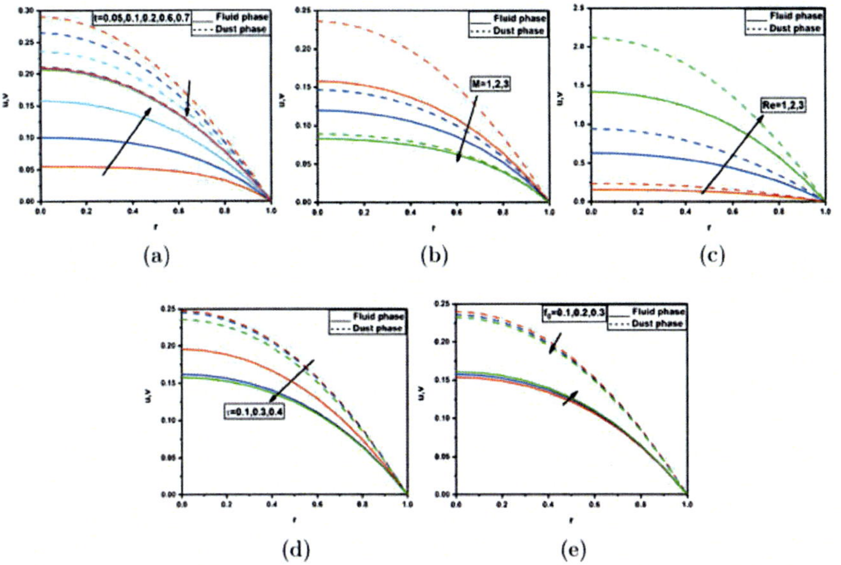


**Fig. 2** Effect of velocity profiles along $r$-axis for different values of **a** time ($t$), **b** magnetic parameter ($M$), **c** Reynolds number ($Re$), **d** relaxation time ($\tau$), **e** mass concentration of the dust ($f_0$)

Fig. 3. It is observed that at steady state (attained nearly at $t = 0.6$), the skin friction of the fluid and dust particles are the same, though initially, the skin friction for clean fluid is greater than that of the dust particles. From Fig. 3a, b, one can observe that the skin friction gradually increases with increase of magnetic parameter and relaxation time of the dust. It is observed from Fig. 3c that the mass concentration of dust has almost no effect on skin friction. This we can expect from the Eq. (11), in which the velocity $v$ is independent of $f_0$ and the Eq. (13) for $u$ contains $f_0$ through $\eta$, in which it is almost constant having no effect of $\eta$.

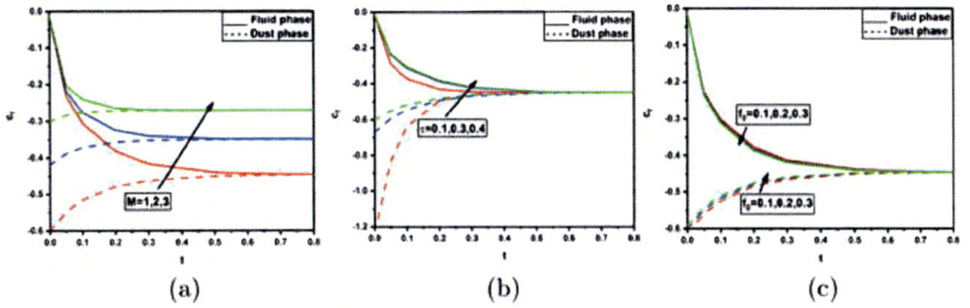


**Fig. 3** Skin friction versus time for diverse values of **a** magnetic parameter ($M$), **b** relaxation time ($\tau$), **c** mass concentration of dust ($f_0$)

## 5 Conclusion

In the present investigation, an unsteady magnetohydrodynamic flow of a dusty fluid through a circular cylindrical pipe has been considered. The fluid particles set in motion from rest by the sudden application of a constant pressure gradient and we obtained the analytical solutions of velocity fields of fluid and dust particles by using the method of separable variables. The graphical visualization of fluid and dust particles is presented with respect to various active parameters. Summarily, the results reveal that

- With increasing magnetic parameter and relaxation time of the dust, the velocity is decelerated in the pipe.
- With increasing time, the fluid velocity is increased whereas it decreased the dust particle velocity.
- Mass concentration of dust has almost no effect on velocity of fluid and dust.

## References

1. Saffman PG (1962) On the stability of laminar flow of dusty gas. J Fluid Mech 13(01):120–128
2. Singh JP, Pathak RS (1976) Unsteady flow of a dusty viscous fluid through a tube with sector of a circle as cross-section. Indian J Pure Appl Math 8(3):314–319
3. Singh SN (1984) Unsteady dusty viscous fluid flow through a circular cylinder. Proc Indian Natl Sci Acad 50(A6):594–598
4. Rukmangadachari E (1986) Unsteady flow of a dusty fluid through a circular cylinder. J Math Phys Sci 20(5):413–427
5. Mandal GC (1990) Unsteady axisymmetric rotational flow of dusty elastico-viscous liquid. Def Sci J 40(2):161–169
6. Gireesha BJ, Madhura KR, Bagewadi CS (2012) Flow of an unsteady dusty fluid through porous media in a uniform pipe with sector of a circle as cross-section. Int J Pure Appl Math 76(1):29–47
7. Singh G, Makinde OD (2014) Axisymmetric slip flow on a vertical cylinder with heat transfer. Sains Malaysiana 43(3):483–489
8. Prakash OM, Makinde OD, Kumar D, Dwivedi YK (2015) Heat transfer to MHD oscillatory dusty fluid flow in a channel filled with a porous medium. Acad Proc Eng Sci 40(4):1273–1282
9. Sandeep N, Sulochana C (2015) MHD flow of a dusty nanofluid over a stretching surface with volume fraction of dust particles. Ain Shams Eng J 7:709–716
10. Attia HA, Abbas W, Aboul-Hassan AL, Abdeen MAM, Ibrahim MA (2016) Unsteady flow of a dusty Bingham fluid through a porous medium in a circular pipe. J Appl Mech Tech Phys 57(4):596–602
11. Dey D (2016) Non-Newtonian effects on hydromagnetic dusty stratified fluid flow through a porous medium with volume fraction. Proc Natl Acad Sci India Sect A Phys Sci 86(1):47–56
12. Attia HA, Ewis KM (2019) Magnetohydrodynamic flow of continuous dusty particles and non-Newtonian Darcy fluids between parallel plates. Adv Mech Eng 11(6):1–11
13. Saqib M, Khan I, Shafie S (2019) Generalized magnetic blood flow in a cylindrical tube with magnetite. J Magn Magn 484:490–496

# Convective Flow over a Vertical Cone Embedded in a Stratified Medium by Group Theoretical Method

**T. Maheshwaran, Bapuji Pullepu, and B. Rushi Kumar**

## 1 Introduction

The group hypothetical strategy is of wide relevance and is very much acknowledged technique to discover the closeness arrangements in numerous actual circumstances. Temperature stratification happens in shut holder and ecological chambers with warmed dividers and is significant in lakes, flows, ocean, condensers of force plants, and different modern units. The actual climate is thermally separated, similar to the sea. This type of flow is also of fundamental important due to its possible applications in heat exchanger gadgets, filtration, and petrol reservoirs. Numerous examinations have been created from the previous 40 years for the free convection flow from different calculations in thermally stratified medium. Moran and Gaggioli [1] isolated for tumbling by one level of non-subordinate parts in express plans which contain, incompletely finished, a Goliath store of dependably differential condition and a monster pile of progress condition. Moran and Gaggioli [2] discussed the similarity analyses of a certain system of PDF and auxillary conditions for a class of compressible boundary-layer flows. Sulochana et al. [3] dissected the stream, warmth, and mass exchange conduct of magnetohydrodynamic stream over a vertical pivoting cone through permeable medium within the sight of warm radiation, compound response, and Soret impacts. Further, Sulochana et al. [4] hypothetically researched the limit layer nature of magnetohydrodynamic nanofluid stream past a vertical extending surface in a turning calculation with gooey scattering, warm radi-

T. Maheshwaran
Department of Statistics, DGVC, Chennai 600 106, Tamil Nadu, India

T. Maheshwaran · B. Pullepu (✉)
Department of Mathematics, SRMIST, Kattankulathur 603 203, Tamil Nadu, India
e-mail: bapujip@yahoo.com

B. Rushi Kumar
School of Advanced Sciences, VIT University, Vellore, TN, India

 
S. Srinivas et al. (eds.), *Recent Advances in Applied Mathematics and Applications to the Dynamics of Fluid Flows*, Lecture Notes in Mechanical Engineering,
https://doi.org/10.1007/978-981-19-1929-9_18

ation, Soret impact, and synthetic response. Kulkarni et al. [5] explored the issue of characteristic convection from an isothermal level plate suspended in a stable straightly delineated liquid medium utilizing the Von karman-Pohlhausen essential strategy. Tripathi and Nath [6] contemplated shaky regular convection stream over an upward level plate inserted in a separated medium. Sandeep et al. [7] examined the impact of different boundaries like substance reaction boundary, warm Grashof number, mass Grashof number, rarefaction boundary, attractive field boundary, radiation boundary, and pull boundary on convective warmth move along a slanted plate installed in permeable medium. Sulochana et al. [8] explored the limit layer investigation of a two-dimensional constrained convection stream along a relentless moving even needle in magnetohydrodynamic radiative nanoliquid. Srinivasan and Angirasa [9] contemplated the flimsy laminar axisymmetric crest that radiate from a wellspring of joined lightness because of synchronous warmth and mass dispersion in thermally separated medium and settled the limit layer condition utilizing express limited distinction technique. Takhar et al. [10] presented an examination on the strong brand name convection limit layer stream over a perseveringly accursed isothermal talk surface gobbled up progress of rising warm air-bound medium. The non-direct coupled fragmentary differential conditions controlling the non-relative stream have been illuminated mathematically using an appreciated proceeded with limit plot. Shapiro et al. [11] beat the transient standard convection of driven stream along a perplexing opposite plate brought down in a development of rising warm air portrayed out medium and considered streams affected by an upsetting change in the splendor of the plate and a befuddling usage of a quality improvement at the plate. Careful systems of the sorting everything out conditions were gotten by the design for Laplace transforms for the circumstance when the Prandtl number is units. Saha et al. [12] confined the two-dimensional solid stream with warmth and mass trade impacts along an isothermal talk level plate in an advancement of rising calm air portrayed fluid. Cutoff layer condition was left on behind by using a specific continued gathering structure in like way as neighborhood non-closeness system. Birkhoff [13] has introduced the social unlawful relationship draws near, as a class of designs which lead to decrease of the degree of self-controlling parts. Tripathi et al. [14] considered the consistent non-comparable free convection limit layer stream over an upward cone installed in separated medium. Mathematical outcomes were introduced for two upsides of the Prandtl number 0.72 (air) and 6.0 (water) and for various upsides of the semi-vertical point and definition boundary. Hossains et al. [15] read non-proportionality answer for the issue of clear smooth, standard convective stream, and warmth move from an upward risky cone brought down in an advancement of rising warm air depicted medium with either a uniform outside warmth or a uniform surface warmth progress. Sandeep and Sugunamma [16] have talked about the impacts of radiation and turn on unstable hydromagnetic free convection stream of a gooey incompressible electrically leading liquid past an imprudently moving vertical plate in a permeable medium by applying slanted attractive field, Bapuji and Immanuel [17] considered the free convection in an improvement of rising warmth air plot along an upward cone. Hossanien and Hamad [18] have developed the micropolar fluid which shows the miniature turn effects and little degree consistency. The difficulty of the appraisal of such fluid issue is nonattendance of cutoff conditions, and existing deformable microelements proportionately as the time become the third free factor. Abd-el-Malek [19] presented the use of the gathering hypothetical strategy to actual issues. Hossanien et al. [20] explored the normal convection limit layer stream of a micropolar liquid along an upward plate in a thermally delineated medium.

## 2 Mathematical Formulation

In a stratified medium, vertical cone set is affected by consistent smooth common convection flow of a thick and incompressible liquid. Viscous dissipation impact and the weight slope are ignored along the breaking point layer. Surface of the cone with encompassing liquid is at respite and maintains the equal heat $T'_{\infty,x}$ at first. The heat of the outside of the cone is raised from $T'_{\infty,x}$ to $T'_w$ when time increases ($t > 0$) and keep up a similar temperature. From Fig. 1, $X$ measures the space alongside the outside of the cone from apex ($x = 0$) and Y measures the space typical to it the external way, semi-vertical point of the cone $\phi$, and nearby radius of the cone is r. Barring thickness varieties, the liquid properties are expected to be consistent. The boundary layer conditions of continuity, momentum, and energy under the Boussinesq estimate are as per the following.

$$\frac{\partial}{\partial x}(ru) + \frac{\partial}{\partial y}(rv) = 0 \tag{1}$$

$$u\frac{\partial u}{\partial x} + v\frac{\partial u}{\partial y} = g\beta\cos\phi(T' - T'_{\infty}) + \frac{\partial^2 u}{\partial y^2} \tag{2}$$

$$u\frac{\partial T'}{\partial x} + v\frac{\partial T'}{\partial y} = \alpha\frac{\partial^2 T'}{\partial y^2}. \tag{3}$$

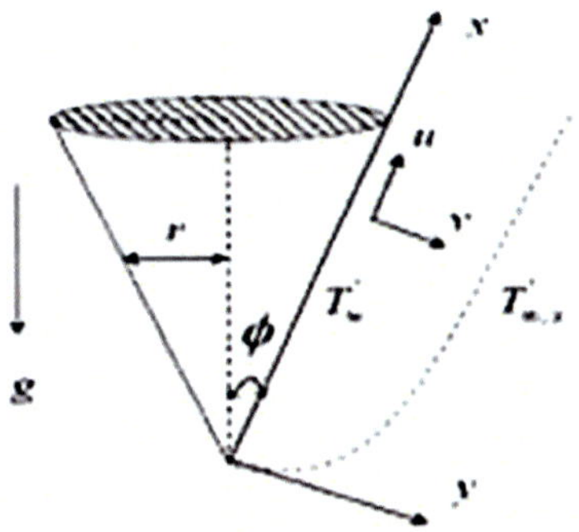

Fig. 1 Physical Model and Coordinate System

The underlying and limit conditions are

$$\left.\begin{array}{l} u(x,0)=0,\, v(x,0)=0,\, T'(x,0)=T'_w \\ u(x,\infty)=0,\, T'(x,\infty)=T'_\infty. \end{array}\right\} \tag{4}$$

Utilizing the accompanying dimensionless quantities

$$\left.\begin{array}{l} x^*=\frac{x}{l},\, y^*=\frac{y}{l}(Gr_l)^{\frac{1}{4}},\, r^*=\frac{r}{l} \text{ where } r=x\sin\phi, \\ u^*=\frac{ul}{\nu}(Gr_l)^{\frac{-1}{2}},\, v^*=\frac{vl}{\nu}(Gr_l)^{\frac{-1}{4}},\, Pr=\frac{\nu}{\alpha}, \\ T=\frac{T'-T'_{\infty,x}}{T'_w-T'_\infty},\, Gr_l=\frac{g\beta(T'_w-T'_{\infty,0})l^3}{\nu^2}. \end{array}\right\} \tag{5}$$

The thermal stratification parameter $S=\frac{1}{\Delta t_0}\left(\frac{dT'_\infty}{dx^*}\right)$ where $\Delta t_0=T'_w-T'_{\infty,0}$.

The linear function of warmth of the delineated medium $T'_{\infty,x}$ is of the Organize X of the form $T'_{\infty,x}=T'_{\infty,0}+SX$.

By using (5), dimensionless form is given as follows after reduction from Eqs. (1) to (3).

$$\frac{\partial}{\partial x^*}(r^*u^*)+\frac{\partial}{\partial y^*}(r^*v^*)=0 \tag{6}$$

$$u^*\frac{\partial u^*}{\partial x^*}+v^*\frac{\partial u^*}{\partial y^*}=T\cos\phi+\frac{\partial^2 u^*}{\partial y^{*2}} \tag{7}$$

$$u^*\frac{\partial T}{\partial x^*}+v^*\frac{\partial T}{\partial y^*}+Su^*=\frac{1}{Pr}\frac{\partial^2 T}{\partial y^{*2}}. \tag{8}$$

With limit conditions are

$$\left.\begin{array}{l} u^*(x,0)=0,\, v^*(x,0)=0,\, T(x,0)=1-Sx \text{ at } y^*=0 \\ u^*(x,\infty)=0,\, T(x,\infty)=0 \text{ as } y^*\to\infty. \end{array}\right\} \tag{9}$$

Stream functions introduced to reduce three equations into two equations $r^*u^*=\frac{\partial\theta}{\partial y^*}$ and $r^*v^*=\frac{\partial\theta}{\partial x^*}$.

By the above stream functions, the following equations are obtained from Eqs. (7) and (8) and satisfy Eq. (6).

$$\frac{\partial\theta}{\partial y^*}\frac{\partial}{\partial x^*}\left(\frac{1}{r^*}\frac{\partial\theta}{\partial y^*}\right)-\frac{1}{r^*}\frac{\partial\theta}{\partial x^*}\frac{\partial^2\theta}{\partial y^{*2}}=Tr^*\cos\phi+\frac{\partial^3\theta}{\partial y^{*3}} \tag{10}$$

$$\frac{1}{r^*}\left(\frac{\partial\theta}{\partial y^*}\frac{\partial T}{\partial x^*}-\frac{\partial\theta}{\partial x^*}\frac{\partial T}{\partial y^*}+S\frac{\partial\theta}{\partial y^*}\right)=\frac{1}{Pr}\frac{\partial^2 T}{\partial y^{*2}}. \tag{11}$$

The limit conditions, condition (9), can be communicated as

$$\left.\begin{array}{l}\lim\limits_{y^*\to 0}\frac{\partial\theta}{\partial y^*}=0,\ \lim\limits_{y^*\to 0}\frac{\partial\theta}{\partial x^*}=0,\ \lim\limits_{y^*\to 0}T=1-Sx^*\\ \lim\limits_{y^*\to\infty}\frac{\partial\theta}{\partial y^*}=0,\ \lim\limits_{y^*\to\infty}T=0.\end{array}\right\}\tag{12}$$

# 3 Solution of the Problem

Our methodology for plan depends upon the usage of one cutoff group change to the PDE (10) and (11). Under this change the two free factors will be diminished by one and the differential conditions (10)–(12) change into an ODE in a solitary variable, which is the likeness variable. $h:\overline{P}=C^{p}(a)P+k^{p}(a)$

A class of one-parameter "a" of the form with group G initiated

$$\left.\begin{array}{l}x=C^{x^*}(a)x^*+k^{x^*}(a);\ y=C^{y^*}(a)y^*+k^{y^*}(a)\\ \overline{\theta}=C^{\theta}(a)\theta+k^{\theta}(a);\ r=C^{r^*}(a)r^*+k^{r^*}(a)\\ \overline{T}=C^{T}(a)T+k^{T}(a).\end{array}\right\}\tag{13}$$

## 3.1 The Invariance Investigation

Through the following chain-rule operations, change of derivative is obtained from G

$$\left.\begin{array}{l}\overline{P}_i=\left(\frac{C^s}{C^i}\right)P_i\\ \overline{P}_{ij}=\left(\frac{C^s}{C^iC^j}\right)P_{ij};\ i,j=x,y,\end{array}\right\}\tag{14}$$

where $P$ stands for $\theta, r^*, T$.

Conditions (10) and (11) changed invariantly, under Eqs. (13) and (14),

$$\begin{aligned}&\frac{\partial\overline{\theta}}{\partial y}\frac{\partial}{\partial x}\left(\frac{1}{r}\frac{\partial\overline{\theta}}{\partial y}\right)-\frac{1}{r}\frac{\partial\overline{\theta}}{\partial x}\frac{\partial^2\overline{\theta}}{\partial y^2}-\overline{rT\cos\phi}-\frac{\partial^3\overline{\theta}}{\partial y^3}\\ &\quad=E_1(a)\left[\frac{\partial\theta}{\partial y^*}\frac{\partial}{\partial x^*}\left(\frac{1}{r^*}\frac{\partial\theta}{\partial y^*}\right)-\frac{1}{r^*}\frac{\partial\theta}{\partial x^*}\frac{\partial^2\theta}{\partial y^{*2}}-r^*T\cos\phi-\frac{\partial^3\overline{\theta}}{\partial y^{*3}}\right]\end{aligned}\tag{15}$$

$$\begin{aligned}&\frac{1}{r}\left[\frac{\partial\overline{\theta}}{\partial y}\frac{\partial\overline{T}}{\partial x}-\frac{\partial\overline{\theta}}{\partial x}\frac{\partial\overline{T}}{\partial y}+S\frac{\partial\overline{\theta}}{\partial y}\right]-\frac{1}{Pr}\frac{\partial^2\overline{T}}{\partial y^2}\\ &\quad=E_2(a)\left[\frac{1}{r^*}\left(\frac{\partial\theta}{\partial y^*}\frac{\partial T}{\partial x^*}-\frac{\partial\theta}{\partial x^*}\frac{\partial T}{\partial y^*}+S\frac{\partial\theta}{\partial y}\right)-\frac{1}{Pr}\frac{\partial^2T}{\partial y^{*2}}\right]\end{aligned}\tag{16}$$

Here $E_1(a)$, $E_2(a)$ are functions (or) may be constants.

$$\frac{C^{\zeta}}{C^{r^*}C^{x^*}C^{y^*}}\left[\frac{1}{r^*}\frac{\partial \theta}{\partial y^*}\frac{\partial^2 \theta}{\partial x^* \partial y^*} - \frac{1}{r^{*2}}\left(\frac{\partial \theta}{\partial y^*}\right)^2 \frac{\partial r^*}{\partial x^*} - \frac{1}{r^*}\frac{\partial \theta}{\partial x^*}\frac{\partial^2 \theta}{\partial y^{*2}}\right]$$
$$- r^* C^{r^*} C^T T \cos\phi - \frac{C^{\theta}}{(C^{y^*})^2}\frac{\partial^3 \theta}{\partial y^{*3}} + A_1(a) \tag{17}$$
$$= V_1(a)\left[\frac{\partial \theta}{\partial y^*}\frac{\partial}{\partial x^*}\left(\frac{1}{r^*}\frac{\partial \theta}{\partial y^*}\right) - \frac{1}{r^*}\frac{\partial \theta}{\partial x^*}\frac{\partial^2 \theta}{\partial y^{*2}} - T r^* \cos\phi - \frac{\partial^3 \theta}{\partial y^{*3}}\right]$$

$$\frac{C^{\zeta}}{C^{r^*}C^{x^*}C^{y^*}}\frac{1}{r^*}\left(\frac{\partial \theta}{\partial y^*}\frac{\partial T}{\partial x^*} - \frac{\partial \theta}{\partial x^*}\frac{\partial T}{\partial y^*}\right) + S\frac{C^{\theta}}{C^{y^*}}\frac{\partial \theta}{\partial y^*} - \frac{1}{Pr}\frac{C^T}{(C^{y^*})^2}\frac{\partial^2 T}{\partial y^{*2}} + A_2(a)$$
$$= V_2(a)\left[\frac{1}{r^*}\left(\frac{\partial \theta}{\partial y^*}\frac{\partial T}{\partial x^*} - \frac{\partial \theta}{\partial x^*}\frac{\partial T}{\partial y^*}\right) + S\frac{\partial \theta}{\partial y^*}\right) - \frac{1}{Pr}\frac{\partial^2 T}{\partial y^{*2}} \tag{18}$$

where

$$A_1(a) = \sum_{n=1}^{\infty}\binom{-1}{n}\left(\frac{k^{r^*}}{C^{r^*}r^*}\right)^m \left(\frac{(C^{\theta})^2}{C^{r^*}C^{x^*}(C^{y^*})^2}\right)\frac{1}{r^*}\left(\frac{\partial \theta}{\partial y^*}\frac{\partial^2 \theta}{\partial x^* \partial y^*} - \frac{\partial \theta}{\partial x^*}\frac{\partial^2 \theta}{\partial x^* \partial y^*}\right)$$
$$- \sum_{n=1}^{\infty}\binom{-2}{n}\left(\frac{k^{r^*}}{C^{r^*}r^*}\right)^m \left(\frac{(C^{\theta})^2}{C^{r^*}C^{x^*}(C^{y^*})^2}\right)$$
$$- \frac{1}{r^{*2}}\left(\frac{\partial \theta}{\partial y^*}\right)^2 \frac{\partial r^*}{\partial x^*} - (C^{r^*}k^{r^*}r^* + k^{r^*}C^{r^*}T)(C^{\cos\phi}\cos\phi + k^{\cos\phi}) \tag{19}$$

$$A_2(a) = \sum_{n=1}^{\infty}\binom{-1}{n}\left(\frac{k^{r^*}}{C^{r^*}r^*}\right)^m \left(\frac{C^{\theta}}{C^{r^*}C^{x^*}(C^{y^*})^2}\right)\frac{1}{r^*}\left(\frac{\partial \theta}{\partial y^*}\frac{\partial T}{\partial x^*} - \frac{\partial \theta}{\partial x^*}\frac{\partial T}{\partial y^*}\right). \tag{20}$$

The invariance of Eqs. (17) and (18)

$\Rightarrow A_1(a) = 0 = A_2(a)$.

Above equations satisfy by substitute

$$k^{r^*} = k^T = k^{y^*} = 0 \tag{21}$$

$$\frac{(C^{\theta})^2}{C^{r^*}C^{x^*}(C^{y^*})^2} = \frac{C^{\theta}}{(C^{y^*})^3} = \frac{C^{\theta}}{C^{y^*}} = C^{r^*}C^T = V_1(a) \tag{22}$$

$$\frac{C^{\theta}C^T}{C^{r^*}C^{x^*}(C^{y^*})^2} = \frac{C^T}{(C^{y^*})^2} = V_2(a). \tag{23}$$

These yields

$$C^{x^*} = (C^{y^*})^2,\ C^{r^*} = \frac{1}{(C^{y^*})^2},\ C^{\theta} = C^{y^*}. \tag{24}$$

Limit Eqs. (19) and (20) are also invariant

$$\Rightarrow k^{r^*}k^T = 0, \;\; C^T = 1. \tag{25}$$

At last, one-boundary exhaustive G which change invariantly obtained.

Conditions (17) and (18) and the most extreme conditions (19) and (20).

From above conditions, we obtain G.

$$G = \begin{cases} x = (C^y)^2 x^* + k^{x^*} \\ y = C^{y^*} y^* \\ r = \frac{r^*}{(C^{y^*})^2} \\ \overline{\theta} = C^{y^*}\theta + k^{\theta} \\ \overline{T} = T. \end{cases} \tag{26}$$

## 3.2 The Total Arrangement of Total Invariants

$$g_j(x, y, \phi, r^*, T) = G_j(\eta(x^*, y^*)); \, j = 1, 2, 3 \tag{27}$$

if the condition that $\eta = \eta(x, y)$ where $x, y$ are independent variables.
Under satisfying the following linear differential equations, one-parameter group function $g_j(x, y, \phi, r^*, T)$ is a complete invariant.

$$\begin{aligned} &\sum_{i=1}^{5}(\alpha_i w_i + \beta_i)\frac{\partial g}{\partial w_i} = 0; \;\; w_i = x^*, y^*, \zeta, r^*, T \\ &\text{where } \alpha_i = \frac{\partial C^{w_i}}{\partial a}(a^0), \;\; \beta_i = \frac{\partial k^{w_i}}{\partial a}(a^0). \end{aligned} \tag{28}$$

Since $k^{r^*} = k^T = k^{y^*} = 0$.

From Eqs. (23) and using (22), we get

$\beta_2 = \frac{\partial k^y}{\partial a}(a^0) = 0, \beta_4 = \frac{\partial k^{r^*}}{\partial a}(a^0) = 0, \beta_5 = \frac{\partial k^T}{\partial a}(a^0) = 0,$

$\beta_3 = \frac{\partial k^\theta}{\partial a}(a^0) = 0,$ (i.e.) $\beta_2 = \beta_3 = \beta_4 = \beta_5 = 0$

By satisfying the first-order linear PDE, $\eta = \eta(x, y)$ is an invariant by Eq. (22).

$$(\alpha_1 x + \beta_1)\frac{\partial \eta}{\partial x^*} + \alpha_2 y^* \frac{\partial \eta}{\partial y^*} = 0. \tag{29}$$

From the above equation, we get

$$\frac{\partial \eta}{\partial x^*} = 0. \tag{30}$$

Therefore Eq. (30)

$$\Rightarrow \eta = y^* \tag{31}$$

Similarly the totally not changing analysis of the dependent variables $\theta, r^*, T$.

$$\theta(x^*, y^*) = \Gamma_1(x)G(\eta),\ r^*(x, y) = \Gamma_2(x)E(\eta),\ T(x, y) = T(\eta), \tag{32}$$

where $\Gamma_1(x)$, $\Gamma_2(x)$, $G(\eta)$, and $E(\eta)$ are functions to be compute. Since $E(\eta)$ must be equal to a constant say $r_0^*$.

$$r^*(x^*, y^*) = r_0^*\Gamma_2(x). \tag{33}$$

## 3.3 The Reduction to ODE

When all is said in done, the supreme not changing has the structure given in condition (31).

Substitute Eq. (32) in to Eq. (17) and subsequent to partitioning, we get

$$\begin{gathered}\Rightarrow \frac{1}{r^*}\frac{\partial\theta}{\partial y^*}\frac{\partial^2\theta}{\partial x^*\partial y^*} - \frac{1}{r^{*2}}\left(\frac{\partial\theta}{\partial y^*}\right)^2\frac{\partial r}{\partial x^*} - \frac{1}{r^*}\frac{\partial\theta}{\partial x^*}\frac{\partial^2\theta}{\partial y^{*2}} - Tr^*\cos\phi - \frac{\partial^3\theta}{\partial y^{*3}} = 0\\ G''' + \frac{1}{r_0^*\Gamma_2}GG''\frac{\partial\Gamma_1}{\partial x^*} - \left(\frac{1}{r_0^*\Gamma_2}\frac{\partial\Gamma_1}{\partial x^*} - \frac{\Gamma_1}{r_0^*\Gamma_2^2}\frac{\partial\Gamma_2}{\partial x^*}\right)G'^2 + \frac{r_0^*\Gamma_2\Gamma_3}{\Gamma_1}T\cos\phi = 0\end{gathered} \tag{34}$$

Likewise, substituting conditions (26)–(28) in condition (13), we get

$$\left.\begin{aligned}&\frac{1}{r^*}\left(\frac{\partial\theta}{\partial y^*}\frac{\partial T}{\partial x^*} - \frac{\partial\theta}{\partial x^*}\frac{\partial T}{\partial y^*}\right) - \frac{1}{Pr}\frac{\partial^2 T}{\partial y^{*2}} = 0\\ &\frac{1}{r_0^*\Gamma_2}\frac{\partial\Gamma_1}{\partial x^*}G(\eta)T' + \frac{1}{Pr}T'' = 0.\end{aligned}\right\} \tag{35}$$

These random coefficients are

$$C_1 = \frac{1}{r_0^*\Gamma_2}\frac{\partial\Gamma_1}{\partial x^*},\ C_2 = \frac{\Gamma_1}{r_0^*\Gamma_2^2}\frac{\partial\Gamma_2}{\partial x^*},\ C_3 = \frac{r_0^*\Gamma_2\Gamma_3}{\Gamma_1}. \tag{36}$$

Conditions (34) and (35) reduced to the following form, by substituting above condition (36)

$$G''' + CGG'' - (C_1 - C_2)G'^2 + C_3\cos\phi T = 0 \tag{37}$$

$$C_1GT' + \frac{1}{Pr}T'' - \frac{SG'}{C_3} = 0. \tag{38}$$

Corresponding limit conditions are

$$G'(0) = 0, T(0) = 1, G'(\infty) = T(\infty) = 0. \tag{39}$$

(i) Putting $C_1 = \frac{3}{4}$, $C_2 = \frac{1}{4}$, and $C_3 = 1$ in Eqs. (37) and (38), we get

$$G''' + \frac{3}{4}GG'' - \frac{1}{2}G'^2 + \cos\phi T = 0 \tag{40}$$

$$\frac{3}{4}GT' + \frac{1}{Pr}T'' - SG' = 0 \text{ (or) } T'' + \frac{3}{4}PrGT' - SG'Pr = 0. \tag{41}$$

(ii) Putting $C_1 = \frac{7}{4}$, $C_2 = \frac{5}{4}$, and $C_3 = 1$ in Eqs. (37) and (38), we get

$$G''' + \frac{7}{4}GG'' - \frac{1}{2}G'^2 + \cos\phi T = 0 \tag{42}$$

$$\frac{7}{4}GT' + \frac{1}{Pr}T'' - SG' = 0 \text{ (or) } T'' + \frac{7}{4}PrGT' - SG'Pr = 0. \tag{43}$$

Corresponding limit conditions are

$$G'(0) = 0, T(0) = 1, G'(\infty) = T(\infty) = 0. \tag{44}$$

## 4 Discussion of Results

Fourth-order R-K technique used to address Eqs. (40)–(43) with limit conditions (44) were fathomed mathematically. Assessments have finished of the step size $\Delta\eta$ and $\eta_\infty$ is edge of the limit level on the framework. Ultimately if we consider $\Delta\eta = 0.035$ and $\eta_\infty = 12$. Toward contact the accuracy of the arithmetical ways of thinking, different appraisals are made with before open work. To ensure the precision of our arithmetical outcomes, the current result in stable state at $x^* = 1.0$. with $\Pr = 0.71$ is connected with existing closeness procedure recorded as a printed adaptation.

Figures 2, 3, 4 and 5 for some values of the parameter Pr and $S$, the velocity and heat exist at $x^* = 1.0$.

In Figs. 2 and 3, for different estimations of Pr with fixed value of $S = 0.4$, the velocity and temperature were plotted.

Likewise saw that, the velocity and temperature profiles will be decreased because of increment of Pr.

From Fig. 3 when Pr is expanding in the external district, it causes negative estimations of the non-dimensional temperature show up as wings. It is additionally observed from Figs. 4 and 5, a few estimations of S with Pr = 7.0 speed and temperature profiles will exist and expansion in the surrounding delineation boundaries constant decreases the warm lightness. Thus, increment in warm definition causes decline in the speed levels. At the point when S is short of what one there exist

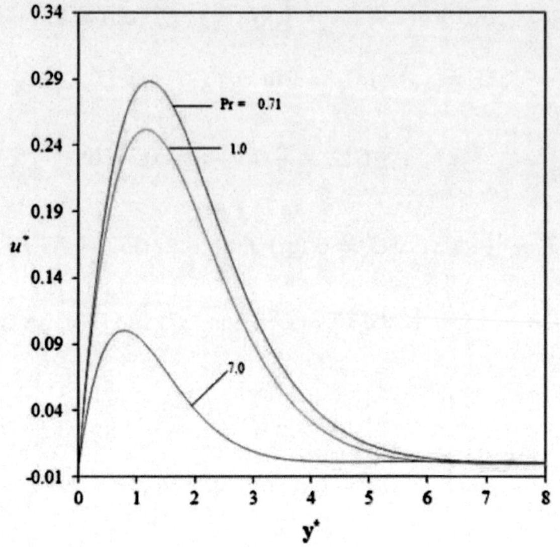


Fig. 2 Velocity profiles at $S = 0.4$ for some values of Pr

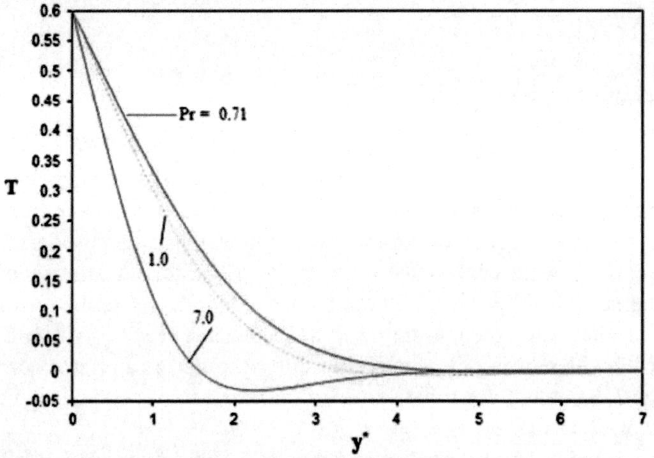


Fig. 3 Temperature profiles at $S = 0.4$ for some values of Pr

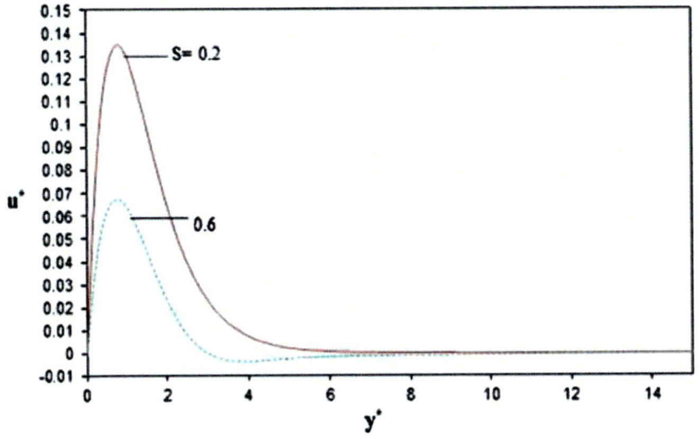


**Fig. 4** Velocity profiles at Pr = 7.0 for some values of *S*

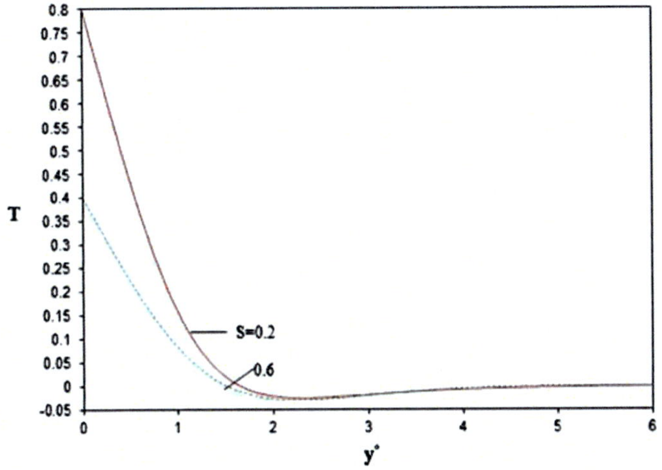


**Fig. 5** Temperature profiles at Pr = 7.0 for some values of *S*

positive lightness at $x^* = 1.0$, however in the external area negative estimations of the non-dimensional heat profiles were noticed. The temperature in the surrounding medium increments quickly with the tallness because of high estimations of S and in the external district of the boundary layer, liquid coming up will be a lot cooler. Consequently, negative upsides of temperature show up in the "wings" of profile. This event is frequently alluded to as "temperature defect" in the flow.

## 5 Conclusions

In a stable flow of rising heat air, stratified medium consistent smooth free convection flow along a perpendicular cone are deals by group theoretical method. The dimensionless regulating limit layer conditions were solved by R-K shooting strategy. The accompanying ends are drawn as follows:

- By increasing impact of parameters S and Pr, the velocity and temperature profiles reduced.
- Due to increasing values of S and Pr, momentum boundary layer will be thinner.
- While Pr is reduced, a flow of rising warm air limit layer becomes thicker.
- Due to effect of increasing values of Pr and S, steady-state values have extended.

## References

1. Moran MJ, Gaggioli RA (1968) Reduction of the number of variables in systems of partial differential equation with auxiliary conditions. SIAM J Appl Math 16:202–215
2. Moran MJ, Gaggioli RA (1968) Similarity analysis via group theory. AIAAJ 6:2014–2016
3. Sulochana C, Samrat SP, Sandeep N (2018) Numerical investigation of magnetohydrodynamic (MHD) radiative flow over a rotating cone in the presence of Soret and chemical reaction. Propuls Power Res 7(1):91–101
4. Sulochana C, Samrat SP, Sandeep N (2017) Magnetohydrodynamic radiative nanofluid flow over a rotating surface with Soret effect. Multidiscip Modell Mater Struct 14(1)
5. Kulkarni AK, Jacobs HR, Hwang JJ (1987) Similarity solution for natural convection flow over an isothermal vertical wall immersed in thermally stratified medium. Int J Heat Mass Trans 30:691–698
6. Tripathi RK, Nath G (1993) Unsteady natural convection flow over a vertical plate embedded in a stratified medium. Int J Heat Mass Trans 36:1125–1128
7. Sandeep N, Vijaya Bhaskar Reddy A, Sugunamma V (2012) Effect of radiation and chemical reaction on transient MHD free convective flow over a vertical plate through porous media. Chem Process Eng Res 12:1–9
8. Sulochana C, Samrat SP, Sandeep N (2017) Boundary layer analysis of an incessant moving needle in MHD radiative nanofluid with joule heating. Int J Mech Sci 128–129:324–331
9. Angirasa D, Srinivasan J (1990) Laminar Axisymmertric multicomponent buoyant plumes in a thermally stratified medium. Int J Heat Mass Trans 33:1751–1757
10. Takhar HS, Chamkha AJ, Nath G (2001) Natural convection flow from a continuously moving vertical surface immersed in a thermally stratified medium. Heat Mass Trans 38:17–24
11. Shapiro A, Fedorovich E (2004) Unsteady convectively driven flow along a vertical plate immersed in a stably stratified fluid. J Fluid Mech 498:333–352
12. Saha SC, Hossain MA (2004) Natural convection flow with combined buoyancy effects due to thermal and mass diffusion in a thermally stratified medium. Non Linear Anal Modell Control 9(1):89–102
13. Birkhoff G (1948) Mathematics for engineers. Elect Eng 67:1185
14. Tripathi RK, Sau A, Nath G (1994) Laminar free convection flow over a cone embedded in a stratified medium. Mech Res Common 21:289–296
15. Hossains MA, Paul SC, Mandal AC (2002) Natural convection flow along a vertical circular cone with uniform surface temperature and surface heat flux in a thermally stratified medium. Int J Num Method Heat Fluid Flow 12:290–305

16. Sandeep N, Sugunamma V (2014) Radiation and inclined magnetic field effects on unsteady hydromagnetic free convection flow past an impulsively moving vertical plate in a porous medium. J Appl Fluid Mech 7(2):275–286
17. Pullepu B, Immanuel Y (2012) Transient free convection flow over a vertical cone embedded in a thermally stratified medium. Bonfring Int J Softw Eng Soft Comput 2(2)
18. Hassanien IA, Hamad MA (2008) Group theoretic method for unsteady free convection flow of a micro polar fluid along a vertical plate in a thermally stratified medium. Appl Math Modell 32:1099–1114
19. Abd-el-Malek MB (1998) Application of the group theoretical method to physical problems. J Nonlinear Math Phys 5:314
20. Hassanien IA, Baker AY, Gorla RSR (1996) Natural convection boundary layer flow of a micro polar fluid along a vertical plate in a thermally stratified medium. Appl Mech Eng 1:381

# Peristaltic Flow of Ferromagnetic Fluid in a Vertical Slot with Mixed Convection

P. Devaki, S. Sreenadh, S. Srinivas, and A. Kavitha

## 1 Introduction

A liquid carrier which is non-conducting mixed with highly concentrated suspension of fine magnetic particles is called a ferrofluid. These fluids have many particle applications in biomedical engineering and technology. Most of the electronic equipment's used in hospitals and industries, kitchen electronic items contain ferrofluids in it. Magnetic and non-magnetic nanoparticles in a vertical channel under the influence of heat transfer were compared by Gul [1]. Rabbi et al. [2] studied on the flow of ferrofluid with the heaters in the bottom and compared results between heat transfer and forced convection. Zeeshan et al. [3] concentrated on velocity and temperature profiles for a ferromagnetic fluid in a stretched sheet with heat transfer using shooting technique. Ferromagnetic fluid and heat transfer effect on external magnetic fluid flow were investigated by Yarahmadi et al. [4]. Lavrova et al. [5] studied on the ferromagnetic particles mixed with ferrofluid, where structure of fluid particles is discussed in detail and obtained interesting results. Study on ferrofluid and its particles under some body forces was done by Asfer et al. [6]. There are many authors [7] concentrating on ferrofluid flow in tubes or channels.

P. Devaki (✉)
Department of Mathematics, SRM Institute of Science and Technology, Ramapuram Campus, Chennai, India 610089
e-mail: devakip@srmist.edu.in

S. Sreenadh
Department of Mathematics, Sri Venkateswara University, Tirupati, India 517502

S. Srinivas
Department of Mathematics, VIT-AP University, Vijayawada 522237, India

A. Kavitha
Department of Mathematics, VIT, Vellore, India



S. Srinivas et al. (eds.), *Recent Advances in Applied Mathematics and Applications to the Dynamics of Fluid Flows*, Lecture Notes in Mechanical Engineering,
https://doi.org/10.1007/978-981-19-1929-9_19

Sadeghinezhad et al. [8] did experimental work on heat transfer effect on hybrid nanofluid and found informative results. Hariri et al. [9] found that Nusselt number increases with increase effect of magnetic field in a non-uniform tube. Closed-form solutions for a non-Newtonian Carreau nanofluid flow in a tapered tube under the influence of magnetic field and peristalsis were investigated by Kothandapani and Prakash [10]. Noreen [11] studied on the flow of Casson fluid under the effect of magnetic field and peristaltic wave in an asymmetric channel. Shit et al. [12] concentrated on couple stress fluid in a magnetohydrodynamic channel with contraction and expansion due to peristaltic wave. Casson fluid flow in a carbon nanotube under the influence of peristalsis and magnetic field was analyzed by Aman et al. [13]. Most of the authors [14–20] are concentrating on the impact of magnetic field in the field of biofluid flows.

In this paper, we study the peristaltic motion of ferromagnetic fluids through a vertical slot with mixed convection under long wavelength and low Reynolds number. The flow is considered in accordance with two boundaries initially at the same temperature and then at different temperatures. The effect of magnetization on the velocity and temperature distributions is understood. The governing equations are highly non-linear and are solved analytically by using the perturbation technique.

## 2 Mathematical Formulation

We consider the flow of a ferromagnetic fluid in a vertical slot with flexible walls in the presence of vertical magnetic field. In the Cartesian coordinate system, $x$-axis is taken along the central line of the channel and a line perpendicular to the $y$-axis (Fig. 1). The wall deformation at any time t is in 'a', which is the half width of the channel. The flow is inherently unsteady in the fixed frame of reference. The flow becomes steady in the wave frame moving with velocity c along the wave. Under the assumption of long wavelength and low Reynolds number, the waveform $(x, y)$ is converted to fixed frame $(x', y')$ with speed $c$ along with the direction of wave. The transformations between the frames are given by $x = x' - ct'$, $y = y'$, $u = u' - c$, $v = v'$.

Basic governing equations, neglecting the viscous dissipation and assuming unidirectional flow in the $x$-direction, take the form

$$\nu \frac{d^2T}{dy^2} + g\beta(T - T_0) - \frac{\mu_0 \alpha}{\rho_0}(T - T_0)\frac{\partial H}{\partial x} = \frac{1}{\rho_0}\frac{\partial p'}{\partial x} \tag{1}$$

$$\frac{\mu_0}{\rho_0 C_0}\left[\frac{\partial M}{\partial T}\right]\left[\frac{\partial H}{\partial x}\right] uT = k\frac{d^2T}{dy^2}, \tag{2}$$

where $p' = p - \mu_0 M_0 H - \frac{1}{2}\mu_0 \chi (H - H_0)^2$ and $k = \frac{k_1}{\rho_0 C_0}$.

**Fig. 1** Physical model

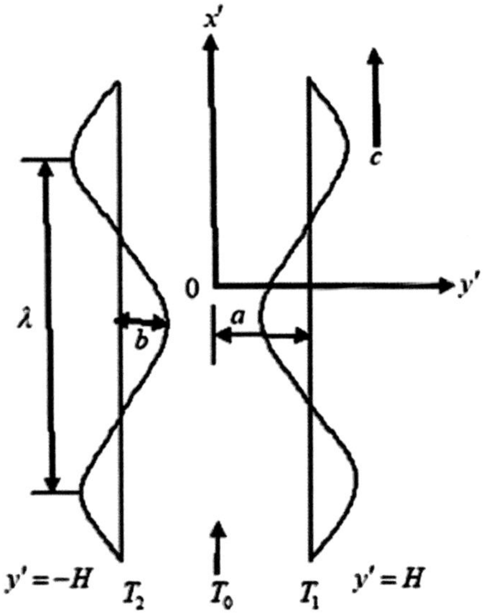


The boundary conditions are

$$u = 0 \;\; at \;\; y = \pm h \tag{3}$$

$$T = T_1 \;\; at \;\; y = h \qquad T = T_2 \;\; at \;\; y = -h. \tag{4}$$

Introducing the following non-dimensional quantities:

$$x^* = \frac{x}{\lambda} \;\; y^* = \frac{y}{a} \;\; U^* = \frac{u}{c} \;\; h^* = \frac{h}{a} \;\; p^* = \frac{pa^2}{\eta c \lambda} \;\; \theta = \frac{T - T_0}{T_1 - T_0}. \tag{5}$$

The non-dimensional governing equations after dropping asterisks (*) become

$$\frac{d^2U}{dy^2} + (BRa + \frac{AG}{\lambda_0})\theta + P = 0 \tag{6}$$

$$\frac{d^2\theta}{dy^2} = AU\theta + A\lambda_0 U, \tag{7}$$

where $b = \frac{K}{AC}$, $G = \frac{c_0 T_0 K}{\nu c^2}$, $\lambda_0 = \frac{T_0}{T_1 - T_0}$, $P = -\frac{\partial p'}{\partial x}$, $Ra = \alpha g(T_1 - T_0)\frac{a^3}{\nu k}$, $A = \frac{\mu_0 \alpha a^2 c}{\rho_0 C_0 k \lambda}\frac{\partial H}{\partial x}$.

The corresponding non-dimensional boundary conditions are

$$U = 0 \;\; at \;\; y = \pm h \quad \theta = 1 \;\; at \;\; y = h \quad \theta = 1 + \alpha \;\; at \;\; y = -h. \tag{8}$$

## 2.1 Method of Solution

Equations (6) and (7) are solved using the boundary conditions (8). Because of the non-linearity in the governing equations, analytical solutions of these equations are difficult. Assuming the magnetization parameter A to be small, we obtain the solutions of Eqs. (6) and (7) in the form

$$\begin{aligned} U &= U_0 + AU_1 + A^2U_2 + ....... \\ \theta &= \theta_0 + A\theta_1 + A^2\theta_2 + .... \end{aligned} \tag{9}$$

Substituting Eq. (9) into Eqs. (6)–(8) and collecting terms of equal powers in $A$ , we obtain the following sets of equations:

**Zeroth-order equations**

$$\frac{d^2U_0}{dy^2} + (BRa)\theta_0 + P = 0 \tag{10}$$

$$\frac{d^2\theta_0}{dy^2} = 0 \tag{11}$$

**First order equation**

$$\frac{d^2U_1}{dy^2} + (BRa)\theta_1 + \frac{G}{\lambda_0}\theta_0 = 0 \tag{12}$$

$$\frac{d^2\theta_1}{dy^2} = U_0\theta_0 + \lambda_0 U_0. \tag{13}$$

The reverent boundary conditions are

$$\begin{aligned} U_0 = 0 \quad U_1 = 0 \quad \theta_0 = 1 \quad \theta_1 = 0 \quad at \quad y = +h \\ U_0 = 0 \quad U_1 = 0 \quad \theta_0 = 1 + \alpha \quad \theta_1 = 0 \quad at \quad y = -h. \end{aligned} \tag{14}$$

The solutions of Eqs. (10)–(13) using the boundary conditions (14) are

$$U_0 = F\left(\frac{h^2 - y^2}{2}\right) - \frac{\alpha BRa}{2}\left[\frac{y^2}{2} - \frac{y^3}{6h} + \frac{hy}{6} - \frac{h^2}{2}\right] \tag{15}$$

$$\theta_0 = 1 + \frac{\alpha}{2}\left[1 - \frac{y}{h}\right] \tag{16}$$

$$U_1 = A_1 + A_2 - \frac{(BRa)^2\,\alpha}{4}[A_3] + \frac{B\alpha Ra}{4}(F)\,[A_4] + A_5 - A_6 + A_7 \tag{17}$$

$$\theta_1 = \frac{F}{2}[B_0] - \frac{\alpha^2 B Ra}{8}[B_1] + \frac{\alpha F}{4}[B_2] + \frac{\lambda_0 F}{2}[B_3] - \left[\frac{\lambda_0 \alpha B Ra}{4} + \frac{\alpha B Ra}{4}\right] B_4, \tag{18}$$

where

$$A_1 = \frac{BRa}{2}(P + BRa)\left[-\frac{h^2y^4}{24} + \frac{y^6}{360} + \frac{5h^5y^2}{24} - \frac{61h^6}{360}\right]$$

$$A_2 = \frac{G}{\lambda_0}\left[\frac{h^2}{2} - \frac{y^2}{2} - \frac{\alpha}{2}\left(\frac{y^2}{2} - \frac{y^3}{6h} + \frac{hy}{6} - \frac{h^2}{2}\right)\right]$$

$$A_3 = -\frac{y^6}{360} + \frac{y^7}{2520h} - \frac{hy^5}{360} + \frac{h^2y^4}{24} + \frac{7h^3y^3}{1080} - \frac{5h^4y^2}{24} + \frac{125h^6y}{756} - \frac{61h^5y}{360} + \frac{61h^6}{360}$$

$$A_4 = -\frac{h^2y^4}{24} + \frac{y^6}{360} + \frac{hy^5}{120} - \frac{y^7}{840h} - \frac{7h^3y^3}{360} + \frac{5h^4y^2}{24} - \frac{11h^5y}{70} + \frac{61h^5y}{360} - \frac{61h^6}{360}$$

$$A_5 = \frac{(B\alpha Ra)^2}{8}\left[-\frac{y^6}{540} + \frac{y^7}{630h} - \frac{5hy^5}{720} + \frac{h^2y^4}{24} - \frac{y^8}{5040h^2} + \frac{13h^3y^3}{1080} - \frac{13h^4y^2}{60} - \frac{635h^5y}{3024} + \frac{2677h^6}{15120}\right]$$

$$A_6 = \frac{\lambda_0\alpha(BRa)^2}{4}\left[-\frac{y^6}{360} + \frac{y^7}{2520h} - \frac{hy^5}{360} + \frac{h^2y^4}{24} + \frac{7h^3y^3}{1080} - \frac{5h^4y^2}{24} + \frac{125h^5y}{756} - \frac{61h^5y}{360} + \frac{61h^6}{360}\right]$$

$$A_7 = \frac{B\lambda_0 Ra}{2}(P + BRa)\left[-\frac{h^2y^4}{24} + \frac{y^6}{360} + \frac{5h^4y^2}{24} - \frac{61h^6}{360}\right]$$

$$F = P + BRa$$

$$B_0 = \frac{h^2y^2}{2} - \frac{y^4}{12} - \frac{5h^4}{12}$$

$$B_1 = \frac{y^4}{18} - \frac{y^5}{15h} + \frac{2hy^3}{9} - \frac{h^2y^2}{2} + \frac{y^6}{90h^2} - \frac{7h^3y}{45} + \frac{13h^4}{30}$$

$$B_2 = \frac{h^2y^2}{2} - \frac{y^4}{12} - \frac{hy^3}{6} + \frac{y^5}{20h} + \frac{7h^3y}{60} - \frac{5h^4}{12}$$

$$B_3 = \frac{h^2y^2}{2} - \frac{y^4}{12} - \frac{5h^4}{12}$$

$$B_4 = \frac{y^4}{12} - \frac{y^5}{60h} - hy^3 18 - h^2y^2 2 - 7h^3y 180 + \frac{5h^4}{12}.$$

## 3 Results and Discussion

Perturbation technique is used to find the velocity and temperature distributions for magnetic fluids in the vertical channel. The magnetization parameter A may be positive or negative depending on the direction of the constant gradient of the magnetic field. The present paper is positive when the gradient of magnetic field acts in the direction of gravity and is negative when it acts opposite to gravity. The velocity and temperature distributions are computed for mixed and free convection.

The velocity and temperature profiles are obtained for different values of between the peristaltic boundaries are shown in Figs. 2 and 3. It is noticed that increment in the values of $\alpha$ result in the increase of velocity and temperature. Figures 4 and 5 depict the fact that the velocity leads to raise with raising the values of magnetization parameter. The temperature distribution obtained for different values of $A$ in the presence and in the absence of peristalsis is investigated through Figs. 6 and 7, respectively. It is noticed that the temperature decreases with increase in magnetization parameter in the presence and in the absence of peristalsis. Also, in the presence of peristaltic wave for fixed values of magnetization parameter, the temperature reduces. Figures 8 and 9 represent the variation of velocity with respect to $y$ for different values of thermal Rayleigh number in the presence and in the absence of peristalsis, respectively.

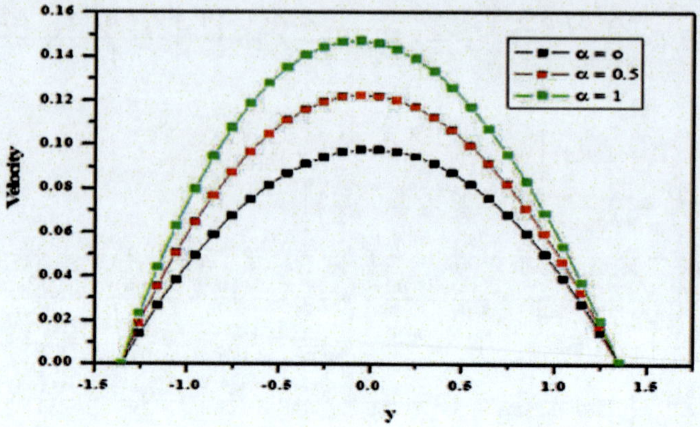


**Fig. 2** Velocity profiles for $\alpha$

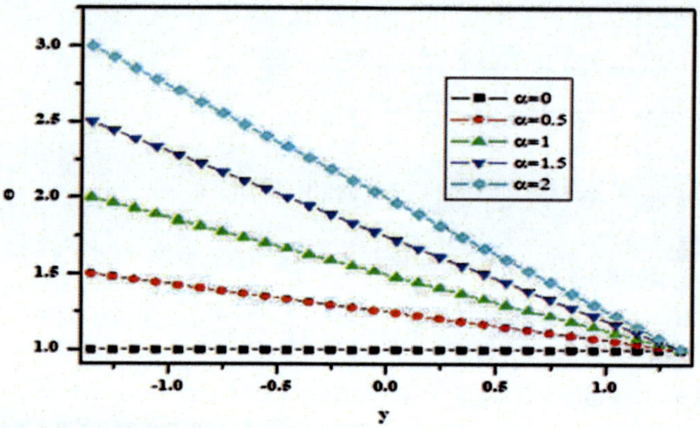


**Fig. 3** Temperature profiles for $\alpha$

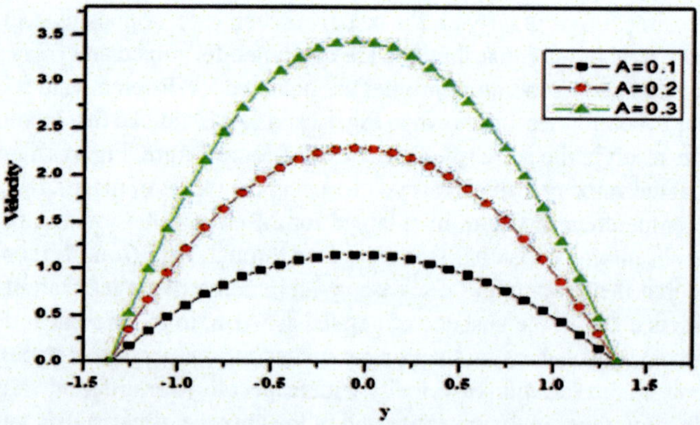


**Fig. 4** Velocity profiles with $h(x)$

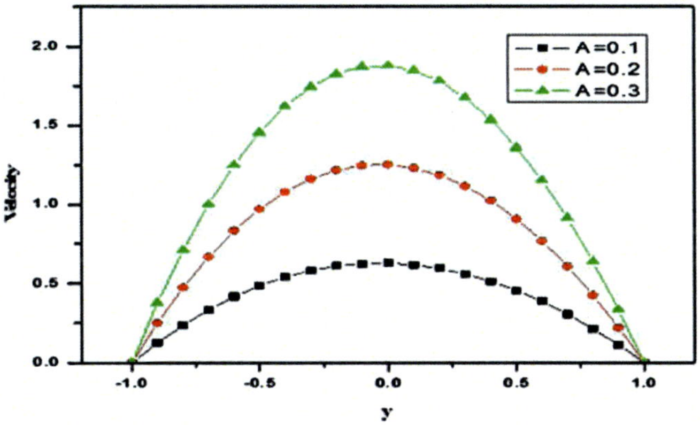


**Fig. 5** Velocity profiles without $h(x)$

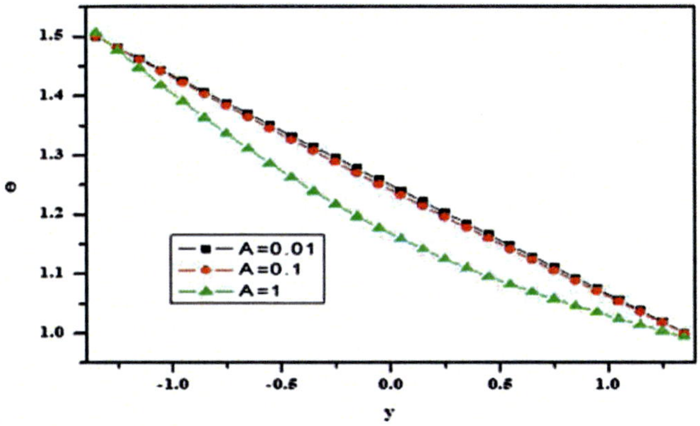


**Fig. 6** Temperature profiles with $h(x)$

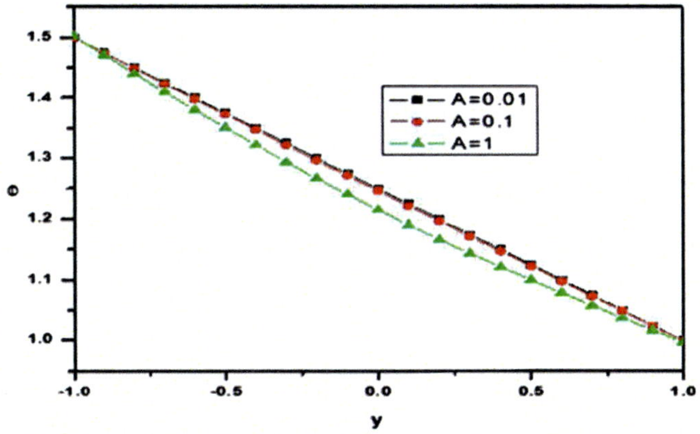


**Fig. 7** Temperature profiles without $h(x)$

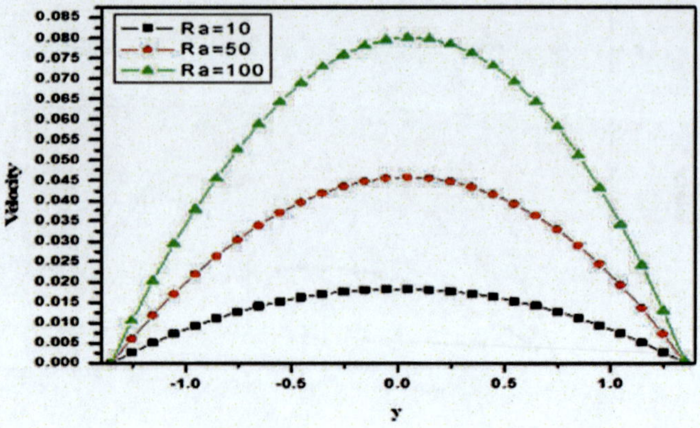


**Fig. 8** Velocity profiles for $Ra$ with $h(x)$

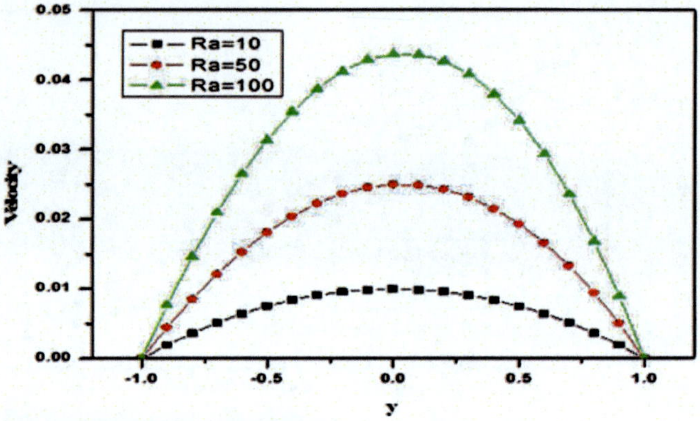


**Fig. 9** Velocity profiles for $Ra$ without $h(x)$

Variations of temperature with $y$ for different values of the Rayleigh number in the presence and in the absence of peristalsis are shown in Figs. 10 and 11, respectively. It is noticed from both the figures that an increasing value of $Ra$, the temperature is decreased. Further, we found that in the presence of peristaltic wave, the temperature behavior is non-linear.

The path of the flow of the fluid in the case of mixed convection is decided by a competitive influence of the pressure gradient and the magnetization parameter. Figures 12 and 13 depict the variation of velocity and temperature profiles with respect to the peristaltic waves for different pressure drops, respectively. From these figures, it is clearly seen that when the values of $P$ increase, then there is an increasing effect on velocity, whereas reverse effects occur in the temperature profile.

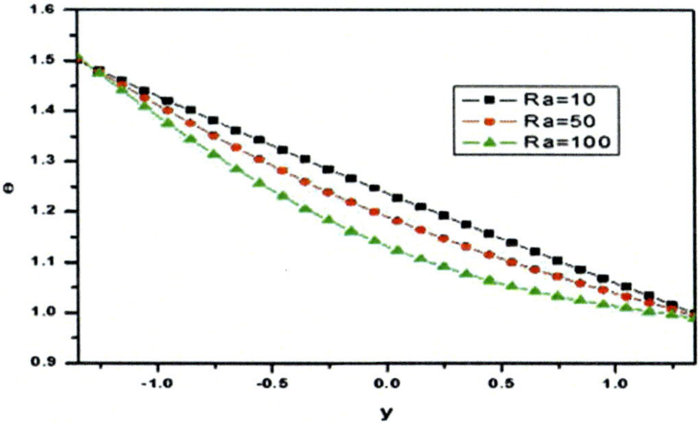


**Fig. 10** Temperature profiles for $Ra$ with $h(x)$

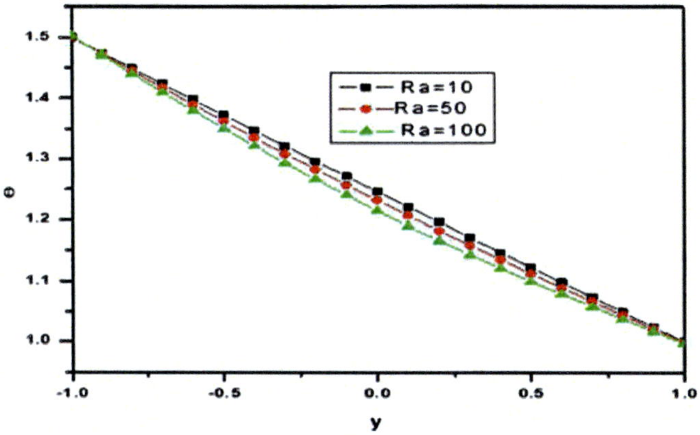


**Fig. 11** Temperature profiles for $Ra$ without $h(x)$

From Fig. 14, we observe that there is a variation of $\delta P$ with $Q$ for different values of $\phi$. It is interesting to note that all curves are intersecting in the pumping region at $Q = 1.0$. For $0 \leq Q \leq 1.0$, we observe that the pressure difference decreases with increase in the amplitude ratio of the peristaltic wave. The same behavior is observed with the increase in $\alpha$ which is shown in Fig. 15.

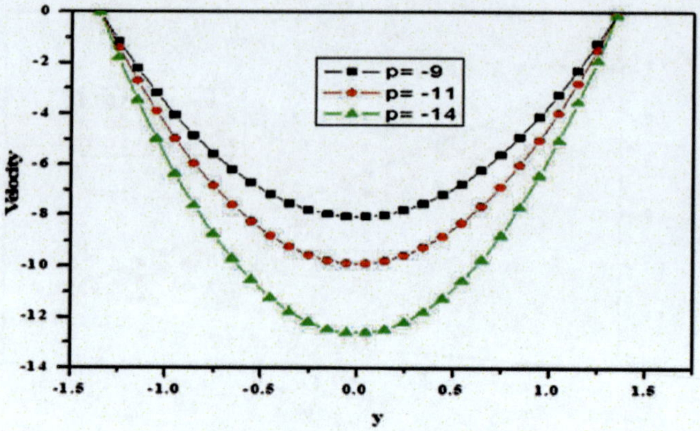


**Fig. 12** Velocity profile for different values of *P*

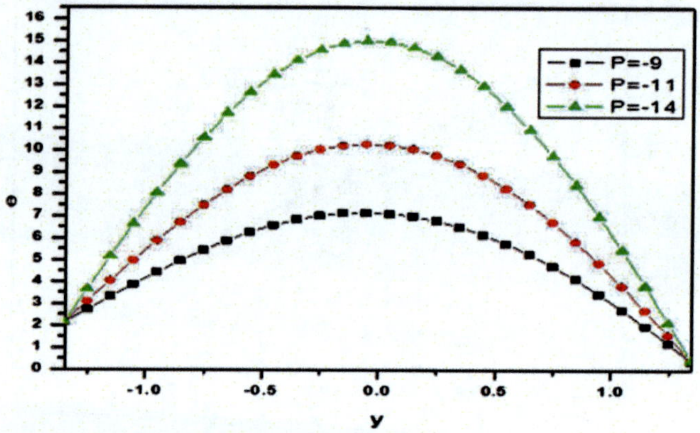


**Fig. 13** Temperature profile for different values of *P*

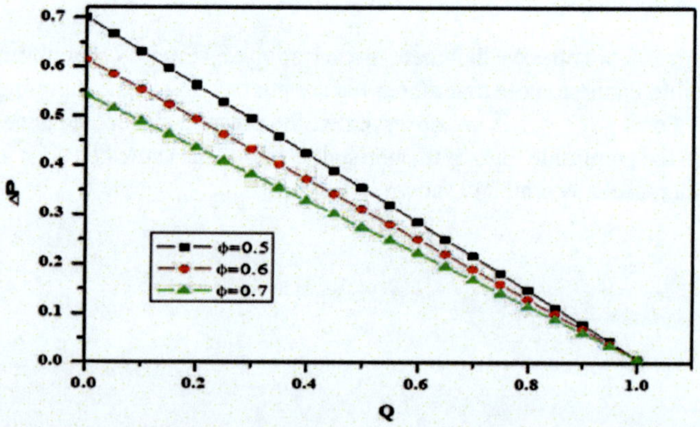


**Fig. 14** Pressure rise versus flux for different $\phi$

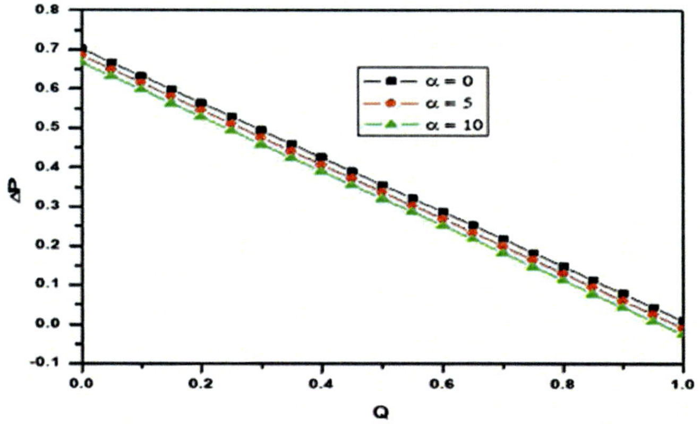


**Fig. 15** Pressure rise versus flux for different $\alpha$

## References


1. Gul A, Khan I, Shafie S, Khalid A, Khan A (2015) Heat transfer in MHD mixed convection flow of a ferrofluid along a vertical channel. PLos ONE 10(11):1–14 (2015)
2. Md. Rabbi K, Saha S, Mojumder S, Rahman MM, Saidur Talaat R, Ibrahim A (2016) Numerical investigation of pure mixed convection in a ferrofluid-filled lid-driven cavity for different heater configurations. Alex Eng J 55(1):127–139
3. Zeeshan A, Majeed A, Ellahi R, Zia QM (2016) Mixed convection flow and heat transfer in ferromagnetic fluid over a stretching sheet with partial slip effects. Therm Sci 268–268
4. Lavrova O, Polevikov V, Tobiska L (2016) Modelling and simulation of magnetic particle diffusion in a ferrofluid layer. Magnetohydrodynamics 52(4):439–452
5. Asfer M, Mehta B, Arun K, Khandekar S, Pradipta Kumar P (2016) Effect of magnetic field on laminar convective heat transfer characteristics of ferrofluid flowing through a circular stainless steel tube. Int J Heat Fluid Fl 59:74–86
6. Goshayeshi HR, Goodarzi M, Safaei MR, Dahari M (2016) Experimental study on the effect of inclination angle on heat transfer enhancement of a ferrofluid in a closed loop oscillating heat pipe under magnetic field. Exp Therm Fluid Sci 74:265–270
7. Shahsavar A, Saghafian M, Salimpour MR, Shafii MB (2016) Experimental investigation on laminar forced convective heat transfer of ferrofluid loaded with carbon nanotubes under constant and alternating magnetic fields. Exp Therm Fluid Sci 76:1–11
8. Sadeghinezhad E, Mehrali M, Akhiani AR, Latibari ST, Dolatshahi-Pirouz AD, Metselaar HC, Mehrali MM (2017) Experimental study on heat transfer augmentation of graphene based ferrofluids in presence of magnetic field. Appl Therm Eng 114:415–427
9. Hariri S, Mokhtari M, Gerdroodbary MB, Fallah K (2017) Numerical investigation of the heat transfer of a ferrofluid inside a tube in the presence of a non-uniform magnetic field. Eur Phys J Plus 132
10. Kothandapani M, Prakash J (2015) Effect of radiation and magnetic field on peristaltic transport of nanofluids through a porous space in a tapered asymmetric channel. J Magn Magn Mater
11. Noreen S (2015) Influence of magnetic field on peristaltic flow of a Casson fluid in an asymmetric channel: application in crude oil refinement. J Magn Magn Mater 378:463–468
12. Shit GC, Ranjit NK, Sinha A (2016) Electro-magnetohydrodynamic flow of biofluid induced by peristaltic wave: a non-newtonian model. J Bionic Eng 13(3):436–448

13. Aman S, Khan I, Ismail Z, Salleh MZ, Ali Alshomrani S, Alghamdi MS (2017) Magnetic field effect on Poiseuille flow and heat transfer of carbon nanotubes along a vertical channel filled with Casson fluid. AIP Adv 7(1)
14. Bhuvaneswari M, Eswaramoorthi S, Sivasankaran S, Hussein A (2019) Cross-diffusion effects on MHD mixed convection over a stretching surface in a porous medium with chemical reaction and convective condition. Eng Trans 67:3–19
15. Niranjan H, Sivasankaran S, Bhuvaneswari M (2017) Chemical reaction, Soret and Dufour effects on MHD mixed convection stagnation point flow with radiation and slip condition Scientia Iranica. Trans Mech Eng B 24:698–706
16. Sivasankaran S, Niranjan H, Bhuvaneswari M (2017) Chemical reaction, radiation and slip effects on MHD mixed convection stagnation-point flow in a porous medium with convective boundary condition. Int J Num Methods Heat Fluid Flow 27:454–470
17. Sivasankaran S, Narrein K (2020) Influence of geometry and magnetic field on convective flow of nanofluids in trapezoidal microchannel heat sink. Iran J Sci Technol Trans Mech Eng 44:373–382
18. Niranjan H, Sivasankaran S, Bhuvaneswari M (2016) Analytical and numerical study on magnetoconvection stagnation-point flow in a porous medium with chemical reaction. Radiat Slip Eff Math Probl Eng 2016:1–12
19. Bindhu R., Sai SundaraKrishnan G, Sivasankaran S, Bhuvaneswari M (2019) Magnetoconvection of water near its maximum density in a cavity with partially thermally active walls. Energy Environ 30(5):833–853
20. Sivasankarana S, Ananthan SS, Abdul Hakeem AK (2016) Mixed convection in a lid-driven cavity with sinusoidal boundary temperature at the bottom wall in the presence of magnetic field. Sci Iranica Trans B Mech Eng 23(3):1027–1036

# Aligned Magnetic Field and Radiation Absorption Effects on Free Convection Chemically Reactive Flow Past An Inclined Surface

**S. Sreelatha, K. S. Balamurugan, and J. Prakash**

## 1 Introduction

The last two decennaries have experienced phenomenal amendments in the exploration concerning the boundary layer flow of Newtonian fluids with reference to heterogeneous conditions across the intercontinental. It is eminence of conviction that the noteworthy researcher augmented the Newtonian fluid on Magnetohydrodynamics (MHD) free convection fluid flow over the proper concentrate of base fluid beneath appropriate conditions. Hydromagnetics is analyzation of magnetic phenomenon of electrically conduction fluids. Various researcher has been examined on MHD free convective fluid flow such as Ping and Minkowycz [1] explored on MHD free convective flow through flat plate embedded in a porous medium by means of relevance to heat transfer as of an embankment. Lai and Kulacki [2] have conversed on MHD mixed convective fluid flow by means of mass and heat transfer through permeable perpendicular isothermal plate. Ramprasad et al. [3] have contemplated unsteady MHD 2D free convection amalgamated head and mass transfer flow through an inclined fluctuating surface by means of heat absorption. Rajakumar et al. [4], Nakayama and Koyama [5], Bestman [6] said various fascinating consequences on MHD flow through vertical plate.

S. Sreelatha (✉)
Department of Mathematics, S.C.W. Degree and PG College, Kothagudem, Telengana, India
e-mail: ssreelathamsc@gmail.com

K. S. Balamurugan
Department of Mathematics, RVR & JC College of Engineering, Guntur, Andhra Pradesh, India
e-mail: muruganbalaks@gmail.com

J. Prakash
Department of Mechanical Engineering Science, University of Johannesburg, Johannesburg, South Africa
e-mail: jagdishp@uj.ac.za



S. Srinivas et al. (eds.), *Recent Advances in Applied Mathematics and Applications to the Dynamics of Fluid Flows*, Lecture Notes in Mechanical Engineering,
https://doi.org/10.1007/978-981-19-1929-9_20

Multifarious chemically affecting regimes amalgamate both heterogeneous and homogeneous reactions, with circumstances experiencing in enzyme catalysis, incineration as well as bio-chemical mechanisms. The interrelatedness among homogenous reactions therein the volume of the fluid and heterogeneous reaction transpiring certain enzyme catalysis surfaces is conventionally remarkably complex and is apprehensive in the generation and pulmonary tuberculosis of responsive species at multifarious rates both in fluid and surfaces of catabolism. Many researches have done a lot of work-related chemical reactions such as Chamkha [7], Muthucumara swamy and Ganesan [8], Ibrahim et al. [9], Makinde [10], Abdelkhalek [11], Arshad Khan et al. [12], and Mythreye et al. [13]. Srinivas et al. [14], Muthuraj et al. [15], Loudu Immaculate et al. [16], Muthuraj et al. [17], Lourdu Immaculate et al. [18], and Muthuraj et al. [19] discussed impact of chemical reaction as well as space porosity on MHD mixed convective flow in a perpendicular asymmetric channel along with peristalsis.

The main objective of current investigation is influence of aligned magnetic field as well as suction velocity on unsteady MHD free convective fluid flow past an inclined plate by means of radiation absorption along with chemical reaction. In this examination, the governing PDEs are determined by employing multiple regular perturbation law. The repercussions of this contemplate have been shown graphically such as "velocity, temperature and concentration" and furthermore the expressions for shear stress along with the rate of heat and mass transfer coefficients are derived in tabular form with dissimilar implications of pertinent parameters.

## 2 Mathematical Fluid Layer with Interpretation of the Problem

Contemplate unsteady MHD 2D flow of heat-absorbing fluid past an inclined plate enclosed in a porous medium in the comportment of chemical reaction. The flow is presumed to be in the direction of $-x^*$, which is occupied along the semi-infinite inclined plate and y-axis upright to it. A magnetic field of uniform strength B0 is interposed at an angle $\xi$ to direction of the flow. In the equation of energy radiation absorption is taken into contemplation (Fig. 1).

From the above presumptions, the flow is governed by the following PDEs considered from Mythreye et al. [13] which are given as follows:

$$v_y^* = 0 \tag{1}$$

$$\left.\begin{aligned} [u_{t^*}^*] + v^*[u_{y^*}^*] &= -\frac{1}{\rho}[p_x^*] + v[u_{y^*y^*}^*] + g\beta_T[T - T_\infty]\,Cos\,(\Psi) \\ &\quad + g\beta_C[C - C_\infty]\,Cos\,(\Psi) - v\left[\frac{u^*}{K^*}\right] - \frac{\sigma}{\rho}B_0^2[u^*]\sin^2\xi \end{aligned}\right\} \tag{2}$$

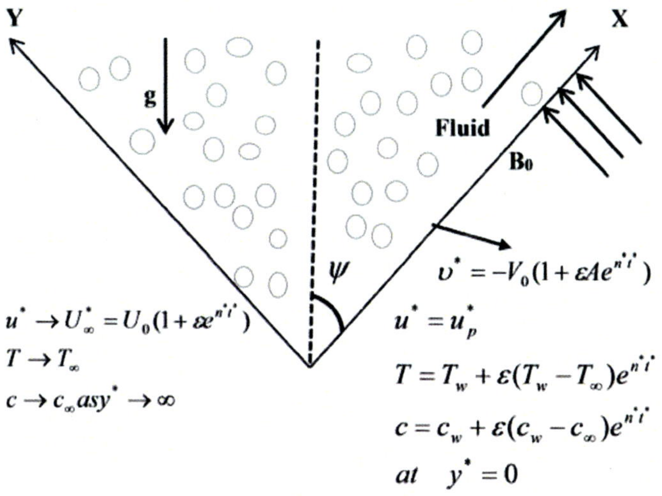


**Fig. 1** Physical model of the problem

$$[T_{t^*}] + v^*\left[T_{y^*}\right] = \alpha\left[T_{y^*y^*}\right] - \frac{Q_0}{\rho C_p}[T - T_\infty] + \frac{R^*}{\rho C_p}[C - C_\infty] \tag{3}$$

$$[C_{t^*}] + v^*\left[C_{y^*}\right] = D\left[C_{y^*y^*}\right] - K\ (C - C_\infty) \tag{4}$$

Under these suppositions, the suitable boundary conditions are

$$\left.\begin{aligned} [u^*] = \left[u_p^*\right],\ [T] = [T_w] + \varepsilon\left[(T_w - T_\infty)e^{n^*t^*}\right], [C] = [C_w] + \varepsilon\left[(C_w - C_\infty)e^{n^*t^*}\right] at\ \ y^* = 0 \\ [u^*] \to [U_\infty^*] = U_0\left[1 + \varepsilon e^{n^*t^*}\right], [T] \to [T_\infty],\ [C] \to [C_\infty]\ \text{as}\ y^* \to \infty \end{aligned}\right\} \tag{5}$$

It is clear that from Eq. (1) the velocity of suction at the surface is the only function of time. Supposing so as to it takes the resulting exponential outlines:

$$v^* = -V_0\left[1 + \varepsilon A\ e^{n^*t^*}\right] \tag{6}$$

where A is constant (non-zero real), $\varepsilon$ and $\varepsilon A < 1$ suction velocity $V_0$ is positive constant. From Eq. (2), we have

$$\left[p_x^*\right] = \left[(U_\infty^*)_t\right] + \frac{\upsilon}{K^*}\left[U_\infty^*\right] + \frac{\sigma}{\rho}B_0^2\left[U_\infty^*\right]. \tag{7}$$

The non-dimensional variables are as follows:

$$\left.\begin{array}{l} u=\left[\dfrac{u^*}{U_0}\right], v=\left[\dfrac{v^*}{V_0}\right], \; y=\left[\dfrac{V_0 y^*}{v}\right], \; U_\infty=\left[\dfrac{U_\infty^*}{U_0}\right] \\ U_p=\left[\dfrac{u_p^*}{U_0}\right], t=\left[\dfrac{V_0^2 y^*}{v}\right], n=\left[\dfrac{n^* v}{V_0^2}\right], K=\left[\dfrac{V_0^2 K^*}{v}\right] \end{array}\right\} \quad (8)$$

According to (6)–(8), Eqs. (2)–(4) are condensed to dimensionless form

$$[u_t]-[1+\varepsilon e^{nt}]u_y=[(U_\infty^*)_x]+[u_{yy}]+G_T[\theta][cos\Psi]+G_C[C][cos\Psi]+N[U_\infty-u] \quad (9)$$

$$[\theta_t]-(1+\varepsilon e^{nt})[\theta_y]=\frac{1}{P_r}[\theta_{yy}]-Q[\theta]+R[C] \quad (10)$$

$$[C_t]-[1+\varepsilon e^{nt}][C_y]=[S_c]^{-1}[C_{yy}]-K_r\;[C] \quad (11)$$

The related boundary conditions are

$$\left.\begin{array}{ll} [u]=[u_p], \theta=[1+\varepsilon e^{nt}], c=[1+\varepsilon e^{nt}] & at \quad y=0 \\ [u]=[u_\infty], [\theta]\rightarrow[0], [C]\rightarrow[0] & as \quad y\rightarrow\infty \end{array}\right\} \quad (12)$$

Here

$$\left.\begin{array}{l} C=\dfrac{(C-C_\infty)}{(C_w-C_\infty)}, G_T=\dfrac{v\beta_T g\,(T_w-T_\infty)}{U_0 V_0^2}, G_c=\dfrac{v\beta c g\,(C_w-C_\infty)}{U_0 V_0^2}, R=\dfrac{vR^*\,(C_w-C_\infty)}{\rho c_p\,(T_w-T_\infty)\,V_0^2} \\ P_r=\dfrac{v\rho c_p}{k}=\dfrac{v}{\alpha}, Sc=\dfrac{v}{D}, Q=\dfrac{vQ_0}{\rho c_p V_0^2}, K_r=\dfrac{kv}{V_0^2}, M=\dfrac{\sigma B_0^2 v}{\rho V_0^2}, \theta=\dfrac{(T-T_\infty)}{(T_w-T_\infty)} \\ N=\left(M\,sin^2\xi+\dfrac{1}{K}\right) \end{array}\right\} \quad (13)$$

## 3 Solution of the Problem

For determining solution of Eqs. (9)–(13), we are employing multiple regular perturbation method so we use the following relations:

$$\left.\begin{array}{l} [u]=u_0[y]+\varepsilon e^{nt}u_1[y]+O[\varepsilon^2]+\ldots \\ [\theta]=\theta_0[y]+\varepsilon e^{nt}\theta_1[y]+O[\varepsilon^2]+\ldots \\ [C]=C_0[y]+\varepsilon e^{nt}C_1[y]+O[\varepsilon^2]+\ldots \end{array}\right\} \quad (14)$$

Substituting Eq. (14) into Eqs. (9)–(11), equating the terms of non-harmonic and harmonic, and neglecting the higher terms of, we obtain

$$\left[u_0''\right] + \left[u_0'\right] - N\left[u_0\right] = -N - G_{T_1}\left[\theta_0\right] - G_{C_1}\left[C_0\right] \tag{15}$$

$$\left[u_1''\right] + \left[u_1'\right] - (N+n)\left[u_1\right] = -A\left[u_0'\right] - G_{T_1}\left[\theta_1\right] - G_{C_1}\left[C_1\right] - \left[N+n\right] \tag{16}$$

$$\left[\theta_0''\right] - \left[\theta_0'\right] Pr - QPr\left[\theta_0\right] = -PrR\left[C_0\right] \tag{17}$$

$$\left[\theta_1''\right] - \left[\theta_1'\right] Pr - (n+Q)Pr\left[\theta_1\right] = -APr\left[\theta_0'\right] - PrR\left[C_1\right] \tag{18}$$

$$\left[C_0''\right] - \left[C_0'\right] Sc - Kr\ Sc\left[C_0\right] = 0 \tag{19}$$

$$\left[C_1''\right] + \left[C_1'\right] Sc - (n+Kr)\left[C_1\right] = \text{"} - \text{"ASc}\left[C_0''\right] \tag{20}$$

$$\left.\begin{array}{l} [u] = \left[U_p\right], [u_1] = [0], [\theta_0] = [1], [\theta_1] = [1], [C_0] = [1], [C_1] = [1] \quad at \quad y = 0 \\ [u] = 1, [u_1] = 1, [\theta_0] \to [1], [\theta_0] \to [0], [C_0] \to [0], [C_1] \to [0] \quad as \quad y \to \infty \end{array}\right\} \tag{21}$$

Solving Eqs. (15)–(20) utilizing Eq. (21), we obtain

$$[u_0] = 1 + c_4 e^{-b_2 y} + S_6 e^{-a_6 y} + S_7 e^{-a_2 y} \tag{22}$$

$$[u_1] = 1 + c_6 e^{-b_4 y} + S_8 e^{-b_2 y} + S_9 e^{-a_6 y} + S_{10} e^{-a_2 y} + S_{11} e^{-a_8 y} + S_{12} e^{-a_4 y} \tag{23}$$

$$[\theta_0] = e^{-a_6 y} + S_2\left(e^{-a_2 y} - e^{-a_6 y}\right) \tag{24}$$

$$[\theta_1] = c_2 e^{-a_8 y} + S_3 e^{-a_4 y} + S_4 e^{-a_2 y} + S_5 e^{-a_4 y} \tag{25}$$

$$[C_0] = e^{-a_2 y} \tag{26}$$

$$[C_1] = e^{-a_4 y} + S_1\left(e^{-a_2 y} - e^{-a_4 y}\right) \tag{27}$$

***Skin friction***:

$$[C_f] = \left(\frac{\partial u}{\partial y}\right)_{y=0} = (-b_2 c_4 - a_6 S_6 - a_2 S_7) + \varepsilon e^{nt}\left(-b_4 c_6 - b_2 S_8 - a_6 S_9 - a_2 S_{10} - a_8 S_{11} - a_4 S_{12}\right) \tag{28}$$

***Nusselt numbers***:

$$[N_u] = \left(\frac{\partial \theta}{\partial y}\right)_{y=0} = (-a_6 - a_2 S_2 + a_6 S_2) + \varepsilon e^{nt}\left(-a_8 c_2 - a_4 S_3 - a_2 S_4 - a_4 S_4\right) \tag{29}$$

***Sherwood numbers***:

$$[S_h] = \left(\frac{\partial C}{\partial y}\right)_{y=0} = (-a_2) + \varepsilon e^{nt}(-a_4 - a_2 S_1 - a_4 S_1) \tag{30}$$

## 4 Results and Discussions

Influence of angle of inclination Ψ on the velocity was illustrated in Fig. 2. From the figure, it was observed that velocity declined with the augmented values of Ψ. Figure 3 represents impact of radiation absorption on the fluid velocity. From the figure, it was found that incremental diverse values of R lead to acceleration in the velocity. Significance of magnetic field M on the velocity was portrayed in Fig. 4: From the figure, it was perceived that various values of M lead to slow down of the flow velocity. As M rises it induces a damping impact on the velocity by creating a drag force that resists the fluid motion, causing the velocity to reduce by means of the augment of Hartman number. Consequence of chemical reaction on the velocity was explained by utilizing Fig. 5: From the figure established on the eventualities, it was evidently revealed that the fluid velocity declined with the incremental values of Kr. Figure 6 epitomizes consequence of radiation absorption on the temperature. From the figure, it was predicate that different incremental values of R lead to enhancement in velocity.

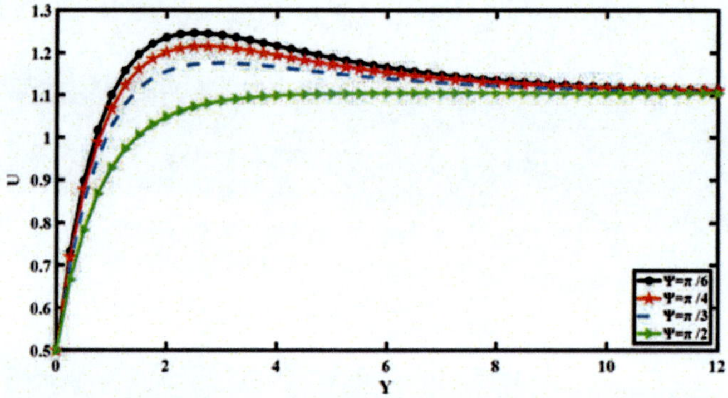


**Fig. 2** Effect of Ψ on velocity

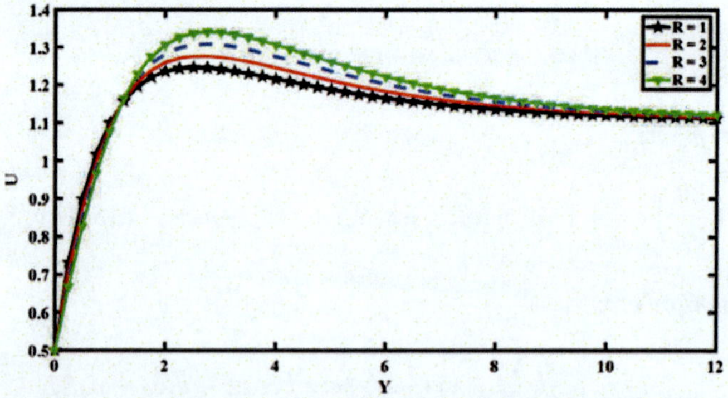


**Fig. 3** Effect of R on velocity

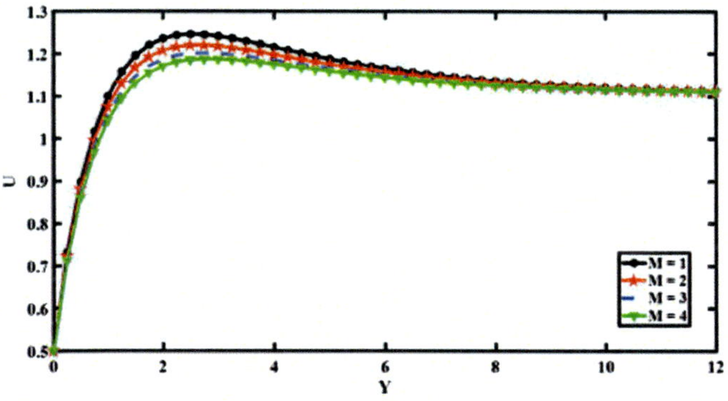


**Fig. 4** Impact of M on velocity

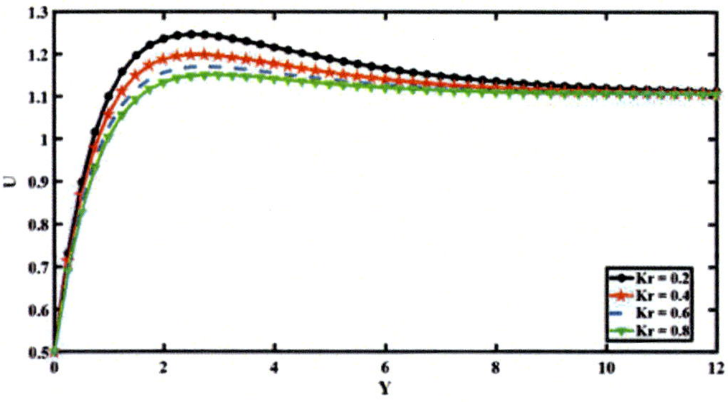


**Fig. 5** Impact of Kr on velocity

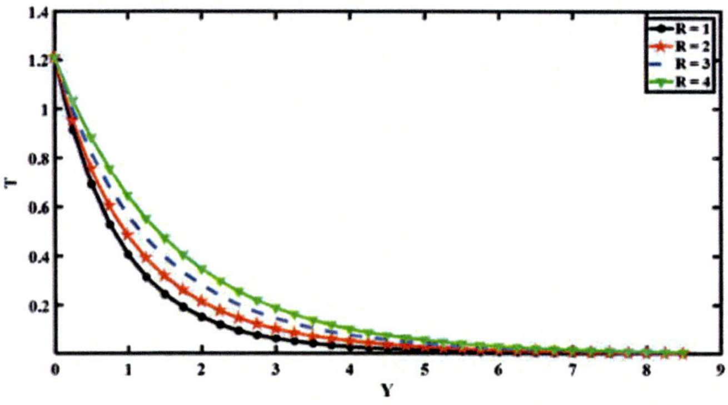


**Fig. 6** Influence of R on temperature

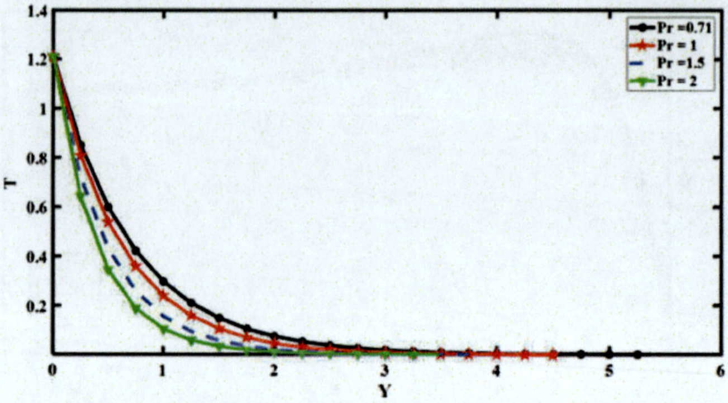


**Fig. 7** Influence of Pr on temperature

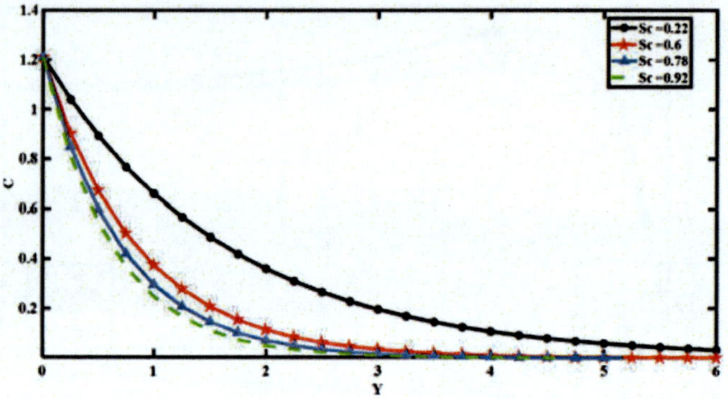


**Fig. 8** Plot of concentration for varying Sc

The significance of Pr on velocity is exemplified in Fig. 7. "It is comprehended that velocity at the surface is restricted on improving Pr values. Owing to the impulse boundary layer thickness is dilute much quick due to the thermal diffusion of the fluid particles, furthermore reinforcement in the viscosity of the fluid which elaborated fluid thickness and consequently causes deterioration in velocity." The concourses of Sc on the concentration are illustrated in Fig. 8: From the figure, it was observed that concentration declined with the accelerated values of Sc. Figure 9 represents the chemical reaction parameter which says that the concentration reduced due to increase of chemical reaction parameter.

Influence of Sc, Kr, Q, Pr , R, M, and Ψ is illustrated in Table 1: From this table, it was observed that skin friction as well as Sherwood number is declined with the incremental values of Sc but contrary impact occurred in case of Nusselt number. However, skin friction along with Nusselt number as well as Sherwood reduced with the accelerated values of Kr and Q; meanwhile, skin friction rises and Nusselt number with augmented values of Pr and M.

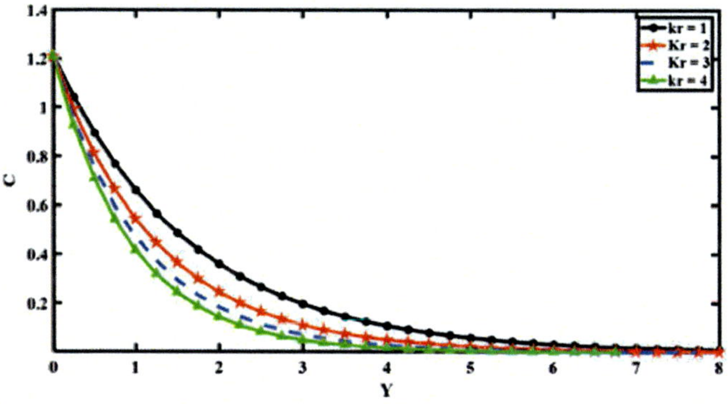


**Fig. 9** Plot of concentration for varying Kr

**Table 1** Influence dissimilar parameters on skin friction, Nusselt number, and Sherwood number

| SC | Kr | Q | Pr | R | M | $\Psi$ | Cf | Nu | Sh |
|---|---|---|---|---|---|---|---|---|---|
| 0.22 | 0.2 | 1 | 0.71 | 1 | 1 | $\pi/6$ | 2.4347 | −0.1269 | −0.7348 |
| 0.60 | 0.2 | 1 | 0.71 | 1 | 1 | $\pi/6$ | 2.3570 | 0.1725 | −1.4178 |
| 0.22 | 0.4 | 1 | 0.71 | 1 | 1 | $\pi/6$ | 2.3596 | −0.1356 | −1.7058 |
| 0.22 | 0.6 | 1 | 0.71 | 1 | 1 | $\pi/6$ | 2.3038 | −0.6277 | −2.0696 |
| 0.22 | 0.2 | 2 | 0.71 | 1 | 1 | $\pi/6$ | 1.9355 | −0.1038 | – |
| 0.22 | 0.2 | 3 | 0.71 | 1 | 1 | $\pi/6$ | 1.8511 | −0.1230 | – |
| 0.22 | 0.2 | 1 | 1 | 1 | 1 | $\pi/6$ | 2.5739 | −0.6768 | – |
| 0.22 | 0.2 | 1 | 5 | 2 | 1 | $\pi/6$ | 2.8770 | −0.18221 | – |
| 0.22 | 0.2 | 1 | 0.71 | 2 | 1 | $\pi/6$ | 2.6671 | 1.4576 | – |
| 0.22 | 0.2 | 1 | 0.71 | 3 | 1 | $\pi/6$ | 2.3194 | 2.9518 | – |
| 0.22 | 0.2 | 1 | 0.71 | 1 | 2 | $\pi/6$ | 2.4067 | – | – |
| 0.22 | 0.2 | 1 | 0.71 | 1 | 3 | $\pi/6$ | 2.3840 | – | – |
| 0.22 | 0.2 | 1 | 0.71 | 1 | 1 | $\pi/4$ | 2.3448 | – | – |
| 0.22 | 0.2 | 1 | 0.71 | 1 | 1 | $\pi/3$ | 2.2266 | – | – |

## 5 Validity of Results

Consider $\frac{R^*}{\rho C_p}[C - C_\infty]$ in Eq. (3) and $\Psi = \pi/2 \& \xi = 0$ in Eq. (2) then our outcomes are in good agreement with the result of Mythreye et al. [13] which is shown in Table 2.

**Table 2** Influence of Q on skin friction

| Q | Mythreye et al. [13] | Present result $C_f$ $when\ R = 0, \Psi = \pi/2, and\ \xi = 0$ |
|---|---|---|
| 0 | 3.7351 | 3.7378 |
| 1 | 3.4707 | 3.4351 |
| 2 | 3.4052 | 3.4729 |
| 3 | 3.5271 | 3.6243 |

## 6 Conclusions

- Velocity depreciated by means of the incremental values of magnetic field.
- When the angle of inclination rises then velocity is atrophied.
- With the accelerated values of $P_r$, temperature gets declined.
- Temperature along with concentration reduced with the accelerated values of $S_c$.
- Temperature as well as concentration depleted with increasing values of $K_r$.

## References

1. Cheng P, Minkowycz WJ (1977) Free convection about a vertical flat plate embedded in a porous medium with application to heat transfer from a dike. J Geophys Res 82(14):2040–2044
2. Lai FC, Kulacki FA (1991) Coupled heat and mass transfer by natural convection from vertical surfaces in porous media. Int J Heat Mass Trans 34(4–5):1189–1194
3. Ramprasad JL, Balamurugan KS, Dharmaiah G (2016) Unsteady MHD convective heat and mass transfer flow past an inclined moving surface with heat absorption. JP J Heat Mass Trans 13(1):33–51
4. Rajakumar KVB, Balamurugan KS, Ramana Murthy CV, Ranganath N (2019) Radiation, dissipation and dufour effects on MHD free convection flow through a vertical oscillatory porous plate with ion slip current. In: Numerical heat transfer and fluid flow, Lecture notes in mechanical engineering, pp 587–596. https://doi.org/10.1007/978-981-13-1903-7-67
5. Nakayama A, Koyama H (1987) A general similarity transformation for combined free and forced convection flows with in a fluid saturated porous medium. J Heat Trans 109(4):1041–1045
6. Bestman AR (1990) Natural convection boundary layer with suction and mass transfer in a porous medium. Int J Energy Res 14(4):389–396
7. Chamkha AJ (1997) Hydromagnetic natural convection from an isothermal inclined surface adjacent to a thermally stratified porous medium. Int J Eng Sci 35(10–11):975–986
8. Muthucumaraswamy R, Ganesan P (2001) Effect of the chemical reaction and injection on flow characteristics in an unsteady upward motion of an isothermal plate. J Appl Mech Tech Phys 42(4):665–671
9. Ibrahim FS, Hassanien IA, Bakr AA (2004) Non classical thermal effects in stokes second problem for micro polar fluids. ASME J Appl Mech 72(4):468–474
10. Makinde OD (2005) Free convection flow with thermal radiation and mass transfer past a moving vertical porous plate. Int Commun Heat Mass Trans 32(10):1411–1419
11. Abdelkhalek MM (2006) The skin friction in the MHD convection stagnation point with mass transfer. Int Commun Heat Mass Trans 33:249–258

12. Khan A, Khan I, Ali F, Khalid A, Shafie S (2015) Exact solutions of heat and mass transfer with MHD flow in a porous medium under time dependent shear stress and temperature. Hindawi Publishing Corporation Abstract and Applied Analysis, Article ID 975201, pp 1–16. https://doi.org/10.1155/2015/97520
13. Mythreye A, Pramod JP, Balamurugan KS (2015) Chemical reaction on unsteady MHD convective heat and mass transfer past a semi-infinite vertical permeable moving plate with heat absorption. Procedia Eng 127:613–620
14. Srinivas S, Muthuraj R (2011) Effect of chemical reaction and space porosity on MHD mixed convective flow in a vertical asymmetric channel with peristalsis. Math Comput Model 123–1227
15. Muthuraj R, Srinivas S, Sakeria J (2013) Combined effects of chemical reactions and wale slip on MHD flow in a vertical wary porous space with Traveling Thermal waves. Walailak J Sci Technol (WJST) 10(4):369–383
16. Immaculate DL, Muthuraj R, Selvi RK, Srinivas S, Shukla AK (2015) The influence of thermo phonetic particle deposition on fully developed MHD mixed convective flow in a vertical channel with thermal-diffusion and diffusion, thermo effect. Ain Shams Eng J 6(2):671–681
17. Muthuraj R, Nirmala K, Srinivas S (2016) Influences of chemical reactions and wall properties on MHD peristaltic transport of Destroy fluid with head and man transfer. Alex Eng J 55:597–611
18. Immaculate DL, Muthuraj R, Shukla AK, Srinivas S (2016) MHD unsteady flow of Williamson Nano fluid in a vertical porous space oscillating wall temperature. Front Heat Mass Trans (FHMT) 7:12
19. Muthuraj R (2021) An analysis Bingham Fluid and Jeffrey fluid flow in a horizontal channel with plug flow and heat transfer. Int J Appl Comput Math 7:35

# Investigating the Risk of Airborne COVID-19 and the Importance of Social Restrictions in Regulating the Transmission

Nidhi S. Vaishnaw and S. Shankar Narayan

## 1 Introduction

The purpose of a mathematical model of an infectious disease is to explain the disease's transmission mechanism, which can generally be described as follows: If within a limited period, the number of cases increases past the normal level, there would be an epidemic of disease. It is an epidemic when the disease spreads swiftly to many people. If fresh susceptibles are introduced to the community, either by conception or relocation, or if reinfection quickly happens, the disease may continue and the virus may stick around to the community for a long time to come. The illness is assumed to be prevalent inside the community in this situation. If the disease spreads to many countries and continents geographically, there is a pandemic [1]. Our present work deals with one such pandemic, COVID-19. SAR COV-2, the virus that is responsible for COVID-19, belongs to the coronavirus family named on their surfaces for the crown spikes [2]. COVID-19, a highly contagious viral infection that primarily attacks the throat and lungs, can cause SAR COV-2 [3]. A series of discussions made over the airborne mode of transmission of the diseases under the supervision of World Health Organization summarized that the spread is possible indoor locations with heavy crowds and poor ventilation are very prone to transmission [4]. A repository containing the well-performed research work can be seen under the WHO COVID-19 Database [5]. Shah et al. [6] proposed a more generalized SEIR model of the disease under study. The model takes account of all potential cases of human-to-human transmission and formulates its reproduction number to examine the reliable transmission dynamics. Mandal et al. [7] devised a mathematical model that implemented initiatives of quarantine category and government involvement to curtail disease transmission. The team analyzed the model's detailed dynamic behavior concerning the basic number of reproductions. The model formulated is

N. S. Vaishnaw · S. Shankar Narayan (✉)
CMR Institute of Technology, Bengaluru 560037, India
e-mail: shankarsubramani3@gmail.com

 
S. Srinivas et al. (eds.), *Recent Advances in Applied Mathematics and Applications to the Dynamics of Fluid Flows*, Lecture Notes in Mechanical Engineering,
https://doi.org/10.1007/978-981-19-1929-9_21

validated using the parameter values considered in Mandal et al. [7]. Similar results are obtained ensuring the correctness of the model formulated.

Section 2 deals with the formulation of a suitable mathematical model to govern the current study conditions. Section 3 outlines the mathematical analysis of the model formulated in Sect. 2. The following Sect. 4 deals with the results obtained on solving the mathematical model considered for the present study with a following conclusions.

## 2 Model Formulation

We consider a mathematical model SEIR that describes the dynamics of COVID-19. We divide the population into four compartments as time-dependent classes. Using first-order differential equations, we form a system that includes interactions between susceptible class (S), Exposed class (E), Infected class (I), and Recovered class (R). The graphical representation of the proposed model is shown in Fig. 1. Based on the best available biological information on the pathogenesis of the disease infection and epidemiology, the following assumptions were drawn about the disease transmission mechanism.

- Transmission occurs horizontally by direct interhost contact.
- The merging of individual hosts is uniform and thus the Law of Mass Action maintains.
- The rate of transmission from a compartment is commensurate with the compartment population size.

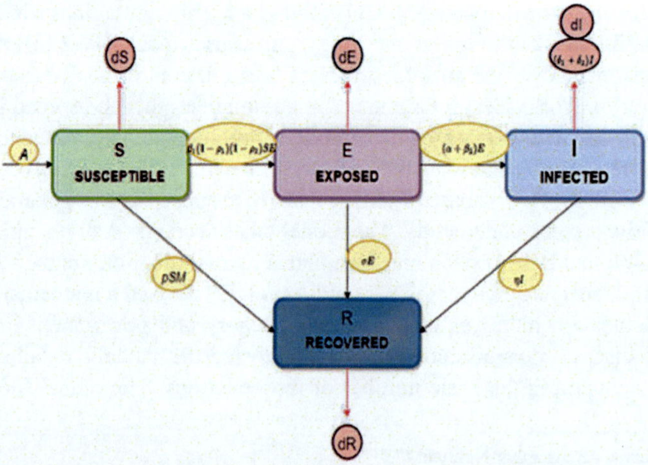


**Fig. 1** Compartment model for disease transmission

- There is no lack of immunity and little risk of reinfection. This implies that the rate of transmission from R back to S is zero.
- At stage 2 of the current study, the mode of transmission is taken to be airborne along with the existing mode of transmission.

A compartment framework is used to model the possibility of airborne spread of the COVID-19 virus within the population. We considered recruitment rate $A$ as constant to probable susceptible class and constant natural death rate in all the classes, $\beta_1$ and $\beta_2$ are the rate of disease transmission through droplet and airborne modes, respectively. The model is formulated by taking the death rate due to infection into two population: one who is primarily infected $\delta_1$ by disease and those who are critically infected $\delta_2$ (with suffering from diabetics, hypertension). The other parameter descriptions can be found in Table 1. The values considered are extracted from various research articles and reports [4, 7].

In the present work, we adopt the same procedure or framework for control policy that has been carried out by Mandal et al. [7].

In this section, we develop a mathematical model by taking into account the above assumptions

$$\frac{dS}{dt} = A - \beta_1(1-\rho_1)(1-\rho_2)SE - pSM - dS \tag{1}$$

$$\frac{dE}{dt} = \beta_1(1-\rho_1)(1-\rho_2)SE - (\alpha + \sigma + d + \beta_2)E \tag{2}$$

**Table 1** Parameters description and values

| Symbols | Description | Values/ranges |
|---|---|---|
| A | Recruitment rate to the susceptible population | 11000 |
| $\beta_1$ | Disease transmission rate | 1 |
| $\beta_2$ | Rate of airborne infection | [0, 1] |
| $\rho_1$ | Population maintain proper precautions | 0.4 |
| $\rho_2$ | Population maintain proper precautions but exposed | 0.45 |
| p | Implemented policy rate | [0, 1] |
| d | Natural death rate | 0.2 |
| $\alpha$ | The transmission rate of infection from exposed class | 0.2 |
| $\sigma$ | The transmission rate of exposure from recovered class | 0.1 |
| $\eta$ | The recovery rate of infection | 0.1 |
| M | Rate of change in p | [0, 1] |
| $\delta_1$ | The death rate due to primary infection | 0.25 |
| $\delta_2$ | The death rate due to conditions like diabetes and hypertension | [0.5, 1] |

$$\frac{dI}{dt} = (\alpha + \beta_2)E - (\eta + \delta_1 + \delta_2 + d)I \tag{3}$$

$$\frac{dR}{dt} = \eta I + \sigma E + pSM - dR \tag{4}$$

where the parameters are described in Table 1, with the initial conditions, $S(0) = S_0 \geq 0$, $E(0) = E_0 \geq 0$, $I(0) = I_0 \geq 0$, $R(0) = R_0 \geq 0$.

## 3 Mathematical Analysis

### 3.1 Boundedness of the System

We observed for boundedness of the system (1–4).

**Lemma 1** *Under the given initial conditions, all solution (S(t), E(t), I(t), R(t)) of the system will be uniformly bounded.*

***Proof*** We assume that N = S + E + I + R.
Therefore, $\frac{dN}{dt} = \frac{dS}{dt} + \frac{dE}{dt} + \frac{dI}{dt} + \frac{dR}{dt}$
$\frac{dN}{dt} = A - (\delta_1 + \delta_2)I - dN$
i.e., $\frac{dN}{dt} + dN \leq A$
By applying the theorem of differential Eq. [8] for integrating the above inequality, we get
$N \leq \frac{A}{d+pM}[1 - e^{-dt}] + N_0 e^{-dt}$
Now for $t \Rightarrow \infty$, we get $0 \leq N \leq \frac{A}{d+pM}$.
Hence, all the solution of (1–4) that are initiating in $R_+^4$ are confined in the region. $\{N \in R_+^4 : 0 \leq N(S, E, I, R) \leq \frac{A}{d+pM} + \epsilon\}$, hence the proof.

### 3.2 Local Stability and Existence of Positive Equilibrium Points

The basic reproduction number plays the most important role to analyze the existence of the unique equilibrium and stability of the governed system of equations. We have various techniques to evaluate $R_0$ for epidemiological models. In our present work, we take the help of the next-generation matrix method, We have $y = \begin{pmatrix} E \\ I \end{pmatrix}$; $\phi(y) = \begin{pmatrix} \beta_1(1-\rho_1)(1-\rho_2)SE \\ 0 \end{pmatrix}$; $\psi(y) = \begin{pmatrix} (\alpha + \sigma + d + \beta_2)E \\ -(\alpha + \beta_2)E + (\eta + \delta_1 + \delta_2 + d)I \end{pmatrix}$ $F = J(\frac{\phi}{E_0}) = \begin{pmatrix} \beta_1(1-\rho_1)(1-\rho_2)S^0 & 0 \\ 0 & 0 \end{pmatrix}$;

$$V = J(\tfrac{\psi}{E_0}) = \begin{pmatrix} (\alpha+\sigma+d+\beta_2) & 0 \\ -(\alpha+\beta_2) & (\eta+\delta_1+\delta_2+d) \end{pmatrix} \text{ Then } V^{-1} = \begin{pmatrix} \frac{\beta_1(1-\rho_1)(1-\rho_2)S^0}{\alpha+\sigma+d+\beta_2} & 0 \\ 0 & 0 \end{pmatrix}$$

The spectral radius of the matrix $FV^{-1}$ is represented as $R_0 = \frac{\beta_1(1-\rho_1)(1-\rho_2)S^0}{\alpha+\sigma+d+\beta_2}$.

**Theorem 1** *The system (1) is locally stable for the virus-free equilibrium point* $E_0$, $R_0 < 1$ *and unstable if* $R_0 > 1$.

***Proof*** The Jacobian at the virus-free equilibrium of the governed system,

$$J_0 = \begin{pmatrix} -(pM+d) & -\beta_1(1-\rho_1)(1-\rho_2)S^0 & 0 & 0 \\ 0 & \beta_1(1-\rho_1)(1-\rho_2)S^0 - (\alpha+\sigma+d+\beta_2) & 0 & 0 \\ 0 & (\alpha+\beta_2) & -(\eta+\delta_1+\delta_2+d) & 0 \\ pM & \sigma & \eta & -d \end{pmatrix}$$ with the characteristic polynomial, $(\lambda+d)(\lambda+d+pM)(\lambda+\eta+d+\delta_1+\delta_2)[\lambda+(\alpha+\sigma+d+\beta_2)(1-R_0)] = 0$.

It is clear that all Eigenvalues are negative if $R_0 < 1$. So, system (1–4) is locally stable if $R_0 < 1$ and unstable if $R_0 > 1$.

**Theorem 2** *The virus equilibrium exists for the system (1), if* $R_0 > 1$.

***Proof*** We equate all the equations of system (1–4) to zero,

$A - \beta_1(1-\rho_1)(1-\rho_2)SE - pSM - dS = 0$ $\beta_1(1-\rho_1)(1-\rho_2)SE - (\alpha+\sigma+d+\beta_2)E = 0$ $(\alpha+\beta_2)E - (\eta+\delta_1+\delta_2+d)I = 0$ $\eta I + \sigma E + pSM - dR = 0$

On solving, we get

$S^* = \frac{\alpha+\sigma+d+\beta_2}{\beta_1(1-\rho_1)(1-\rho_2)}$

$E^* = \frac{A\beta_1(1-\rho_1)(1-\rho_2)-(d+pM)(\alpha+\sigma+d+\beta_2)}{[\beta_1(1-\rho_1)(1-\rho_2)]^2}$

$I^* = \frac{(\alpha+\beta_2)[A\beta_1(1-\rho_1)(1-\rho_2)-(d+pM)(\alpha+\sigma+d+\beta_2)]}{(\eta+\delta_1+\delta_2+d)[\beta_1(1-\rho_1)(1-\rho_2)]^2}$

$R^* = \frac{(\eta I^*+\sigma E^*+pS^*M)}{d}$

The Jacobian for endemic equilibrium is given by

$$J_* = \begin{pmatrix} a_{11} & -\beta_1(1-\rho_1)(1-\rho_2)S^* & 0 & 0 \\ \beta_1(1-\rho_1)(1-\rho_2) & a_{22} & 0 & 0 \\ 0 & (\alpha+\beta_2) & -(\eta+\delta_1+\delta_2+d) & 0 \\ pM & \sigma & \eta & -d \end{pmatrix}$$

where

$a_{11} = -\beta_1(1-\rho_1)(1-\rho_2)E^* - (pM+d)$

$a_{22} = -\beta_1(1-\rho_1)(1-\rho_2)S^* - (\alpha+\sigma+d+\beta_2)$

Then the characteristic equation around the endemic equilibrium points is given by

$[-(\eta+\delta_1+\delta_2+d)-\lambda][-d-\lambda][(a_{11}-\lambda)(a_{22}-\lambda)+\beta^2(1-\rho_1)^2(1-\rho_2)^2S^*] = 0$

$[\lambda+(\eta+\delta_1+\delta_2+d)][\lambda+d][\lambda^2+K_1\lambda+K_2] = 0.$

where

$K_1 = \beta_1(1-\rho_1)(1-\rho_2)(E^*+S^*) + (pM+d) + (\alpha+\sigma+d+\beta_2)$

$K_2 = \beta^2(1-\rho_1)^2(1-\rho_2)^2S^*E^* + [\beta_1(1-\rho_1)(1-\rho_2)((\alpha+\sigma+d+\beta_2)E^* + (pM+d)S^*)] + (pM+d)(\alpha+\sigma+d+\beta_2) + \beta^2(1-\rho_1)^2(1-\rho_2)^2S^*E^*$

By the above equation, it is very clear that at least two eigenvalues are negative and remaining roots are the roots of the quadratic polynomial and ($S^*$, $E^*$, $R^*$) are giving positive values. Hence, according to Routh–Hurwitz criteria for second-order polynomial, both coefficients satisfy $K_i > 0$, so we can conclude that the system (1–4) is locally asymptotically stable for endemic equilibrium.

## 4 Results and Discussion

Table 2 demonstrates the importance of the control policy on regulating the number of infected people for 30 days. The sample data taken here corresponds to a state of India (Karnataka) [9]. It can be seen very clearly that the rate of increase in the number of infected people is more when there are no control strategies applied by the government. It can also be seen that the equilibrium is reached at a rapid rate when there is no control policy applied.

Figure 2 shows the graphical representation of the infected and the recovered population densities for the case in which the control policy is applied and Fig. 3 is for the situation when the government unlocks completely.

Table 3 and Fig. 4 demonstrate the scenario when the disease caused by the deadly coronavirus becomes airborne, with [5] as a base. The researchers led by Lidia Morawska of the International Laboratory for Air Quality and Health, WHO Collaborating Center, Queensland University of Technology, Brisbane, Australia, consulted the clinical network and the applicable national and global bodies to perceive the potential for airborne spread of COVID-19 [10]. The statement made by the team mentions the severity caused when the disease becomes airborne. As per the assumptions made at the beginning of the work, we assume that the rate of airborne transmission varies directly as the proportion of droplets in one's environment. Plot 4 clearly shows that as the transmission rate increases gradually, the number of infectious cases show an abrupt jump in the values, in other words, the airborne mode is the most dangerous mode of transmission of any disease, in particular COVID-19.

**Table 2** Infected population with and without control policy

| Day | Infected population with no control policy | Infected population with control policy |
|---|---|---|
| 1 | 2551 | 2541 |
| 5 | 9469 | 9498 |
| 10 | 10007 | 10018 |
| 15 | 9995 | 10008 |
| 20 | 10005 | 10000 |
| 25 | 10004 | 9999 |
| 30 | 10003 | 10000 |

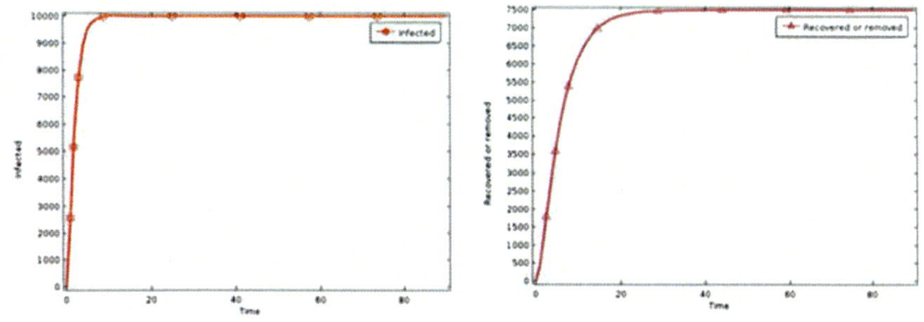


**Fig. 2** Population densities of infected and recovered compartments with control policy adopted

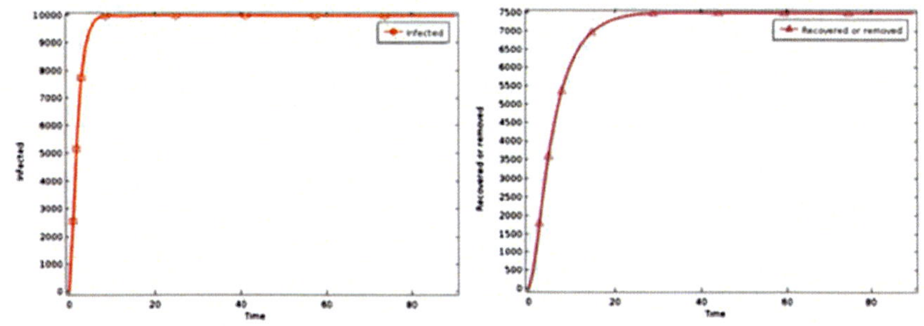


**Fig. 3** Population densities of infected and recovered compartments without control policy

**Table 3** Number of infected people with different airborne rate

| Day | $\beta_2 = 0$ | $\beta_2 = 0.1$ | $\beta_2 = 0.5$ | $\beta_2 = 0.9$ |
|---|---|---|---|---|
| 0 | 1 | 1 | 1 | 1 |
| 1 | 2540 | 2609 | 2918 | 3289 |
| 5 | 9494 | 9916 | 12009 | 15470 |
| 10 | 10015 | 10378 | 12040 | 14367 |
| 15 | 10004 | 10356 | 12053 | 14419 |
| 20 | 9996 | 10349 | 12051 | 14425 |
| 25 | 9996 | 10349 | 12051 | 14422 |
| 30 | 9997 | 10350 | 12051 | 14422 |

The slope of a straight line represents the rate at which the y-quantity varies concerning a unit change in the x-quantity [11]. Figure 5 displays the slope of trendlines for the case of an airborne simulation study of COVID-19. The change in the slope values are in the order: 889.06, 920.74, 1063.5, 1265.3 persons/day for the case of the airborne rate $\beta_2$: 0, 0.1, 0.5, and 0.9, respectively. These slope scatterplots envisage that, as the rate $\beta_2$ is increased, for every 0.1 rises, there can be an increase in

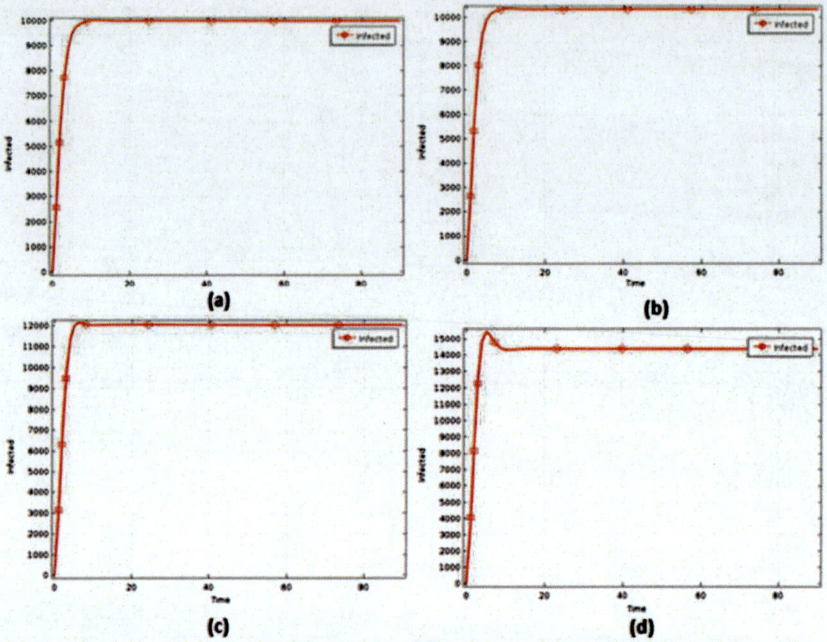


**Fig. 4** Infection population densities for $\beta_2 = 0$, $\beta_2 = 0.1$, $\beta_2 = 0.5$, $\beta_2 = 0.9$, respectively

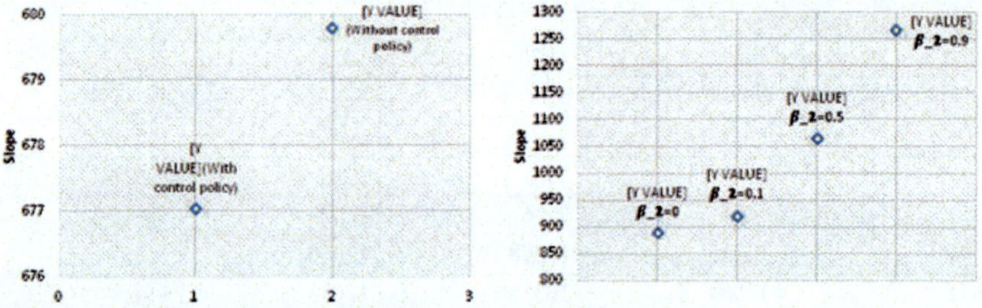


**Fig. 5** Plots depicting the slope values of the various trendlines

the infected population by 100. Thus, the evolution of the virus to airborne puts the entire globe under threat.

## 5 Conclusions

In the present work, we investigated the importance of control policies to avoid gatecrashing of coronavirus. Most airborne illnesses can run their course within a couple of weeks. Also, the obtained results promise us that when the airborne transmission is triggered, the infectious number will suppress the recovered number to a considerable extent. In our future work, we will consider age as one of the factors causing heterogeneity in the host population.

## References


1. Li MY (2018) An introduction to mathematical modeling of infectious diseases. Springer International Publishing
2. Naming the coronavirus disease (COVID-19) and the virus that causes it. World Health Organization (2020). https://www.who.int/emergencies/diseases/novel-coronavirus-2019/technical-guidance/naming-the-coronavirus-disease-(covid-2019)-and-the-virus-that-causes-it. Accessed 07 Aug 2020
3. Gorbalenya AE et al (2020) The species Severe acute respiratory syndrome-related coronavirus: classifying 2019-nCoV and naming it SARS-CoV-2. Nat Microbiol, Nat Res 5(4):536–544. https://doi.org/10.1038/s41564-020-0695-z
4. Coronavirus | WHO: Indoor airborne spread of virus possible—The Hindu. https://www.thehindu.com/news/international/who-indoor-airborne-spread-of-coronavirus-possible/article32037196.ece. Accessed 07 Aug 2020
5. COVID-19. https://search.bvsalud.org/global-literature-on-novel-coronavirus-2019-ncov/. Accessed 07 Aug 2020
6. Shah NH, Suthar AH, Jayswal EN (2020) Control strategies to curtail transmission of COVID-19. Int J Math Math Sci. https://doi.org/10.1155/2020/2649514
7. Mandal M, Jana S, Nandi SK, Khatua A, Adak S, Kar TK (2020) A model based study on the dynamics of COVID-19: prediction and control. Chaos Solitons Fractals 136:109889. https://doi.org/10.1016/j.chaos.2020.109889
8. Birkoff G , Rota GC (1982) Ordinary differential equations. Boston, Ginn
9. Home—COVID-19 INFORMATION PORTAL. https://covid19.karnataka.gov.in/english. Accessed 13 Aug 2020
10. The Hindu Explains—Is airborne transmission of COVID-19 a risk?—The Hindu. https://www.thehindu.com/sci-tech/health/the-hindu-explains-is-airborne-transmission-of-covid-19-a-risk/article32053924.ece. Accessed 13 Aug 2020
11. Xie M, Chen Q (2020) Insight into 2019 novel coronavirus—An updated interim review and lessons from SARS-CoV and MERS-CoV. Int J Infect Dis 94:119–124. https://doi.org/10.1016/j.ijid.2020.03.071

# A Numerical Algorithm Based on Tension Spline Scheme for Solving Singularly Perturbed Differential Difference Equations with Shifts

P. Murali Mohan Kumar and A. S. V. Ravi Kanth

## 1 Introduction

A general model problem of singularly perturbed differential difference equation (SPDDE) consists of positive and negative shifts is as follows:

$$\varepsilon \mathrm{v}''(\varsigma) + p(\varsigma)\mathrm{v}'(\varsigma) + q(\varsigma)\mathrm{v}(\varsigma - \delta) + r(\varsigma)\mathrm{v}(\varsigma) + s(\varsigma)\mathrm{v}(\varsigma + \eta) = \mathrm{g}(\varsigma), \quad (1)$$

with

$$\mathrm{v}(\varsigma) = \psi(\varsigma) \ \text{ on } \ -\delta \leqslant \varsigma \leqslant 0, \quad (2)$$
$$\mathrm{v}(\varsigma) = \gamma(\varsigma) \ \text{ on } \ 1 \leqslant \varsigma \leqslant 1 + \eta, \quad (3)$$

where the constants $\varepsilon$, $\delta$, $\eta$ are singular perturbation, delay, and advance parameters, respectively. The coefficient of the terms in (1) $p(\varsigma), q(\varsigma), r(\varsigma), s(\varsigma), \gamma(\varsigma), \psi(\varsigma)$, $\mathrm{g}(\varsigma)$ are functions in $\varsigma \in \Omega = (0, 1)$ and sufficiently smooth through out the interval.

Differential difference equation with respect to the singular perturbation parameter plays a significant role in discrete fields of science and engineering. The investigation of numerical analysis for SPDDE has made considerable progress for many authors. The applications of these problems often arise in engineering and science such as micro-scale heat transfer [5], optically bistable device [3], reaction–diffusion

P. M. Mohan Kumar (✉)
Department of Basic Sciences & Humanities, GMR Institute of Technology,
Rajam, Andhra Pradesh 532 127, India
e-mail: muralimohan.p@gmrit.edu.in

A. S. V. R. Kanth
Department of Mathematics, National Institute of Technology Kurukshetra,
Haryana 136 119, India
e-mail: asvravikanth@yahoo.com

 
S. Srinivas et al. (eds.), *Recent Advances in Applied Mathematics and Applications to the Dynamics of Fluid Flows*, Lecture Notes in Mechanical Engineering,
https://doi.org/10.1007/978-981-19-1929-9_22

equations [1], hydrodynamics of liquid helium [4], stability [2], etc. The solutions of SPDDEs exhibits boundary layers when $\varepsilon$ is goes to zero. Classical numerical schemes fail to generate precise alternatives because of the incidence of these boundary layers. Special methods are therefore necessary to achieve appropriate numerical method to SPDDEs. There are distinct volumes available on the impact of shift parameters on boundary layers. The asymptotic analysis is initiated and studied the impact of small parameter in the boundary layer region [6, 7]. There have been several numerical techniques to solve SPDDEs such as finite difference scheme [8, 9], non-standard finite difference scheme [10], optimized B-Spline collocation method [11], compact implicit method [13], and B-Spline collocation method [12]. The aim of this paper is to develop uniform convergent hybrid numerical scheme based on non-uniform Shishkin mesh for SPDDEs with positive shift ($\delta = o(\varepsilon)$) and negative shift ($\delta = o(\varepsilon)$). The hybrid numerical scheme combines tension spline numerical scheme on Shishkin mesh and mid-point numerical scheme on uniform mesh.

The content of the paper is as follows. Some priori results are given for continuous problem in Sect. 2. Section 3 describes the description of piece-wise uniform mesh and derivation of the hybrid numerical scheme. Error analysis is discussed in Sect. 4. Numerical results are presented in Sect. 5 to validate theoretical findings. Conclusions are given in Sect. 6.

## 2 Continuous Problem

By considering application of Taylor series, Eqs. (1)–(3) becomes

$$\mathrm{L} \equiv \varepsilon \mathrm{v}''(\varsigma) + \mathrm{a}(\varsigma)\mathrm{v}'(\varsigma) + \mathrm{b}(\varsigma)\mathrm{v}(\varsigma) = \mathrm{g}(\varsigma), \tag{4}$$

$$\text{with } \mathrm{v}(0) = \psi(0) = \psi_0,\ \mathrm{v}(1) = \gamma(1) = \gamma_1, \tag{5}$$

where $\mathrm{a} = p - \delta q + \eta r \geqslant \mathrm{a}^* > 0, \mathrm{b} = q + r + s \leqslant -\theta < 0$, for positive value of $\theta$. The solution of (4)–(5) exhibits the boundary layer in the left side of the interval.

**Lemma 1** *A function with smoothness $\varkappa(x)$ that satisfies $\varkappa(0) \geqslant 0$ and $\varkappa(1) \geqslant 0$. Then $\mathrm{L}\varkappa(\varsigma) \leqslant 0, \forall\, \varsigma \in (0,1)$ implies $\varkappa(\varsigma) \geqslant 0, \forall\, \varsigma \in [0, 1]$.*

***Proof*** Suppose $m^* \in \overline{\Omega} \ni \varkappa(m^*) < 0$ and $\varkappa(m^*) = \min_{x \in \overline{\Omega}} \varkappa(m)$. Clearly, $m^* \neq 0$ and $m^* \neq 1$; therefore, $\varkappa'(m^*) = 0$ and $\varkappa''(m^*) \geqslant 0$ implies $\mathrm{L}\varkappa(m^*) = \varepsilon\varkappa''(m^*) + \mathrm{a}(m^*)\varkappa'(m^*) + \mathrm{b}(m^*)\varkappa(m^*) > 0$, that is, in contradiction with our assumption. It follows that $\varkappa(m^*) \geqslant 0$ and thus $\varkappa(\varsigma) \geqslant 0\ \forall\, \varsigma \in \overline{\Omega}$.

**Lemma 2** *Suppose $\mathrm{v}(\varsigma)$ be the solution of (4) and (5) then $\|\mathrm{v}\| \leqslant max\,(|\psi_0|, |\gamma_1|) + \theta^{-1}\,\|\mathrm{g}\|$.*

***Proof*** Suppose $\varkappa^{\pm}(x)$ are two barrier functions and defined by $\varkappa^{\pm}(x) = \max(|\psi_0|, |\gamma_1|) + \|g\|\theta^{-1} \pm v(x)$. Then this implies, $\varkappa^{\pm}(0) \geqslant 0$, $\varkappa^{\pm}(1) \geqslant 0$ and

$$L\varkappa^{\pm}(x) = b(x)\left\{\|g\|\,\theta^{-1} + \max(|\psi_0|, |\gamma_1|)\right\} \pm g(x).$$

We have $L\varkappa \leqslant 0 \ \forall\ x \in \overline{\Omega}$ as $b(x)\theta^{-1} \leqslant -1$ and $\|g\| \geqslant g(x)$. We get required inequality by using Lemma 1.

**Lemma 3** *Assume* $v(\varsigma) \in C(R, j)$ *be the solution for the equations* (4) *and* (5), *then* $\left|v^{(i)}(\varsigma)\right| \leqslant \Theta\left[1 + \frac{exp\left(\frac{-a^*\varsigma}{\varepsilon}\right)}{\varepsilon^i}\right], 0 \leqslant i \leqslant j.$

***Proof*** One can refer in [14].

## 3 Description of the Method

Firstly, we partition the domain [0,1] by the knots $\varsigma_i = \sum_{w=0}^{i-1} h_w,\ h_w = \varsigma_{w+1} - \varsigma_w,\ i = 1(1)N-1$. A function as $\mathbf{S}(\varsigma, \varpi) \in C^2[0, 1]$ interpolates at each point in $v(\varsigma)$ and depends on a parameter $\varpi > 0$. The relation $\mathbf{S}(\varsigma, \varpi) = \mathbf{S}(\varsigma)$ satisfying in a sub-interval $[\varsigma_i, \varsigma_{i+1}]$, then

$$\frac{(\varsigma_{i+1} - \varsigma)}{h_i}\left[\mathbf{S}''(\varsigma_i) + \varpi\mathbf{S}(\varsigma_i)\right] + \frac{(\varsigma - \varsigma_i)}{h_i}\left[\mathbf{S}''(\varsigma_{i+1}) + \varpi\mathbf{S}(\varsigma_{i+1})\right] = \mathbf{S}''(\varsigma) + \varpi\mathbf{S}(\varsigma), \tag{6}$$

where $\mathbf{S}(\varsigma_i) = v_i$ is called as tension spline. Following the steps in [15], we have the following relation through the knots $i = 1(1)N-1$:

$$\theta_1 h_{i-1}\mathcal{M}_{i-1} + \theta_2(h_{i-1} + h_i)\mathcal{M}_i + \theta_1 h_i\mathcal{M}_{i+1} = \frac{(v_{i+1} - v_i)}{h_i} - \frac{(v_i - v_{i-1})}{h_{i-1}}, \tag{7}$$

where

$$\theta_1 = \frac{1}{\theta^2}\left(1 - \frac{\theta}{\sinh\theta}\right),\quad \theta_2 = \frac{1}{\theta^2}\left(\theta\coth\theta - 1\right), \tag{8}$$

and

$$\theta = h\varpi^{1/2},\quad \mathcal{M}_j = v''(\varsigma_j),\quad j = i, i \pm 1.$$

Rewrite Eq. (4) as

$$\varepsilon\mathcal{M}_k = g(\varsigma_k) - a(\varsigma_k)v'_k - b(\varsigma_k)v_k,\quad k = i, i \pm 1. \tag{9}$$

We have the following first-order derivative approximations:

$$v'_i \cong \frac{1}{\chi_i h_{i-1} h_i}\left(h_{i-1}^2 v_{i+1} + (h_i - h_{i-1})\chi_i v_i - h_i^2 v_{i-1}\right), \tag{10}$$

$$v'_{i+1} \cong \frac{1}{\chi_i h_i h_{i-1}}\left[(h_{i-1}^2 + 2h_i h_{i-1})v_{i+1} - (\chi_i)^2 v_i + h_i^2 v_{i-1}\right], \tag{11}$$

$$v'_{i-1} \cong \frac{1}{\chi_i h_i h_{i-1}}\left[-h_{i-1}^2 v_{i+1} + (\chi_i)^2 v_i - (h_i^2 + 2h_i h_{i-1})v_{i-1}\right], \tag{12}$$

where $\chi_i = h_{i-1} + h_i$. Substituting Eq. (9) in Eq. (7) and using (10)–(12), we obtain

$$\begin{aligned}
&\left[\frac{\varepsilon}{h_{i-1}\chi_i} - \theta_1\frac{(h_i + 2h_{i-1})}{\chi_i^2}a_{i-1} - \theta_2\frac{h_i}{h_{i-1}\chi_i}a_i\right.\\
&\left.+\theta_1\frac{h_i^2}{h_{i-1}\chi_i^2}a_{i+1} + \theta_1\frac{h_{i-1}}{\chi_i}b_{i-1}\right]v_{i-1}\\
&+\left[-\frac{\varepsilon}{h_i h_{i-1}} + \theta_1\frac{1}{h_i}a_{i-1} + \theta_2\frac{(h_i - h_{i-1})}{h_i h_{i-1}}a_i - \theta_1\frac{1}{h_{i-1}}a_{i+1} + \theta_2 b_i\right]v_i\\
&+\left[\frac{\varepsilon}{h_{i-1}\chi_i} - \theta_1\frac{h_{i-1}^2}{\chi_i^2}a_{i-1} + \theta_2\frac{h_{i-1}}{h_i\chi_i}a_i\right.\\
&\left.+\theta_1\frac{(h_{i-1} + 2h_i)}{\chi_i^2}a_{i+1} + \theta_1\frac{h_i}{\chi_i}b_{i+1}\right]v_{i+1}\\
&= \theta_1\frac{\chi_i}{\chi_i}g_{i-1} + \theta_2 g_i + \theta_1\frac{h_i}{\chi_i}g_{i+1},\ i = 1(1)N-1.
\end{aligned} \tag{13}$$

## 3.1 Piece-Wise Uniform Shishkin Mesh

In the boundary layer region, we generate more mesh points than in the outer region, i.e., the domain split in to two sub-intervals by a transition parameter ($\omega$) and is defined as

$$\omega = \min\left(0.5, \frac{2\,\varepsilon \ln N}{\sqrt{\theta}}\right).$$

We denote

$$h_i = \begin{cases} 2\omega/N, & \text{if } i = 1, 2, \ldots, \frac{N}{2},\\ 2(1-\omega)/N, & \text{if } i = \frac{N}{2}+1, \frac{N}{2}+2, \cdots, N. \end{cases}$$

and

$$\varsigma_i = \begin{cases} i\mathrm{h}_i, & \text{if } i = 0, 1, \ldots, \frac{N}{2}, \\ \omega + (i - N/2)\mathrm{h}_i, & \text{if } i = \frac{N}{2}+1, \frac{N}{2}+2, \cdots, N. \end{cases}$$

## 3.2 *Hybrid Scheme*

In this section, we introduce a hybrid numerical scheme that combines tension spline scheme and mid-point scheme [16]. For smaller values of singular perturbation parameter to the problem (4)–(5), tension spline may lead to oscillations in the outer region. To preserve its stability in the solution at the boundary layer region, we use mid-point scheme on uniform mesh and tension spline scheme on Shishkin mesh. We have

$$\mathrm{L}^N \equiv e_i^- \mathrm{v}_{i-1} + e_i^c \mathrm{v}_i + e_i^+ \mathrm{v}_{i+1} = \tilde{f},\ i = 1(1)N-1 \tag{14}$$

$$\text{with } \mathrm{v}_0(0) = \psi_0,\ \mathrm{v}_N(1) = \gamma_1, \tag{15}$$

where $e_i^-, e_i^c, e_i^+$ are as follows in the boundary layer region as well as outer region correspondingly. The values in the boundary layer region (For $i = 1, 2, \ldots, N/2$)

$$\begin{aligned}
e_i^- &= \left[\frac{\varepsilon}{\mathrm{h}_{i-1}\chi_i} + \theta_1 \frac{\mathrm{h}_i^2}{\mathrm{h}_{i-1}\chi_i^2}\mathrm{a}_{i+1} - \theta_2 \frac{\mathrm{h}_i}{\mathrm{h}_{i-1}\chi_i}\mathrm{a}_i \right.\\
&\quad \left. + -\theta_1 \frac{(\mathrm{h}_i + 2\mathrm{h}_{i-1})}{\chi_i^2}\mathrm{a}_{i-1} + \theta_1 \frac{\mathrm{h}_{i-1}}{\chi_i}\mathrm{b}_{i-1}\right],\\
e_i^c &= \left[-\frac{\varepsilon}{\mathrm{h}_i\mathrm{h}_{i-1}} - \theta_1 \frac{1}{\mathrm{h}_{i-1}}\mathrm{a}_{i+1} + \theta_2 \frac{(\mathrm{h}_i - \mathrm{h}_{i-1})}{\mathrm{h}_i\mathrm{h}_{i-1}}\mathrm{a}_i + \theta_1 \frac{1}{\mathrm{h}_i}\mathrm{a}_{i-1} + \theta_2 \mathrm{b}_i\right],\\
e_i^+ &= \left[\frac{\varepsilon}{\mathrm{h}_i\chi_i} + \theta_1 \frac{(\mathrm{h}_{i-1} + 2\mathrm{h}_i)}{\chi_i^2}\mathrm{a}_{i+1} + \theta_2 \frac{\mathrm{h}_{i-1}}{\mathrm{h}_i\chi_i}\mathrm{a}_i - \theta_1 \frac{h_{i-1}^2}{\mathrm{h}_i\chi_i^2}\mathrm{a}_{i-1} + \theta_1 \frac{\mathrm{h}_i}{\chi_i}\mathrm{b}_{i+1}\right],\\
\tilde{f} &= \theta_1 \frac{\mathrm{h}_i}{\chi_i}\mathrm{g}_{i+1} + \theta_2 \mathrm{g}_i + \theta_1 \frac{\mathrm{h}_{i-1}}{\chi_i}\mathrm{g}_{i-1}.
\end{aligned}$$

The values in the outer region (For $i = N/2+1, \cdots, N-1$)

$$\begin{aligned}
e_i^- &= \frac{2\varepsilon}{\mathrm{h}_{i-1}\chi_i},\\
e_i^c &= -\frac{2\varepsilon}{\mathrm{h}_{i-1}\chi_i} - \frac{2\varepsilon}{\mathrm{h}_i\chi_i} - \frac{1}{2\mathrm{h}_i}(\mathrm{a}_i + \mathrm{a}_{i+1}) + 0.5(\mathrm{b}_i + \mathrm{b}_{i+1}),
\end{aligned}$$

$$e_i^+ = \frac{2\varepsilon}{h_i \chi_i} + \frac{1}{2h_i}(a_i + a_{i+1}) + 0.5(b_i + b_{i+1}),$$
$$\tilde{f} = 0.5g_i + 0.5g_{i+1}.$$

The above system of Eqs. (14) with (15) can be solved by known algorithm.

## 4 Error Estimate

In this part, we estimate the error in truncation of the proposed approach, we calculate in the boundary layer region is given by

$$\begin{aligned}\Pi_i &= e_i^- v(\varsigma_{i-1}) + e_i^c v(\varsigma_i) + e_i^+ v(\varsigma_{i+1}) \\ &\quad - \left[\theta_1 \frac{h_i}{\chi_i} g(\varsigma_{i+1}) + \theta_2 g(\varsigma_i) + \theta_1 \frac{h_{i-1}}{\chi_i} g(\varsigma_{i-1})\right].\end{aligned} \tag{16}$$

Using the Eq. (4) for $g(\varsigma_k),\ k = i, i \pm 1$ in (16) and apply Taylor series expansion for $v(\varsigma_{i+1})$ and $v(\varsigma_{i-1})$, we obtain

$$\Pi_i = \Pi_{0,i} v(\varsigma_i) + \Pi_{1,i} v'(\varsigma_i) + \Pi_{2,i} v''(\varsigma_i) + \Pi_{3,i} v^{(iii)}(\varsigma_i) + \Pi_{4,i} v^{(iv)}(\varsigma_i) + \text{h.o.t}, \tag{17}$$

where

$$\Pi_{0,i} = e_i^- + e_i^c + e_i^+ - \theta_1 \frac{h_{i-1}}{\chi_i} b_{i-1} - \theta_2 b_i - \theta_1 \frac{h_i}{\chi_i} b_{i+1},$$

$$\begin{aligned}\Pi_{1,i} =& - h_{i-1} e_i^- + h_i e_i^+ - \theta_1 \frac{h_{i-1}}{\chi_i} a_{i-1} - \theta_2 a_i - \theta_1 \frac{h_i}{\chi_i} a_{i+1} \\ &+ \theta_1 \frac{h_{i-1}^2}{\chi_i} b_{i-1} - \theta_1 \frac{h_i^2}{\chi_i} b_{i+1},\end{aligned}$$

$$\begin{aligned}\Pi_{2,i} =& - \varepsilon(\theta_1 + \theta_2) + \frac{h_{i-1}^2}{2!} e_i^- + \frac{h_i^2}{2!} e_i^+ + \theta_1 \frac{h_{i-1}^2}{\chi_i} a_{i-1} - \theta_1 \frac{h_i^2}{\chi_i} a_{i+1} \\ &- \theta_1 \frac{h_{i-1}^3}{2!\chi_i} b_{i-1} - \theta_1 \frac{h_i^3}{2!\chi_i} b_{i+1},\end{aligned}$$

$$\begin{aligned}\Pi_{3,i} =& - \frac{h_{i-1}^3}{3!} e_i^- + \frac{h_i^3}{3!} e_i^+ + \varepsilon \frac{\theta_1}{\chi_i}\left(h_{i-1}^2 - h_i^2\right) \\ &- \frac{\theta_1}{2!\chi_i}(h_{i-1}^3 a_{i-1} + h_i^3 a_{i+1}) + \frac{\theta_1}{3!\chi_i}(h_{i-1}^4 b_{i-1} - h_i^4 b_{i+1}),\end{aligned}$$

$$\Pi_{4,i} = \frac{h_{i-1}^4}{4!}e_i^- + \frac{h_i^4}{4!}e_i^+ - \varepsilon\theta_1\frac{(h_{i-1}^3 + h_i^3)}{2!\chi_i} + \theta_1\frac{(h_{i-1}^4 a_{i-1} + h_i^4 a_{i+1})}{3!\chi_i} - \frac{\theta_1}{4!\chi_i}(h_{i-1}^5 b_{i-1} + h_i^5 b_{i+1}).$$

It is obvious,

$$\Pi_{0,i} = \Pi_{1,i} = 0,\ \Pi_{2,i} = \varepsilon\left(1/2 - (\theta_1 + \theta_2)\right),$$

$$\Pi_{3,i} = \varepsilon\frac{(h_i - h_{i-1})}{3!} - \theta_1\varepsilon(h_i - h_{i-1}), \qquad \Pi_{4,i} = \varepsilon\left(\frac{1}{4!} - \frac{\theta_1}{2!}\right)\frac{(h_i^3 + h_{i-1}^3)}{\chi_i}.$$

Also, $\Pi_{2,i} = 0$ for $\theta_1 + \theta_2 = 1/2$. Thus, we have

$$\Pi_i = \varepsilon\left(\frac{1}{4!} - \frac{\theta_1}{2!}\right)\frac{(h_i^3 + h_{i-1}^3)}{\chi_i}v^{(iv)}(\varsigma_i) + O(N^{-3}). \tag{18}$$

Similarly, the truncation error in outer region is given by

$$\Pi_i = \frac{\varepsilon}{3}\Big(h_{i-1} - h_i\Big)v^{(iii)}(\varsigma_i) + \frac{2\varepsilon}{4!}\left(\frac{h_i^3 + h_{i-1}^3}{\chi_i}\right)v^{(iv)}(\varsigma_i) + O(N^{-3}). \tag{19}$$

By using the Lemma 3, we obtain the following proposition which gives the error estimation for the proposed hybrid numerical scheme.

**Theorem 1** *Let $v(\varsigma)$, $\varsigma \in \overline{\Omega}$ be the solution of (1)–(3) and let $U(\varsigma_i)$, $\varsigma \in \overline{\Omega}^N$ be the solution of (14)–(15), respectively. The local truncation error then satisfies the estimation of the following*

$$\sup_{0<\varepsilon\leqslant 1} \|U - v\|_{\Omega^N} \leqslant \lambda N^{-2}(\ln N)^3,$$

*where $\lambda$ is a constant.*

## 5 Numerical Results

We present numerical examples in this section to solve SPDDE with shifts to show the accuracy of the proposed hybrid numerical scheme and discuss the accuracy of the method. Some error formulas are listed below using the double mesh principle [17]. Maximum absolute error (MAE) and convergence rates are calculated by

$$E_\varepsilon^N = \max_{0 \leqslant i \leqslant N} \left| \mathrm{v}_i^N - \mathrm{v}_{2i}^{2N} \right|,$$

$$r_\varepsilon^N = \ln_2(E_\varepsilon^N / E_\varepsilon^{2N}).$$

**Example 1** Consider the SPDDE (1)–(3) with $p(\varsigma) = 1$, $q(\varsigma) = 2$, $r(\varsigma) = -3$, $s(\varsigma) = 0$, $g(\varsigma) = 0$, $\psi(\varsigma) = 1$, and $\gamma(\varsigma) = 1$ in the interval [0,1]. Computational results different values of $\delta$, $\eta$, $\varepsilon$, and $N$ are listed in Table 1. For $\theta_1 = 1/12$, $\theta_2 = 5/12$, it can be seen that the hybrid numerical scheme is uniformly convergent. The width of the boundary layer decreases as $\delta$ decreases in Fig. 1.

**Example 2** Consider the SPDDE (1)–(3) with $p(\varsigma) = 1$, $q(\varsigma) = -2$, $r(\varsigma) = -5$, $s(\varsigma) = 1$, $g(\varsigma) = 0$, $\psi(\varsigma) = 1$, and $\gamma(\varsigma) = 1$ in the interval [0,1]. Hybrid numerical scheme is uniformly convergent by observing the results in Table 2 for $\theta_1 = 1/12$, $\theta_2 = 5/12$. The results with method [9] in Tables 3 and 4 are compared and in good agreement. Figures 2 presents the approximate solution for various values of $\delta$ and $\eta$, respectively.

**Table 1** MAE ($E_\varepsilon^N$) for $\delta = 0.5 \times \varepsilon = \eta$ of Example 1.

| $\varepsilon \downarrow$ | $N = 2^8$ | $N = 2^9$ | $N = 2^{10}$ |
|---|---|---|---|
| $2^{-2}$ | 1.62e-04 | 8.10e-05 | 4.05e-05 |
| $2^{-4}$ | 1.47e-04 | 7.34e-05 | 3.67e-05 |
| $2^{-8}$ | 1.93e-04 | 6.28e-05 | 3.48e-05 |
| $2^{-16}$ | 1.87e-04 | 5.90e-05 | 1.82e-05 |
| $2^{-24}$ | 1.87e-04 | 5.90e-05 | 1.82e-05 |
| $2^{-32}$ | 1.87e-04 | 5.90e-05 | 1.82e-05 |

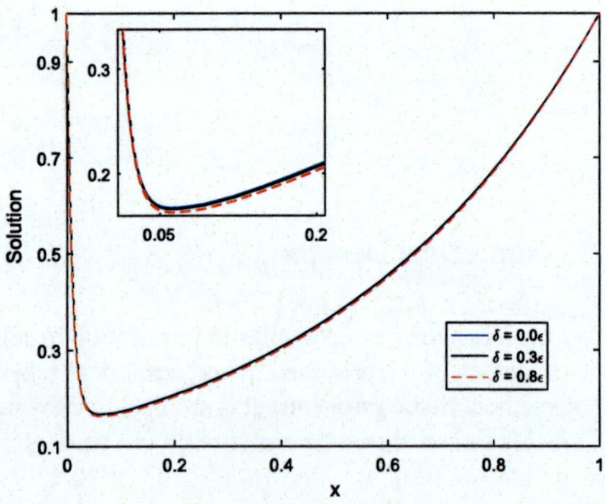


**Fig. 1** Approximate solution for $\varepsilon = 0.001$ of Example 1

**Table 2** MAE ($E_\varepsilon^N$) for $\delta = 0.5 \times \varepsilon = \eta$ of Example 1

| $\varepsilon \downarrow$ | $N = 2^8$ | $N = 2^9$ | $N = 2^{10}$ |
|---|---|---|---|
| $2^{-2}$ | 8.35e-04 | 4.15e-04 | 2.06e-04 |
| $2^{-4}$ | 8.25e-04 | 4.08e-04 | 2.03e-04 |
| $2^{-8}$ | 4.80e-04 | 2.00e-04 | 9.02e-05 |
| $2^{-16}$ | 2.13e-04 | 6.74e-05 | 2.08e-05 |
| $2^{-24}$ | 2.13e-04 | 6.74e-05 | 2.08e-05 |
| $2^{-32}$ | 2.13e-04 | 6.74e-05 | 2.08e-05 |

**Table 3** MAE for $\varepsilon = 0.1,\ \eta = 0.05$ of Example 2

| | $\delta \downarrow$ | Proposed method | Method in [8] |
|---|---|---|---|
| $N = 2^5$ | 0.00 | 6.40e-03 | 3.45e-02 |
| | 0.05 | 5.81e-03 | 3.82e-02 |
| | 0.09 | 5.38e-03 | 4.11e-02 |
| $N = 2^7$ | 0.00 | 1.45e-03 | 1.16e-02 |
| | 0.05 | 1.31e-03 | 1.29e-02 |
| | 0.09 | 1.21e-03 | 1.40e-02 |
| $N = 2^9$ | 0.00 | 3.54e-04 | 3.00e-03 |
| | 0.05 | 3.20e-04 | 3.35e-03 |
| | 0.09 | 2.96e-04 | 3.62e-03 |

**Table 4** MAE for $\varepsilon = 0.1,\ \delta = 0.05$ of Example 2

| | $\eta \downarrow$ | Proposed method | Method in [8] |
|---|---|---|---|
| $N = 2^5$ | 0.00 | 6.10e-03 | 3.64e-02 |
| | 0.05 | 5.81e-03 | 3.82e-02 |
| | 0.09 | 5.59e-03 | 3.96e-02 |
| $N = 2^7$ | 0.00 | 1.38e-03 | 1.22e-02 |
| | 0.05 | 1.31e-03 | 1.29e-02 |
| | 0.09 | 1.26e-03 | 1.34e-02 |
| $N = 2^9$ | 0.00 | 3.37e-04 | 3.17e-03 |
| | 0.05 | 3.20e-04 | 3.35e-03 |
| | 0.09 | 3.08e-04 | 3.49e-03 |

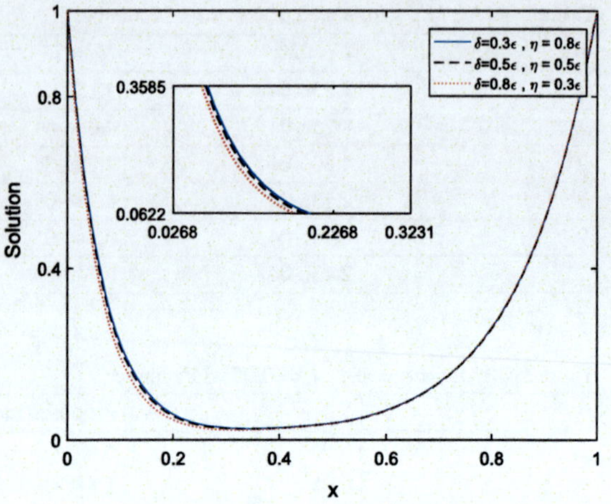


**Fig. 2** Approximate solution for $\varepsilon = 0.001$ of Example 2

# 6 Conclusions

In this study, we have employed hybrid numerical technique scheme for the solution of singularly perturbed differential difference equation with shifts. By employing this technique for the problem on non-uniform Shishkin mesh, we obtain the parameter uniform convergence. Finally, numerical examples are experimented for to validate the theoretical applicability. Numerical computations were provided by using Scilab software package.

# References


1. Bestehorn M, Grigorieva EV (2004) Formation and propagation of localized states in extended systems. Ann Phys 13:7–8. https://doi.org/10.1002/andp.200410085
2. Burton TA (2005) Fixed points, stability and exact linearization. Nonlinear Anal Theory Methods Appl 61:857–8708. https://doi.org/10.1016/j.na.2005.01.079
3. Derstine MW, Gibbs HM, Hopf FA, Kaplan DA (1982) Bifurcation gap in a hybrid optically bistable system. Phys Rev A 26:3720–3722. https://doi.org/10.1103/PhysRevA.26.3720
4. Joseph DD, Preziosi L (1989) Heat waves. Rev Mod Phys 61:41–73. https://doi.org/10.1103/RevModPhys.61.41
5. Tzoun DY (1997) Micro-to-macroscale heat transfer. Taylor & Francis, Washington, D.C.
6. Lange CG, Miura RM (1994) Singular perturbation analysis of boundary value problems for differential difference equations. V. Small shifts with layer behavior. SIAM J Appl Math 54:249–272. https://doi.org/10.1137/S0036139992228120
7. Lange CG, Miura RM (1994) Singular perturbation analysis of boundary value problems for differential difference equations. VI. Small shifts with rapid oscillations. SIAM J Appl Math 54:273–283. https://doi.org/10.1137/S0036139992228119
8. Kadalbajoo MK, Sharma KK (2005) Numerical treatment of a mathematical model arising from a model of neuronal variability. J Math Anal Appl 307:606–627. https://doi.org/10.1016/j.jmaa.2005.02.014

9. Kadalbajoo MK, Patidar KC, Sharma KK (2006) $\varepsilon$-uniformly convergent fitted methods for the numerical solution of the problems arising from singularly perturbed general DDEs. Appl Math Comput 182:119–139. https://doi.org/10.1016/j.amc.2006.03.043
10. Patidar KC, Sharma KK (2006) Uniformly convergent non-standard finite difference methods for singularly perturbed differential-difference equations with delay and advance. Int J Num Methods Eng 66:272–296. https://doi.org/10.1002/nme.1555
11. Vivek KA, Sharma KK (2008) An optimized B-spline method for solving singularly perturbed differential difference equations with delay as well as advance. Neural Parall Sci Comput 16:371–385
12. Kadalbajoo MK, Kumar D (2008) Fitted mesh B-spline collocation method for singularly perturbed differential-difference equations with small delay. Appl Math Comput 204:90–98. https://doi.org/10.1016/j.amc.2008.05.140
13. Salama AA, Al-Amery DG (2013) High order method for singularly perturbed differential-difference equations with small shifts. Int J Pure Appl Math 88:273–295. https://doi.org/10.12732/ijpam.v88i2.10
14. Miller JJH, O'Riordan RE, Shishkin GI (1996) Fitted numerical methods for singular perturbation problems. World Scientific, Singapore
15. Khan A, Khan I, Aziz T, Stojanovic MA (2004) Variable-mesh approximation method for singularly perturbed boundary-value problems using cubic spline in tension. Int J Comput Math 81:1513–1518. https://doi.org/10.1080/00207160412331284169
16. Stynes M, Roos HG (1997) The midpoint upwind scheme. Appl Num Math 23:361–374. https://doi.org/10.1016/S0168-9274(96)00071-2
17. Doolan EP, Miller JJH, Schilderse WHA (1980) Uniform numerical methods for problems with initial and boundary layers. Boole Press, Dublin

# Splines for Atmospheric Data

**P. Venkata Ramana Moorthy, B. Sarojamma, and S. Venkatramana Reddy**

## 1 Introduction

Wind speed is an atmospheric quantity measure and air moving from high pressure to low pressure. Higher wind speeds generate wind energy when strong wind blows the blades to rotate in higher speeds. The major advantages of wind are that it is a renewable, sustainable, and free resource. Wind is an environmentally friendly source of energy for human beings. Wind turbines occupy small places and towers are high above the ground. Wind turbines are both industrial and domestic. These wind turbines produce energy of power to remote locations. In 1888, wind turbines were operated and, nowadays, wind power industry has going upward. Wind turbines have low maintenance cost, low running costs, and it has huge potential. Wind energy is low cost for generating electricity. By using local renewable energy sources, a country can reduce its dependency on external supplies of natural resources. By use of this, a country can increase its energy security. Nowadays, wind energy industry creates jobs in industry.

Impact of wind direction and speed on dusty days was given by [1] and was analyzed by Yazd in Iran constantly exposed to the dust particles in the air, which impact human life and causes health issues. In this paper, they used Pearson correlation coefficient to study the relationship between wind direction and wind speed, and wind speed and visibility. They concluded that there exists a negative correlation between wind speed and visibility and positive correlation between wind speed and direction.

P. V. R. Moorthy · S. V. Reddy (✉)
Department of Physics, Sri Venkateswara University, Tirupati 517 502, A.P., India
e-mail: drsvreddy123@gmail.com

B. Sarojamma
Department of Statistics, Sri Venkateswara University, Tirupati 517 502, A.P., India



S. Srinivas et al. (eds.), *Recent Advances in Applied Mathematics and Applications to the Dynamics of Fluid Flows*, Lecture Notes in Mechanical Engineering,
https://doi.org/10.1007/978-981-19-1929-9_23

Time-series prediction of wind speed using machine learning algorithms: A case study of Osorio wind farm in Brazil was given by [2]. In this paper, they predicted wind speed data of Osorio wind farm in South Brazil using machine learning algorithms like multi-layer feed-forward neural network (MLFFNN), support vector regression (SVR), fuzzy inference system (FIS), adaptive neuro fuzzy inference system (ANFIS), group method of data handling (GMDH)-type neural network, ANFIS optimized with particle swarm optimization algorithm (ANFIS-PSO), and ANFIS optimized with genetic algorithms (ANFIS-GA).

A review on the forecasting of wind speed and generated power was given by [3]. In this paper, they discussed about physical models, conventional statistic models, spatial correlation models, and artificial intelligence and new models. Probabilistic wind speed forecasting using ensembles and Bayesian model averaging was given by [4]. They used Bayesian model averaging for processing numerical weather prediction. An hourly average wind speed and direction and average hourly power measured from January to April 2002 with a total of 2800 data points for prediction of wind energy using adaptive estimation methods and dynamic models by Ismael Sanchez in the paper short-term prediction of wind energy production [5]. Short-term prediction of local wind conditions was given by [6]. Reddy et al. [7] published a research paper for wind power density analysis for micro-scale wind turbines using various distributions.

## 2 Methodology

There are many regression models for wide usage of wind speed. Some of the most popular regression models are simple linear regression, multiple linear regression, polynomial regression models, etc. In the present paper by taking temperature, visibility, and day-wise "time" as independent variables and wind speed as dependent variable, we are fitting piece-wise regression models. The different piece-wise regression models fitted are linear spline, polynomial spline, and cubic spline.

### *2.1 Linear Spline or Piece-wise Linear Regression*

If we have huge data, data contains highest value almost in one situation, then data is divided into two parts by taking the highest value as "knot" and we will fit two linear regression equations for left and right data sets by forcing knot to be continuous. If we have $x_1, x_2, \ldots, x_n \ldots, x_s$ are independent variables and $y$ is dependent variable. If $x_n$ is a higher value of '$x$' variable, consider it as knot. Now data is divided into two parts, and for each part, we are fitting linear model separately.

| Data | Model |
|---|---|
| Subdivision1 : $x_1, \ldots, x_n$ | $y_i = a + b_i x_i$ |
| Subdivision2 : $x_n, \ldots, x_s$ | $y_j = a_j + b_j x_i$ |

According to data, we may fit different linear regression models for various subdivisions by taking different knots.

## 2.2 Polynomial Spline (or) Piece-Wise Polynomial Regression Model

In linear regression model, we fitted linear regression equation for every subdivision data by continuous knots. In polynomial splines, we fitted polynomial regression for each subdivision of data. If data is $t_{01}, t_{11}, \ldots, t_{k1}, t_{02}, t_{12}, \ldots, t_{k2}, t_{03}, t_{13}, \ldots, t_{k3}, t_{04}, t_{14}, \ldots, t_{k4}, \ldots, t_{0k}, t_{1k}, \ldots, t_{kk}$. Here, in the data, we have $k$ subgroups with '$k$' knots. Therefore, the model fitted is

| | | |
|---|---|---|
| sub division1 | $t_{01}, t_{11}, t_{k1}$ | $y_{1i} = a_{1i} + b_{1i} t_{j1} \quad j = 0 \le t \le k$ |
| sub division2 | $t_{02}, t_{12}, t_{k2}$ | $y_{2i} = a_{2i} + b_{2i} t_{j2} \quad j = 0 \le t \le k$ |
| ...... | ...... | ...... |
| ...... | ...... | ...... |
| sub division k | $t_{0k}, t_{1k}, \ldots, t_{kk}$ | $y_{ki} = a_{ki} + b_{ki} t_{jk} \quad j = 0 \le t \le k$ |

## 2.3 Cubic Regression Model or Cubic Spline

The cubic regression equation used in general is

$$Y = a + bx + cx^2 + dx^3$$

where $a, b, c$, and $d$ are constants, $x$ is the independent variable, and $y$ is the dependent variable. Cubic spline is as similar as linear spline and polynomial spline. For each subdivision, cubic spline is fitted for subdivided data. If data is divided by two knots, i.e., three groups,

| | | |
|---|---|---|
| $x_1, x_2, \ldots, x_n$ | $y_1 = a_1 + b_1 x_i + c_1 x_i^2$ | $i$ is $1 \le j \le n$ |
| $x_n, x_{n+1}, x_{n+2}, \ldots, x_s$ | $y_2 = a_2 + b_2 x_j + c_2 x_j^2$ | $n$ is $1 \le j \le s$ |
| $x_s, x_{s+1}, x_{s+2}, \ldots, x_r$ | $y_3 = a_3 + b_3 x_t + c_2 x_t^2 + d_1 x_t^3$ | $s$ is $1 \le j \le r$ |

**Multiple R squared Value:** It is the ratio of sum of area by total sum of squares subtracted from 1.

$$R^2 = 1 - \frac{ESS}{TSS}$$

Generally, $R^2$ value lies between 0 and 1 and commonly value is multiplied with 100 and expressed performance over the independent variable or dependent variable.
**Root Mean Square Error (RMSE):** RMSE is the measure of difference between actually observed from the thing which is being modeled or estimates and values predicted by a model.

## 3 Empirical Investigations

Wind speed is dependent variable and various independent variables are temperature, day-wise time, and visibility. The three piece-wise regression models are linear spline or linear piece-wise regression model, quadratic spline or quadratic piece-wise regression model, and cubic spline or cubic piece-wise regression model. Data chosen for fitting splines is from 1 January 2014 to 31 December 2018 [8]. We selected knots of the data by high and low temperatures as tabulated in Table 1.

For the above ten data sets, we fitted linear, quadratic, and cubic splines separately by taking wind speed with temperature, wind speed with day-wise time, and wind speed with visibility. By taking wind speed as dependent variable and temperature as independent variable, the fitted linear spline, their $R^2$ values, and RMSE values are listed in Table 2. The quadratic splines fitted for the temperature as independent variable and wind speed as dependent variable, their multiple $R^2$ values, and RMSE values are listed in Table 3. Cubic splines fitted for the wind speed as dependent variable and temperature as independent variable and their RMSE and multiple $R^2$ values are listed in Table 4. The fitted linear temperature spline, quadratic temperature spline, and cubic temperature spline and wind speed according to time with temperature are presented in Fig. 1.

By taking visibility as independent variable and wind speed as dependent variable, we fitted linear piece-wise regression model, quadratic piece-wise regression model,

**Table 1** Data set subdivisions

| | |
|---|---|
| Data set Subdivision 1 | 01.01.2014 to 11.06.2014 |
| Data set Subdivision 2 | 11.06.2014 to 01.09.2014 |
| Data set Subdivision 3 | 01.09.2014 to 26.05.2015 |
| Data set Subdivision 4 | 26.05.2015 to 22.01.2016 |
| Data set Subdivision 5 | 22.01.2016 to 04.06.2016 |
| Data set Subdivision 6 | 04.06.2016 to 31.12.2016 |
| Data set Subdivision 7 | 31.12.2016 to 05.06.2017 |
| Data set Subdivision 8 | 05.06.2017 to 01.01.2018 |
| Data set Subdivision 9 | 01.01.2018 to 12.06.2018 |
| Data set Subdivision 10 | 12.06.2018 to 01.01.2019 |

**Table 2** Fitted linear splines for temperature data set subdivisions

| Subdivision | Model | $R^2$ | RMSE |
|---|---|---|---|
| 1 | $1.4584 + 0.0514T$ | 0.1332 | 1.9599 |
| 2 | $10.6749 + 0.1862T$ | 0.1671 | 1.6953 |
| 3 | $2.1822 + 0.03177T$ | 0.0377 | 2.0738 |
| 4 | $-2.9229 + 0.0938T$ | 0.2805 | 1.9265 |
| 5 | $2.6252 + 0.0218T$ | 0.0098 | 1.9910 |
| 6 | $-2.2723 + 0.0801T$ | 0.1645 | 1.9651 |
| 7 | $2.2925 + 0.0356T$ | 0.0659 | 1.9355 |
| 8 | $-0.917 + 0.0620T$ | 0.1025 | 2.0588 |
| 9 | $1.7699 + 0.0457T$ | 0.1125 | 1.8569 |
| 10 | $-3.3338 + 15.9144T$ | 0.2288 | 2.1423 |

**Table 3** Fitted quadratic splines for temperature data set subdivisions

| Subdivision | Model | $R^2$ | RMSE |
|---|---|---|---|
| 1 | $-9.6925 + 0.3643T - 0.0021T^2$ | 0.1663 | 1.9221 |
| 2 | $-29.5239 + 0.6025T - 0.0023T^2$ | 0.1678 | 1.7752 |
| 3 | $10.9762 - 0.2212T + 0.0018T^2$ | 0.0605 | 2.0318 |
| 4 | $14.9192 - 0.4042T + 0.0034T^2$ | 0.3266 | 1.8637 |
| 5 | $-10.2852 + 0.3946T - 0.0026T^2$ | 0.0274 | 1.9603 |
| 6 | $12.5409 - 0.3140T + 0.0025T^2$ | 0.1841 | 1.9419 |
| 7 | $-6.1826 + 0.2676T - 0.0015T^2$ | 0.0814 | 1.9194 |
| 8 | $35.5765 - 0.9360T + 0.0067T^2$ | 0.2130 | 1.9280 |
| 9 | $-0.7438 + 0.1148T - 0.0005T^2$ | 0.1141 | 1.8553 |
| 10 | $5.9865 - 0.1590T + 0.0017T^2$ | 0.2410 | 2.1252 |

**Table 4** Fitted cubic splines for temperature data set subdivisions

| Sub division | Model | $R^2$ | RMSE |
|---|---|---|---|
| 1 | $-1.9045 + 0.0411T + 0.0023T^2 - 0.0004T^3$ | 0.1668 | 1.9215 |
| 2 | $-1829.4152 + 60.5157T - 0.6658T^2 + 0.0024T^3$ | 0.1895 | 1.6953 |
| 3 | $-25.7564 + 1.3682T - 0.0206T^2 + 0.0001T^3$ | 0.0758 | 2.0318 |
| 4 | $-42.7718 - 1.5660T + 0.019T^2 + 0.407T^3$ | 0.3306 | 1.8581 |
| 5 | $22.5542 - 1.0347T + 0.0178T^2 - 0.0001T^3$ | 0.0302 | 1.9703 |
| 6 | $65.6836 - 2.4278T + 0.0301T^2 - 0.0001T^3$ | 0.1918 | 1.9328 |
| 7 | $1.0816 - 0.0342T + 0.0026T^2 - 0.0002T^3$ | 0.0819 | 1.9188 |
| 8 | $14.1648 - 0.0618T - 0.0051T^2 + 0.00001T^3$ | 0.2143 | 1.9263 |
| 9 | $23.6830 - 0.8912T + 0.0130T^2 - 0.0001T^3$ | 0.1178 | 1.8514 |
| 10 | $8.9865 - 0.2832T + 0.0034T^2 - 0.00001T^3$ | 0.2411 | 2.125 |

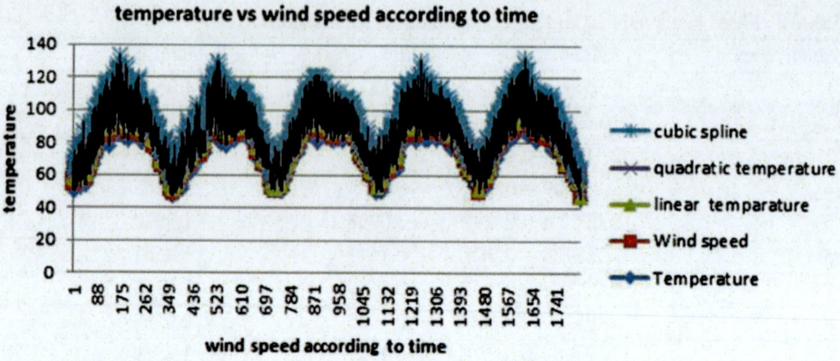


Fig. 1 Temperature versus wind speed according to time fitted for various models

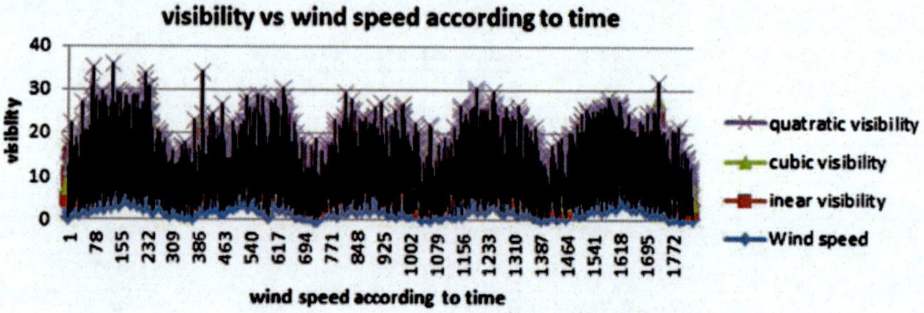


Fig. 2 Visibility versus wind speed according to time fitted for various models

and cubic piece-wise regression model. $R^2$ and RMSE values for each regression model are listed in Table 5.

In Table 5, first column represent the subdivision using knots, second column explain various models used for that subdivision like linear '1', quadratic '2', and cubic '3', third column describes the $R^2$ values, and fourth column explain RMSE values.

The fitted linear visibility spline, quadratic visibility spline, and cubic visibility spline and wind speed according to time with visibility are presented in Fig. 2.

By taking day-wise time as independent variable and wind speed as dependent variable for the day-wise data from 1 January 2014 to 31 December 2018, i.e., 5 years. These 5 years data is divided into ten subdivisions by taking nine knots. The fitted linear spline models, cubic spline models, and quadratic splines, their multiple $R^2$ and RMSE values are listed in Table 6.

The fitted linear model, quadratic model, and cubic model for time as independent variable and wind speed as dependent variables are shown in Fig. 3.

**Table 5** Fitted linear, quadratic and cubic splines for visibility data

| Subdivision | Model | $R^2$ | RMSE |
|---|---|---|---|
| 1 | $2.2711 + 2.1829V$ | 0.3086 | 1.7505 |
| | $2.8538 + 1.0638V + 0.4397V^2$ | 0.314 | 1.7436 |
| | $3.1808 + 0.0612V + 1.2403V^2 - 0.1828V^3$ | 0.3147 | 1.7427 |
| 2 | $0.0822 + 3.2677V$ | 0.219 | 1.6641 |
| | $-3.5277 + 7.3553V - 1.1312V^2$ | 0.2244 | 1.6583 |
| | $9.6618 - 16.2623V + 12.5015V^2 - 2.5462V^3$ | 0.2301 | 1.6522 |
| 3 | $2.6159 + 1.4276V$ | 0.1288 | 1.9727 |
| | $2.8212 + 1.0422V + 0.1499V^2$ | 0.1293 | 1.9721 |
| | $2.7320 + 1.3274V - 0.0959V^2 + 0.0617V^3$ | 0.1294 | 1.9721 |
| 4 | $1.0374 + 2.5254V$ | 0.3583 | 1.8192 |
| | $1.556 + 1.5390V + 0.3910V^2$ | 0.3607 | 1.8158 |
| | $3.7695 - 5.3176V + 6.2207V^2 - 1.4708V^3$ | 0.3717 | 1.8001 |
| 5 | $1.1079 + 1.8721V$ | 0.1713 | 1.8214 |
| | $2.8833 - 0.4521V + 0.9216V^2$ | 0.1824 | 1.8091 |
| | $2.2275 + 1.6241V - 0.9298V^2 + 0.4944V^3$ | 0.1831 | 1.8083 |
| 6 | $0.9947 + 2.1349V$ | 0.2622 | 1.8467 |
| | $1.4261 + 1.3462V + 0.2978V^2$ | 0.2643 | 1.8441 |
| | $2.9253 - 3.0874V + 3.7359V^2 - 0.7830V^3$ | 0.2754 | 1.8301 |
| 7 | $2.3516 + 1.6733V$ | 0.201 | 1.7901 |
| | $4.6452 - 2.2005V + 1.3736V^2$ | 0.2265 | 1.7612 |
| | $4.1331 - 0.8944V + 0.4006V^2 + 0.2221V^3$ | 0.2267 | 1.761 |
| 8 | $0.8387 + 2.0120V$ | 0.2143 | 1.9263 |
| | $2.8278 - 1.2734V + 1.1668V^2$ | 0.2357 | 1.8999 |
| | $0.7042 + 4.9935V - 3.9632V^2 + 1.2573V^3$ | 0.2429 | 1.891 |
| 9 | $2.9927 + 1.4566V$ | 0.1244 | 1.8445 |
| | $3.5218 + 0.5164V + 0.3508V^2$ | 0.1265 | 1.8423 |
| | $4.1850 - 1.5411V + 2.0827V^2 - 0.4335V^3$ | 0.1273 | 1.8415 |
| 10 | $0.6690 + 2.3033V$ | 0.1359 | 2.2675 |
| | $0.2101 + 2.9747V - 0.2298V^2$ | 0.1363 | 2.2671 |
| | $2.7151 - 2.9136V + 3.9927V^2 - 0.9449V^3$ | 0.1385 | 2.2642 |

**Table 6** Fitted linear, quadratic and cubic splines for visibility data

| Subdivision | Model | $R^2$ | RMSE |
|---|---|---|---|
| 1 | $3.7423 + 0.0183t$ | 0.1653 | 21.9233 |
| | $3.1755 + 0.0390t - 0.0001t^2$ | 0.1793 | 1.9072 |
| | $3.2813 + 0.0314t - 0.00001t^2 - 0.00001t^3$ | 0.1795 | 1.9068 |
| 2 | $10.3733 - 0.0213t$ | 0.0701 | 1.8158 |
| | $50.1949 - 0.4189t + 0.00097t^2$ | 0.1348 | 1.7515 |
| | $492.1285 - 7.0865t + 0.0339t^2 - 0.0001t^3$ | 0.2179 | 1.665 |
| 3 | $2.412 + 0.0052t$ | 0.0364 | 2.0747 |
| | $12.5457 - 0.0504t + 0.0001t^2$ | 0.0712 | 2.0369 |
| | $83.1678 - 0.6422t + 0.0017t^2 - 0.0001t^3$ | 0.0309 | 1.9703 |
| 4 | $15.800 - 0.018t$ | 0.3041 | 1.8945 |
| | $23.7586 - 0.0435t + 0.0001t^2$ | 0.3056 | 1.8925 |
| | $-261.48 + 1.3304t - 0.0022t^2 - 0.0001t^3$ | 0.3237 | 1.8677 |
| 5 | $-12.0155 + 0.0205t$ | 0.0491 | 1.9511 |
| | $-298.716 + 0.7468t - 0.0005t^2$ | 0.0583 | 1.9416 |
| | $4966.083 - 19.2641t + 0.0249t^2 - 0.0001t^3$ | 0.0602 | 1.9397 |
| 6 | $17.2261 - 0.0135t$ | 0.2377 | 1.8771 |
| | $29.0079 - 0.0381t + 0.0001t^2$ | 0.2387 | 1.8759 |
| | $-832.141 + 2.6686t - 0.0028t^2 - 0.0001t^3$ | 0.2667 | 1.841 |
| 7 | $-10.7585 + 0.0134t$ | 0.0911 | 1.9092 |
| | $-310.88 + 0.5252t - 0.0002t^2$ | 0.13 | 1.8679 |
| | $4400.02 + 10.9885t - 0.0091t^2 + 0.0001t^3$ | 0.1382 | 1.859 |
| 8 | $26.9404 - 0.0169t$ | 0.2221 | 1.916 |
| | $238.2359 - 0.3288t + 0.0001t^2$ | 0.2523 | 1.8792 |
| | $-6509.55 + 14.6192t - 0.0109t^2 + 0.00002t^3$ | 0.2998 | 1.8185 |
| 9 | $-21.6172 + 0.0174t$ | 0.1712 | 1.7946 |
| | $78.3650 - 0.2122t + 0.00004t^2$ | 0.1729 | 1.7927 |
| | $-2850.51 + 5.5867t - 0.0036t^2 + 0.0001t^3$ | 0.1739 | 1.7915 |
| 10 | $41.5676 - 0.0216t$ | 0.2692 | 2.0854 |
| | $91.8659 - 0.0797t + 0.00002t^2$ | 0.2696 | 2.0847 |
| | $2592.698 - 4.4328t + 0.003t^2 - 0.00001t^3$ | 0.2706 | 2.0833 |

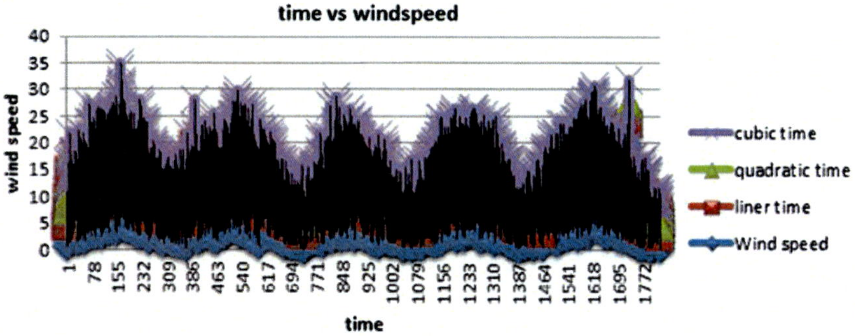


**Fig. 3** Time versus wind speed for various fitted models

**Table 7** Summary of the best fitted models for Temperature, Visibility and time

| Subdivision | Temperature | Visibility | Time |
|---|---|---|---|
| 1 | $-1.9045 + 0.0411T+$ | $3.1808 + 0.0612V+$ | $3.2813 + 0.0314t-$ |
| | $0.0023T^2 - 0.0004T^3$ | $1.2403V^2 - 0.1828V^3$ | $0.0001t^2 - 0.00001t^3$ |
| 2 | $-1829.4152 + 60.5157T-$ | $9.6618 - 16.2623V+$ | $492.1285 - 7.0865t+$ |
| | $0.6658T^2 + 0.0024T^3$ | $12.5015V^2 - 2.5462V^3$ | $0.0339t^2 - 0.0001t^3$ |
| 3 | $-25.7564 + 1.3682T-$ | $2.7320 + 1.3274V-$ | $83.1678 - 0.6422t+$ |
| | $0.0206T^2 + 0.0001T^3$ | $0.0959V^2 + 0.0617V^3$ | $0.0017t^2 - 0.0001t^3$ |
| 4 | $-42.7718 - 1.5660T+$ | $3.7695 - 5.3176V+$ | $-261.48 + 1.3304t-$ |
| | $0.019T^2 + 0.407T^3$ | $6.2207V^2 - 1.4708V^3$ | $0.0022t^2 - 0.0001t^3$ |
| 5 | $22.5542 - 1.0347T+$ | $2.2275 + 1.6241V-$ | $4966.083 - 19.2641t+$ |
| | $0.0178T^2 - 0.0001T^3$ | $0.9298V^2 + 0.4944V^3$ | $0.0249t^2 - 0.0001t^3$ |
| 6 | $65.6836 - 2.4278T+$ | $2.9253 - 3.0874V+$ | $-832.141 + 2.6686t-$ |
| | $0.0301T^2 - 0.0001T^3$ | $3.7359V^2 - 0.7830V^3$ | $0.0028t^2 - 0.0001t^3$ |
| 7 | $1.0816 - 0.0342T+$ | $4.1331 + 0.8944V+$ | $4400.02 + 10.9885t-$ |
| | $0.0026T^2 - 0.0002T^3$ | $0.4006V^2 + 0.2221V^3$ | $0.0091t^2 + 0.0001t^3$ |
| 8 | $14.1648 - 0.0618T-$ | $0.7042 + 4.9935V-$ | $-6509.55 + 14.6192t-$ |
| | $0.0051T^2 + 0.00001T^3$ | $3.9632V^2 + 1.2573V^3$ | $0.0109t^2 + 0.0002t^3$ |
| 9 | $23.6830 - 0.8912T+$ | $4.1850 - 1.5411V+$ | $-2850.51 + 5.5867t-$ |
| | $0.0130T^2 - 0.0001T^3$ | $2.0827V^2 - 0.4335V^3$ | $0.0036t^2 + 0.0001t^3$ |
| 10 | $8.9865 - 0.2832T+$ | $2.7151 - 2.9136V+$ | $2592.698 - 4.4328t+$ |
| | $0.0034T^2 - 0.00001T^3$ | $3.9927V^2 - 0.9449V^3$ | $0.003t^2 - 0.00001t^3$ |

## 4 Summary and Conclusions

Splines are playing vital role and, in this paper, we have used linear splines, quadratic splines, and cubic splines for day-to-day data of temperature, visibility, and time. Temperature plays an important role in blowing of wind. When temperature is high, wind speed is also high and when temperature is low, wind speed is also low. We fitted linear, quadratic, and cubic spline models for wind speed as dependent variable and temperature, visibility, and time as independent variables. The highest and lowest temperature points as knots and fitted corresponding models are considered for ten subdivisions of data. The best model is using RMSE and $R^2$ values. The best-fitted models for temperature, visibility, and time are summarized in Table 7.

## References


1. Dehghanpour AR, Halabian AH, Fallahpour M (2014) Impact of wind direction and speed on dusty days. Int J Adv Biol Biomed Res 2(5):1742–1749
2. Khosravi A, Machado L, Nunes RO (2018) Time-series prediction of wind speed using machine learning algorithms: a case study Osorio wind farm, Brazil. Appl Energy 224(C):550–566
3. Lei Ma, Shiyan Luan, Chuanwen Jiang, Hongling Liu, Yan Zhang (2009) A review on the forecasting of wind speed and generated power. Renew Sustain Energy Rev 13:915–920
4. McLean Sloughter J, Tilmann G, Raftery AE (2010) Probabilistic wind speed forecasting using ensembles and Bayesian model averaging. J Am Stat Assoc 105(489):25–35
5. Sanchez Ismael (2006) Short-term prediction of wind energy production. Int J Forecast 22:43–56
6. Landberg L, Watson SJ (1994) Short-term prediction of local wind conditions. Boundary-Layer Meteorol 70:171–195
7. Reddy KKG, Reddy SV, Ramkumar TK, Sarojamma B (2014) Wind power density analysis for micro-scale wind Turbines. Int J Eng Sci (IJES) 3(2):53–60
8. www.visualcrossing.com

# Dynamics of SEIR Model of Nipah Virus

**V. S. V. Naga Soundarya Lakshmi and A. Sabarmathi**

## 1 Introduction

A mathematical model can translate the real-world problems into mathematical problems. It became an important tool of epidemiology to measure different strategies to restrict the disease. Recently, mathematical models on epidemiology of infectious disease has increased its influence on both theory and practice to manage and control diseases [1]. Particularly, SIR model is important to represent the flow of an infectious disease. SEIR is a model developed from SIR model which has an exposed class, that means the individuals in the class have been infected by the disease but not yet infectious.

Infectious diseases caused by living organisms such as viruses and bacteria. They can be passed from human to human through secretion, insects or by other means. These diseases can be spread by direct contact or indirect contact. Nipah virus is a 'bat borne virus' with the scientific name 'Nipah henipa virus'. The first identification of Nipah virus as a cause of outbreak was reported in Meherpur District of Bangladesh. The outbreaks have been reported in Malaysia, Singapore, Bangladesh and India, in which Bangladesh has the highest mortality rate due to Nipah virus [1]. In India, Nipah virus outbreak was reported in Siliguri of West Bengal in 2001 [2] and in Nadia District of West Bengal in 2007 [3]. Nipah virus was reported in Perambra near Calicut in Kerala in 2018.

Haider Ali Biswas [4] has investigated the Nipah virus infected in Bangladesh with the basic SIR model and proposed the control strategies for Nipah virus. Biswas [5, 6] has discussed the spread of Nipah virus among host and pathogen in Bangladesh and the control strategies were numerically studied by optimal control theory and studied the infectious diseases by SEIR model with vaccination. Zhang [7] has analysed the global dynamics of SEIR model for infectious diseases when the treatment

V. S. V. Naga Soundarya Lakshmi (✉) · A. Sabarmathi
Department of Mathematics, Auxilium College, Vellore, India
e-mail: san9sak14@gmail.com

 
S. Srinivas et al. (eds.), *Recent Advances in Applied Mathematics and Applications to the Dynamics of Fluid Flows*, Lecture Notes in Mechanical Engineering,
https://doi.org/10.1007/978-981-19-1929-9_24

discontinuous. Sultana and Podder [8] has analysed the Nipah virus by a SIR model with two control strategies. Mondal [9] has proposed the model for controlling the spread of Nipah virus with vital dynamics. Shah et al. [10, 11] have investigated SEI model of Nipah virus in bat and human population and also with control strategies for Nipah virus. Sinha and Sinha [12] have analysed the SEIR model for Zoonotic Nipah virus in South-East Asia region, also investigated the recovery rate with vaccination and without vaccination. In this paper, we have investigated the Nipah virus by SEIR model with vital dynamics and numerically analysed the infection of Nipah virus in Kerala.

## 2 Mathematical Model

The SEIR model is formulated for Nipah virus. Let $S(t)$, $E(t)$, $I(t)$, $R(t)$ be the number of Susceptible, Exposed, Infectious and Recovered class at time '$t$', respectively.

The transition of Nipah virus is described with the parameters $\omega$—Average birth rate, $\mu$—Average death rate, $\beta$—Transistion infectious rate, $\alpha$—Exposed rate, $\gamma$—Recovery rate, $N$—Total Population. The following diagram represents the SEIR model of Nipah virus (Fig. 1).

The following system of Ordinary Differential equation represents the SEIR model of Nipah virus.

$$\begin{aligned} \frac{dS}{dt} &= \omega N(t) - \mu S(t) - \beta S(t) I(t) \\ \frac{dE}{dt} &= \beta S(t) I(t) - \mu E(t) - \alpha E(t) \\ \frac{dI}{dt} &= \alpha E(t) - \gamma I(t) - \mu I(t) \\ \frac{dR}{dt} &= \gamma I(t) - \mu R(t). \end{aligned} \tag{1}$$

With the initial conditions as $S(t), E(t), I(t), R(t) \geq 0$, $\alpha, \beta, \gamma, \mu, \omega > 0$, $S(t) + E(t) + I(t) + R(t) = N(t)$.

Fig. 1 SEIR model for transition of Nipah virus

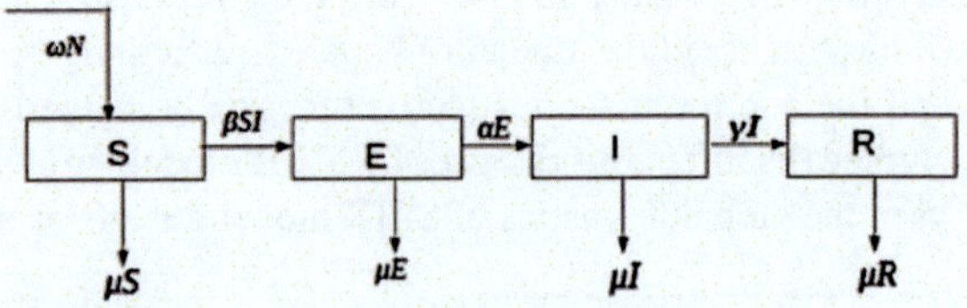

# 3 Equilibrium Analysis

Now set $s(t) = \frac{S(t)}{N}$, $e(t) = \frac{E(t)}{N}$, $i(t) = \frac{I(t)}{N}$, $r(t) = \frac{R(t)}{N}$ and $r(t) = 1 - s(t) - e(t) - i(t)$.

Hence, (1) is reduced to the following equations:

$$\frac{ds}{dt} = \omega - \mu s - \beta si \tag{2}$$

$$\frac{de}{dt} = \beta si - \mu e - \alpha e \tag{3}$$

$$\frac{di}{dt} = \alpha e - \gamma i - \mu i \tag{4}$$

The steady states are $G_0(0, 0, 0)$, $G_1(\overline{s}, \overline{e}, 0)$, $G_2(s^*, e^*, i^*)$.
Case(1): Always the trivial steady state $G_0(0, 0, 0)$ exists.
Case(2): For $G_1(\overline{s}, \overline{e}, 0)$ (i.e. in the absence of infectious class)
Let $\overline{s}, \overline{e}$ be the positive solutions of $\frac{ds}{dt} = 0$, $\frac{de}{dt} = 0$.

From (2) and (3), $G_1(\overline{s}, \overline{e}, 0) = G_1(\frac{\omega}{\mu}, 0, 0)$, Disease-free equilibrium
Case(3): For $G_2(s^*, e^*, i^*)$ (i.e. endemic equilibrium)
Let $s^*, e^*, i^*$ be the positive solutions of $\frac{ds}{dt} = 0$, $\frac{de}{dt} = 0$, $\frac{di}{dt} = 0$

From (2),

$$s^* = \frac{\omega}{(\mu + \beta i^*)} \tag{5}$$

From (3),

$$s^* = \frac{(\mu + \alpha)e^*}{\beta i^*} \tag{6}$$

From (4),

$$e^* = \frac{(\gamma + \mu)i^*}{\alpha} \tag{7}$$

Using (7) in (6), $s^* = \frac{(\mu+\alpha)(\gamma+\mu)}{\alpha\beta}$.
Hence, from (5), $i^* = \frac{-\mu}{\beta} + \frac{\omega\alpha}{(\mu+\alpha)(\gamma+\mu)}$

and from (7), $e^* = \frac{(\gamma+\mu)}{\alpha}\{\frac{\omega\alpha}{(\mu+\alpha)(\gamma+\mu)} - \frac{\mu}{\beta}\}$.

Hence, the endemic equilibrium is
$(s^*, e^*, i^*) = (\frac{(\mu+\alpha)(\gamma+\mu)}{\alpha\beta}, \frac{(\gamma+\mu)}{\alpha}(\frac{\omega\alpha}{(\mu+\alpha)(\gamma+\mu)} - \frac{\mu}{\beta}), (\frac{\omega\alpha}{(\mu+\alpha)(\gamma+\mu)} - \frac{\mu}{\beta}))$.

## 4 Stability Analysis

To find the local stability of (2)–(4), we use the Routh–Hurwitz criteria. The system is locally stable for $S^3 + a_2S^2 + a_1S + a_0 = 0$ if and only if $a_2 > 0$, $a_0 > 0$, $a_2a_1 > 0$.

The Jacobian matrix for the system (2)–(4) is

$$\begin{pmatrix} -\beta i - \mu & 0 & -\beta s \\ \beta i & -\mu - \alpha & \beta s \\ 0 & \alpha & -\gamma - \mu \end{pmatrix} \tag{8}$$

At the interior equilibrium (8) becomes

$$\begin{pmatrix} \frac{-\omega}{s} & 0 & -\beta s \\ \beta i & \frac{-\beta si}{e} & \beta s \\ 0 & \alpha & \frac{-\alpha e}{i} \end{pmatrix} \tag{9}$$

The characteristic equation of (9) is given by

$$\begin{vmatrix} \frac{-\omega}{s} - \lambda & 0 & -\beta s \\ \beta i & \frac{-\beta si}{e} - \lambda & \beta s \\ 0 & \alpha & \frac{-\alpha e}{i} - \lambda \end{vmatrix} = 0$$

$$\lambda^3 + \left(\frac{\alpha e}{i} + \frac{\beta si}{e} + \frac{\omega}{s}\right)\lambda^2 + \left(\frac{\alpha e\omega}{si} + \frac{\omega\beta i}{e}\right)\lambda + \alpha\beta^2 is = 0 \tag{10}$$

which is of the form $S^3 + a_2S^2 + a_1S + a_0 = 0$,
where $a_0 = \alpha\beta^2 is$, $a_1 = \frac{\alpha e\omega}{si} + \frac{\omega\beta i}{e}$, $a_2 = \frac{\alpha e}{i} + \frac{\beta si}{e} + \frac{\omega}{s}$
Here, $a_0 > 0$; $a_2 > 0$ as $\alpha, \beta, \omega, i, s, e$ are positive.

$$a_2a_1 = \frac{\alpha^2e^2\omega}{si^2} + \frac{\alpha e\omega^2}{s^2i} + 2\alpha\omega\beta + \frac{\omega\beta^2si^2}{e^2} + \frac{\omega^2\beta i}{se}$$

$\therefore$, $a_2a_1 > 0$, as $\alpha, \beta, \omega, i, s, e$ are positive.
Hence, (2)–(4) satisfies the condition for local stability by Routh–Hurwitz criteria.

To find the global stability at $(s^*, e^*, i^*)$, we construct the following Lyapunov function.

$$V(s, e, i) = \left[(s - s^*) - s^* \ln\frac{s}{s^*}\right] + l_1\left[(e - e^*) - e^* \ln\frac{e}{e^*}\right] + l_2\left[(i - i^*) - i^* \ln\frac{i}{i^*}\right] \tag{11}$$

Differentiate (11) with respect to t,

$$\frac{dV}{dt} = \left(\frac{s - s^*}{s}\right)\frac{ds}{dt} + l_1\left(\frac{e - e^*}{e}\right)\frac{de}{dt} + l_2\left(\frac{i - i^*}{i}\right)\frac{di}{dt}$$

By using the model equations (2)–(4),

$$= (s - s^*)\left[\frac{\omega}{s} - \mu - \beta i\right] + l_1(e - e^*)\left[\frac{\beta si}{e} - \mu - \alpha\right] + l_2(i - i^*)\left[\frac{\alpha e}{i} - \gamma - \mu\right]$$

At $(s^*, e^*, i^*)$, we have

$$\frac{dV}{dt} = (s - s^*)\left[\frac{\omega}{s} - \frac{\omega}{s^*}\right] + l_1(e - e^*)\left[\frac{\beta si}{e} - \frac{\beta s^* i^*}{e^*}\right] + l_2(i - i^*)\left[\frac{\alpha e}{i} - \frac{\alpha e^*}{i^*}\right]$$

$$= -\frac{\omega}{ss^*}(s - s^*)^2 + l_1\frac{\beta(e - e^*)}{ee^*}(e^* si - es^* i^*) + l_2\frac{\alpha(i - i^*)}{ii^*}(ei^* - ie^*)$$

Choosing $l_1 = \frac{1}{\beta}, l_2 = \frac{1}{\alpha}$,

$$\frac{dV}{dt} = -\frac{\omega}{ss^*}(s - s^*)^2 + \frac{(e-e^*)}{ee^*}(e^* si - es^* i^*) + \frac{(i-i^*)}{ii^*}(ei^* - ie^*)$$

$$= -\frac{\omega}{ss^*}(s - s^*)^2 + \left[(si + s^* i^*) - \left(\frac{e^*}{e}si + \frac{e}{e^*}s^* i^*\right)\right] + \left[(e + e^*) - \left(\frac{e}{i}i^* + \frac{i}{i^*}e^*\right)\right]$$

$\therefore$, $\frac{dV}{dt} < 0$, as all the terms in R.H.S. are negative.
By Lyapunov theorem, the system (2)–(4) is globally asymptotically stable.

## 5 Basic Reproduction Number

Basic reproduction number $R_0$ is defined as the number of new cases of an infection caused by one infected individual in a population of susceptible. In epidemiology, to find $R_0$ the next-generation matrix method is used. $FV^{-1}$ is known as the next-generation matrix, where $F$ represents the rate of appearance of new infections and $V$ represents the rate of transfer of individuals. The largest eigen value of $FV^{-1}$ is the basic reproduction number [13].

From (2)–(4), we have

$$F = \begin{bmatrix} 0 & 0 & 0 \\ 0 & 0 & \beta si \\ 0 & 0 & 0 \end{bmatrix}, \quad V = \begin{bmatrix} -\mu & 0 & -\beta s \\ 0 & -(\mu + \alpha) & 0 \\ 0 & \alpha & -(\gamma + \mu) \end{bmatrix},$$

$$FV^{-1} = \begin{bmatrix} 0 & 0 & 0 \\ 0 & \frac{\alpha\beta si}{(\mu+\alpha)(\gamma+\mu)} & \frac{\beta si}{(\gamma+\mu)} \\ 0 & 0 & 0 \end{bmatrix}$$

We have

$$R_0 = \frac{\alpha\beta}{(\mu + \alpha)(\gamma + \mu)} \tag{12}$$

We define

$$\gamma = \frac{1}{\text{mean infectious period}},$$
$$\alpha = \frac{1}{\text{mean latent period}}, \tag{13}$$
$$\beta = \frac{\text{Effective contact}}{\text{Total contact}}$$

Stability analysis by the basic reproduction number $R_0$ [14]:

1. The disease-free equilibrium is stable if $R_0 < 1$, otherwise unstable.
2. The endemic equilibrium is stable if $R_0 > 1$, otherwise unstable.

## 6 Numerical Analysis

The latency period of Nipah virus is 5–14 days [15], from (13), $\alpha = 0.1053$. The infectious period of Nipah virus is 3–14 days, from (13), $\gamma = 0.1176$.

The population of Kerala in 2012 is N = 34800000. By the natural birth rate and death rate of Kerala in 2012, $\omega = 0.0408$, $\mu = 0.01890$. Let us assume $\beta = 0.75$.

From (12), $R_0 = 4.66$, Which is unstable in disease-free equilibrium and stable in endemic equilibrium.

The following table gives the value of $R_0$ for different values of $\gamma$ and $\beta$.

| $\beta$ | $\gamma$ | $R_0$ | Disease-Free Equilibrium | Endemic Equilibrium |
|---|---|---|---|---|
| 0.35 | 0.18 | 1.4919 | Unstable | Stable |
| 0.3 | 0.2 | 1.1619 | Unstable | Stable |
| 0.25 | 0.22 | 0.8872 | Stable | unstable |
| 0.2 | 0.23 | 0.6813 | Stable | unstable |
| 0.55 | 0.4474 | 1 | Critical value | Critical value |

Figure 2 shows the flow of variables with respect to time. It is clear from Fig. 3 that the number of susceptibles which increases day by day. Figure 4 shows clearly that the exposure of the disease starts from day 3.

It is clear from Fig. 5 that the spread of infection of disease starts from day 4. Figure 6 represents the recovery individuals which starts from day 4. Figure 7 represents that as the transition infectious rate $\beta$ increases the susceptible individuals decreases.

Figure 8 gives a clear image that the individuals in the exposed class increases as the exposed rate $\alpha$. Figure 9 clears that as the recovery rate $\gamma$ increases, the individuals in the infectious class decreases. Figure 10 represents that the individuals in the recovered class increases as the recovery rate $\gamma$ increases.

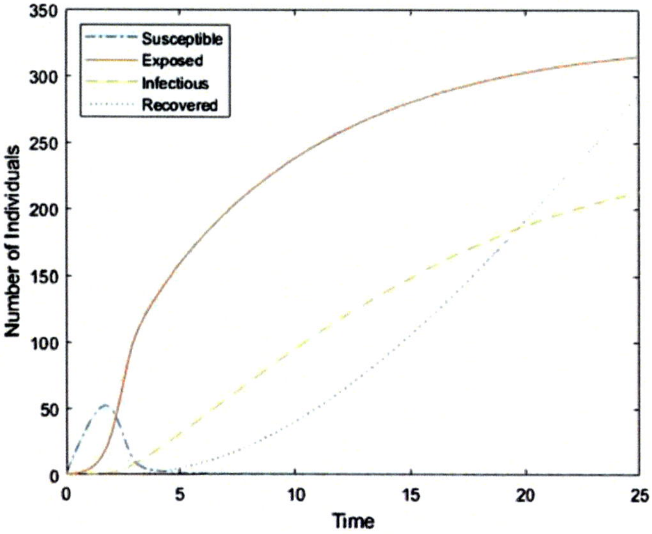


**Fig. 2** Flow of variables with respect to time

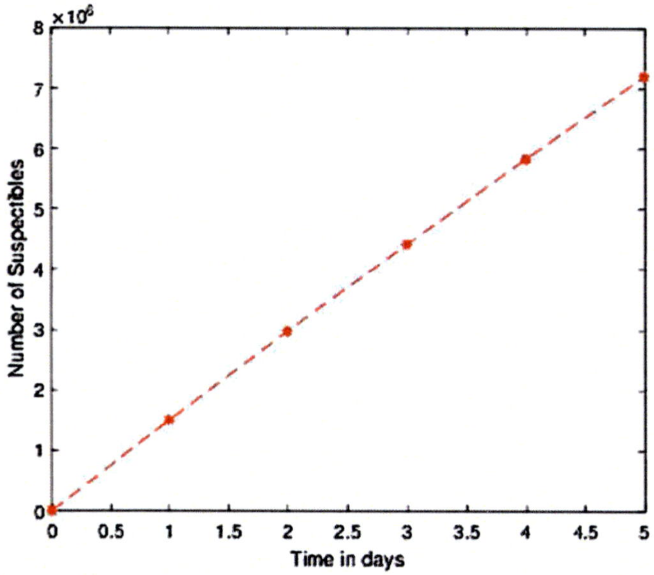


**Fig. 3** Susceptible class with respect to time

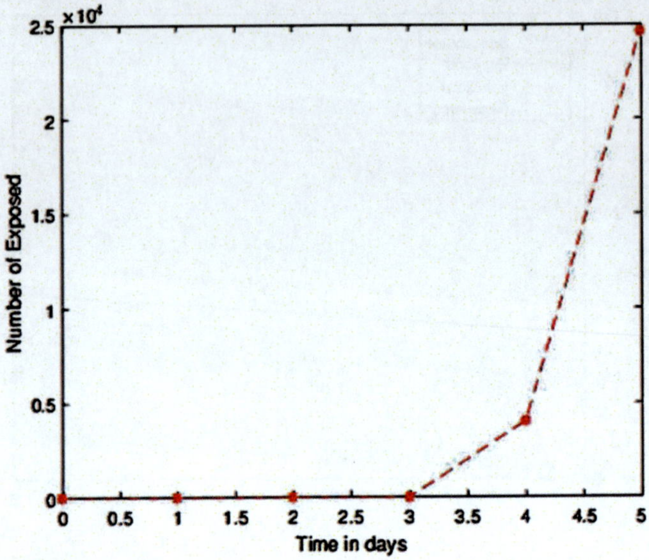


**Fig. 4** Exposed class with respect to time

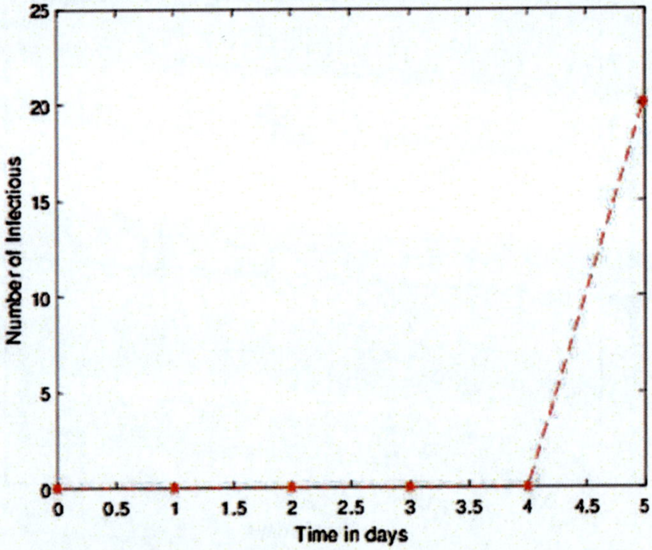


**Fig. 5** Infectious class with respect to time

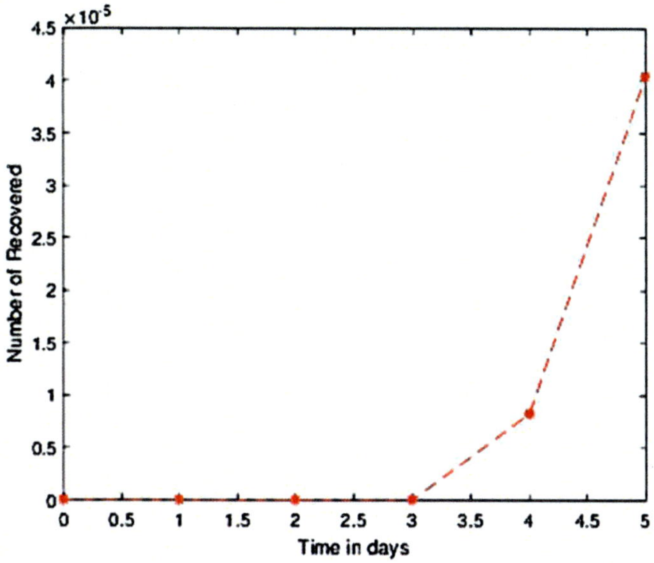


**Fig. 6** Recovered class with respect to time

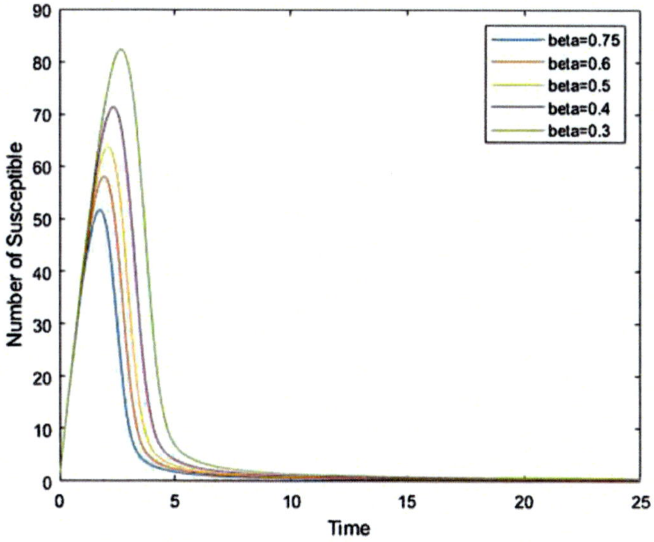


**Fig. 7** Susceptible class for different values of $\beta$

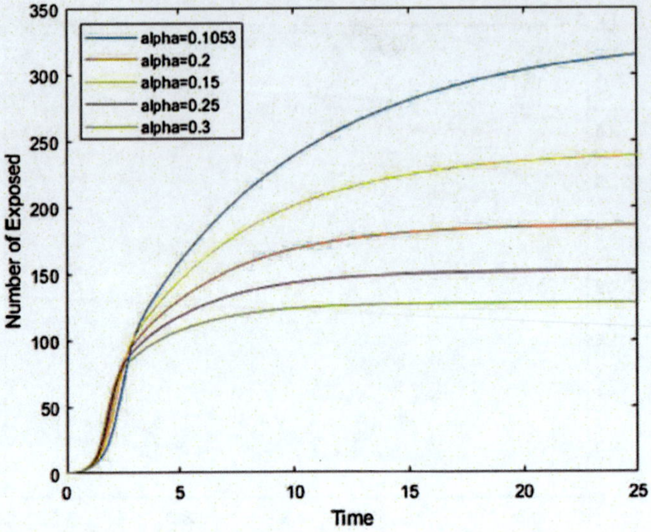


**Fig. 8** Exposed class for different values of $\alpha$

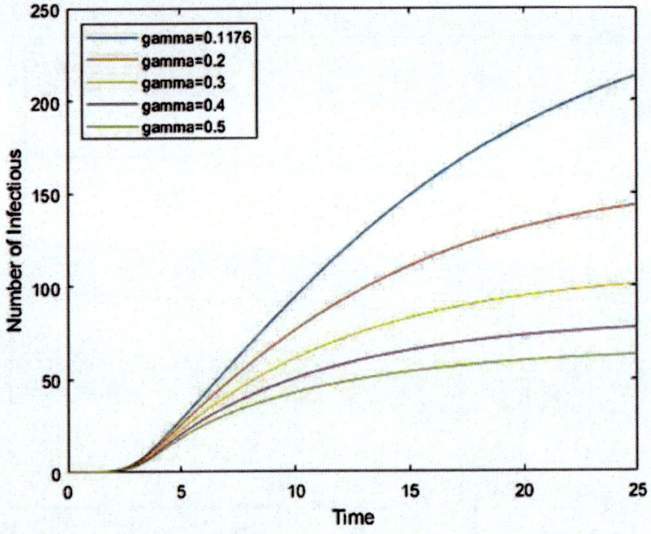


**Fig. 9** Infectious class for different values of $\gamma$

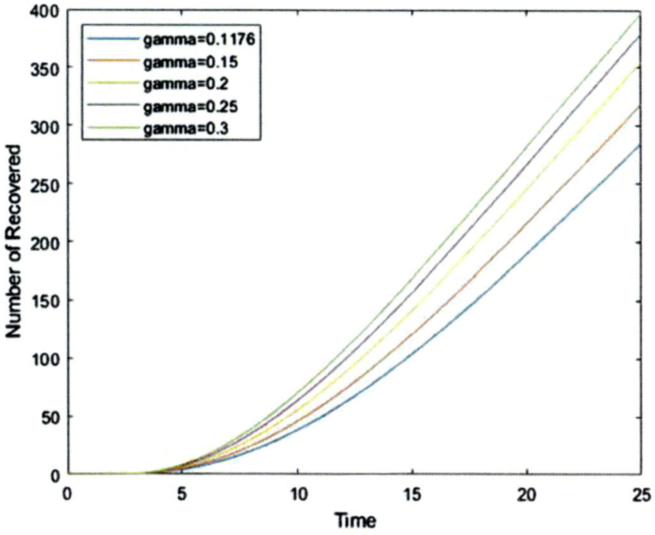


**Fig. 10** Recovered class for different values of $\gamma$

# 7 Conclusion

The local and the global stability of the model for the system is analysed. The basic reproduction number of Nipah virus in Kerala is 4.66. That is the maximum number of people get the Nipah virus from the contagious person is 5. From numerical simulations, the exposure of disease is from day 3, spread of infection of disease from day 4 and the recovery of the disease from day 3. The graphs of susceptible, exposed, infectious and recovered were shown for different values of $\beta, \alpha$ and $\gamma$, respectively.

# References


1. Chong HT, Hossain MJ, Tan CT (2008) Differences in epidemiology and clinical features of Nipah virus encephalitis between Malaysian and Bangladesh outbreaks. Neurol Asia 13:23–26
2. Harit AK, Ichhpujani RL, Gupta S, Gill KS, Lal S, Ganguly NK, Agarwal SP (2006) Nipah/Hendra virus outbreak in Siliguri, West Bengal, India in 2001, Indian. J Med Res 123:553–560
3. Arankalle VA, Bandyopadhyay BT, Ramdasi AY, Jadi R, Patil DR, Rahman M, Majumdar M, Banerjee PS, Hati AK, Goswami RP, Neogi DK (2011) Genomic characterization of Nipah virus,West Bengal, India. Emerg Infect Dis 17(5)
4. Haider Ali Biswas (2012) Model and control strategy of the deadly Nipah virus (NiV) infections in Bangladesh. BioSci RRBS 6(12):370–377
5. Biswas MHA, Haque MM, Duvvuru G (2015) A mathematical model for understanding the spread of Nipah fever epidemic in Bangladesh. In: International conference on industrial engi-

neering and operations management (IEOM), pp 1–8

6. Biswas MHA, Paiva LT, de Pinho MdR (2014) A SEIR model for control of infectious diseases with constraints. Math Biosci Eng 11(4):761–784. https://doi.org/10.3934/mbe.2014.11.761
7. Zhang T, Kang R, Wang K, Liu J (2015) Global dynamics of an SEIR epidemic model with discontinuous treatment. Adv Differ Equ 2015(361). 10.1186/s13662-015-0695-0
8. Sultana J, Podder CN (2016) Mathematical analysis of nipah virus infections using optimal control theory. J Appl Math Phys 4:1099–1111
9. Mondal MK, Hanif M, Biswas MHA (2017) A Mathematical analysis for controlling the spread of Nipah virus infection. Int J Modell Simul 37(3):185–197
10. Shah NH, Suthar AH, Thakkar FA, Satia MH (2018) SEI-model for transmission of Nipah virus. J Math Comput Sci 8(6):714–730
11. Shah NH, Trivedi ND, Thakkar FA, Satia MH (2018) Control strategies for Nipah virus. Int J Appl Eng Res 13(21):15149–15163
12. Sinha D, Sinha A (2019) Mathematical model of zoonotic NIPAH virus in South-East Asia region. Acta Sci Microbiol 2(7):82–89
13. Diekmann O, Heesterbeek JAP, Roberts MG (2010) The construction of next-generation matrices for compartmental epidemic models. J R Soc Interface 7:873–885
14. Inaba H, Nishiura H (2008) The basic reproduction number of an infectious disease in a stable population: the impact of population growth rate, on the eradication threshold. Math Model Nat Phenom 3(7):194–228
15. Prarthana MS (2018) Nipah virus in India: past, present and future. Int J Community Med Public Health 5(9):3653–3658

# Thermodynamic Analysis of Cross-Flow Heat Exchanger with Organic Blends As Substitute to Ionic Coolants

Vidya Ch., G. Ravi Kiran Sastry, P. Phani Prasanthi, and Ch. Lakshmi Kanth

## Nomenclature

| | |
|---|---|
| $H_2O$ | water |
| $P_r$ | Prandtl number |
| $EG$ | ethylene glycol |
| $N_u$ | Nusselt number |
| $PG$ | propylene glycol |
| $h_a$ | heat transfer coefficient of air |
| $GLY$ | glycerin |
| $h_c$ | heat transfer coefficient of coolant |
| $0.25EG$ | 25% concentration ethylene glycol solution |
| $A_a$ | heat transfer area of fins |
| $0.5EG$ | 50% concentration ethylene glycol solution |
| $A_c$ | heat transfer area of tube bank |
| $0.6EG$ | 60% concentration ethylene glycol solution |
| $A_o$ | overall heat transfer area |
| $0.25EG$ | 25% concentration propylene glycol solution |
| $U_o$ | overall heat transfer coefficient |
| $0.5EG$ | 50% concentration propylene glycol solution |
| $\eta_{fin}$ | fin efficiency |
| $0.6EG$ | 60% concentration propylene glycol solution |
| $m_a$ | mass flow rate of air |
| $0.25GLY$ | 25% concentration glycerol solution |

Vidya Ch. (✉) · G. R. K. Sastry
National Institute of Technology, Tadepalligudem, Andhra Pradesh, India
e-mail: vidyanitap@gmail.com

P. P. Prasanthi · Ch. Lakshmi Kanth
P.V.P. Siddhartha Institute of Technology, Vijayawada, Andhra Pradesh, India



S. Srinivas et al. (eds.), *Recent Advances in Applied Mathematics and Applications to the Dynamics of Fluid Flows*, Lecture Notes in Mechanical Engineering,
https://doi.org/10.1007/978-981-19-1929-9_25

| | |
|---|---|
| $m_c$ | mass flowrate of coolant |
| $0.25EG$ | 50% concentration glycerol solution |
| $C_r$ | heat capacity ratio |
| $0.25EG$ | 60% concentration glycerol solution |
| $T_o$ | dead state temperature |
| $\rho_a$ | density of air |
| $T_{c,i}$ | coolant inlet temperature |
| $\rho_c$ | density of coolant |
| $T_{c,o}$ | coolant outlet temperature |
| $C_{p,a}$ | specific heat of air |
| $T_{a,i}$ | air inlet temperature |
| $C_{p,c}$ | specific heat of coolant fluid |
| $\Delta P_c$ | pressure drop by coolant |
| $K_a$ | thermal conductivity of air |
| $P_{a,i}$ | inlet air pressure |
| $K_c$ | thermal conductivity of coolant |
| $P_{a,o}$ | outlet air pressure |
| $D_{h,a}$ | hydraulic diameter of air side |
| $T_{a,i}$ | air inlet temperature |
| $D_{h,v}$ | coolant side hydraulic diameter |
| $T_{a,o}$ | air outlet temperature |
| $V_c$ | flow rate of coolant |
| $S_g$ | entropy generation rate |
| $R_e$ | Reynolds number |
| $T_{c,mean}$ | mean temperature of coolant fluid |

## 1 Introduction

Rapid technological developments and miniaturization of scientific applications demand for a compact system than the existing size. One such demand gaining application is the cooling system of automotive engines. Automotive engines are arranged with cross-flow heat exchangers for the sake of rejecting waste heat from the atmosphere. Here compactness of the application demands for a reduction in size, weight and supporting structure which leads to a reduction in energy requirement and operational cost. However, the optimum thermal performance of cross-flow heat exchangers gets influenced by large pressure drops which demand for more pumping energy. Therefore, it often turns out to be necessary to take steps to bring an acceptable tradeoff between the enhanced heat transfer rates to the pumping power due to unavoidable pressure drops with the existing resources. Also, the analysis concerned with exergy and irreversibility may be applied for that purpose which would be the best deal for the situation. In the past few decades ionic liquids attracted the attention of research groups as an eco-friendly substitute for the organic compounds because of their favourable properties. The prominent characteristics of these liquids such as non-volatility and better solving ability made them to be thought of as an alternative.

The best cooling arrangement facilitates the engine to warm up quickly so as to attain optimum working temperatures [1]. A huge number of objectionable pollutants get released during cruising of automobile engine because of incomplete combustion of fuel [2]. The geometric area ratio of the core structure influences the exergetic performance of the heat pump predominantly in order to produce simultaneous heating as well as cooling effects [3]. The number of rows in tubes staggered leads to pronounced improvement in heat transfer rate while causing little influence on frictional penalty [4]. Parametric studies on automotive radiators reported a significant influence of fin geometric parameters towards heat transfer [5]. Entropy principle by being basic has a good agreement in the evaluation of heat exchangers [6]. The inlet position of the fluid streams has an influence on the relative distribution of two-phase flow, and the magnitude of air flow velocity affects more than the fluid velocity. Parallel flow configuration permits best flow distribution for both the streams [7].

Propylene glycol is found as the best alternative to ethylene glycol to address the issues relating to safety and operational life of the equipment while improving the performance of the same simultaneously [8]. Literature review reports that in the last two decades the research community has attempted to address the issues related to operating temperatures of the engine cooling system by adding ethylene glycol or propylene glycol as antifreeze to water. By utilizing glycol-based water solutions as coolant the freezing point temperature could be depressed significantly. But glycol family solutions are toxic in nature. A very limited research has been conducted to study the possibility of glycerol-based solution as a coolant fluid. This is identified as a research gap and the final work is planned on glycerol-based solution as a coolant to bring the benefits of these coolants in heat exchange applications using numerical studies.

Glycerol is found to be a naturally available non-toxic compound with considerable anti-corrosive properties. Hence, the present investigation analyses the performance parameters of cross-flow heat exchanger using glycerol solutions as alternative antifreeze additive to glycols.

## 2 Mathematical Modelling

The cross-flow heat exchanger considered for the present analysis is of wavy fin-type flat tube radiator of a truck, as shown in Fig. 1.

The core structure of heat exchanger consists of parallel rectangular flat tubes with continuous wavy fin layers arranged in between as shown in Fig. 1. The geometric dimensions of the respective test unit are provided in Table 1.

The present study focuses on the comparison of the performance of cross-flow heat exchangers with various cold fluid blends for exergy and energy analysis along with heat transfer rate and irreversibility effects. Thermo physical properties of the blend constituents are taken from the literature (Tables 2 and 3).

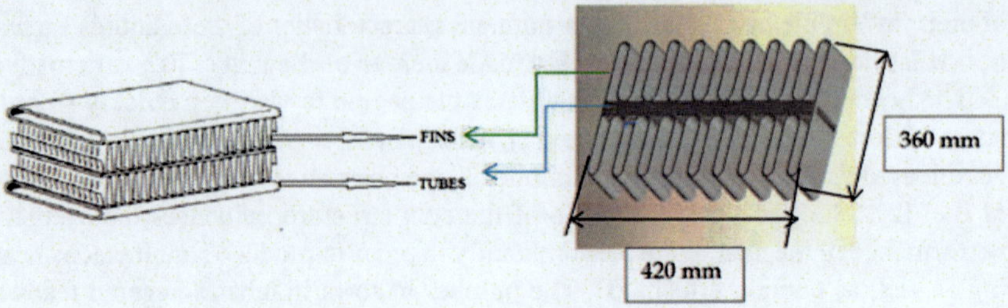


**Fig. 1** Schematic of tube and fin arrangement in cross-flow heat exchanger

**Table 1** Heat exchanger dimensions

| Parameter | Dimension |
|---|---|
| Width of heat exchanger core Wcore | 360 mm |
| Height of heat exchanger core Hcore | 420 mm |
| Depth of heat exchanger core Bcore | 30 mm |
| No. of coolant flow tubes | 37 |
| Fin width wf | 8 mm |
| Fin layer thickness | 0.75 mm |
| No. of fin layers | 38 |
| Fin pitch | 3 mm |
| Air-side heat transfer area/unit volume Aa | 916 $m^2/m^3$ |
| Coolant-side heat transfer area/unit volume Ac | 190 $m^2/m^3$ |

**Table 2** Boiling and freezing points data of base fluid and blend constituents [8]

| Type of fluid | Freezing point temperature (°C) | Boiling point temperature (°C) |
|---|---|---|
| Water | 0 | 100 |
| Ethylene glycol | –59 | 187.4 |
| Propylene glycol | –12.9 | 197.3 |

Below mentioned assumptions have been considered to carry out the analysis.

- Steady flow condition prevails for both hot as well as cold fluids.
- The whole heat exchanger setup acts as an adiabatic system. This assumption permits energy conservation across hot stream to cold stream fluids.
- All fluid thermo physical properties have been correlated with respect to the bulk mean temperature of the fluid.
- Air is assumed to be compressible fluid while coolant fluids as incompressible.
- Air thermo physical properties recorded at ambient temperature.
- Air inlet temperature taken to be constant.

**Table 3** Thermo physical property data of various blend constituents [9–11]

| Type of fluid | Freezing point temperature (°C) | Boiling point temperature (°C) | Density (kg/m$^3$) | Thermal conductivity (W/m.K) | Specific heat (kJ/kg.K) |
|---|---|---|---|---|---|
| 25% EG-H2O blend | −10.8 | 103.3 | 1029 | 0.495 | 3.917 |
| 50% EG-H2O blend | -36.8 | 107.2 | 1067 | 0.442 | 3.534 |
| 60% EG-H2O blend | -52.8 | 111.1 | 1082 | 0.452 | 3.354 |
| 25% PG-H2O blend | −11 | 87.4 | 1007 | 0.411 | 0.947 |
| 50% PG-H2O blend | -34 | 91.3 | 1023 | 0.299 | 0.85 |
| 60% PG-H2O blend | −48 | 93 | 1029 | 0.262 | 0.805 |
| 25% GLY-H2O blend | -7.2 | 102 | 1063 | 0.482 | 3.718 |
| 50% GLY-H2O blend | -22.8 | 106 | 1138 | 0.354 | 3.017 |
| 60% GLY-H2O blend | -34.4 | 109 | 1176 | 0.335 | 2.814 |

Different thermo dynamic performance parameters have been analysed as follows: The heat absorption rate of air has been calculated by

$$Q_a := \rho_a \times \mathrm{H_{core}} \times \mathrm{W_{core}} \times \mathrm{C_{p,a}} \tag{1}$$

Fin-side heat transfer coefficient has been calculated by using

$$h_a := \mathrm{N_{u_a}} \times \mathrm{K_a} \div \mathrm{D_{h.a}} \tag{2}$$

where the Nusselt number value of air for laminar flow zone is expressed by

$$Nu := 0.664 \times \mathrm{R_e}^{0.5} \times \mathrm{P_r}^{0.333} \tag{3}$$

Now, the heat capacity rate of fluid is given by

$$Q_c := \rho_c \times \mathrm{V_c} \times \mathrm{C}_{p,c} \tag{4}$$

Tube-side heat transfer coefficient could be given by

$$h_c := \frac{N_{u_c} \times K_c}{D_{h_c}} \tag{5}$$

Nusselt number for turbulent flow of cold fluid is expressed as

$$N_{u,c} := \frac{\left(\frac{C_f}{2}\right) \times R_{e_f} \times P_{r_f}}{\left[1 + 12.8 \times \left(\frac{C_f}{2}\right)^{0.5} \times \left({P_{r_f}}^{0.68} - 1\right)\right]} \tag{6}$$

where the friction factor has been calculated from

$$C_f := 0.37\left(\ln R_{e_c}\right)^{-2.584} \tag{7}$$

Overall heat transfer coefficient of the heat exchanger would be expressed as

$$\frac{1}{U_o A_o} := \frac{1}{\eta_a h_a A_a} + \frac{1}{h_f A_f} \tag{8}$$

where surface temperature effectiveness of core is calculated as

$$\eta_a := 1 - \frac{A_{fin}}{A_a} \times (1 - \eta_{fin}) \tag{9}$$

The effectiveness of non-mixing type cross-flow heat exchanger is expressed as

$$\varepsilon := 1 - e^{\left[\frac{NTU^{0.22}}{C_r} \times e^{\left(-C_r NTU^{0.78} - 1\right)}\right]} \tag{10}$$

Then, the actual heat transfer rate of heat exchanger can be expressed as

$$Q_{act} := \varepsilon \times C_{min} \times \left(T_{c,i} - Ta, i\right) \tag{11}$$

Hence, exergy loss by cooling medium is expressed by

$$\Delta E_c := Q_{act} - T_0\left[m_c \times C_{p,c} \times \ln\left(\frac{T_{c,i}}{T_{c,o}}\right) - \frac{m_c \times \Delta P_c}{\rho_c \times T_{c'mean}}\right] \tag{12}$$

This implies, exergy gain by rushing air stream may be given as

$$\Delta E_a := Q_{act} - T_0\left[m_a \times C_{p,a} \times \ln\left(\frac{T_{a,o}}{T_{a,i}}\right) + m_a \times R_a \times \ln\left(\frac{P_{a,i}}{P_{a,o}}\right)\right] \tag{13}$$

Now, the irreversibility of the system can be expressed by

$$I = \Delta E_c - \Delta E_a = T_0 S_g \tag{14}$$

For a cross-flow heat exchanger, the liquid coolant flowing through the tube bank can be assumed as an incompressible fluid stream, while the rushing air stream may be taken as compressible. Hence, the entropy generation rate of the system Sg can be given as

$$S_g := \mathrm{m_a}\left[C_{p,a}\ln\left(\frac{T_{a,o}}{T_{a,i}}\right) - R_a\ln\left(\frac{P_{a,o}}{P_{a,i}}\right)\right] + \mathrm{m_c}\left[C_{p,c}\ln\left(\frac{T_{c,o}}{T_{c,i}}\right) - \frac{P_{a,o}-P_{a,i}}{\rho_c T_{c,mean}}\right] \quad (15)$$

Now, irreversibility of the system can be expressed by EEQ Solver code developed to solve the set of analytical equations of the study. The average temperature has been chosen as a reference to correlate thermo physical properties of both the fluid streams. For the simulation, the inlet temperature of the cooling agent has been taken as a variable with particular values of 50, 60, 70, 80 and 90 °C, whereas the inlet atmospheric air average temperature of 30 °C was taken. Various parametric calculations are performed for coolant flow rates ranging from 10 to 60 LPM.

## 3 Results and Discussions

The influence of water-based polyethylene glycol, polypropylene glycol as well as glycerol blends on various parameters such as heat transfer rate, effectiveness, frictional loss, exergetic efficiency, pumping power and entropy generation rate of the studied heat exchanger are plotted in Figs. 2, 3, 4, 5, 6, 7 and 8. From the plot of coolant flow rate versus heat capacity rate, it has been observed that for all blends of

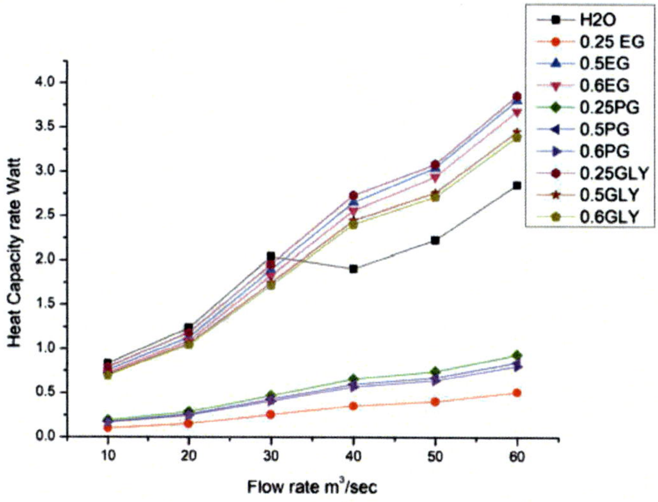


**Fig. 2** Variation of heat capacity rate with coolant flow rate

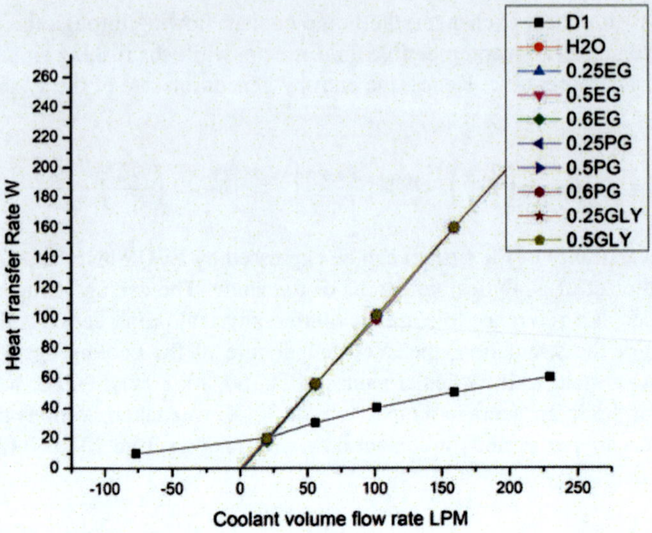


**Fig. 3** Variation of heat transfer rate with coolant flow rate

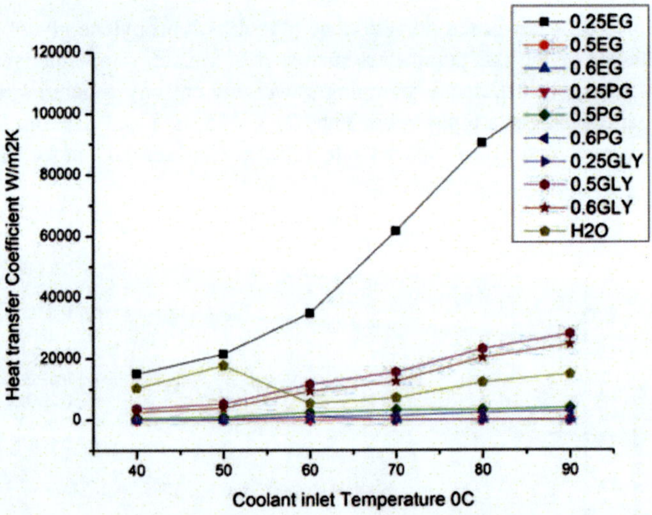


**Fig. 4** Variation of heat transfer coefficient with coolant inlet temperature

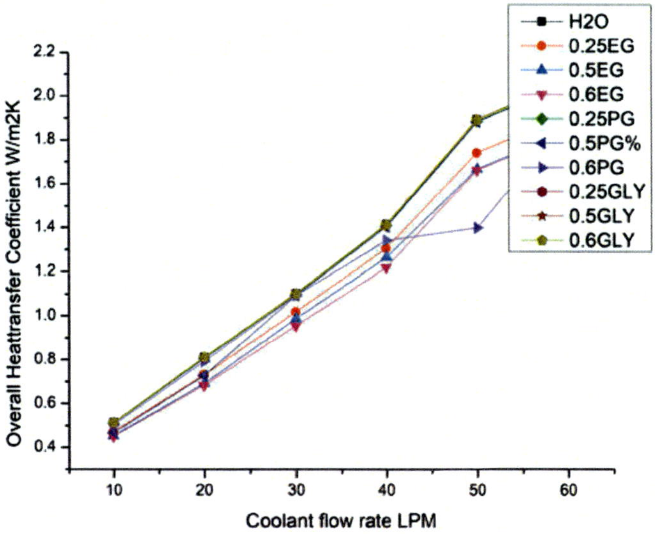


**Fig. 5** Variation of overall heat transfer coefficient with coolant flow rate

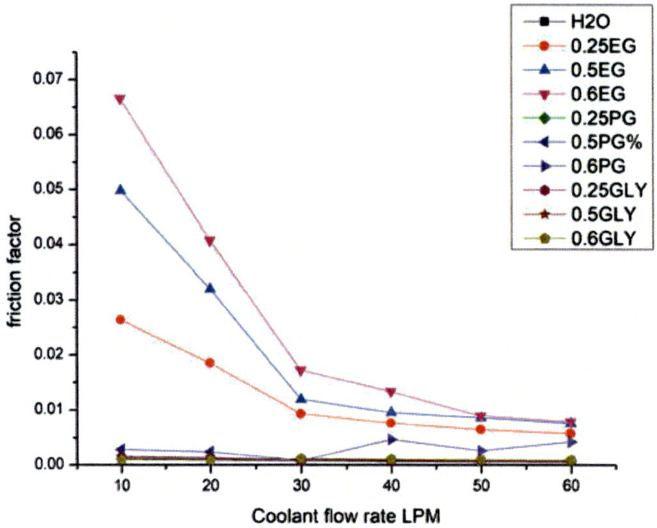


**Fig. 6** Variation of friction factor with coolant flow rate

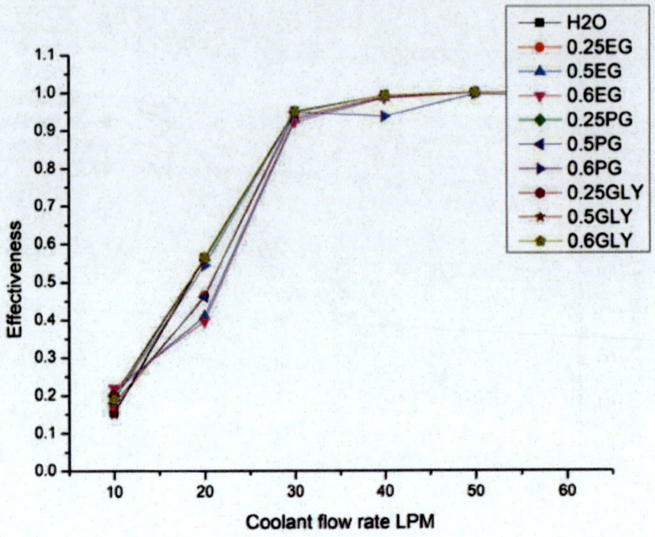


**Fig. 7** Variation of effectiveness with coolant flow rate

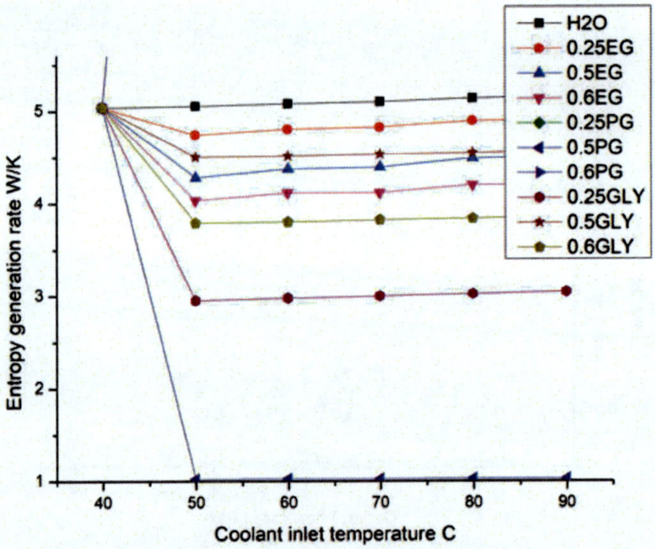


**Fig. 8** Variation of entropy generation rate with coolant inlet temperature

study heat capacity rate increases progressively with flow rate. However, 25% glycerol solution reported a dominant heat capacity rate as compared to glycol solutions. From this plot it may be concluded that due to its high heat capacity rate glycerol solution can absorb or reject more heat for the same flow rate. Glycerol solution reported an enhancement of maximum 15%. For the same coolant flow rate and inlet temperature all considered blends have reported nearly similar heat transfer rate. In comparison to the remaining samples polypropylene glycol solution of 60% concentration in the base fluid has reported an improved heat transfer rate of around 7% on an average.

Variation of heat transfer coefficient with respect to cooling fluid inlet temperature reported that 25% polyethylene glycol blend leads to significant enhancement than other samples. Superior thermal conductivity of polyethylene glycol facilitates minimum temperature gradient existence between cooling agent and air stream which leads to reduced internal irreversibility enhancing the value of Prandtl number. However, 50% glycerol solution also reported consistent improvement of the same. However, it was observed that the heat transfer coefficient value using pure water as coolant followed an irregular trend because of its reducing viscosity value which causes the Prandtl number to decrease. As a net effect the heat transfer coefficient reports non-uniform variation. 0.6PG solution also reported an abnormal trend as compared to other blends for its influence on the overall heat transfer coefficient. Being non-toxic clear solvent of kosmotropic nature, glycerol is found to be the best substitute for polyethylene glycol and polypropylene glycol blends ensuring a stable solution in order to achieve enhanced heat transfer coefficient. As for as the overall heat transfer coefficient is concerned, empirical calculations reported that water and 25% polypropylene glycol solution produce the highest value than remaining samples.

On studying the behaviour of friction factor with concentration of blends variation in coolant flow rates is drawn and plotted as shown in Fig. 5. This shows clearly that the friction factor varies drastically for ethylene glycol solution in comparison to propylene glycol and glycerol solutions. 60% ethylene glycol blends have reported an enhancement in friction factor of around 70% in comparison to other samples. Glycerol blend reported more or less constant friction factor irrespective of flow rate. An average friction factor of 0.000848 was obtained for glycerol blends which were observed to be in the range reported by pure water.

The effect of coolant flow rate on the effectiveness of the heat exchanger reveals that increasing trend for all blends in comparison to pure base fluid. 25% glycerol solution recorded better than ethylene glycol and propylene glycol solutions due to its lowest dynamic viscosity. For polypropylene glycol, as the concentration increases the graph shows a fluctuating trend which would be due to abnormal changes in viscosity, which in turn leads to thermal irreversibility.

From the variation of entropy generation rate with respect to coolant flow rate, it has been observed that the lowest exergy destruction with 60% concentration of polypropylene glycol. On the other hand, 25% glycerol solution also reported comparatively least entropy generation than other blends which has been accounted to be around 47%.

## 4 Conclusions

Thermodynamic exergetic performance analysis of cross-flow heat exchangers with wavy fins and rectangular flat tube arrangement has been carried out while applying various blends in base cooling fluid. The analysis results in the following conclusions.

- Polypropylene glycol being already used as antifreeze in heat exchangers has reported enhanced heat transfer rate along with overall heat transfer coefficient.
- Heat transfer rate, heat transfer coefficient and effectiveness of the heat exchanger keep increasing with the flow rate of coolant, whereas the friction factor is observed to be declining.
- Glycerol solution reported enhanced heat capacity rate along with dominant overall heat transfer coefficient while reporting constant friction factor. Being non-toxic and metallic surface-friendly in comparison to other blends, glycerol solution is found as the best substitute for glycol blends without compromising performance.
- By attempting glycerol solution as a coolant in a cross-flow heat exchanger, the system becomes compact in size which leads to a reduction in weight of the same. The life span of the flow tubes gets extended because of their non-corrosive nature.
- As the engine accessories turn lightweight, a propulsive effort is required to overcome the inertia that gets reduced, which in turn influences specific fuel consumption.

## References

1. Torregrosa A et al (2008) assessment of the influence of different cooling system configurations on engine warm-up, emissions and fuel consumption. Int J Automot Technol 9(4):447–458
2. Broatch A et al (2008) measurement of hydrocarbon and carbon monoxide emissions during the starting of automotive di diesel engines. Int J Automot Technol 9(2):129–140
3. Sarkar J et al (2005) transcritical $co_2$ heat pump systems: exergy analysis including heat transfer and fluid flow effects. Energy Convers Manag 46:2053–2067
4. Wang C-C et al (2000) parametric studies on automotive radiators. part II: correlation. Int J Heat Mass Transfer 43:2693–2700
5. Oliet C et al (2007) Parametric studies on automotive radiators. Appl Thermal Eng 27:2033–2043
6. Yilmaz M et al (2001) Performance evaluation criteria for heat exchangers based on second law analysis. Exergy Int J 1(4):278–294
7. Saad SB et al (2011) Experimental distribution of phases and pressure drop in a two-phase offset strip fin type compact heat exchanger. Int J Multiphase Flow 37:576–584
8. Sahoo RR et al (2017) Energy and exergy comparisons of water based optimum brines as coolants for rectangular fin automotive radiator. Int J Heat Mass Transfer 105:690–696
9. kormann et al (2011) Glycerine-containing antifreezing agent concentrates with corrosion protection. US 7927,505 B2
10. Sreedhara Rao B (2016) Development of correlation for fluid flow in wavy corrugated plate heat exchangers. Int J Innovat Res Sci Eng Technol 5(3)
11. Sarkar J et al (2013) Performance analysis of louvered fin tube automotive radiator using nanofluids as coolants. Int J Nanomanuf 9(1)

# Modified Image Processing Technique for Measurement of Intermittent Flow Characteristics

**Kiran Mansuriya, Gandham Yedukondalu, Jignesh Thaker, and Jyotirmay Banerjee**

## 1 Introduction

The usage of image processing techniques for studying the physical behavior of transport phenomena is raising in various research fields. Two-phase flow is one such field, which is benefited with the rise of image analyzing techniques. Understanding the transport phenomena of gas–liquid flow is complicated as it shows the compressibility of one of the phases and the characteristics of the deformable interface. In gas–liquid flow, the intermittent flow patterns such as slug and plug flow occur in many industrial processes such as the geothermal steam production, oil and gas production, the condensation and boiling processes, the emergency cooling of nuclear reactors and the transport and handling of cryogenic fluids, designing the slug catchers for the two-phase hydrocarbon's transportation, etc. [1]. The presence of the intermittent flow patterns inside the piping system causes degradation of material or thinning of wall due to corrosion and erosion in piping components. In past decades, many destructive failures with serious injuries and fatalities have been observed in industrial applications due to such material degradation inside the piping system. To prevent such failures, it is strongly required to estimate the dynamics of intermittent flow and their associated flow characteristics accurately [2]. Therefore, the assessment of this flow type during the design stages must be considered to protect piping equipment. The detailed analysis of liquid plug/slug velocity, plug/slug frequency, plug/slug bubble length, and liquid plug/slug length for intermittent flow subpatterns

K. Mansuriya (✉) · J. Thaker
Department of Mechanical Engineering, Adani Institute of Infrastructure Engineering, Ahmedabad, India
e-mail: kiranmansuriya@gmail.com

G. Yedukondalu · J. Banerjee
Department of Mechanical Engineering, Sardar Vallabhbhai National Institute of Technology, Surat, India

 
S. Srinivas et al. (eds.), *Recent Advances in Applied Mathematics and Applications to the Dynamics of Fluid Flows*, Lecture Notes in Mechanical Engineering,
https://doi.org/10.1007/978-981-19-1929-9_26

are important for pressure drop and holdup calculations. Moreover, for the designer of pipeline, the examination of the intermittent characteristics for various inlet flow conditions is necessary for the processing equipment design and for predicting the real corrosion–erosion rates in piping systems.

For the above study and detailed analysis, the optical visual-based observation techniques using high-end and high-speed camera are being used by the researchers nowadays due to its non-intrusiveness, availability, moderate cost, and safe operation. The images obtained from this optical visual-based observation techniques and high-speed camera need to be processed for the estimation of intermittent flow characteristics. Mayor et al. [3] proposed an image processing technique, which includes the operations such as image contrast enhancement, image conversion, image filtering, and image object labeling. The authors have used the technique for studying of gas and liquid slug flow characteristics in vertical pipes. Ahmed [4] used digital image processing to compute the characteristics of slug flow in which the binary images were obtained from the captured images by applying the gray level threshold from the full white–black histogram and all binary images were unified using processing algorithm to get the time-averaged values of slug flow characteristics. Amaral et al. [5] used image analysis techniques like top-hat filtering, watershed segmentation, and H-minima transform to examine bubble contour and velocities from the images. Limitations of this method are difficulty in bubble rear detection and high image processing time. Abdalellah et al. [6] used the approximate median method for the foreground extraction followed by the blob analysis for the computation of flow characteristics. The examination was bound to slug translational velocity and slug length. Computation of slug frequency and slug tail velocity was not calculated. There were some limitations to the approximate median method used in their study as it was slow to change its estimated background model. Also, the image processing time can be reduced with better foreground extraction technique.

The present work's focus is to develop an image processing technique, which can measure intermittent flow characteristics accurately and reduce the processing time significantly. In what follows in this paper, Sect. 2 describes the experimental test facility. The proposed modified image processing technique is demonstrated in Sect. 3. Section 4 deals with the estimation and validation of intermittent flow characteristics measured using the proposed technique. Finally, the key observations from the present work are discussed in Sect. 6.

## 2 Experimental Facility

The flow visualization experiments are performed on two-phase experimental flow facility, developed in the laboratory of Advanced Fluid Dynamics at SVNIT, Surat, India. The two-phase flow experimental facility is shown in Fig. 1 majorly includes two-phase flow test setup and high-speed photography system. Details about this test facility are explained elsewhere [7]. Flow visualization images are captured using high-speed Photron made FASTCAM camera for various inlet flow conditions in

**Fig. 1** Experimental test setup for two-phase flow measurement [7]

**Table 1** Experimental conditions

| | |
|---|---|
| System | Air–Water |
| Pipe internal diameter | 25 mm |
| Pipe inclination | Horizontal |
| Range for $Re_{SL}$ | 4120.5–13,735 |
| Range for $Re_{SG}$ | 1414.7–2829.42 |

terms of different combinations of superficial Reynolds number of liquid ($Re_{SL}$) and gas ($Re_{SG}$). The range of $Re_{SL}$ and $Re_{SG}$ for intermittent flow regime is selected from the sub-regime map for intermittent flow in horizontal pipe proposed by Thaker and Banerjee [1]. Table 1 summarizes the selected experimental conditions for the present work. Then, captured images are processed further for estimating intermittent flow characteristics using the proposed methodology of modified image processing technique.

## 3 Modified Image Processing Technique

The proposed image processing technique is developed using MATLAB tool for accurate examination of the characteristics of the intermittent flow from the captured time-series images of intermittent flow at different inflow conditions. Figure 2 represents the flowchart of the developed image processing technique. The primary six stages are: image pre-processing, background subtraction, Binary Large Object (BLOB) extraction analysis, morphological operations, motion tracking, and extraction of flow characteristics.

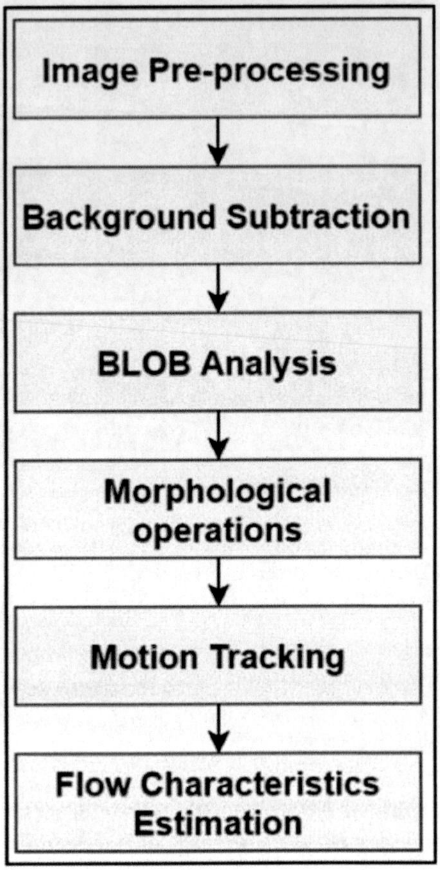


**Fig. 2** Flowchart of the developed image processing technique

Image pre-processing has been carried out to reduce the noise and enhance the contrast of captured digital images. Moreover, these images are in form of true color (i.e. RGB images), processing of these RGB images is computationally inefficient. Thus, RGB images are converted into greyscale images in the pre-processing step. This is done by using the 'rgb2gray' function in the MATLAB. Image cropping is also performed to eliminate unwanted portions from the captured images (Fig. 3).

In background subtraction (or foreground detection) stage, a Modified Frame Difference Method (MFDM) is implemented, which gives better foreground image extraction compared to approximate median method. It focuses on selecting the proper step length for different foreground movements. For slower foreground movement, larger step length selection, and for faster foreground movement, the smaller step length selection is recommended. The step length of four is selected in this study. In this method, the background image for the next time step is considered to be actual time-series image of previous time step. Figure 4 shows the foreground extracted images using modified frame difference method. These processing results

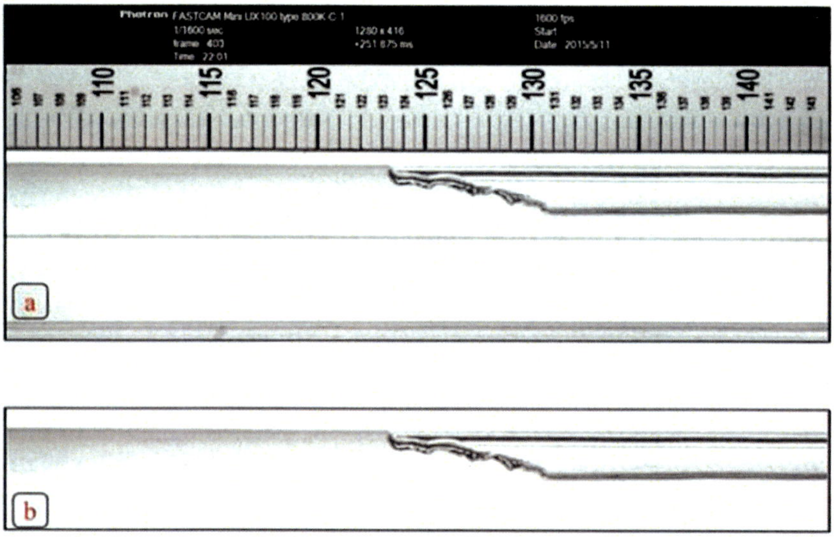


**Fig. 3** Captured plug flow image: **a** before pre-processing; **b** After pre-processing

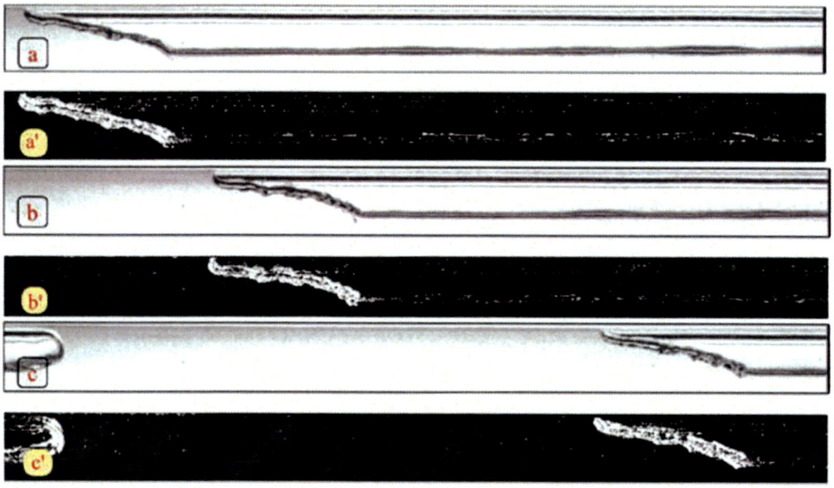


**Fig. 4** Foreground extraction using MFDM: **a**, **b**, **c** are captured plug flow images for different time intervals; **a′**, **b′**, **c′** are corresponding foreground extracted images of (**a**, **b**, **c**) using MFDM

are found better than the extraction using Approximate Median Method (AMM) and Frame Difference Method (FDM).

For qualitative comparison, foreground extracted images using the AMM, FDM, and MFDM are shown in Fig. 5a′, b′, a″, b″, a‴ and b‴, in which Fig. 5a′ and b′ shows the foreground extracted using images AMM; Fig. 5a″ and b″ shows the

**Fig. 5** Comparison of background subtraction methods: **a** and **b** images after pre-processing; **a′** and **b′** foreground extracted images using AMM; **a″** and **b″** foreground extracted images using FDM; **a‴** and **b‴** foreground extracted images using MFDM

foreground extracted images using FDM; Fig. 5a‴ and b‴ shows the foreground extracted images using MFDM. With the better estimation of background model, it can be observed that quality foreground extraction in MFDM is superior to that by FDM and AMM. Even though the complete contour of elongated plug bubble is not achieved, but the exact interface between the liquid plug and elongated plug bubble is obtained accurately.

The above background subtraction is followed by the Binary Large Object (BLOB) extraction. The purpose of this stage is to isolate the large objects from the binary image. Binary large object (BLOB) extraction analysis is carried out in place of normal edge detection as the latter is having the difficulties in slug movement tracking. Fig. 6a″ and b″ shows the extracted objects after BLOB implementation. After this stage, morphological operations (dilation and erosion) are performed for uniformity of region and to favor the motion tracking as shown in Fig. 6a‴ and b‴. The BLOBs (liquid slug/plug front and liquid slug/plug tail) after this step are tracked by defining a rectangular bounding box over the region of interest. It is explained as covering all pixels for a BLOB and extracting the four pixels with the maximum x-value, minimum x-value, maximum y-value, and minimum y-value, respectively. For the estimation of flow parameters, the position of the bounding box in each step is saved in an array of a matrix.

**Fig. 6** **a** and **b** represent images after pre-processing; **a′** and **b′** represent foreground extracted images using MFDM; **a″** and **b″** represent images after BLOB analysis; **a‴** and **b‴** represent images after morphological operations

# 4 Estimation and Validation of Intermittent Flow Characteristics

Intermittent flow characteristics are computed by developed image processing technique that requires length scale transformation. Using the fixed-point data analysis approach, three characteristics, namely, slug/plug frequency, liquid slug/plug velocity, and liquid slug/plug length are computed. According to this approach, two fixed points are selected in such a way that it covers the maximum field span of the captured image (shown in Fig. 7). The pixel-based axial position for the interface of the elongated bubble is determined at the time of entering ($pixel_i$) and the exit ($pixel_j$) of the camera view field. The pixel size (mm/pixel) is calculated by dividing the camera view field (mm) by the pixel numbers in the x-direction. In the present study, pixel size is 38.5 (mm)/1280 (pixel).

The average liquid slug/plug velocity is formulated as:

**Fig. 7** Bubble interface tracing using fixed-point analysis approach

$$V_{Liquid\ plug/slug} = 0.5 \times (V_{Liquid\ plug/slug\ front} + V_{Liquid\ plug/slug\ tail}) \quad (1)$$

$$V_{Liquid\ plug/slug\ front}/V_{Liquid\ plug/slug\ tail} = \left(\frac{pixel\ j - pixel\ i}{frame\ j - frame\ i}\right) \times pixel\ size \times Vacq \quad (2)$$

where, $Vacq$ is the velocity of acquisition of digital images (FPS).

The liquid slug/plug length is calculated using the mean liquid slug/plug velocity and the travel time of liquid slug/plug from reference 1 to reference 2, and it is calculated as:

$$L_{Liquid\ plug/slug} = V_{Liquid\ plug/slug} \times (t_{Liquid\ plug/slug\ front} - t_{following\ Liquid\ plug/slug\ tail}) \quad (3)$$

The liquid slug/plug frequency is computed by summing the number of the liquid slugs/plugs crossing a reference point in a specific time period and is formulated as:

$$f = N/(t_2 - t_1) \quad (4)$$

where $f$ is the liquid slug/plug frequency; $N$ is the number of the liquid slugs/plugs crossing a reference point in a specific time period; $t_2$ refers to the time at which the last liquid slug/plug passes the reference point 1 or 2; $t_1$ refers to the time at which the first liquid slug/plug passes the reference point 1 or 2.

The above-proposed technique is formulated as an algorithm in MATLAB tool. In an average of 14 min processing period, 10,000 frames of 504*1280 pixels are processed on a PC with the specifications as Windows 8.1 (64 bit), 8 GB RAM, Intel Corei7. Intermittent flow characteristics computed by this modified technique are compared with the same characteristic's values obtained from the conventional method. These comparison results are represented in Figs. 8 and 9. Figure 8a shows the obtained result of plug/slug velocity; Fig. 8b represents the obtained result of liquid plug/slug length while Fig. 9a and b represents the obtained results of maximum and minimum liquid plug/slug lengths.

A good agreement is achieved between conventional method estimations and the current technique's estimations, with the average errors of 0.92% and 0.50% for the liquid slug/plug length and liquid slug/plug velocity, respectively. Table 2 represents the image processing time required to estimate the intermittent flow characteristics by various researchers. The present method is found to be more advantageous in terms of less image processing time and accurate estimation of intermittent flow characteristics.

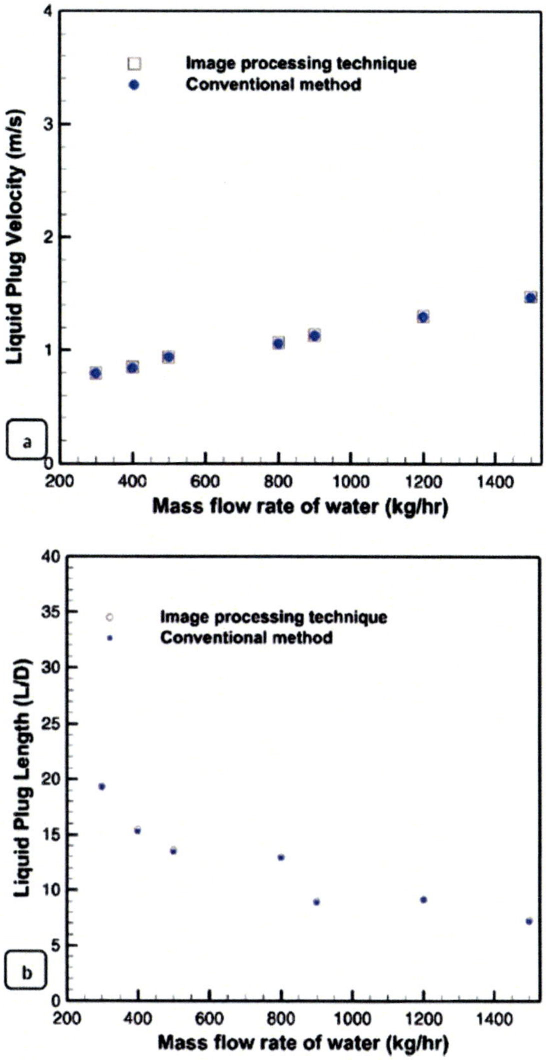


**Fig. 8** Comparison of obtained results of image processing data to the data from conventional method: **a** Liquid plug/slug velocity; **b** Liquid plug/slug length

## 5 Conclusions

In the present work, a modified image processing technique development using MATLAB is attempted to examine the characteristics of the intermittent flow in a horizontal air–water pipe flow. In this technique, the major modification is applied in the stage of background subtraction. In this stage, a Modified Frame Difference Method (MFDM) is implemented, which gives better foreground image extraction compared to approximate median method and frame difference method. Further, Binary large

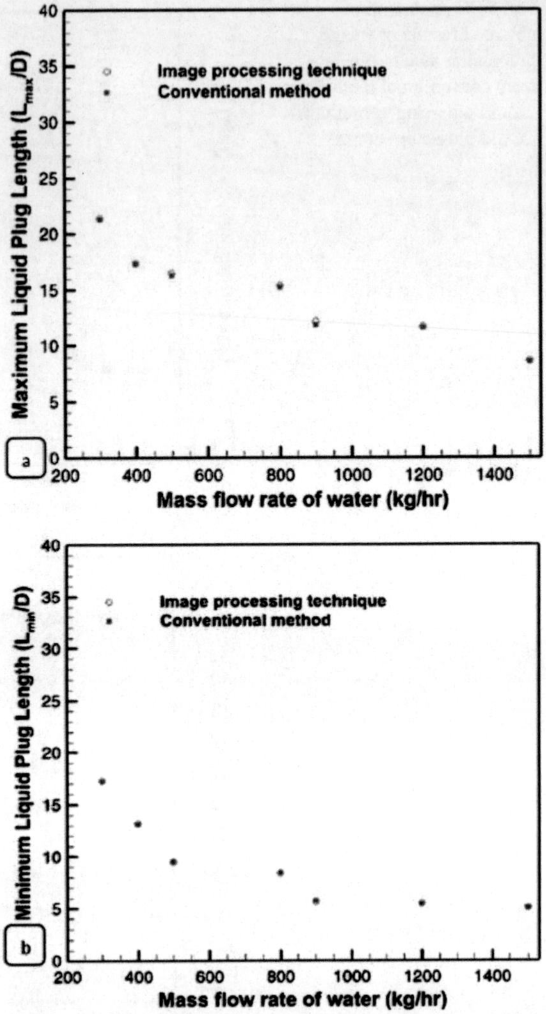


**Fig. 9** Comparison of obtained results of image processing data to the data from conventional method: **a** Maximum liquid plug length; **b** Minimum liquid plug length

object (BLOB) extraction analysis is carried out in place of normal edge detection as the latter is having difficulties in slug movement tracking. The intermittent flow characteristics evaluated are namely: Slug/plug length, liquid slug/plug velocity, and liquid slug/plug velocity. A good agreement is achieved between conventional method estimations and the current technique's estimations, within the average errors of 0.92% and 0.50% for the liquid plug/slug length and liquid plug/slug velocity, respectively.

**Table 2** Comparison of image processing time of present technique with existing techniques

| Authors/Year | Image pixels | Image processing time in minutes (For 5000 frames) | Computational resource details |
|---|---|---|---|
| Amaral et al. [5] | 232 × 500 | 37.5 | Intel Quad Core 2 (@2.836 GHz) with 4 GB of memory and running Windows 7, 64bits |
| Abdalellah et al. [6] | 480 × 960 | 12 | Intel Core i7 (@3.4 GHz) with 8 GB RAM and a Windows 7 (64 bit) operating system |
| Present Technique | 504 × 1280 | 7 | Windows 8.1 (64 bit), 8 GB RAM, Intel Core i7 |

## References


1. Thaker J, Banerjee J (2015) Characterization of two-phase slug flow sub-regimes using flow visualization. J Pet Sci Eng 135:561–576
2. Ahmed WH (2012) Nuclear power—practical aspects. InTech, New York, USA, Chapter 6:153–178
3. Mayor TS, Pinto AMFR, Campos JBLM (2007) An image analysis technique for the study of gas-liquid slug flow along vertical pipes–associated uncertainty. Flow Meas Inst 18:139–147
4. Ahmed WH (2011) Experimental investigation of air-oil slug flow using capacitance probes, hot-film anemometer, and image processing. Int J Multiphase Flow 37:876–887
5. Amaral CEF, Alves RF, da Silva MJ, Arruda LVR, Dorini L, Morales REM, Pipa DR (2013) Image processing techniques for high-speed videometry in horizontal two-phase slug flows. Flow Meas Inst 33:257–264
6. Abdalellah OM, Mohammad SN, Hussain HA, Rune WT (2016) Measurements of translational slug velocity and slug length using an image processing technique. Flow Meas Inst 50:112–120
7. Thaker J, Banerjee J (2016) Influence of intermittent flow sub-patterns on erosion–corrosion in horizontal pipe. J Pet Sci Eng 145:298–320

# Modification of Surface Properties of AA7075 by Friction Stir Processing

**Koona Bhavani, V. S. N. Venkata Ramana, R. Rahul, Ch. Lakshmi Kanth, K. Sri Ram Vikas, and Ch. Kishore Reddy**

## Nomenclature

| | |
|---|---|
| FSW | Friction Stir Welding |
| FSP | Friction Stir Processing |
| NZ | Nugget Zone |
| OC | Open Current |
| PFZ | Precipitate Free Zone |
| SCE | Saturated calomel electrode |

## 1 Introduction

Out of all aluminium alloys, AA7075 is one of the candidate materials that possesses high strength, fatigue, toughness, and ductility. It is commonly used in aircraft structural parts. The primary alloying element in this alloy is zinc, and other alloying elements include magnesium and copper. This alloy is not resistant to general as well as stress corrosion. Joining by fusion welding is very difficult and a lot of problems are associated with this alloy [1, 2]. Solid-state welding techniques are successfully

K. Bhavani · V. S. N. Venkata Ramana
Department of Mechanical Engineering, GITAM Institute of Technology, GITAM (Deemed to be University), Visakhapatnam 530045, Andhra Pradesh, India

R. Rahul (✉)
Department of Mechanical Engineering, Chaitanya Bharathi Institute of Technology(A), Gandipet, Telangana 500075, Hyderabad, India
e-mail: luhar6100@gmail.com

Ch. Lakshmi Kanth · K. Sri Ram Vikas · Ch. Kishore Reddy
Department of Mechanical Engineering, Prasad V. Potluri Siddhartha Institute of Technology, Vijayawada 520007, Andhra Pradesh, India



S. Srinivas et al. (eds.), *Recent Advances in Applied Mathematics and Applications to the Dynamics of Fluid Flows*, Lecture Notes in Mechanical Engineering,
https://doi.org/10.1007/978-981-19-1929-9_27

implemented to join this alloy. Venugopal et al. [3] have worked extensively in joining AA7075 by friction stir welding (FSW) technique. Even though the joining of AA7075 is made successful but corrosion is one of the key problems which is getting pivoted every time. The main reason associated with less corrosion resistance is its precipitates, mainly $MgZn_2$ which are highly anodic, and another reason is precipitate free zones (PFZ's), because of this, preferential dissolution takes place [4]. Many researchers have been trying to improve the corrosion resistance of this alloy by adding Nanopowders and modifying the surface of this alloy [5–7]. Friction stir processing (FSP) is a novel technique which is originated from the FSW technique. It can be extensively used for modifying surfaces. Typically, tool materials that are used for processing possess high hardness, and a variety of tool materials and their effects are available in reference [8]. Our main intention in this investigation is to improve pitting corrosion resistance by modifying the surface of this alloy. Usually, hard tools are commonly employed for this purpose, here an attempt has been made by using a softer tool. In this study, AA60601 is used as a tool for the surface modification of AA7075. As we know AA6061 is precipitation hardened which contains magnesium and silicon. It is not strong as AA7075 but it is good at corrosion resistance and it also possesses good weldability and formability. It is also used in aerospace applications. This FSP is a very versatile technique, which can be used for modifying the surfaces as well as it can be used for depositing another material over the other. Madhusudhan Reddy et al. [9] modified the surface of commercial A356 alloy using the FSP technique and found wear and corrosion resistance improvement. Researchers are using FSP for coating the substrate, Tokisue et al. [10] performed friction surfacing using AA2017 on AA5052 and observed the hardness of the coated surface on substrate to be equivalent to AA2017 and even tensile strength of the coated substrate has been improved.

FSP has been successfully utilised to improve the wear and ballistic resistance of AA7075 because the microstructure after processing becomes finer, Sudhakar et al. [11] induced $B_4C$ and $MOS_2$ in AA7075 by FSP and observed improvement in these properties, and the same researcher has established both analytical and experimental approach on surface modification of AA7075 [12]. Even in situ AA7075 composites are successfully processed through this technique [13]. Fusion-welded joint properties can be enhanced by modifying the surface of the joint, recently dissimilar AA6061 and AA7075 fusion welded joint property has been enhanced by FSP, tensile strength and hardness properties were enhanced due to refinement of grains in the nugget zone (NZ) [14]. Recently different filler materials such as copper, multiwalled carbon nanotubes, and silicon carbide powders have been utilised to process AA7075 and found significant improvement in properties [15]. FSP parameters and tool shape also play a vital role in enhancing the surface properties [16, 17]. AA6061 is a pseudo-binary alloy [18] that contains an ageing precipitate $Mg_2Si$, it possesses good corrosion resistance, solid-state joining by FSW enhances stress corrosion resistance in this alloy [19]. In the present AA7075-T6 plate has been FSPed using AA6061-T6 rod.

## 2 Experimental Procedure

The present investigation focuses on improving the pitting corrosion resistance of AA7075 alloy by depositing a layer of AA6061 on it by friction stir processing. Two base materials viz. wrought AA7075-T6 and wrought AA6061-T6 were used. The chemical composition of these materials is presented in Table 1. AA6061-T6 rod (length 75 mm, pin diameter 10mm) is used as a consumable mechtrode for processing on AA7075-T6 plate (60 × 100 × 25 mm) on a horizontal Milling machine with vertical head attachment. Processing operation was made at a tool rotational speed 1000 rpm with transverse speed 180 mm/min and force 900N, schematic representation of this process has been presented in Figs. 1 and 2 that shows the FSPed AA7075-T6 plate using AA6061-T6 tool.

After FSP operation samples were polished using a different set of emery papers followed by disc polishing. Keller's etchant is used for etching the surface and samples were dried. Microstructural examination of the samples was observed using an optical metallurgical microscope (Olympus GX51 Inverted Metallurgical Micro-

**Table 1** Chemical composition of the commercial AA7075 and AA6061 used in this investigation

| Material | Cu | Mg | Si | Fe | Mn | Cr | Zn | Ti | Al |
|---|---|---|---|---|---|---|---|---|---|
| AA7075 | 2 | 2.9 | 0.4 | 0.5 | 0.3 | 0.25 | 5.7 | 0.2 | Balance |
| AA6061 | 0.31 | 0.69 | 0.53 | 0.23 | 0.33 | – | – | – | Balance |

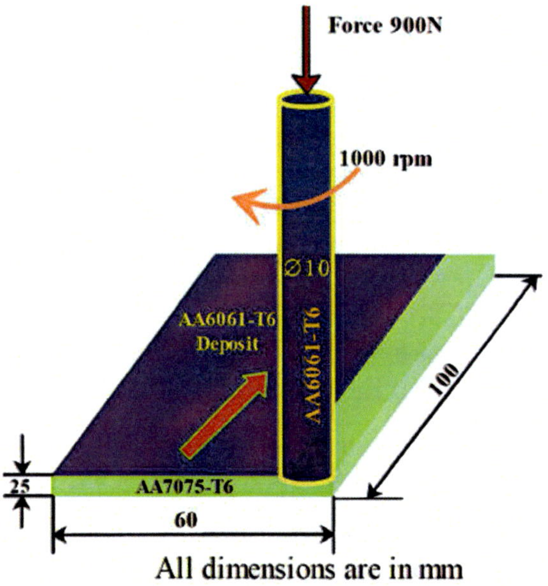


**Fig. 1** Schematic representation of FSP used in this study

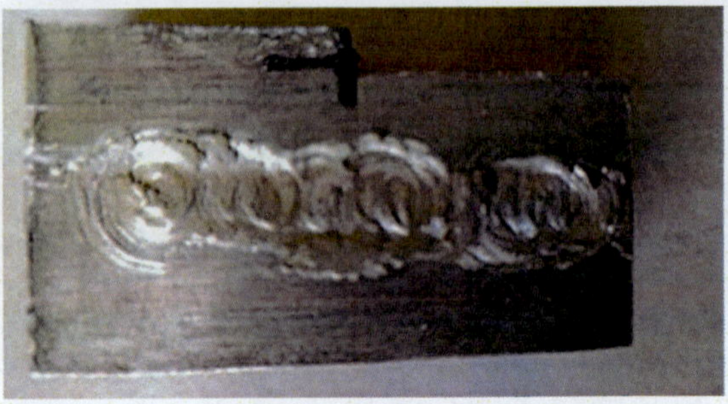

**Fig. 2** Friction stir Processed AA7075-T6 plate using AA6061-T6 Tool

scope). The specifications of the optical microscope are: it's a Binocular Head microscope with 10x/22 mm eyepieces and with 5 positions nosepiece. Pitting corrosion tests were done as per ASTMG69 standard in 3.5%NaCl, the potential scanning was carried at 0.166 mV/s with an initial potential −0.25V open current (OC) Saturated calomel electrode (SCE) on a 1 $cm^2$ area of the sample, Gill AC basic electrochemical system Software is used to study the pitting corrosion behaviour. Hardness test was conducted using micro-Vickers hardness tester as per ASTM E 384 standard, a load of 5 Kgf is applied with dwell 15 s, readings were taken five times and an average value of hardness has been presented in the results. Wear test is conducted using a pin on disc apparatus to make DUCOM TR-20 as per ASTMG99 standard. Samples of AA7075-T6, AA6061-T6 and FSPed condition were prepared to study dry sliding wear behaviour, all samples were made 10 mm diameter and height 20 mm, later samples were polished and cleaned using acetone. Wear test was performed at 1 Kgf load, 460 rpm for 10 min, the diameter of the sliding track on the disc surface is 40 mm.

## 3 Results and Discussions

### *3.1 Microstructure Studies*

Optical micrograph of the mechtrode AA6061-T6 is shown in Fig. 3. The black particles in the microstructure are soluble $Mg_2Si$ particles which are responsible for strengthening, the $Mg_2Si$ are uniformly distributed and the grey script like particles are $Fe_3SiAl_{12}$ is one of the stable phases [20]. From the aluminium-magnesium silicide system (Fig. 4) also it is evident that these phases are identified

Optical micrograph of AA7075-T6 is presented in Fig. 5. The grain structure is well developed. The optical micrograph shows the intermetallic precipitates of $MgZn_2$(dark) dispersed uniformly in and around grain boundaries [21]. Figure 6 shows the optical micrograph of the deposited AA6061-T6 on AA7075-T6, coating thickness has been measured using image analyzer software which is inbuilt with

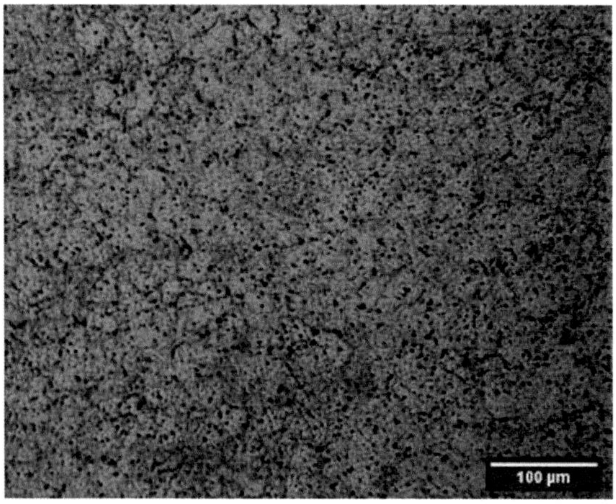


**Fig. 3** Optical micrograph of AA6061-T6 showing $Mg_2Si$ (black) and $Fe_3SiAl_{12}$(grey) particles

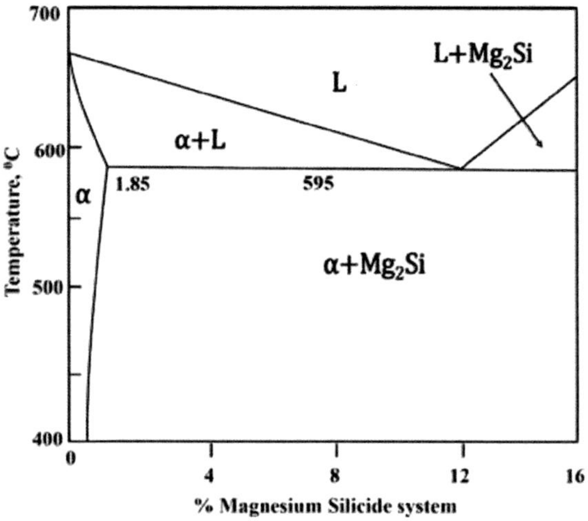


**Fig. 4** Aluminium-magnesium silicide system

the microscope, the coating thickness of 29.44 μm has been observed. Figure 7 shows the optical micrograph of FSPed AA7075-T6 with AA6061-T6, micrograph clearly reveals the recrystallized grains that are formed after processing, and it also shows the uniform distribution of secondary phase particles throughout the matrix, this can be attributed to the extensive plastic deformation caused due to stirring action of AA6061-T6 tool and also temperature generated which is responsible for recrystallization.

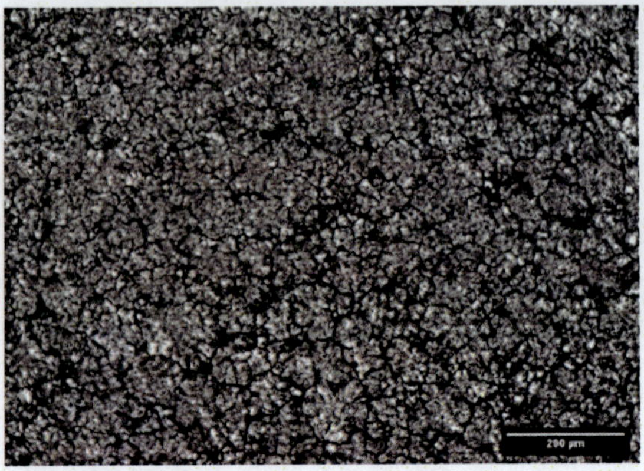


**Fig. 5** Optical micrograph of AA7075

**Fig. 6** Coating thickness of friction stir processed AA7075-T6 by AA6061-T6

**Fig. 7** Optical micrograph of FSPed AA7075-T6 with AA6061-T6 showing $Mg_2Si$ particles(dark)

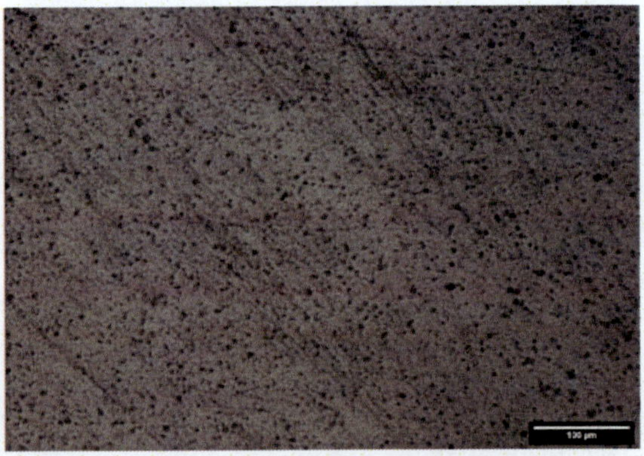

### 3.2 Hardness Studies

The main intention of studying the hardness of the material is to understand the hardness changes that occur after processing. It is observed from the hardness studies that there is a substantial improvement in hardness after processing. The hardness of the processed surface has been improved as compared with AA6061-T6. This is mainly because of the refinement of the microstructure and also beneath the deposit high strength AA7075-T6 is offering resistance towards indentation [22], overall hardness has been improved by 12.5% with respect to the AA6061-T6 hardness. Table 2 and Fig. 8 show the hardness values of both alloys and as well as FSPed alloy.

### 3.3 Pitting Corrosion Studies

Corrosion study is very important in this alloy because, as already discussed in the introduction, this alloy is used in structural parts in aerospace applications and there are many more applications of this alloy. Improvement of corrosion resistance is very important in this alloy. Basically, many of the aluminium alloys are susceptible to corrosion due to their secondary phase particles [23]. These secondary phase

**Table 2** Mirco Vickers hardness of AA7075-T6, AA6061-T6 and FSPed AA7075-T6

| Condition | VHN |
|---|---|
| Base Metal AA7075-T6 | 160 |
| Mechtrode AA606-T6 | 104 |
| Friction stir processed AA7075-T6 by AA6061-T6 | 117 |

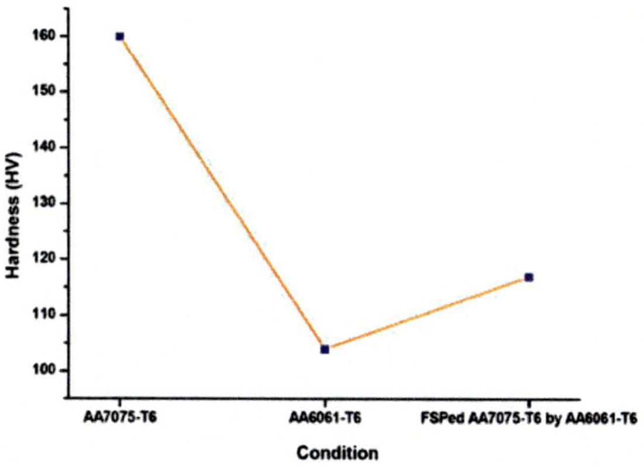


**Fig. 8** Mirco Vickers hardness of AA7075-T6, AA6061-T6 and FSPed AA7075-T6

particles decrease corrosion potential because of potential difference with the matrix [24]. Pitting corrosion studies have been done by understanding dynamic polarisation under 3.5% NaCl. Solution pH is maintained to 10 by adding KOH. Table 3 shows the $E_{corr}$ values of AA7075-T6 and AA6061-T6. Figures 9 and 10 show the potentiodynamic polarisation curves of AA7075-T6 and AA6061-T6. These curves also confirm the lower corrosion resistance of AA7075-T6 compared with AA6061-T6. Figure 11 shows the overall comparison of the potentiodynamic polarisation curves of individual alloys and FSPed alloy.

In the above graphs, it is observed negative open circuit potential (OCP), this means to raise in corrosion can be attributed to increasing potential. As we know, the oxide film growth is inversely proportional to the electric field. During the polarisation of aluminium alloys, $Al^{3+}$ and $OH^{-}$ ions migrate by the electric field and

**Table 3** Corrosion potentials, $E_{corr}$ (mV), SCE of AA7075 alloy in -T6 condition

| Condition | $E_{corr}$ |
|---|---|
| AA7075-T6 | −1029 |
| AA6061-T6 | −703 |
| FSP of AA7075 by AA6061 | −787 |

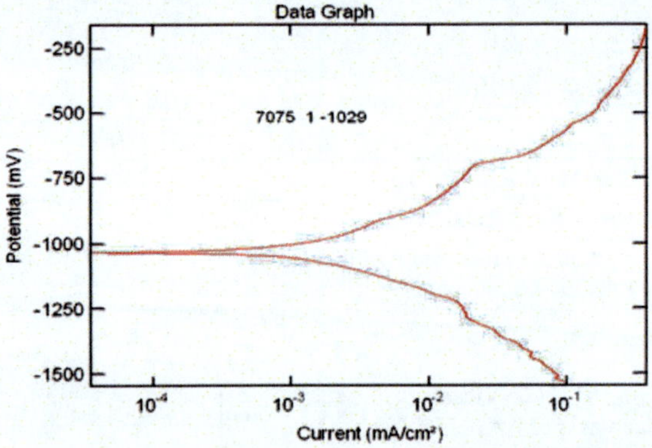


**Fig. 9** Potentiodynamic polarisation curves of AA7075-T6

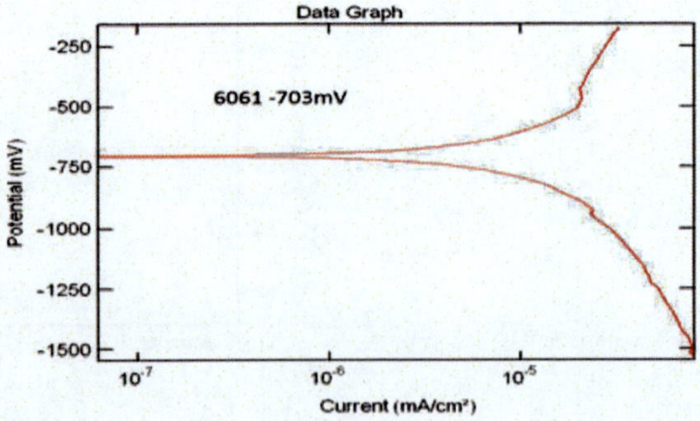


**Fig. 10** Potentiodynamic polarisation curves of AA6061-T6

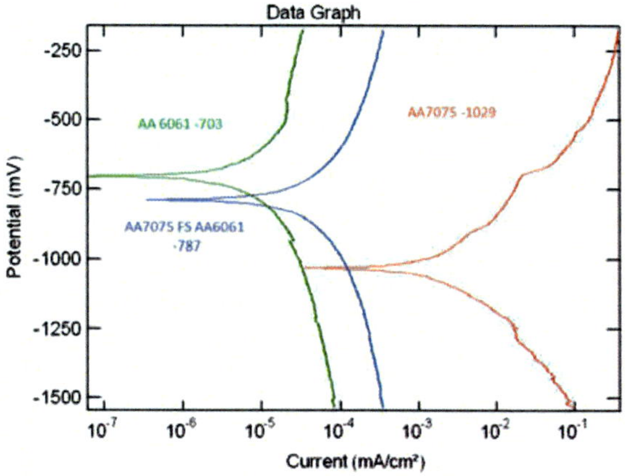


**Fig. 11** Comparison of corrosion resistance of AA7075-T6, AA6061-T6 and AA7075-T6 FSPed with AA6061-T6

from oxide film [25]. In AA6061, the low corrosion tendency could be attributed to the less galvanic action between the $Mg_2Si$ and the surrounding matrix. Solutions with pH range 8 to 14 do not affect the $Mg_2Si$ phase, but as the pH decreases from Fig. 8, magnesium starts decoupling in $Mg_2Si$, and this Mg dissolution in this phase does not depend on the potential [26]. Even though $Mg_2Si$ is active in the matrix of AA6061 [27], this phase has higher corrosion potential than $MgZn_2$, i.e. the M-phase that is present in AA7075, due to this corrosion resistance is higher in AA6061 than AA7075 alloy [28]. The above $E_{corr}$ values presented in Table 3, show the same trend. Of the two alloys, AA7075 is more corrosive, whereas, in the friction stir processed condition, the pitting corrosion resistance of AA6061 is slightly decreased due to intermixing of AA6061 with AA7075. But the pitting potential at the surface is more compared with AA7075 alloy. It implies that the corrosion resistance on the surface of AA7075 substrate after friction stir processing is far better than in its original form.

### *3.4 Wear Studies*

Comparison of variation of wear with the time between friction stir processed AA7075 with non-consumable rod AA6061-T6 and individual alloys are shown in Fig. 12. From the graph, it is clear that the wear rate is lower when AA7075-T6 is friction stir processed with AA6061 compared to the base metal AA6061-T6. Individually AA6061-T6 wear rate is high as compared with AA7075-T6. It is reported that wear rate depends on different factors like grain refinement, material mixability and hardness [29, 30], and also it depends on asperities between pin and disc [31]. AA6061-T6 is soft as compared with AA7075-T6. In the case of AA7075-T6 high resistance towards wear is because of $MgZn_2$ strengthening precipitates. In this condition, these M-phase precipitates distribute uniformly where both hardness and

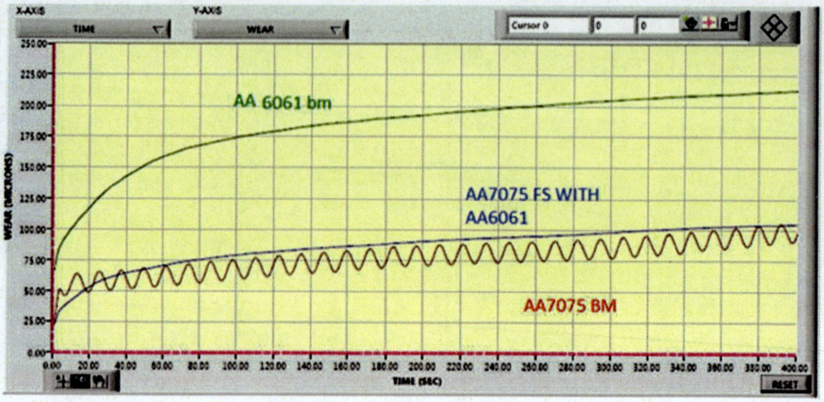


**Fig. 12** Comparison of wear of AA7075-T6, AA6061-T6, AA7075-T6 friction stir processed with AA6061-T6 with Time

wear resistance is enhanced. Similar observations were observed by Deore et al. [15]. Higher wear resistance in AA7075-T6 as compared with AA6061-T6 is due to the presence of Zn [32]. In the case of FSPed AA7075-T6 with AA6061-T6, wear resistance is closer to AA7075-T6. This may be due to better inter mixability and good bonding of AA6061-T6 with AA7075-T6. To analyse this further, SEM studies of worn samples should be done and further wear test by varying parameters such as load, sliding distance and sliding velocity should be analysed. On these conditions wear rate of FSPed AA7075-T6 shows better wear resistance with respect to AA6061-T6 and it is very close to AA7075-T6.

## 4 Conclusions

From the present investigation, the following conclusions were made.

1. Optical micrograph of FSPed AA7075-T6 with AA6061-T6, clearly reveals the recrystallized grains that are formed after processing, and it also shows the uniform distribution of secondary phase particles throughout the matrix.
2. Hardness of AA7075-T6 base metal is modified with friction stir processing by using AA6061-T6 as mechtrode.
3. Friction stir processing with AA6061-T6 increased pitting corrosion resistance of AA7075-T6.
4. Wear resistance of AA7075-T6 has been improved as compared with AA6061-T6 and the wear resistance of FSPed AA7075-T6 with AA6061-T6 is nearer to the AA7075-T6.

## References


1. Nicolas M, Deschamps A (2003) Characterisation and modelling of precipitate evolution in an Al-Zn-Mg alloy during non-isothermal heat treatments. Acta Mater 51(20):6077–6094
2. Alatorre N, Ambriz RR, Noureddine B, Amrouche A, Talha A, Jaramillo D (2014) Tensile properties and fusion zone hardening for GMAW and MIEA welds of a 7075-T651 aluminum alloy. Acta Metall Sin (English Lett) 27(4):694–704
3. Venugopal T, Rao KS, Rao KP (2004) Studies on friction stir welded AA 7075 aluminium alloy. Trans Indian Inst Met 57(6):559–663
4. Shikama T, Yoshihara S (2017) Highly SCC resistant 7000-series aluminum alloy extrusion. Kobelco Technol Rev 35:65–68
5. Kumar PV, Reddy GM, Rao KS (2015) Microstructure and pitting corrosion of armor grade AA7075 aluminum alloy friction stir weld nugget zone - effect of post weld heat treatment and addition of boron carbide. Def Technol 11(2):166–173
6. Kumar S, Kumar A, Vanitha C (2019) Corrosion behaviour of Al 7075 /TiC composites processed through friction stir processing. Mater Today Proc 15:21–29
7. Ikumapayi OM, Akinlabi ET (2019) Efficacy of $\alpha$ - $\beta$ grade titanium alloy powder (Ti-6Al-2Sn-2Zr-2Mo-2Cr-0.25Si) in surface modification and corrosion mitigation in 3.5% NaCl on friction stir processed armour grade 7075-T651 aluminum alloys-insight in defence applications. Mater Res Express 6(7):076546
8. Rai R et al (2011) friction stir welding tools. Sci Technol weld Join 16(4):325–342
9. Madhusudhan Reddy G, Srinivasa Rao K (2010) Enhancement of wear and corrosion resistance of cast A356 aluminium alloy using friction stir processing. Trans Indian Inst Metals 63(5):793–798
10. Tokisue H et al (2006) Mechanical properties of 5052/2017 dissimilar aluminum alloys deposit by friction surfacing. Mater Trans 47(3):874–882
11. Sudhakar I et al (2015) Enhancement of wear and ballistic resistance of armour grade AA7075 aluminium alloy using friction stir processing. Def Technol 11(1):10–17
12. Sudhakar I, Reddy GM, Rao KS (2016) Ballistic behavior of boron carbide reinforced AA7075 aluminium alloy using friction stir processing-an experimental study and analytical approach. Defence Technol 12(1):25–31
13. Rajan, Michael HB et al (2016) Influence of friction stir processing on microstructure and properties of AA7075/TiB2 in situ composite. J Alloys Compd 657:250–260
14. Mehdi H, Mishra RS (2020) Effect of friction stir processing on microstructure and mechanical properties of TIG welded joint of AA6061 and AA7075. Metallogr Microstruct Anal 9(3):403–418
15. Deore HA et al (2019) Effect of filler material and post process ageing treatment on microstructure, mechanical properties and wear behaviour of friction stir processed AA 7075 surface composites. Surf Coat Technol 374:52–64
16. Girish G, Anandakrishnan V (2020) Determination of friction stir processing window for AA7075. Mater Today: Proc 21:557–562
17. Patel Vivek V, Badheka Vishvesh, Kumar Abhishek (2017) Effect of polygonal pin profiles on friction stir processed superplasticity of AA7075 alloy. J Mater Process Technol 240:68–76
18. Dean WA (1967) Effects of alloying elements and impurities on properties. Paper from Aluminum, V. 1. ASM, METALS PARK, OHIO, 1967, pp 163–208
19. Lim S et al. (2005) Stress corrosion cracking behavior of friction-stir-welded Al 6061-T651.' Metall Mater Trans A 36(7):1977–1980
20. Bedard BA et al (2018) Microstructure and micromechanical response in gas-atomized Al 6061 alloy powder and cold-sprayed splats. J Therm Spray Technol 27(8):1563–1578
21. Aluminum association. Aluminum: properties and physical metallurgy. ASM international (1984)
22. Khodabakhshi F et al (2017) Friction-stir processing of a cold sprayed AA7075 coating layer on the AZ31B substrate: structural homogeneity, microstructures and hardness. Surf Coat Technol 331:116–128

23. Aballe A et al (2001) Localized alkaline corrosion of alloy AA5083 in neutral 3.5% NaCl solution. Corros Sci 43(9):1657–1674
24. Liao Chi-Min, Wei Robert P (1999) Galvanic coupling of model alloys to aluminum-a foundation for understanding particle-induced pitting in aluminum alloys. Electrochim Acta 45(6):881–888
25. Srinivasa Rao K, Prasad Rao K (2006) Corrosion resistance of AA2219 aluminium alloy: electrochemical polarisation and impedance study. Mater Sci Technol 22(1):97–104
26. Linardi E, Haddad R, Lanzani L (2012) Stability analysis of the $Mg_2Si$ phase in AA 6061 aluminum alloy. Procedia Mater Sci 1:550–557
27. Rao KS, Rao KP (2004) Pitting corrosion of heat-treatable aluminium alloys and welds: a review. Trans Indian Inst Met 57(6):593–610
28. Buchheit RG (1995) A compilation of corrosion potentials reported for intermetallic phases in aluminum alloys. J Electrochem Soc 142(11):3994
29. Ahmadnia M et al (2015) Determining influence of ultrasonic-assisted friction stir welding parameters on mechanical and tribological properties of AA6061 joints. Int J Adv Manuf Technol 78(9-12):2009–2024
30. Michael Rajan HB et al (2014) Effect of TiB 2 content and temperature on sliding wear behavior of AA7075/TiB 2 in situ aluminum cast composites. Arch Civil Mech Eng 14:72–79
31. Dhanashekar M et al (2020) Mechanical and wear behaviour of AA6061/SiC composites fabricated by powder metallurgy method. Mater Today: Proc 21:1008–1012
32. Baradeswaran A et al (2014) Experimental investigation on mechanical behaviour, modelling and optimization of wear parameters of $B_4C$ and graphite reinforced aluminium hybrid composites. Mater Design 63:620–632

# CFD Simulation of a Closed Turn Pulsating Heat Pipe with Hydrocarbon As a Working Fluid with Varying Fill Ratios for the Enhancement of Heat Transfer

E. Revanth and Dillip Kumar Mohanty

## 1 Introduction

Heat pipe is a heat absorbing and transferring device that has an efficient thermal conductivity. Heat pipe has closed evaporator-condenser system that performs the phenomenon of absorbing heat from the hot source and dissipating it to the colder surrounding. This process is achievable due to the working fluid that flows inside the heat pipe container. Working fluid undergoes phase change in the evaporator to become vapor from liquid and in the condenser region the vapor condensates into liquid form. Over the past 30 years, the technology of standard heat pipes had been adapted successfully, for the thermal management of numerous applications such as electronic cooling, spacecrafts, thermal control systems, heat recovery systems, heat exchangers, etc.

In a single turn closed-loop oscillating heat pipe, experiments like transient and steady state were performed by Naik et al. [1] where wall temperatures of evaporator, condenser are measured and for both the sections with ID 1.95 mm and OD 3 mm the material of the tube utilized here is the copper, respectively. With the fill ratios varying from 60 to 80% the working fluids utilized here are acetone, methyl alcohol, and ethyl alcohol. Results revealed that acetone is identified to be optimum at a fill ratio of 60% among other fluids. However, abundant designs of ordinary heat pipes are available, and current industry trends show the limitations of these conventional designs regularly which resulted in the evolution of modern technology. Pulsating Heat Pipe (PHP) proposed by Akachi and co-workers [2, 3] which originated in the year 1990 is a heat exchange apparatus which possesses optimum outcomes in electronic cooling devices, in the area of green energy which is utilized for cooling

E. Revanth (✉) · D. K. Mohanty
School of Mechanical Engineering, VIT-AP University, Inavolu, 522 237 Near Vijayawada, Andhra Pradesh, India
e-mail: revanth.e@vitap.ac.in



S. Srinivas et al. (eds.), *Recent Advances in Applied Mathematics and Applications to the Dynamics of Fluid Flows*, Lecture Notes in Mechanical Engineering,
https://doi.org/10.1007/978-981-19-1929-9_28

or heating buildings utilizing the geothermal heat sources or solar heating. PHP also termed as oscillating heat pipe comprises three sections, namely, evaporator, adiabatic, and condenser. Heat is being transferred to the condenser section from the evaporator section. This type of heat pipe is generally comprised of several bends or turns with capillary dimensions. At the beginning, the PHP is evacuated and filled partially with working fluid so the evaporation of working fluid begins when the heat has been applied to the evaporator region, which leads to rise in the vapor pressure in the tube and makes the bubbles to grow inside the evaporator region and the liquid is propelled to condenser region. The vapor pressure is decreased and bubble condensation takes place at the condenser part. The entire continual procedure across evaporation region, condenser region leads to a pulsating motion inside the tube. Transportation of heat is done by latent heat of vaporization, which is transferred through sensible heat carried by the liquid slugs.

Shafii et al. [4] explained about the basic study of PHP and also manufactured a closed-loop PHP made of Cu tubes with internal diameter of 1.8mm. The volumetric filling ratios varying from 30–80% for the working fluids $H_2O$ and ethyl alcohol had been introduced into the closed turn pulsating heat pipe, by varying heat inputs (5–70 W) and the results exhibited that the CLPHP has optimum heat transfer performance for $H_2O$ and ethyl alcohol at 40 and 50% fill ratios, respectively. An experimental investigation was conducted on a closed-loop PHP by Clement et al. [5], who designed a multi-turn PHP which comprises 15 turns, which was constructed using cu tube with 1.375 mm diameter. With the fill ratios from 30 to 70%, the working fluids used in this investigation are methyl alcohol, ethyl alcohol, acetone, and deionized water. Results indicated that pulsating heat pipe with methyl alcohol at a fill ratio of 45% has better thermal performance comparatively to the other fluids. Rudresha et al. [6] explained about the thermal performance of a heat pipe charged with nanofluids, and performed the CFD analysis in which $SiO_2$/DI water and $Al_2O_3$/DI water and $DI$ water served as the working fluids with concentrations by different masses 10, 20 g/lit in the heat pipe. The heat pipe is constructed by copper tube of length 1785 mm, thickness 1mm, and outer diameter 3 mm. CFD results in comparison with the experimental results indicated that the thermal resistance, coefficient of heat transfer, thermal conduction, and efficiency of closed turn pulsating heat pipe charged with $SiO_2$/DI water and $Al_2O_3$/DI water nanofluids are better than using DI water as the working fluid.

Baitule et al. [7] conducted another study on a CLPHP with two turns. Copper is used as the material with OD of 3 mm and ID of 2 mm. The working fluids filled with ratios ranging 0–100% utilized here are acetone, water, methyl alcohol, and ethyl alcohol. Therefore, the results from the experiment displayed that at 60% fill ratio for disparate heat inputs, the heat transfer performance characteristics reduce thermal resistance and escalate the coefficient of heat transfer of pulsating heat pipe. Pachgare [8] had performed experiments on the performance of heat transfer in a closed turn pulsating heat pipe, the copper tube used for the manufacture of PHP having internal diameter 2 mm and external diameter 3.6 mm and the filling ratio was 50%. And the number of turns used in this experimentation is 10. The working fluids introduced here are methyl alcohol, ethyl alcohol, acetone, $H_2O$, and different binary mixtures. The results demonstrate that with increase in the heat input supplied, the thermal resistance reduces immediately. Out of the various operating fluids, acetone has the most effective heat transfer performance.

An experimental investigation was performed [9] on a multi-turn open-loop pulsating heat pipe (five turns) in startup thermal conditions and steady-state thermal conditions, respectively. Different parameters, namely, working fluid, heat input, charge ratio, and concentration of ferrofluid were explored. From the experimental results, it shows that using ferrofluid improves the quality of thermal performance when compared to distilled water as working fluid. Also observed that the efficiency of thermal performance was improved to a great extent, when the magnetic field is applied to an open-loop PHP which is charged with ferrofluid. By using CFD modeling, Fadhl et al. [10] explained about the application to simulate the two-phase heat transfer mechanism in a wickless heat pipe called the thermosyphon. R134a and R404a were used as working fluids for thermosyphon. The obtained CFD results were compared with experimental results with good agreement obtained between predicted temperature profiles and experimental temperature data. With R134a and R404a charged thermosyphon it is proved that CFD modeling was efficient enough in reproducing the heat and mass transfer process. Qu et al. [11] have inspected on silicon-based micro-pulsating heat pipe by analyzing about how the measurement of temperature takes place. And the flow visualization was also implemented to examine the thermal and flow behavior of Si-based micro-PHP with hydraulic diameter of 352 $\mu$m. The working fluids used are the FC72 and R113. Variations in temperature versus time graph of micro-PHP under different power input and flow patterns in micro-channels were recorded. Heat input of 6.3w is supplied to the evaporator wall temperature of 42.1 C and 41.9 C which are obtained for micro-PHP when charged with R113 at filling ratios of 41 and 58%. In case of micro-PHP charged with FC72, a maximum power input of 9.5w is associated, respectively. Finally, he achieved a result that the visualization study exhibited at the evaporation, adiabatic, and condenser sections was largely occupied by annular slug slug-bubbly flows. The experimental studies are carried out in a closed-loop PHP with three turns which were elucidated by Naik et al. [12], who had conducted the experimental analysis on steady state for discrete heat loads, working fluids. The working fluids introduced in PHP are ethyl alcohol, methyl alcohol, heptane, and distilled water. The execution parameters like thermal resistance and heat exchange coefficient are assessed. The outcomes demonstrated that heptane and acetone display optimum heat exchange qualities of pulsating heat pipe. Goshayeshi et al. [13] have examined the experimental study on the use of $Fe_2O_3$/kerosene, nanofluid for the copper OHP in the presence of applied magnetic field. In a PHP to assess the augmentation of heat transfer, a study was conducted experimentally on iron oxide ($Fe_2O_3$) dispersed nanoparticles into kerosene (base fluid). Furthermore, the results indicated that the inclusion of iron oxide nanoparticles may improve the performance of heat transfer particularly under the effect of magnetic field. Saba et al. [14] investigated experimentally about the effect of perpendicular, uniform, gradient magnetic field on convective heat transfer coefficient of $Fe_3O_4/H_2O$ nanofluids. The heat transfer coefficient was measured at volume concentration of 0.5,1,2,3 % and temperatures of 20, 30, 40 °C. Thus, it is noticed that, with increase in the nanofluid concentration $Fe_3O_4/H_2O$, the convective heat transfer coefficient also increases. Further observations were made by the author that without magnetic field the convective heat transfer coefficient of $Fe_3O_4/H_2O$ nanofluids was improved by 5.2% at volume fraction of 3% and with the magnetic field applied to $Fe_3O_4/H_2O$ nanofluids the enhancement of convective heat transfer coefficient was improved by 4.2 and 8.1%, respectively.

The available research on CFD simulations of two-phase heat transfer/flow within a wickless heat pipe clearly shows a gap. As a result, the objective of this study is to develop a CFD model that encompasses all aspects of two-phase flow and heat transfer during the operation of a wickless heat pipe charged with acetone and how the various parameters such as geometry, fill ratio, and working fluid affect its performance. The CFD model which is created was validated with the experimental paper [1] with good consensus.

## 2 Governing Equations

The common procedure to model a two-phase flow which employs the formulation of a single fluid, supported volume of fluid method (VOF) are going to come after modeling the motion of slug flow in capillary tube. The hydro-dynamical movement is frequently characterized by the Navier–Stokes equations to figure out the two-phase flow interface, called Volume of Fluid (VOF).

Conservation of mass:

$$\nabla \cdot \mathbf{u} = 0 \tag{1}$$

Conservation of momentum:

$$\partial (\rho \mathbf{u})\, \partial t + \nabla \cdot (\mathbf{u} \cdot \rho \mathbf{u}) = -\nabla p + \nabla \cdot [\mu (\nabla \cdot \mathbf{u} + (\nabla \cdot \mathbf{u})\, T)] + \mathbf{f}_\sigma \tag{2}$$

Conservation of energy:

$$\partial (\rho c T)\, \partial t + \nabla \cdot (\rho c \mathbf{u} T) = \nabla \cdot (k \nabla T) \tag{3}$$

From the above equations $T,\ k$ represents temperature and thermal conductivity. $\rho,\ \mathbf{u},\ p,\ \mu,\ c$ represents density, vector velocity, pressure, dynamic viscosity, and specific heat capacity, respectively. In the momentum equation, the term $\mathbf{f}_\sigma$ is the surface tension force. With the VOF technique the interfaces are represented and furthermore the governing equations are solved in the phases correspondingly.

# 3 Simulation of CFD

The main purpose is to analyze the heat transfer performance in an oscillating heat pipe aka pulsating heat pipe by changing the load conditions in an evaporator section. The working fluid used here is acetone. The complete study is performed in CFD ANSYS software.

## *3.1 Methodology*

Firstly, based on the performance and operational analysis of pulsating heat pipe the literature survey is done. Many suggestions have been taken into consideration on journals and papers which are examined to occur with defined problem statement. On this basis, the reference paper by Naik et al. [1] for basic geometry, performance parameters, and operating conditions is taken into consideration as this paper has given the scope to perform the CFD analysis. The tools required for the model to be solved are set and worked on. From the results, the data obtained have been characterized and compared to our basic reference paper. The results of the CFD analysis are then verified analytically.

## *3.2 Geometry and Meshing of PHP*

In this work to create a geometry of a single turn PHP, the reference of geometric design is to be considered from Naik et al. [1], for each model independent study of grid is done and also ensured that the solution is also independent of mesh solution (Fig. 1).

### 3.2.1 Geometry

A CLPHP with single turn is made of copper has ID of 2 mm and OD of 3 mm, respectively. In the geometric modeling, the copper domain is not considered as the fluid domain is only the matter of criteria.

### 3.2.2 Meshing

For the shape of the mesh, the automatic method is selected and the mesh sizing is to be completed manually. The maximum, minimum sizes of the meshes are selected to be 0.0002 m as shown below (Fig. 2). The mesh has been generated with 6,17874 number of elements and 6,64427 number of nodes with this control settings.

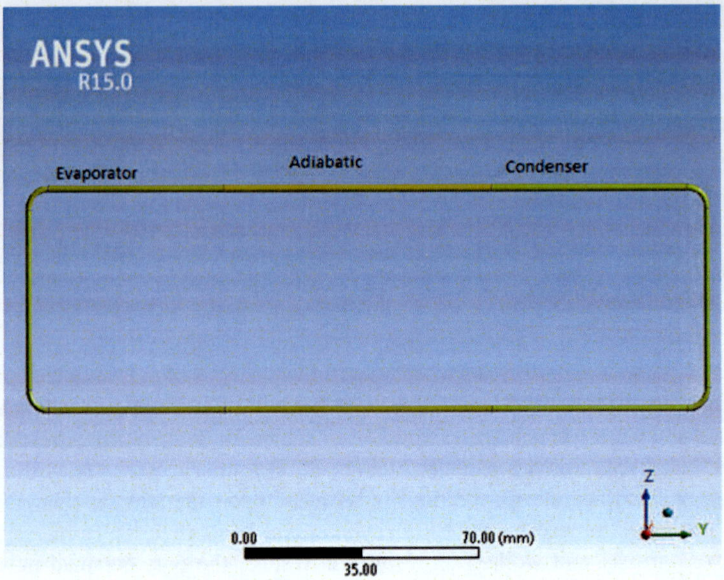


**Fig. 1** Illustration of a Geometric model of single turn PHP

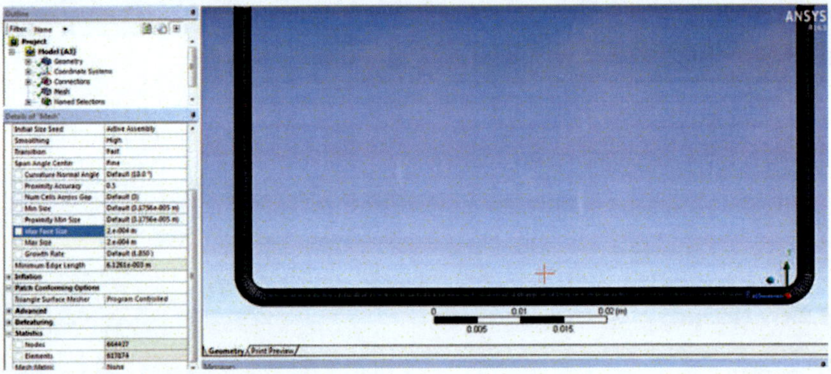


**Fig. 2** Meshing (ANSYS CFX image)

## 3.3 *Domain Settings*

Primarily three domains were defined for the entire geometry of pulsating heat pipe such as evaporator, adiabatic, and condenser region. Besides, the adiabatic region is separated to include the increase of acetone level in PHP from evaporator at base. A charge ratio of 60% by volume was filled with acetone working fluid in the evaporator section.

### 3.4 Boundary Conditions

In evaporator region, heat fluxes such as 9, 11, 13, and 15 W were specified, respectively. The value of heat flux is zero in case of adiabatic region. The heat flux which was given negative in the condenser region has to be calculated from reference paper taken. For the CFD analysis, VOF model has been employed.

The constant temperature of 302 K was given at the condenser region. Correspondingly, the rate of heat flux is zero Wm$^2$ K$^{-1}$ as the condition is adiabatic in the adiabatic region. Equivalently, the value of total constant heat flux is 7945 Wm$^2$ K$^{-1}$ as the external source is attached in evaporator region.

### 3.5 Turbulence Model

K-epsilon model has been one of the familiar turbulence models which have been executed in this analysis. In most of the CFD codes, it is implemented and is ought to be the industry standard model. For common purpose simulations, the K-epsilon offers a better understanding with regard to accuracy and robustness. Within the CFX the standard two equation model was applied by the K-epsilon turbulence model. The two equations are the transport equations one for dissipation and the other for turbulent kinetic energy ($k$) and to improve accuracy and robustness it utilizes the scalable wall function approach. Therefore, for the flow of fluids in the domain it provides good predictions.

## 4 Results and Discussions

The simulation output is generally examined in numerous ways such as charts or graphs, temperature contours, and streamlines. Different parameter values over the time steps are taken down and they are plotted to make graphs. The temperatures of acetone, and the volumetric fraction of acetone vapor and acetone at condenser region and at evaporator region are plotted to obtain the comprehensible output picture. The simulation results are often analyzed by examining these graphs (Tables 1 and 2).

**(i) Variation of Thermal Resistance**: The thermal resistance of pulsating heat pipe:

$$R = \frac{T_e - T_c}{Q} \, (K/W)$$

where $Q$ = heat flux in W, $T_e$ = temperature of the evaporator in k, and $T_c$ = temperature of the condenser in k (Table 3).

It is indicated from Fig. 3 there is a depletion in the thermal resistance at 60% volume fraction as compared with 70 and 80% volume fractions. Thus, acetone with

**Table 1** Heat input effects on evaporator temperature, condenser temperature: acetone at 60% volume fraction

| Heat flux (W) | Temperature of the evaporator (K) | Temperature of the condenser (K) |
|---|---|---|
| 9 | 312 | 292 |
| 11 | 316 | 292.5 |
| 13 | 320 | 293 |
| 15 | 324 | 294 |

**Table 2** Effects of fill ratio on evaporator temperature

| Heat flux (W) | 80% vof | 70% vof | 60% vof |
|---|---|---|---|
| 9 | 320 | 317 | 312 |
| 11 | 322 | 321 | 316 |
| 13 | 325 | 324 | 320 |
| 15 | 327 | 326 | 324 |

**Table 3** Variation of thermal resistance for CFD and experimental at 60% volume fraction (vof)

| Q (W) | Thermal resistance (K/W) at 60% vof | |
|---|---|---|
| | CFD | Experimental |
| 9 | 2.33 | 2.8 |
| 11 | 2.2 | 2.4 |
| 13 | 2.07 | 1.9 |
| 15 | 2 | 1.8 |

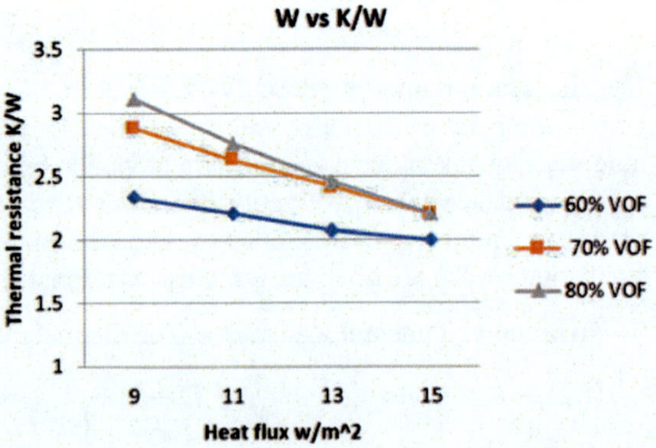


**Fig. 3** Comparison graph for variation in thermal resistance with heat flux at different volume fractions

**Table 4** Text

| Heat flux in W | Heat transfer coefficient in W/m$^2$k at 60% vof | |
|---|---|---|
| | CFD | Experimental |
| 9 | 379 | 316 |
| 11 | 402 | 368 |
| 13 | 427 | 465 |
| 15 | 442 | 491 |

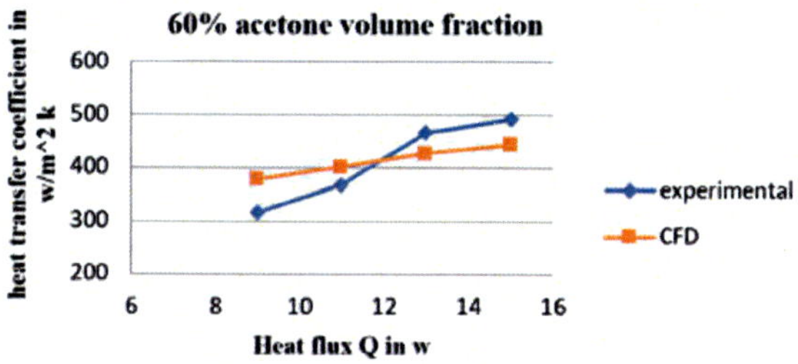


**Fig. 4** Comparison graph between CFD and experimental results for variation in heat transfer coefficient with heat flux at 60% volume fraction

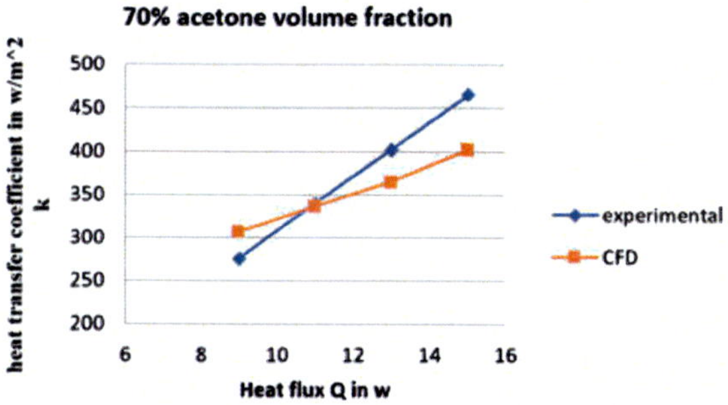


**Fig. 5** Comparison graph between CFD and experimental results for variation in heat transfer coefficient with heat flux at 70% volume fraction

60% volume fraction shows the better result, i.e., decrease in the thermal resistance is more with respect to heat flux, respectively (Table 4).

**(ii) Variation of heat transfer coefficient**: The coefficient of heat transfer for a pulsating heat pipe:

$$h = \frac{Q}{A_s\,(T_e - T_c)}\ (\mathrm{W/m^2K})$$

The heat transfer coefficient increases as the heat flux increases, as shown in Fig. 4. The maximum difference between experimental and CFD analysis was found to be 16 percent. And w.r.t the value of increasing heat flux, the deviation is decreased, respectively (Fig. 5).

The heat transfer coefficient increases as the heat flux increases, as seen in the maximum difference between experimental and CFD analysis observed is 15%. When compared to a 60 percent volume fraction, this deviation is less.

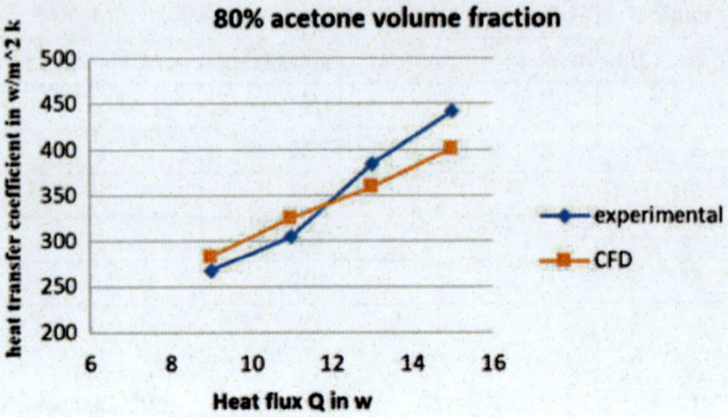


**Fig. 6** Comparison graph between CFD and experimental results for variation in heat transfer coefficient with heat flux at 80% volume fraction

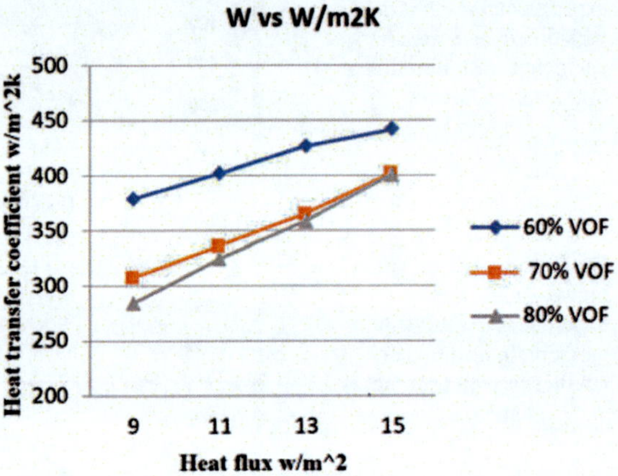


**Fig. 7** Comparison graph for variation in heat transfer coefficient with heat flux at different volume fractions

The heat transfer coefficient increases as the heat flux increases, as seen in Fig. 6. The maximum difference between experimental and CFD analysis was found to be 10%.

It is inferred from Fig. 7 that the heat transfer coefficient is high at 60% volume fraction as compared with 70% and 80% volume fractions.

The temperature contours of PHP at varied volume fractions, different time step size, and varying heat load are shown in Figs. 8, 9, and 10. However, Fig. 10a, b illustrates temperature contours of 70 and 80% volume fractions for 15 and 13 W heat loads at various time steps. As the time step rises, the volume fraction of acetone in liquid decreases to zero, leaving only acetone vapor to flow in the pulsating heat pipe. At this period, the evaporator portion's temperature is steadily increasing and the evaporator section is kept at the same temperature. The temperature contours are depicted in Fig. 12, respectively

**Contours of PHP by varying Volume fraction**:
The volume fractions of acetone liquid, acetone vapor, and air in a single-loop pulsing heat pipe are displayed for various time increments. The volume fraction contours of acetone liquid, acetone vapor, and air in PHP at the initial condition for 60, 70% filling ratio are shown in the first three contours (Figs. 11 and 12).

**Temperature Contours of PHP:**

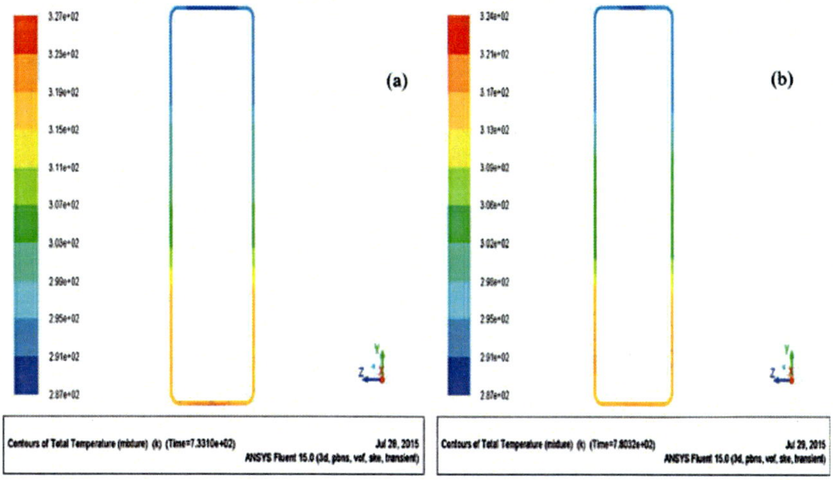


**Fig. 8** **a** and **b** represent temperature contours of PHP at 9 W at 60 and 70% volume fraction for acetone working fluid

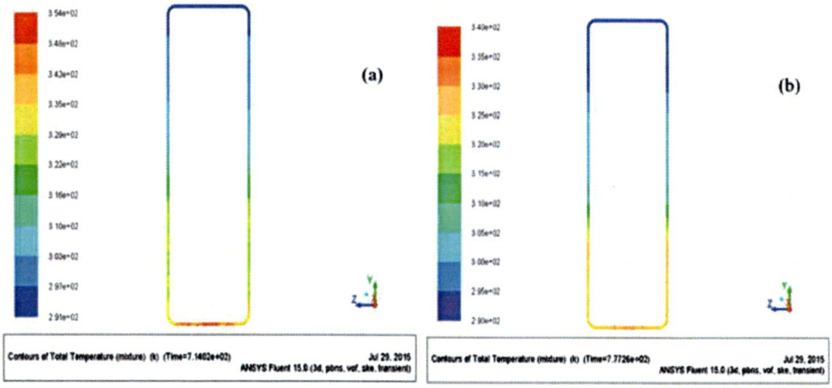


**Fig. 9** **a** and **b** represent temperature contours of PHP at 15 W for 60 and 70% volume fraction for acetone working fluid

This clearly demonstrates that the PHP is filled with 60% acetone fluid before the heat load is applied to the evaporator, with the remaining 40% of the PHP containing air in it. And at initial condition there is no acetone vapor in the PHP, these acetone-vapor contours are shown in Fig. 12b. These volume fraction contours at initial condition are same for different heat loads.

Figures 11 and 12 show the volume fraction contours of acetone liquid at different heat loads, i.e., 9, 13, and 15 W, respectively. It is observed from these volume fraction

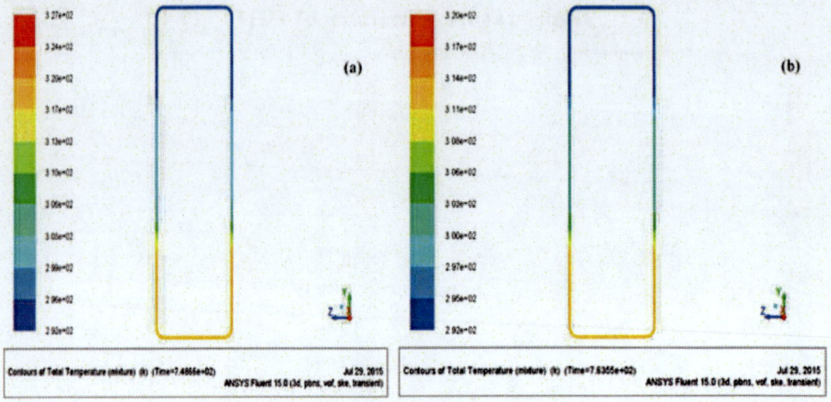


**Fig. 10** **a** and **b** represent temperature contours of PHP at 13 W and 15 W at 70 and 80% volume fraction for acetone working fluid

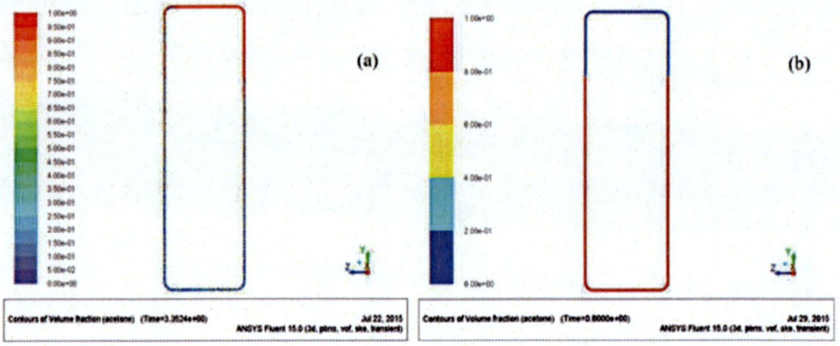


**Fig. 11** **a** and **b** represent contours of PHP for 60 and 70% volume fraction at initial condition

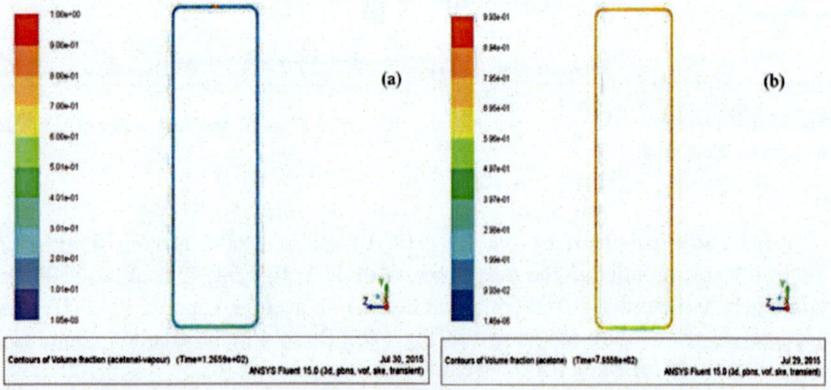


**Fig. 12** **a** and **b** represent contours of PHP for 60 and 70% volume fraction at initial condition

contours as time step size increases the volume fraction of acetone liquid decreases when the heat load is given at evaporator section. And also it is observed that the acetone-liquid volume fraction decreases as heat load increases in the PHP.

## 5 Future Scope

1. Number of turns can be increased which improves the better thermal performance of a pulsating heat pipe.
2. Use of nanofluids can improve thermal performance better than the ordinary working fluids.
3. Proper inclination angle and gravity may also affect the performance of a PHP with good thermal performance and increase in heat transfer, respectively.
4. By using the ferrofluid (magnetic nanofluid) which has greater thermal conductivity, it improves the thermal performance of a PHP significantly.

## 6 Conclusions

- In this CFD analysis of pulsating heat pipe, we can conclude that simulation of two-phase flow is simulated successfully in FLUENT 15.0.
- It can be examined that with raising the heat flux, the value of thermal resistance goes down and the convective heat transfer coefficient is increased.
- In this analysis, the difference in wall temperatures of condenser and evaporator with respect to flow time is detected.
- The PHP can display favorable heat transfer characteristics at a fill ratio of 60%, respectively.
- Among three fill ratios 60, 70, and 80%, it is noted that acetone at 60% is found to be a better fluid with better charging ratio for the augmentation of heat transfer rate.

## References


1. Naik R, Varadarajan V, Pundarika G, Narasimha KR (2013) Experimental investigation and performance evaluation of a closed loop pulsating heat pipe. J Appl Fluid Mech 6(2):267–275
2. Akachi H (1990) Structure of Heat Pipe. US patent, 4921041
3. Akachi H, Polasek F, Stulc P (1996) Pulsating Heat pipes. In: Proceeding's 5th international heat pipe symposium, pp 208–217
4. Shafii MB, Arabnejad S, Saboohi Y, Jamshidi H (2010) Experimental investigation of pulsating heat pipes and a proposed correlation. Heat Trans Eng 31(10):854–861
5. Clement J, Xia W (2013) Experimental investigation of pulsating heat pipe performance with regard to fuel cell cooling application. Appl Thermal Eng 50(1):268–274

6. Rudresha S, Kumar V (2014) CFD analysis and experimental investigation on thermal performance of closed loop pulsating heat pipe using different nanofluids. Int J Adv Res 2(8): 753–776
7. Baitule A, Pramod R (2013) Experimental Analysis of Closed loop Pulsating Heat Pipe with variable Filling ratio. Int J Mech Eng Robot Res 2(3):113–121
8. Pramod R, Ashish M (2013) Effect of pure and binary fluids on closed loop pulsating heat pipe thermal performance. Procedia Eng 51:624–629
9. Taslimifar M, Maziar M, Hossein A, Hassan S, Behshad S (2013) Overall thermal performance of ferrofluidic open loop pulsating heat pipes: an experimental approach. Int J Thermal Sci 65:234–241
10. Fadhl B, Wrobel C, Hussam J (2015) CFD modelling of a two-phase closed thermosyphon charged with R134a and R404a. Appl Thermal Eng 78:482–490
11. Jian Q, Hui-Ying W, Qian W (2012) Experiment investigation of Si based micro pulsating heat pipe. Nanoscale Micro scale Thermo Phys Eng 16(1):37–49
12. Naik R, Varadarajan V, Pundarika G, Narasimha KR (2013) Experimental investigation and performance evaluation of a closed loop pulsating heat pipe. J Appl Fluid Mech 6(2):267–275
13. Goshayeshi HR, Goodarzi M, Daharic M (2015) Effect of magnetic field on the heat transfer rate of kerosene/$Fe_2O_3$ nanofluid in a copper oscillating heat pipe. Exp Therm Fluid Sci 68:663–668
14. Fitnat S, Naveed A, Umar Khan S, Tauseef M (2019) A novel coupling of (CNT-$Fe_3O_4/H_2O$) hybrid nanofluid for improvements in heat transfer for flow in an asymmetric channel with dilating/squeezing walls. International Journal of Heat and Mass Transfer 136:186–195

# Thermo-Hydraulic Performance Analysis of a Shell and Tube Heat Exchanger with Different Single Segmental Baffle Configurations

**K. Anjineyulu and Dillip Kumar Mohanty**

## 1 Introduction

Heat exchangers have always been considered as pivotal in the processing, power generation, HVAC/R, transportation, and manufacturing industries. Due to the low-cost maintenance, limitation ranges of pressure and temperature designs this STHX is vastly used say 35–40% among all types of heat exchangers [1]. Among the various components, the baffles play a significant role in increasing the turbulence in case of the shell side flow which consequently adds to the overall performance of the heat exchanger. With helical baffles on the shell side Shinde and Chayan [2] has investigated the flow of heat, fluid in a STHX. Impacts of Different baffle configurations in STHX were examined by Ambekar et al. [3] on coefficient of heat transfer, pressure drop respectively. The design of ladder type fold baffle which has an ability to obstruct the leakage zones (triangular) in actual heat exchangers with helical baffles had been demonstrated by Xiao et al. [4]. Wen et al. [5] have investigated the improvement in performance and enhancement of heat transfer with installing the sealers in the shell side and the results show that overall heat transfer coefficient, exergy efficiency, pressure losses were increased on shell side. The influence of friction characteristics on a heat exchanger model using disparate swirl vane numbers at various locations across the length of the pipe to augment the rate of heat transfer was inspected by Yehia et al. [6].

The computational analysis for study of performance parameters with different geometrical configurations of a STHX has been applied by various researchers in last few decades. In a STHX Leoni et al. [7] have analyzed the flow in a shell side using CFD. The numerical model of a STHX pertained to the conviction of porosity and permeability to procure the thermo-hydraulic performances in a shell side had

K. Anjineyulu (✉) · D. K. Mohanty
School of Mechanical Engineering, VIT-AP University, Inavolu, Near Vijayawada 522 237, Andhra Pradesh, India
e-mail: anjineyulu.k@vitap.ac.in



S. Srinivas et al. (eds.), *Recent Advances in Applied Mathematics and Applications to the Dynamics of Fluid Flows*, Lecture Notes in Mechanical Engineering,
https://doi.org/10.1007/978-981-19-1929-9_29

been studied by You et al. [8]. Pal et al. [9] have inspected the behavior of a fluid flow, the temperature pattern in a STHX with and without the presence of baffle and concluded that the cross flow at the nozzle part has a remarkable contribution in the direction of the heat transfer. The rate of heat transfer in the shell side of a STHX and the 3D numerical simulation of a turbulent fluid flow had been examined by Mellal et al. [10].

In a STHX the steady-state numerical simulations have been carried out accompanied by twisted tapes on a shell side was investigated by Yang et al. [11]. Ozden and Tari [12] have studied the design of a shell side in a STHX, eventually the results are susceptible to the selection of a turbulence model. The numerical inspections on a combined multi-pass STHX with continual helical baffles had done by Wang et al. [13]. Bhutta et al. [14] have done the review which is mainly focused on the applications of CFD and observed that CFD has been implemented in following: maldistribution of fluid flow, to identify the fouling factor, pressure drop, thermal analysis in the phases of design and optimization. Wang [15] had investigated that the heat transfer enhancement can be done in a STHX by implanting the sealers in the shell side. The results revealed that the coefficient of heat transfer at shell side was increased from 18.2 to 25.5%, the overall heat transfer coefficient was raised to 15.6 to 19.7% and the exergy efficiency was increased to 12.9–14.1%, respectively.

The significant factors for the evaluation of shell and tube heat exchangers which can give more advantageous results with helical baffles are the performance of rate of heat transfer and pressure drop [16]. Comparatively to the conventional segmental baffles the usage of continual helical baffles has resulted in 10% increase in the coefficient of heat transfer. The use of continuous helical baffles results in nearly 10% increase in heat transfer coefficient compared with that of conventional segmental baffles [17]. With the help of theoretical and numerical methods Chit et al. [18] had studied the flow analysis and the influence of baffle spacing of a STHX.

The present work has investigated computationally the impact of baffles on the thermo-hydraulic performances in a STHX. The overall coefficient of heat transfer, pressure drop has been estimated in the heat exchangers without baffles with parameters of single segmental baffles keeping constant. The baffle cut has been considered to be 22.5% for the single segmental baffles.

## 2 Methodology

Computational fluid dynamics (CFD) is a significant tool which has been quite useful for study of fluid flow behavior, rate of heat transfer, etc. by solving a set of mathematical equations with the help of numerical simulations [14]. Shell and tube heat exchangers have a quite complex geometry for which an exact numerical analysis is too difficult. However, CFD employs a quite simple and effective technique by resolving the entire heat exchanger system into a number of small cells or grids. Certain parameters like distribution of pressure, difference in temperature gradients and flow parameters can be obtained by applying the governing equations on these particular distinct elements.

## 2.1 Modeling

The geometric model was designed in CATIA V5R20 and later it was exported to ANSYS- FLUENT version to procure the finite element analysis. Figure 1 represents the geometric model of single segmental heat exchanger baffle. Shell side of the heat exchanger is made up of stainless steel with internal and external shell diameters as 153 and 167 mm, respectively. The exchanger consists of 24 tubes arranged in square pitch arrangement. The internal and external diameters of the tubes are 9.5 and 12 mm respectively while the length of each tube is 1200 mm. In case of the exchanger with single segmental baffle, 4 mm thick baffles with 22.5% baffle cut were used. The working fluid considered for shell side of a heat exchanger is a hot fluid.

## 2.2 Governing Equations

The shell side flow and the rate of heat transfer are considered to be steady and the working fluid gravity is ignored completely. The governing equations pertained to the conservation of mass, momentum, and energy for numerical simulations can be expressed as follows [13]

Continuity Equation

$$\frac{\partial u_i}{\partial x_i} = 0 \tag{1}$$

Momentum equation

$$\frac{\partial u_i u_j}{\partial x_i} = -\frac{\partial p}{\rho \partial x_i} + \frac{\partial}{\partial x_j}\left((\nu + \nu_t)\left(\frac{\partial u_j}{\partial x_i} + \frac{\partial u_i}{\partial x_j}\right)\right) \tag{2}$$

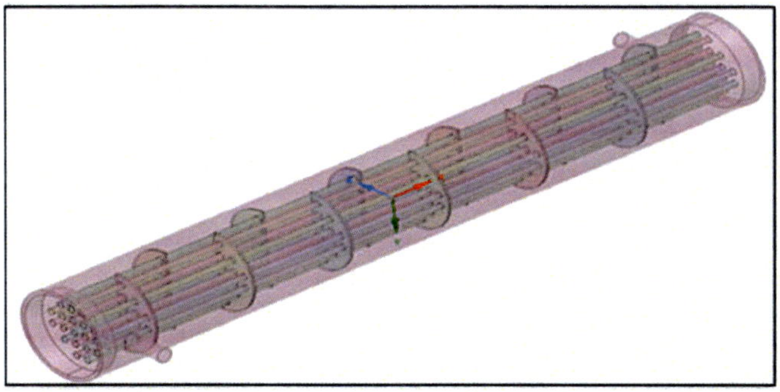

**Fig. 1** Geometric model of a single segmental baffle in a shell and tube heat exchanger

Energy equation

$$\frac{\partial u_i T}{\partial x_i} = \rho \frac{\partial}{\partial x_i}\left(\left(\frac{\nu}{Pr} + \frac{\nu_t}{Pr}\right)\left(\frac{\partial T}{\partial x_i}\right)\right) \tag{3}$$

Turbulent kinetic energy $k$ equation

$$\frac{\partial u_i k}{\partial x_i} = \frac{\partial}{\partial x_i}\left(\left(\nu + \frac{\nu_t}{\sigma_k}\right)\left(\frac{\partial k}{\partial x_i}\right)\right) + \Gamma - \varepsilon \tag{4}$$

Rate of energy dissipation $\varepsilon$ equation

$$\frac{\partial (\rho u_j)}{\partial x_j} = \frac{\partial}{\partial x_i}\left(\left(\nu + \frac{\nu_t}{\sigma_\varepsilon}\right)\left(\frac{\partial \varepsilon}{\partial x_i}\right)\right) + c_1 \Gamma \varepsilon - c_2 \frac{\varepsilon^2}{k + \sqrt{\nu \varepsilon}} \tag{5}$$

where $\Gamma$ denotes the production rate of $k$ and is given by

$$\Gamma = -\overline{u'_i u'_j} \frac{\partial u_i}{\partial x_i} = \nu_t \left(\frac{\partial u_i}{\partial x_j} + \frac{\partial u_j}{\partial x_i}\right) \frac{\partial u_i}{\partial x_i} \tag{6}$$

$$\nu_t = c_\mu \frac{k^2}{\varepsilon} \tag{7}$$

## 2.3 Meshing

The modeling of the test section is meshed with ANSYS. The fine mesh is created throughout the effective length of the tube and then the domains were meshed with polyhedral. Due to its excellent merit of managing very complex 3D geometries. Figure 2 represents the heat exchanger with single segmental baffle. The smooth meshing was created such that edges as well as regions of temperature constraints meshed. For carrying out this CFD analysis both the hot and cold fluid are considered to be incompressible and in single phase only. The effect of radiation and net convection are neglected. The thermophysical properties are assumed to be independent of temperature. The flow is hydrodynamic and constant heat flux is used. Grid independence tests have been performed and later came to a conclusion that the total number of cells procured around 1.33 millions with single segmental baffles, respectively.

## 2.4 Thermophysical Properties

The hot and cold fluids introduced in the shell and tube side are considered to be water for the analysis of heat transfer performances. The thermophysical properties of hot, cold fluids are tabulated below (Table 1).

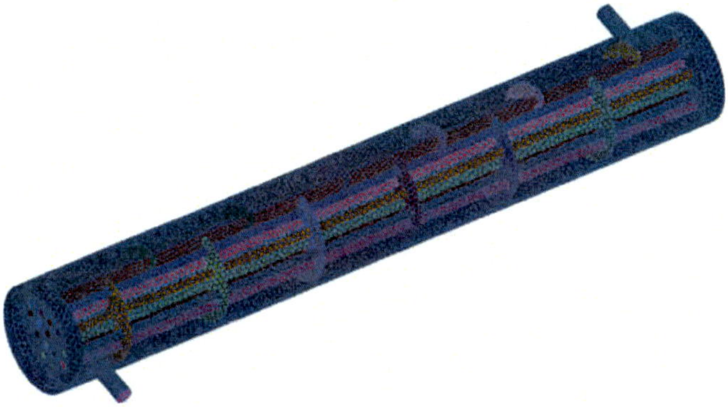

**Fig. 2** Meshing of shell and tube heat exchanger with single segmental baffle

**Table 1** Thermo-physical properties of hot and cold fluids

| S. No. | Fluid | Density (kg/m$^3$) | Thermal conductivity (W/m–K) | Viscosity (Ns/m$^2$) | Specific heat $C_p$ (J/kg K) |
|---|---|---|---|---|---|
| 1 | Cold $H_2O$ | 997 | 0.6207 | 0.00089 | 4182 |
| 2 | Hot $H_2O$ | 985 | 0.6493 | 0.000504 | 4181 |

## *2.5 Boundary Conditions*

The boundary conditions of the test section in the working fluids within shell and tubes for STHX are hot fluid and cold fluids. The 3D, double precision, pressure based, and steady solver are acquired for numerical simulations by utilizing ANSYS-FLUENT. The desired mass flow rate of shell side is 3 kg/s and tube side is 0.5 kg/s. The temperature values assigned to the shell side inlet are 25 °C and tube side inlet are 55 °C. The gauge pressure values of the shell side and tube side are zero. No slip boundary condition is set to STHX wall surfaces. The thermal boundary layer of the single segmental STHX wall is set to be adiabatic. The constant wall temperature of 27 °C is assigned to tube walls. The second-order upwind scheme is set to energy, momentum equations, and SIMPLE algorithm is allotted for the coupling between pressure-velocity fields.

## *2.6 Numerical Analysis*

The numerical method adopted particularly for this analysis is the explicit analysis also known as second order upwind scheme method. In this analysis Mass/inertia and damping are included. One of the main advantages using this method is no iteration required as the nodal accelerations are solved directly. Once accelerations are known at time $n$, velocities are calculated at time $n + 1/2$ and displacements at time $n + 1$. The strain is derived from the displacements and stress is obtained from strain. And the entire cycle is repeated this way, respectively.

## 3 Results and Discussion

The CFD simulation of various profiles like temperature, pressure, and velocity in single segmental STHX for disparate baffle cut configurations has been considered. Parameters like effectiveness, pressure drop, overall coefficient of heat transfer have been considered for the measurement of performance and efficiency of heat exchangers. A steady-state CFD model was simulated in ANSYS-FLUENT. The parameters such as overall coefficient of heat transfer, pressure drop when compared to the mass flow rate it is observed that both these parameters are directly proportional to the mass flow rate. The outcome of the computational analysis along the shell side flow has been described with regard to temperature, velocity contours, and pressure distribution of the heat exchangers with single segmental baffles, respectively.

### 3.1 *Effect on Temperature*

The temperature distributions of shell and tube side heat exchanger of segmental baffle are shown in Fig. 3. It is examined that in the outer shell pass of the shell and tube heat exchanger, the temperature increases evenly and smoothly. From the temperature contours shown in Fig. 4 it can be observed that temperature of the water at the cold pipes in the single segmental STHX is 22.5% is higher than remaining baffle cuts. From the temperature contours it is noticed that temperature has been decreased continuously from the shell side inlet to shell side outlet. Conversely the temperature increases continuously from cold side inlet to cold side outlet.

From the graph shown in Fig. 5 it is examined that temperature distribution varies with mass flow rate which implies mass flow rate is increased with increase in the temperature distribution.

### 3.2 *Effect on Velocity Contours*

The velocity distribution of a STHX with single segmental baffle is shown in Fig. 6. The orientation of the baffle cut types and the fluid flow pattern has been represented in the form of path lines. The formation of a zigzag flow pattern in the shell side of a STHX is observed. From different single segmental baffle cuts, it can be seen that velocity distribution of hot water in a single segmental model is more uniform.

**Fig. 3** Shell side heat exchanger of a single segmental baffle

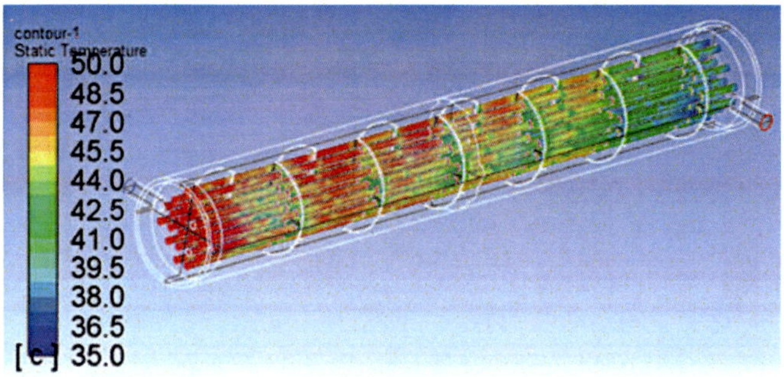


**Fig. 4** Tube side heat exchanger of a single segmental baffle

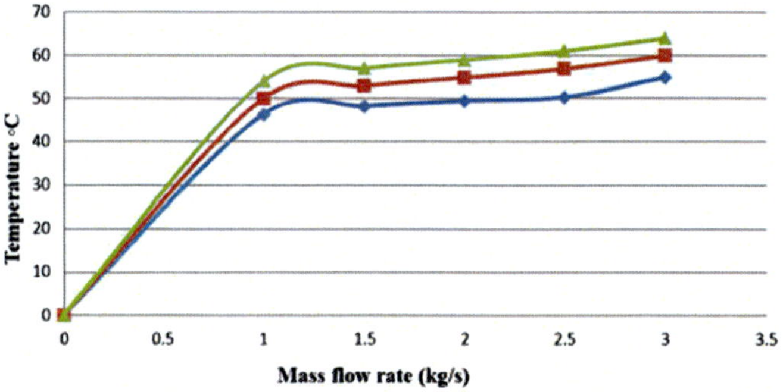


**Fig. 5** Comparison of mass flow rate versus temperature distribution with different baffle cuts

## 3.3 *Effect on Pressure*

For the design of a STHX, the pressure drop is considered to be a significant factor because it is associated with the parameters like operating cost and efficiency. If the pressures drop is low resulting in low pumping power, which leads to the increase in

**Fig. 6** Heat exchanger with single segmental Baffle

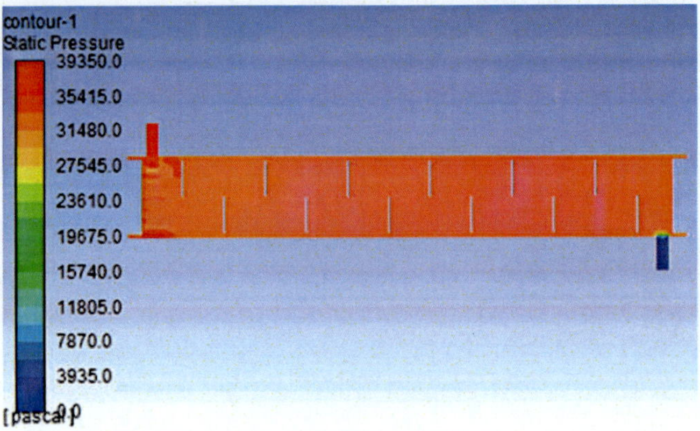


**Fig. 7** Heat exchanger with single segmental Baffle

the overall system efficiency. From Figs. 7 and 8, it can be observed that the required pressure drop for a fluid needs to be higher at inlet stage to reach the outlet.

From the above graph which represents mass flow rate to pressure drop it is noticed that the mass flow rate is directly proportional to pressure drop. In case of baffle cut at 22.5% the pressure drop increases and with baffle cut at 42.75% the pressure drop decreases gradually. Apparently, it can be concluded that the pressure drop is inversely proportional to baffle cut, respectively.

## 4 Conclusions

The overall heat transfer rate and pressure drop for heat exchangers with single segmental baffles at disparate inlet mass flow rates have been analyzed using ANSYS-FLUENT. The conclusions that can be taken into consideration from the study are as follows:

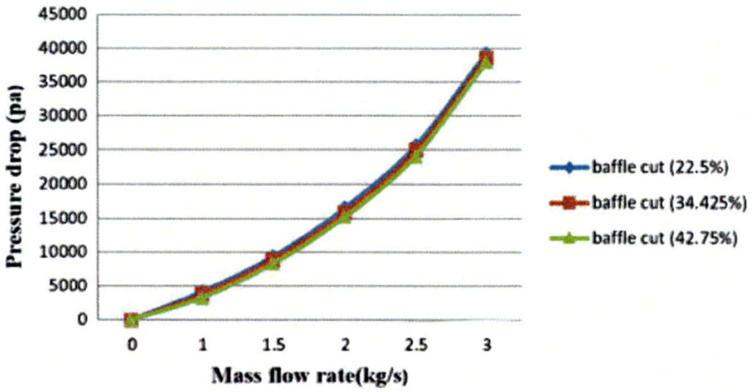


**Fig. 8** Comparison of mass flow rate versus pressure drop with different baffle cuts

1. Single segmental baffles with 22.5% baffle cut obtain the better overall coefficient of heat transfer compared to other baffle cuts.
2. This work is intended to develop an optimal configuration of a shell and tube heat exchanger which can provide maximum possible coefficient of heat transfer with minimum possible pressure drop.
3. The CFD estimated pressure drop is compatible within ±10% with those estimated by the Bell-Delaware method.

## References

1. Mohanty DK (2016) Gravitational search algorithm for economic optimization design of a shell and tube heat exchanger. Appl Thermal Eng 107:184–193
2. Shinde S, Chayan U (2018) Numerical and experimental analysis on shell side thermo-hydraulic performance of shell and tube heat exchanger with continuous helical FRP baffles. Thermal Sci Eng Progress 5:158–171
3. Ambekar AS, Sivakumar R, Anantharaman N, Vivekanandan M (2016) CFD simulation study of shell and tube heat exchangers with different baffle segment configurations. Appl Thermal Eng 108:999–1007
4. Xiao J, wang S, Ye S, wang J, Tu J (2019) Experimental investigation on pre-heating technology of coal water slurry with different concentration in shell and tube heat exchangers with ladder type fold baffles. Int J Heat Mass Transfer 132:1116–1125
5. Wen J, Yang H, wang S, Xue Y, Tong X (2015) Experimental investigation on performance comparison for shell and tube heat exchangers with different baffles. Int J Heat Mass Transfer 84:990–997
6. Yehia MG, Attia AAA, Abdelatif OE, Khalii EE (2016) Heat transfer and friction characteristics of Shell and tube heat exchanger with multi-inserted swirl vanes. Appl Thermal Eng 102:1481–1491
7. Leoni GB, Klein TS, Medronho RA (2017) Assessment with computational fluid dynamics of the effects of baffle clearances on the shell side flow in a shell and tube heat exchanger. Appl Thermal Eng 112:497–506

8. You Y, Fan A, Huang S, Liu W (2012) Numerical modelling and experimental validation of heat transfer and flow resistance on shell side of a shell and tube heat exchanger with flower baffles. Int J Heat Mass Transfer 55:7561–7569
9. Pal E, Kumar I, Joshi JB, Maheshwari NK (2016) CFD simulations of shell-side flow in a shell-and-tube type heat exchanger with and without baffles. Chem Eng Sci 143:314–340
10. Mellal M, Benzeguir R, Sahel D, Ameur H (2017) Hydro-thermal shell side performance evaluation of a shell and tube heat exchanger under different baffle arrangement and orientation. Int J Thermal Sci 121:138–149
11. Yang D, Khan TS, Al-Hajri E, Ayub ZH, Ayub AH (2019) Geometric optimization of shell and tube heat exchanger with interstitial twisted tapes outside the tubes applying CFD techniques. Appl Thermal Eng 152:559–572
12. Ozden E, Tari I (2010) Shell side CFD analysis of a small shell and tube heat exchanger. Energy Convers Manag 51:1004–1014
13. Wang Q, Chen Q, Chen G, Zeng M (2009) Numerical investigation on combined multiple shell pass shell and tube heat exchanger with continuous helical baffles. Int J Heat Mass Transf 52:1214–1222
14. Bhutta MMA, Hayat N, Bashir MH, Khan AR, Ahmad KN, Khan S (2012) CFD applications in various heat exchangers design: A review. Appl Therm Eng 32:1–12
15. Wang S, Wen J, Li Y (2009) An experimental investigation of heat transfer enhancement for a shell-and-tube heat exchanger. Appl Thermal Eng 29:2433–2438
16. Bichkar P, Dandgaval O, Dalvi P, Godase R, Dey T (2018) Study of shell and tube heat exchanger with the effect of types of baffles. Procedia Manuf 20:195–200
17. Li H, Peng L (2018) Numerical investigation on double tube-pass shell-and-tube heat exchangers with different baffle configurations. Appl Therm Eng 143:561–569
18. Chit SP, San NA, Soe MM (2015) Flow analysis in shell side on the effect of baffle spacing of shell and tube heat exchanger. Int J Sci, Technol Soc 3:254–259

# Influence of Dry Sliding Wear Parameters on the Rate of Wear Al–Mg Matrix Composites Reinforced with Ternary Alloy Particulate

**Ch. Kishore Reddy, V. S. N. Venkata Ramana, Rahul, K. Sri Ram Vikas, and Dasari Madhusudhan**

## Nomenculture

| | |
|---|---|
| HSAp | High strength ternary alloy particulate |
| ANOVA | Analysis of Variance |
| AMCs | Aluminium-based metal matrix composites |
| MMCs | Metal Matrix Composites |
| GRG | Grey Relational Grade |
| GRA | Grey Relational Analysis |
| GRC | Grey Relational Coefficient |

Ch. Kishore Reddy (✉) · K. Sri Ram Vikas
Department of Mechanical Engineering, Prasad V. Potluri Siddhartha Institute of Technology, Vijayawada 520007, Andhra Pradesh, India
e-mail: kishorereddymtech@gmail.com

V. S. N. Venkata Ramana
Department of Mechanical Engineering, GITAM Institute of Technology, GITAM (Deemed to be University), Visakhapatnam, Andhra Pradesh 530045, India

Rahul
Department of Mechanical Engineering, Chaitanya Bharathi Institute of Technology(A), Gandipet, Hyderabad 500075, Telangana, India

D. Madhusudhan
Department of Metallurgical and Materials Engineering, RAjiv Gandhi University of Knowledge Technologies, Nuzivid 521202, Andhra Pradesh, India



S. Srinivas et al. (eds.), *Recent Advances in Applied Mathematics and Applications to the Dynamics of Fluid Flows*, Lecture Notes in Mechanical Engineering,
https://doi.org/10.1007/978-981-19-1929-9_30

# 1 Introduction

'Wear' is contemporary and existed issue to discuss on both ends beneficiary and malicious for various industrial applications depending on input tribological factors especially in the case of aluminium-based metal matrix composites (AMCs), which occupied about 70% of applications [1, 2]. These AMCs are feeble, particularly in the dry sliding conditions against the wear [3]. When the soft graphite particulate at various proportions was incorporated in AA6063, the wear rate of the resultant composite is greatly reduced [4]. Similar output was observed when $TiO_2$ particulate was incorporated in aluminium-based alloy [5]. Most of the aluminium-based AMMCs were weak in tribological behaviour [6]. The tribological behaviour of composites AA7075 with the inclusion of SiC+$Al_2O_3$ particulate is optimised using the ANOVA method and is concluded that the optimised parameters are 550 rpm stirring speed and 12% of inclusion [7]. The $Al_2O_3$ reinforced particulates greatly influence the properties of aluminium metal matrix composites. By the inhibition of silicon carbide (SiC) particulate in the aluminium-based matrix, by varying proportions, at 20 wt.%, the composite has the highest tensile strength as well as highest hardness [8]. For the optimization of the wear behaviour of composites, many statistical tools are available. Statistical techniques like Taguchi techniques Grey Relational Grade (GRG) method are some of the effective techniques optimizations of the parameters like rate of wear (RoW), coefficient of friction (CoF) concerning input parameters like the sliding velocity, distance of sliding, etc. When SiC particulate was reinforced in aluminium 2xxx series alloy, the velocity of sliding was found to be the most affected input parameter on the composite's dry sliding wear property. Statistical analysis for optimising wear behaviour using Taguchi techniques was conducted on the composites of Al2219 with SiC and graphite as reinforcements. The velocity of sliding load and sliding distance was taken as input parameters and observed that the hybrid composite fabricated with the combination of SiC and graphite particulates shown better resistance to wear [9]. In the present investigation, an attempt was made to accomplish better mechanical properties coupled with good wear properties for the selected materials in the conventional stir casting route followed by a secondary process hot-extrusion.

# 2 Materials and Methods

In the present research study, the matrix material is Al4Mg which is prepared using the ordinary casting method. When mixed with copper, this Al4Mg binary alloy possessed enriched properties which insinuation for significant applications in aircraft, automobile industries [10]. Another base material, the ternary alloy Al20Cu20Mg is also casted and particulate is prepared from this ternary alloy using filing technique. The average size of particulate is 200 microns, obtained after ball milling and sieving. Using the stir casting technique, the composites are prepared by the addition of 5, 10, and 15% of particulate in the base alloy which was melted at 700 °C. In order to avoid oxidation, a continuous flow of argon gas was maintained during the addition of particulate into the melt of base alloy. The particulate was preheated at 100 °C before addition into the vortex of melt. A continuous vortex is maintained using the stirrer at a speed of 750 rpm. The melt was casted into 60 mm diameter billets. These billets of composites are extruded into rods of 16mm in diameter using a hot-extrusion press of (200T) capacity.

## 3 Wear Test

The specimens under ASTM: G99 standards were prepared using Wire cut Electric Discharge Machine (WEDM). Dry sliding wear test was conducted using pin-on-disc apparatus—DUCOM TR-20 wear testing machine. The standard wear specimens of 4 mm dia. × 30 mm long. The cut specimens are as shown in Fig. 1. The pin-on-disc test was performed at normal room temperature.

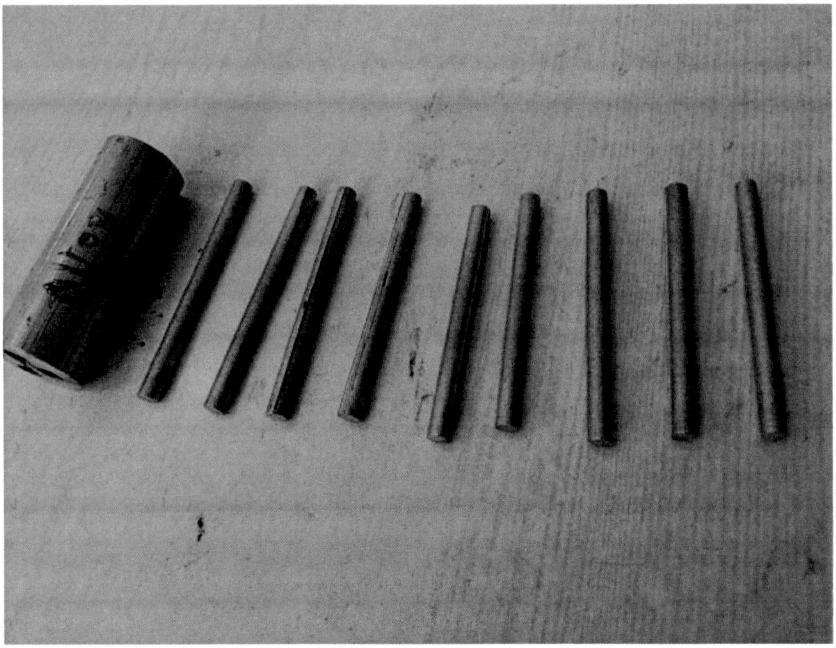

**Fig. 1** Wear cut specimen using WEDM

# 4 Results and Discussions

## 4.1 Grey Relational Analysis (GRA)

The wear rate outcomes from the pin-on-disc test are projected, for optimising through GRA. The designed experiment is successfully conducted as assumed. Three input factors each at three levels are tabulated in Table 1. Table 2 shows the analytical responses.

The grey rational coefficients (GRCs) were calculated from the Signal to Noise ratio (S/N ratio) of all response parameters and Grey Rational Grades (GRGs). The above GRCs were premeditated for all combinations of experiments, and then ranks are allotted.

$$\eta = -10\log_{10}\left(\frac{1}{n}\sum_{i=1}^{n} {y_{ij}}^2\right) \quad (1)$$

$$\eta = -10\log_{10}\left(\frac{1}{n}\sum_{i=1}^{n} \frac{1}{{y_{ij}}^2}\right) \quad (2)$$

where $\eta$ = Signal to noise ratio, $y_{ij}$ = Observations, $n$ = Number of replicates.

**Table 1** The three assumed levels of the process parameters

| Parameter type | Level: 1 | Level: 2 | Level: 3 |
|---|---|---|---|
| Velocity (m/min) | 100.00 | 120.00 | 140.00 |
| Applied load (N) | 4.91 | 9.81 | 14.71 |
| Time (min) | 15 | 30 | 45 |

**Table 2** Investigative table of RoW responses

| Experiment No. | Velocity (m/min) | Load (N) | Time (min) | Rate of Wear (RoW) (mg/min) | | | |
|---|---|---|---|---|---|---|---|
| | | | | Alloy | 5% | 10% | 15% |
| 1 | 1 | 1 | 1 | 4.61 | 4.28 | 1.94 | 1.28 |
| 2 | 1 | 2 | 2 | 0.61 | 1.66 | 0.67 | 0.92 |
| 3 | 1 | 3 | 3 | 3.54 | 4.18 | 1.53 | 1.1 |
| 4 | 2 | 1 | 2 | 0.87 | 1.2 | 0.47 | 0.52 |
| 5 | 2 | 2 | 3 | 0.73 | 1.23 | 0.57 | 1.43 |
| 6 | 2 | 3 | 1 | 1.19 | 1.98 | 0.98 | 0.87 |
| 7 | 3 | 1 | 3 | 1.33 | 0.5 | 0.89 | 1.91 |
| 8 | 3 | 2 | 1 | 0.98 | 0.77 | 0.85 | 0.81 |
| 9 | 3 | 3 | 2 | 1.89 | 0.98 | 1.05 | 1.78 |

## 4.2 Grey Relational Analysis (GRA) for Rate of Wear (RoW)

The grey system theory (GRA) is the influential measurement technique. This technique examines the uncertainty relations between major factor with other factors in a given system. When the experimental trials can't be completed precisely, GRA assists to compensate for the shortcomings in statistical regression. GRA is a measurement of the absolute value of the data differences between sequences and it can be used to measure the estimated correlation between sequences [11, 12].

$$N - S/Nratio = \eta_n = \frac{\eta_i - min(\eta_i)}{max\,(\eta_i) - \; min\,(\eta_i)} \quad (3)$$

$$f = \left[\max\left[\eta_{ni-n_{in}}\right]\right] - \eta_{ni} \quad (4)$$

$$GRC = \frac{min\,(f Dc_i) + f * max\,(Dc_i)}{(Dc_i) + f * max\,(Dc_i)} \quad (5)$$

$$GRG = \frac{1}{m}\sum_{k=1}^{m} GRC_{ik} \quad (6)$$

where $m$ = number of output parameters, $f$ = quality loss function, GRC = grey relational coefficient, Dc = Distinguishing coefficient (range 0Dc1), GRG = grey relational grade, S/N ratio = Signal to noise ratio, N S/N ratio = n = Normalised Signal to Noise ratio.

The S/N ratios were calculated and tabulated in Table 3. The Grey relational coefficients, Grey relational grades, and ranks are tabulated in Table 4. The optimal conditions on the basis of level means tabulated in Table 5 are A2-B3-C3.

**Table 3** S/N ratios for all experiments

| Experiment no. | Rate of Wear (RoW) (mg/min) | | | |
|---|---|---|---|---|
| | Alloy | 5P | 10P | 15P |
| 1 | –6.0639 | –7.1967 | –6.4444 | –6.0639 |
| 2 | –6.8485 | –8.9432 | –7.7478 | –6.8485 |
| 3 | –7.3471 | –9.1571 | –8.0280 | –7.3471 |
| 4 | –7.2346 | –9.1273 | –7.6042 | –7.2346 |
| 5 | –9.3077 | –9.4258 | –9.2480 | –9.3077 |
| 6 | –9.4843 | –9.8272 | –9.3669 | –9.4843 |
| 7 | –7.6042 | –7.7478 | –6.0206 | –7.6042 |
| 8 | –6.9271 | –8.2995 | –6.8485 | –6.9271 |
| 9 | –8.2995 | –8.5627 | –6.8485 | –8.2995 |

**Table 4** Calculation of Grey relational coefficients (GRCs)

| Experiment no. | GRC | | | | GRC | Rank |
|---|---|---|---|---|---|---|
| | Alloy | 5P | 10P | 15P | | |
| 1 | 0.333 | 0.333 | 0.364 | 0.333 | 0.341 | 9 |
| 2 | 0.394 | 0.549 | 0.508 | 0.394 | 0.461 | 6 |
| 3 | 0.445 | 1.000 | 0.555 | 0.445 | 0.611 | 3 |
| 4 | 0.432 | 0.590 | 0.487 | 0.432 | 0.485 | 5 |
| 5 | 0.906 | 0.669 | 0.934 | 0.906 | 0.854 | 2 |
| 6 | 1.000 | 0.818 | 1.000 | 1.000 | 0.954 | 1 |
| 7 | 0.476 | 0.381 | 0.333 | 0.476 | 0.417 | 7 |
| 8 | 0.401 | 0.443 | 0.399 | 0.401 | 0.411 | 8 |
| 9 | 0.591 | 0.481 | 0.399 | 0.591 | 0.516 | 4 |

**Table 5** Level means of (GRG) for Rate of wear

| Parameter type | Level: 1 | Level: 2 | Level: 3 | Delta | Rank | Optimal level |
|---|---|---|---|---|---|---|
| Velocity—A | 0.471 | 0.764 | 0.448 | 0.316 | 1st | L:2 |
| Load—B | 0.414 | 0.575 | 0.694 | 0.279 | 2nd | L:3 |
| Time—C | 0.569 | 0.487 | 0.627 | 0.058 | 3rd | L:3 |

## *4.3 ANOVA*

### 4.3.1 Checking of Data and Adequacy/Tolerability of the Model

The average probability plot of the residuals is shown in Fig. 2. From this plot, it can be declared that the experimental outputs are nearer to the line fit in the normal plot. Hence the validity of operative parameters holds good for all cases.

The contribution on effectiveness of all process parameters, namely, the sliding velocity, load applied, and time of sliding on wear rate outputs, is premeditated using ANOVA technique.

The ANOVA for the rate of wear (RoW) was conducted as tabulated in Table 6. It can be stated that the most significant factor is velocity followed by load and time.

## *4.4 Microstructural Results*

In Fig. 3, the SEM micrograph, the microstructure of matrix alloy Al4Mg is shown. In the SEM micrograph of AL4Mg growth of dendrites of small size is observed., and the uniform distribution of magnesium is seen as whitewashed-like spots all over

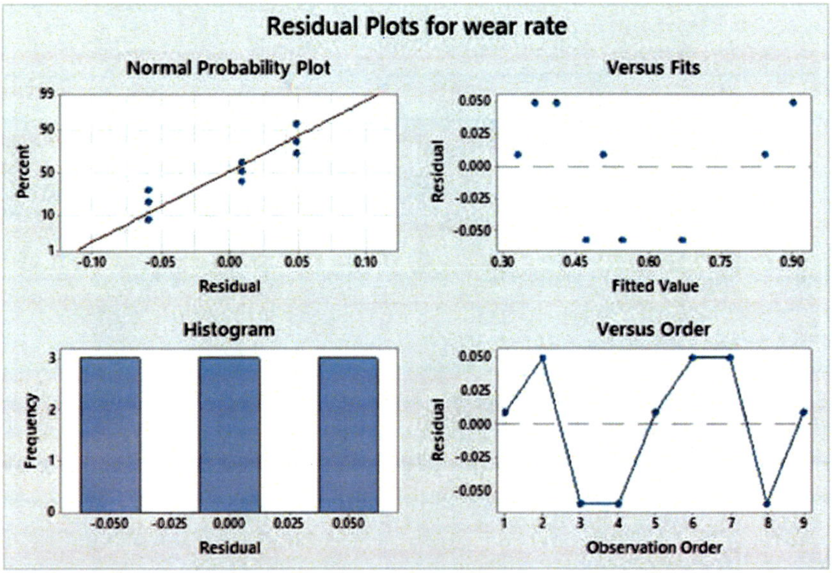


**Fig. 2** Normal probability—Residual plots for the Rate of Wear

**Table 6** ANOVA output for GRGs

| Source | D.F | S.S | M.S | F value | % Contribution |
|---|---|---|---|---|---|
| Velocity | 2 | 0.18664 | 0.093320 | 10.38 | 52.9866 |
| Load | 2 | 0.11795 | 0.058975 | 6.56 | 33.48569 |
| Time | 2 | 0.02966 | 0.014828 | 1.65 | 8.41926 |
| Error | 2 | 0.01799 | 0.008993 | | 5.106178 |
| Total | 8 | 0.35223 | | | |

the surface. In Fig. 4, the SEM micrograph of the composite reinforced with 10% particulate shows uniform distribution of high strength ternary alloy particulates (HSAp) in the Al4Mg alloy. This can clearly be evidenced throughout the grain boundaries which are thin white strands, similar observations were made by Praveen Kumar et al., in AA2024 which is reinforced with High Entropy Alloy particulates [13], and similar white clusters of Al-20Cu-10Mg were observed in A356 alloy matrix composites [14].

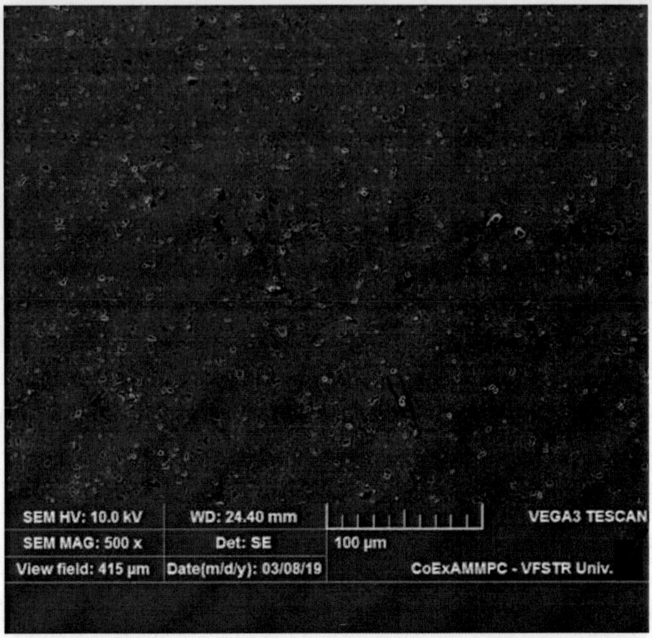


**Fig. 3** Microstructure of base alloy Al4Mg showing uniform distribution of Magnesium

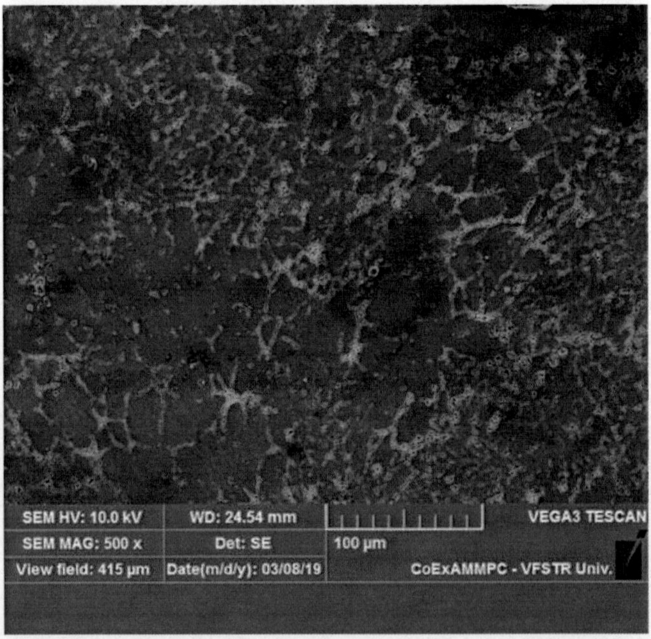


**Fig. 4** Microstructure of wt.% 10 composite showing uniform distribution of HSAp in Al4Mg

## 4.5 SEM Observation of Worn-out Specimens

The wear behaviour of the samples was analysed by examining the SEM micrographs of the worn-out surfaces. The micrographs revealed that the samples were abraded due to the adhesive and abrasive wear mechanism, from the SEM micrograph of worn-out Al4Mg alloy (Fig. 5) reveals more ploughing and broader wear tracks, which indicates wider grooves adjacent to the ploughed region and more debris is observed which is adhered on the surface, broadening of wear tracks and deep grooves are due to plastic deformation, similar observations were reported by Kumar et al. [15].

In Fig. 6, the SEM image of the composite with wt. % 10 showing reduced delamination and uniform wear track, which indicates increased wear resistance with an increase in reinforcement content. Improvement in wear resistance can be ascribed to the strong interfacial bonding strength between the matrix and reinforcement particulate, which resists the delamination and debonding of particulate. Kumar et al. [16] reported similar observations in aluminium composites.

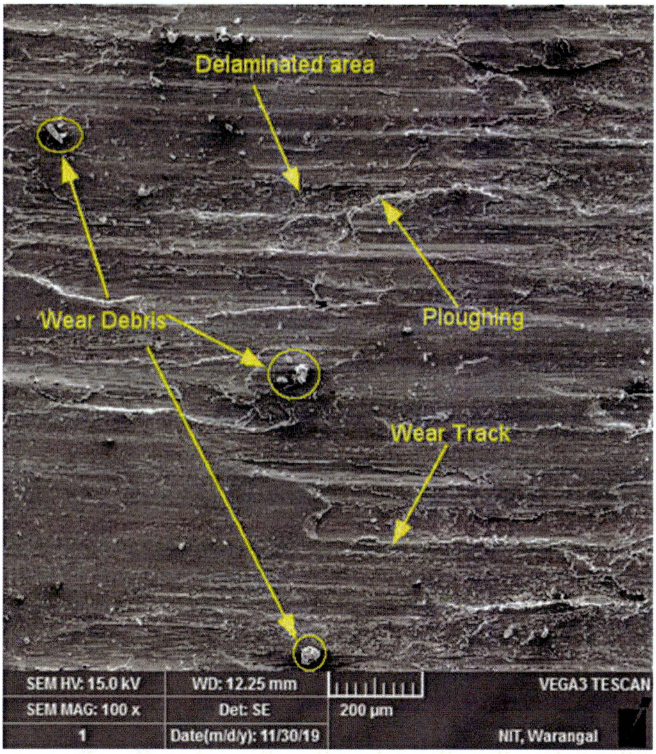


**Fig. 5** Worn-out surface of alloy showing delamination with broad wear tracks with a significant amount of Ploughing at 1.5 kgf load with 120 m/min velocity for 45 min

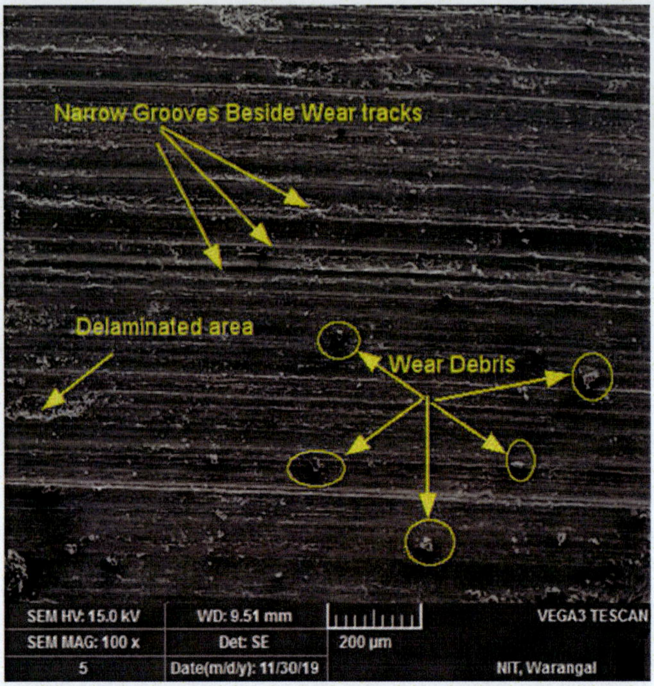


**Fig. 6** Worn-out surface of 10 wt.% particulate composite showing thin wear tracks with less delamination and wear debris at 1.5 kgf load with 120 m/min velocity for 45 min

## 5 Conclusions

- The wear rate of the composites gradually enhanced with the increase in the reinforcement of particulate.
- Wear rate is optimised with moderate velocity and larger values of load and time.
- From GRA, the optimal combination of process parameters with the assumed levels is determined as A2–B3–C3, i.e. the desired parameters are moderate Velocity, and higher levels of load and time.
- The highest influential contribution is by velocity (V) and it is 52%, followed by Load (L) and Time (T), with contributions of 33% and 8%, respectively.
- The worn-out micrographs reveal plastic deformation in the alloy which resulted in wider grooves and delamination with a high amount of ploughing along the sliding direction and narrower wear tracks, less delamination, and debris formation in the alloy with 10wt.% particulate reinforcement.
- Wear track thickness is less from the microstructural observations for the alloy with 10wt.% particulate reinforcement manifests good agreement with the grey relational analysis GRA.

## References


1. Prakash JU, Ananth S, Kumar GS, Moorthy TV (2018) Multi-objective optimization of wear parameters for aluminium matrix composites (413/B4C) using grey relational analysis. Mater Today 5:7207–7216
2. Mohan Kumar S, Pramod R, Govindaraju HK (2017) Evaluation of mechanical and wear properties of aluminium AA430 reinforced with SiC and MgO. Mater Today 4: 509–518
3. Surappa MK (2003) Aluminium matrix composites: challenges and opportunities. Sadhana 28(1 & 2):319–334
4. Alaneme KK, Sanusi KO (2015) Microstructural characteristics, mechanical and wear behaviour of aluminium matrix hybrid composites reinforced with alumina, rice husk ash, and graphite. Eng Sci Technol Int J 1–7
5. Antony vasantha Kumar C, Selvin Rajadurai J (2016) Influence of rutile (TiO2) content on wear and microhardness characteristics of aluminum-based hybrid composites synthesized by powder metallurgy. Trans Nonferrous Met Soc 26: 63–73
6. Bahmani E, Abouei V, Shajari Y, Razavb SH, Bayat O (2018) Investigation on microstructure, wear behavior and microhardness of AlSi/SiC nanocomposite. Surf Eng Appl Electrochem 54(4):350–354
7. Rajeswar B, Amirthagadeswaran KS, Barasu KGA (2015) Investigation on mechanical properties of A7075-SiC-Al h-composite using the Taguchi method. Australian J Mech Eng 13:127–135
8. Mohsin M, Mohd A, Suhaib M, Arif S, Siddiqui MA (2017) Investigation of the effect of alumina and compaction pressure on physical, electrical and tribological properties of Al-Fe-Cr-Al2O3 powder composites. Mater Res Exp 4(10):106510
9. Basavarajappa S, Chandramohan G (2006) Dry sliding wear behavior of metal matrix composites: a statistical approach. J Mater Eng Perform 15(6):656–60
10. NIshida, Y (2013) Introduction to metal matrix composites: fabrication and recycling. Springer Science & Business Media
11. Tosun N (2006) Determination of optimum parameters for multi-performance characteristics in drilling by using grey relational analysis. Int J Adv Manuf Technol 28(5):450–455
12. Ho C-Y, Lin Z-C (2003) Analysis and application of grey relation and ANOVA in chemical-mechanical polishing process parameters. Int J Adv Manuf Technol 21.1: 10–14
13. Kumar KP (2015) Fabrication and characterization of 2024 aluminium-high entropy alloy composites. J Alloys Compd 640(2015):421–427
14. Gopi KM et al (2017) Metal-metal composites-an innovative way for multiple strengthening. Mater Today: Proc 4.8: 8085–8095
15. Kumar SS, Pandian RS, Pitchipoo P (2020) A study on tribological behavior of Al-4% Mg incorporated with MoS2. Mater Res Exp 7.1: 016578
16. Kumara, TS et al (2016) Synthesis, microstructure and mechanical properties of Al-Si-Mg alloy hybrid (zircon+ alumina) composite. Indian J Eng Maters Sci 23:20–26

# POD—ANN Reduced Order Macromodel of Nonlinear Autonomous Dissipative System Using Machine Learning

S. Nagaraj, D. Seshachalam, and G. Jayalatha

## 1 Introduction

Many physical world systems are represented by highly nonlinear dynamical system models. Modeling mathematically and simulating such nonlinear autonomous dissipative (NAD) systems require higher degrees of freedom in accurately describing data flow. NAD system governed by partial differential equation (PDE) needs discretization to analyze in the domain of computation. Application of finite difference (FD) scheme in discretizing the system results in a set of nonlinear algebraic equations. Employing numerical techniques like Runge–Kutta (RK) or Newton–Raphson (NR), these equations are solved. If viewed in terms of processing time and memory, these methods are observed to be inefficient.

Size and complexity of time variant/invariant system are significantly reduced using various MOR techniques [1]. These techniques include Passive Reduced Order Interconnect Macromodeling Algorithm method (PRIMA), Arnoldi, Laguerre, optimal Hankel norm and proper orthogonal decomposition (POD) [20].

Nonlinear system MOR exposes to totally distinct confrontations which have not been experienced in linearizing the model. Due to large condition numbers, wide range of time constants embracing different orders of dimension, nonlinear system show a highly analogue nature [23].

S. Nagaraj (✉) · D. Seshachalam
BMS College of Engineering, Bengaluru 560019, India
e-mail: nagaraj.research@gmail.com

D. Seshachalam
e-mail: dschalam.ece@bmsce.ac.in

G. Jayalatha
RV College of Engineering, Bengaluru 560059, India
e-mail: jayalathag@rvce.edu.in



S. Srinivas et al. (eds.), *Recent Advances in Applied Mathematics and Applications to the Dynamics of Fluid Flows*, Lecture Notes in Mechanical Engineering,
https://doi.org/10.1007/978-981-19-1929-9_31

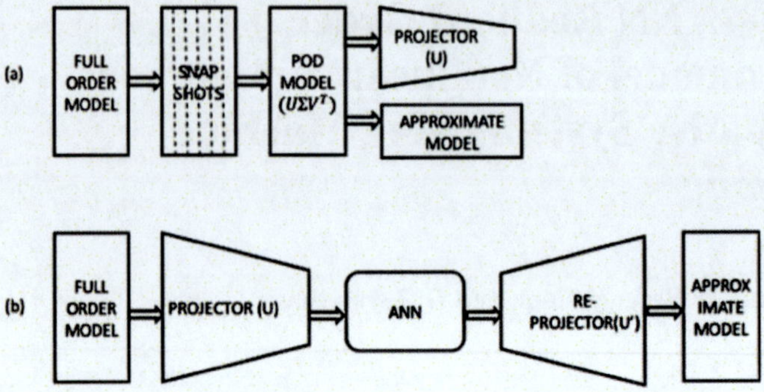


**Fig. 1** POD—ANN approximation model

POD has cost effectively and efficiently solved nonlinear problems overcoming the above challenges [8]. It has successfully reduced order of the system model describing the nonlinear behavior. This method is also known as Karhune-Loéve decomposition. POD has solved many applications in various mechanical and physical problems effectively since 1990s [12]. Recently POD has also been applied in nonlinear system macromodeling [10].

However nonlinear evaluation of the model found disadvantage due to significant CPU cost [21]. This has been overcome by developing macromodel of nonlinear system using ANN [6, 9, 11, 25]. Neural network models were found to be promising in reducing memory and processing time with good accuracy.

Neural network technology has become popular in recent years. They are applied in various fields such as dynamical system modeling, control system modeling and signal processing [5, 19, 22]. ANN learning models predict the system behavior in all engineering fields and provide promising approach to overcome modeling bottleneck. In past few decades, there has been a solid resurgence in the field of ANNs including specialists from numerous differing disciplines, examining the reasonableness of AI methods, with the end goal of full scale demonstrating of complex dynamical systems utilizing neural system [2, 3, 16–18]. The machine learning framework of the current work utilizes Bayesian Regularization [4].

By observing the advantages of POD in reducing the nonlinear model order and ANN in macromodeling of nonlinear algebraic equation, POD-ANN combination is generated to gain the benefit from both [11, 16, 17]. The novelty of the proposed work is applying POD-ANN technique to nonlinear autonomous dissipative systems. The POD-ANN reduced order macromodel of nonlinear complex systems has been developed as shown in Fig. 1.

Here, preliminary work on using ANN to assist modeling of nonlinear autonomous dynamic system has been presented. Data-driven supervised machine learning framework is utilized to learn the nonlinear behavior of the autonomous dynamic system.

The presented work is ordered as follows. Section 2 narrates the very nature of nonlinear autonomous dynamical system. The method of extracting linearity of the nonlinear behavior is explained in Sect. 3. Sections 4 and 5 discuss about the MOR of autonomous dynamical system using POD. Section 6 leads to the explanation of neural network model being used and applied. Finally numerical results obtained and conclusions of this work have been brought out in Sects. 7 and 8, respectively.

## 2 Nonlinear Autonomous Dynamical System

Nonlinear dynamic system has been described using the following state space model:

$$\begin{aligned} \dot{v}(t) &= Av(t) + f(v(t)) + Bu(t), \\ y(t) &= C^T v(t), \end{aligned} \tag{1}$$

where $v(t) \in R^N$ is a state vector, $f : R^N \to R^N$ is a function of nonlinear vector values, $A$ is $N \times N$ dimensioned system matrix, $B$ is $N \times M$ dimensioned input matrix, $C$ is $N \times K$ dimensioned output matrix, $y : R \to R^K$ is the output signal and $u : R \to R^M$ is an input signal.

When force input signal to the system is zero, the system is called autonomous system and is described by

$$\begin{aligned} \dot{v}(t) &= Av(t) + f(v(t)), \\ y(t) &= C^T v(t). \end{aligned} \tag{2}$$

In present paper, autonomous nonlinear dissipative system described by equation (2) has been explained through fluid flow model using 1-D Burgers' equation.

### 2.1 *Fluid Flow Model Using 1-D Burgers' Equation*

One of the basic, important and widely used fluid flow model is the Navier–Stokes equation given by

$$\begin{aligned} \nabla \cdot q &= 0, \\ (\rho\, \mathbf{q})_t + \nabla \cdot (\mathbf{q}\, \mathbf{q}) + \nabla p - \tau \nabla^2 \mathbf{q} &= 0, \end{aligned} \tag{3}$$

where $\mathbf{q}$ is the velocity vector, $\rho$ is the density, viscosity represented as v and pressure is $p$. The equation set (3) defines the dynamic behavior of an incompressible divergence free flow. The gravitational effects are negligible. Considering and accounting only for $x$ component of the velocity vector with zero pressure gradient, the equation is given by

$$q_t + q\, q_x - v\, q_{xx} = 0. \tag{4}$$

Hence it is shown here that the Navier–Stokes equation is qualitatively approximated as Burgers' equation and is given by Eq. (4). Specially discretized form with $x_i = i\Delta x$ and $q_i = q(x_i)$ of Eq. (4) is given by

$$\frac{\partial q_i}{\partial t} = \frac{v}{(\Delta x)^2}\left[q_{i-1} - 2q_i + q_{i+1}\right] - \frac{q_i}{2\Delta x}\left[q_{i+1} - q_{i-1}\right]. \tag{5}$$

Therefore Burgers' equation (5) can be written in the state space form (2) and solved numerically by Runge–Kutta method.

## 3 Linearizing Method

When $N$ is defined as model order, full order model evaluation has the complexity of $O(N)$. This complexity of evaluation includes both linear and nonlinear components of the model. To separate the linear component and nonlinear component, polynomial expansion with zero initial condition of the mathematical model is considered. The function $\mathcal{F}(k)$ is expressed as infinite polynomial series as:

$$\mathcal{F}(k) = \mathcal{F}(k_0) + J_0(k - k_0) + \frac{1}{2}W_0(x - x_0) \otimes (x - x_0) + \ldots, \tag{6}$$

where $W_0$ and $J_0$ represent Hessian and Jacobian for $x = x_0$ and $\mathcal{F}(.)$, Kronecker product is represented by$\otimes$. Linearly approximating $\mathcal{F}(.)$ as:

$$\begin{aligned} \dot{\mathcal{F}}(k) &= J_0(k - k_0), \\ \mathcal{G} &= C^T\mathcal{F}(k). \end{aligned} \tag{7}$$

Linearized model of the nonlinear system is presented as Eq. (7). This model is still reduced to form reduced order linear model. Due to less evaluation complexity, execution of reduced linear model is faster compared to full model. Output of reduced linear model is fed to the trained ANN model to get approximated full model output. This methodology has been applied on fluid flow model.

## 4 Model Order Reduction of Nonlinear Autonomous Dynamic System

Nonlinear autonomous dynamic system is represented by equation (1) which has $N$ states defined here as $N$th order system. The primary goal of MOR is to build low dimension approximation with $q$ order which is very much less than the full order $N$ of the system, while precisely safeguarding the input/output behavior of the original system.

In this work, POD has been applied as a MOR technique for generating lower order approximation which is typical subspace of Hilbert space. Observations from experimental model or numerical solution set from mathematical model at different time intervals are defined as snapshots. Optimal orthogonal basis set is extracted using snapshot set $\{y_1 \dots y_n\} \subset \mathbb{R}^n$.

### 4.1 Approximation Using POD

It is observed that, by exercising singular value decomposition (SVD) to snapshot results of nonlinear dynamic system sample space matrix $Y = [y^1, \dots y^{n_s}] \in \mathbb{R}^{n \times n_s}$, left singular vector set $U$ with orthogonal singular basis vector represented by $U \in \mathbb{R}^{n \times r}$, is obtained such that,

$$U \Sigma W^T = Y. \tag{8}$$

Singular values with $\sigma_1 \geqslant \sigma_2 \cdots \geqslant \sigma_r > 0$ are presented as diagonal matrix $\Sigma$. Right orthonormal singular basis vectors set is $W \in \mathbb{R}^{n_s \times r}$. Then Eq. (8) has the optimal solution $\{u_i\}_{i=1}^{k}$. Using the POD bases, snapshot matrix is approximated and 2-norm for minimum error is represented by

$$\sum_{i=1}^{n_s} \|y_i - \sum_{j=1}^{K} (Y_i^T v_j) v_j \|_2^2 = \sum_{j=k+1}^{r} \sigma_j^2. \tag{9}$$

It is also observed that choosing the snapshot ensemble is a crucial factor for constructing the POD basis.

## 5 Macro Model Using Machine Learning

The elementary structure of the basic feed-forward ANN as shown in Fig. 2a is composed of $M$ layers, each having number of predefined neuron cells. Every ANN layer is amalgamated with transfer function and each neuron with bias [5]. The process is described by the equation,

$$\mathbf{X}_m = G_2(\mathbf{W}_m \mathbf{X}_{m-1} + \mathbf{B}_m). \tag{10}$$

Here $m$th layer biasing parameter matrix is represented by $B_m$. The $m$ and $m-1$ layer linking weight matrix is represented by $W_m$ and input for the $m$th layer is given by $X_{m-1}$. Initial input layer and weight matrix are, respectively, given by

$$\mathbf{X}^0 = [\mathbf{x}_1^0, \mathbf{x}_2^0, \dots, \mathbf{x}_{n_s}^0], \;\; \mathbf{W}^1 = [\mathbf{w}_1^{1^T}, \mathbf{w}_2^{1^T}, \dots, \mathbf{w}_{n_q}^{1^T}]. \tag{11}$$

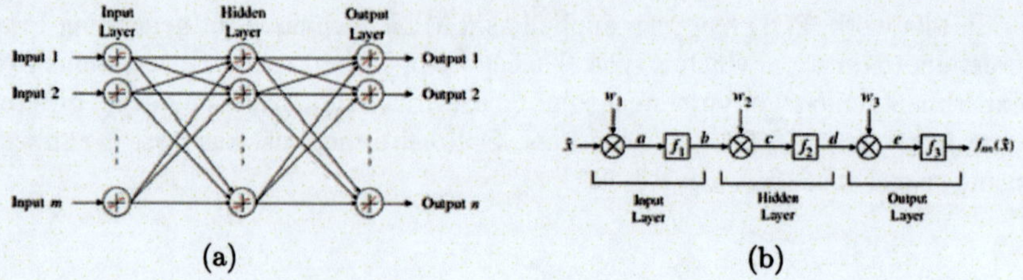


**Fig. 2** **a** Neural network abstraction. **b** Internal structure of feed-forward neural network

Every unit cell with transfer function $G_m$ at a particular layer $m$ operates on bias with input, realizing an output. Across the network, this output is further 'fed forward'. The neurons of first layer connected with input data set pass the data without computation. The elements of succeeding layers called hidden layers (see Fig. 2b) perform computation using inputs and respective biases. The output layer is the final layer which generally consists of linear activation function with a bias.

In general, any differential function can be considered as activation function of layer other than input layer [24]. Usually monotonically increasing, bounded and differentiable sigmoidal functions are used as activation function [7]. Here Tan-Sigmoid activation is used in present investigation which is represented as

$$G_m(s) = \frac{2}{1 + e^{(-2s)}} - 1. \tag{12}$$

Biases and linear weights that are used to predict the target are called best fit parameters. To regulate the set of best fit parameters, the ANN is trained, minimizing the error between the output and target. A single layer feed-forward ANN is used for training objective function. In Bayesian regularization, the network errors are minimized with penalization of sum of all squared weights [4, 14].

Gradient-based optimization is used to set the weight co-efficients through back propagation. To obtain the gradient-based optimization, the combination fo Gauss–Newton's and steepest descent-based fast search technique called Levenberg- Marquardt (LM) technique is utilized [13, 15]. The initialization of layer weights is according to the Nguyen-Widrow methodology [22].

## 6 Accuracy and Performance Index

This section focuses on accuracy of the system's response approximated by neural network reduced models. Let full order system response be $\mathbf{y}$ and POD-ANN reduced model obtained with a POD basis of dimension $\ell$ be $\mathbf{y}^{(\ell)} = \Phi_\ell \tilde{\mathbf{y}}$. Then with reference to the full order system's response $\mathbf{y}$, approximate response $\bar{\mathbf{y}}$ with relative error $L_2(\Omega)$ is given by:

$$\frac{\|\mathbf{y}-\bar{\mathbf{y}}\|_{L_2(\Omega)}}{\|\mathbf{y}\|_{L_2(\Omega)}} := \bar{e}, \tag{13}$$

where $\Omega = [0, T] \times [0, L]$, in time and space respectively. $\bar{\mathbf{y}} = \mathbf{y}^{(\ell)}$ is the approximate response for POD-ANN in (13).

Performance of the presented models is also evaluated by mean squared error (MSE). This is also referred to as the risk function corresponding to full order solution. For computing MSE, neural network approximation of reduced order model and numerical evaluation of full order model are considered. Difference between the above attributes are calculated to find the error. Mean of the square of the above error is calculated to get the MSE as given by

$$MSE = \frac{1}{N}\sum_{i=1}^{N}\left(\hat{y}_i - y_i\right)^2. \tag{14}$$

## 7 Methodology

Neural network is modeled using the "learning through experience" technique. Here, the supervised learning methodology is utilized by providing appropriate inputs and corresponding outputs. The network gets trained with proper machine learning algorithms and self-validates with untrained set of data. The summary of the logical steps brief the methodology as followed:

1. Construct the reduced order nonlinear model for the fluid flow governed by 1-D Burgers' equation using POD and linear model using Arnoldi algorithm
2. Construct the neural network with a feed-forward network model considering one input, one output and one hidden layer. Also, in the hidden layer consider two neurons in input and one neuron each in output layers
3. For different $\nu$ points get the output data for reduced order nonlinear and linear Burgers' model and store the same as output and input data sets respectively
4. The neural network is trained using 80% of data set and same is validated with 20% of data set, with supervised learning method
5. Check for accuracy of the output providing new input data. If the accuracy is not satisfactory, increase the neurons in the hidden layer. When the result is found satisfactory, deploy the neural network model for real-time execution.

# 8 Numerical Results

In this section, performance of the devised novel method, applied on fluid flow model using Burgers' equation is discussed. In the present research work, it is intended to study whether reduction in model order preserves accuracy compared with full order FD model. One-dimensional Burgers' equation with homogeneous Dirichlet boundary condition has been considered. A step function $y_0(x)$ as initial condition has been considered as

$$q_t + \left(\frac{1}{2}q^2 - \nu q_x\right)_x = f,\ (x,t) \in (0,L) \times (0,T),\ q(t,0) = q(t,L) = 0, \\ t \in (0,T),\ q(0,x) = q_0(x), \quad x \in (0,L), \tag{15}$$

and $q_0(x)$ is given by

$$q_0(x) = \begin{cases} 1, & \text{if } 0 \le x \le \frac{L}{2} \\ 0, & \text{if } \frac{L}{2} < x \le L \end{cases} \tag{16}$$

Finite difference scheme is applied on the nonlinear equation (4), with 80 grid points in space and 79 grid points in time. As shown in Fig. 3, eigenvalues in descending order with respect to the POD are obtained. Reduced order $r = 10$ is selected and POD is applied on full order FDM model to get reduced order data used as output to ANN. Reduced linear model data are used as input to ANN. Considering different $\nu$ values, 100 data sample sets are constructed which are used to train the ANN

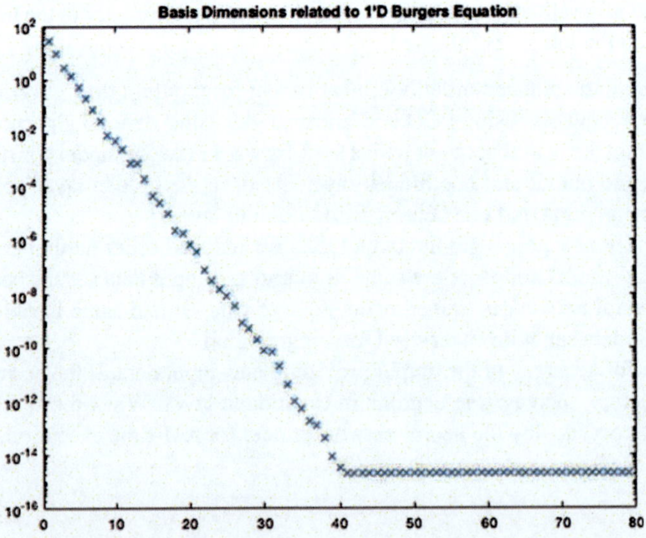


**Fig. 3** Singular values of Burger equation for $\nu = 0.1$

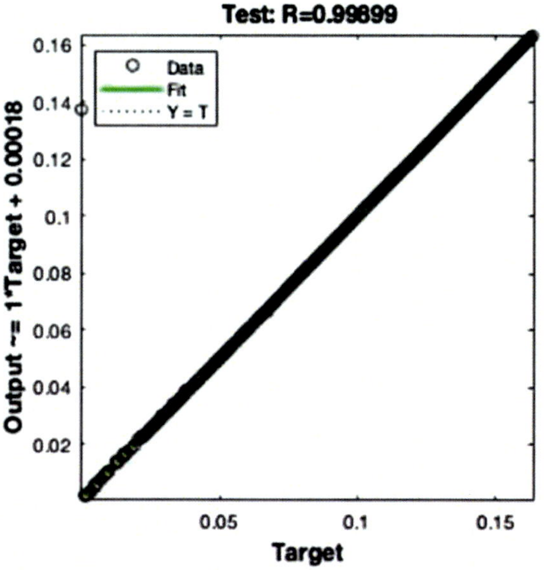


**Fig. 4** Regression plot of neural network observed for $\nu = 0.01$

model. Trained POD-ANN model is verified for $\nu = 0.01, 0.06$ and $0.1$ respectively as depicted in Fig. 5. In this case also feed-forward neural network with 10 neurons on single hidden layer is used for ANN model. As it is observed from the plots that trained network can successfully trace the nonlinear behavior from the linear model, given variation in the system parameter. Performance of the model is verified with full order FDM model using $L_2(\Omega)$ error and mean square error. It is observed from Tables 1 and 2 that the errors are in an acceptable limit and the model shows the better performance. It is also observed from Table 3, that the very less time has been taken by POD-ANN model solution compared to full order model solution. The accuracy of the model is also confirmed using the regression plot. By the regression plot, (Fig. 4), it is observed that the output test data is along the 45 degree line. This proves that the ANN approximation is much close to the original model.

## 9 Conclusion

A Neural network model using proper orthogonal decomposition and machine learning for nonlinear dissipative system has been introduced in the present work. The generic nature of the network makes it suitable for real-world problems. The capability of the machine learnt network to assimilate the temporally sparse measurements of state variables is depicted in the results shown above.

Multiple of such temporal patterns with variations in parameters for each sample could be fed to the network. The neural network is also tested by interpolating with a

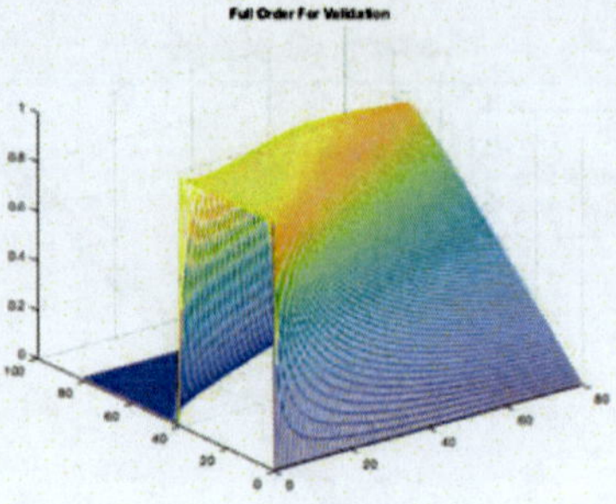


(a) Full order solution for $\nu = 0.01$

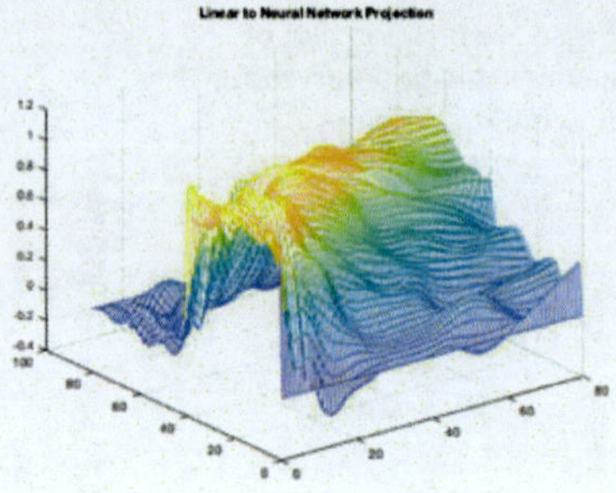


(b) ANN solution for $\nu = 0.01$

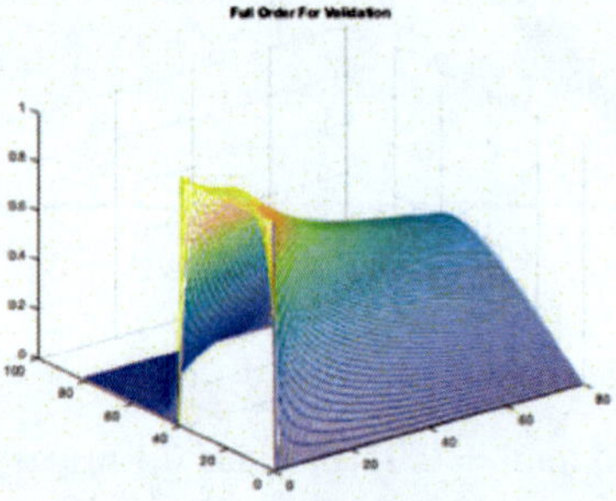


(c) Full order solution for $\nu = 0.06$

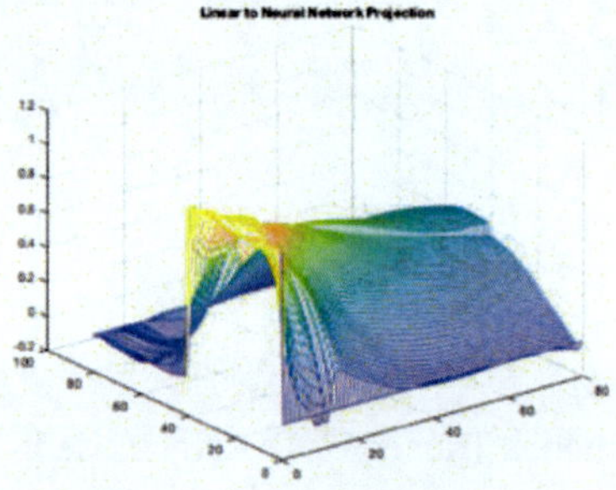


(d) ANN solution for $\nu = 0.06$

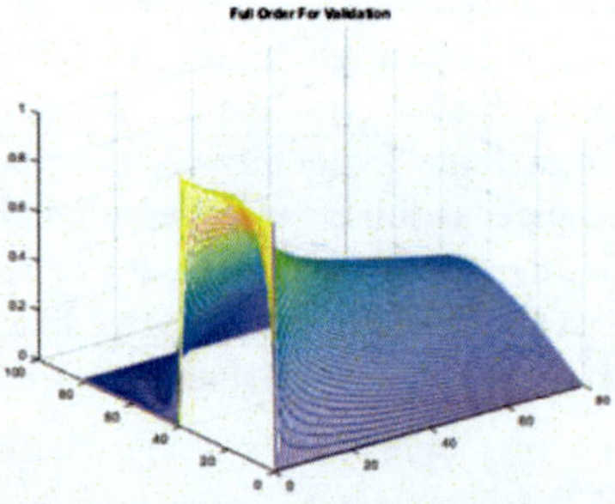


(e) Full order solution for $\nu = 0.1$

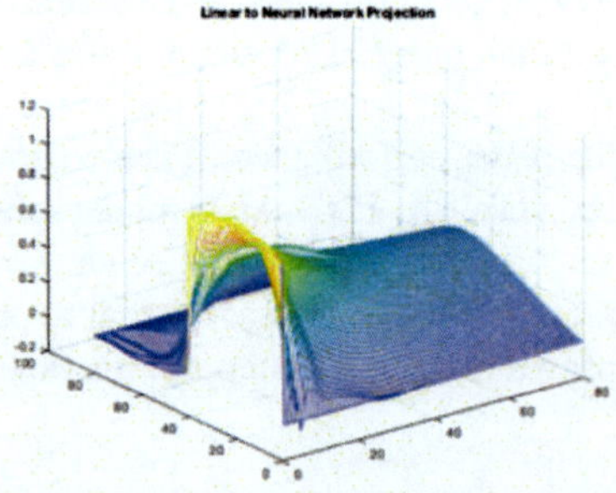


(f) ANN solution for $\nu = 0.1$

**Fig. 5** Full order model solution and neural network approximation solution of Burger equation for $\nu = 0.01, 0.06$ and $0.1$, respectively

**Table 1** Performance index—Mean Square Error (MSE)

| $\nu$ | Order ($q$) | MSE (Linear model) | MSE reduced nonlinear model | $\nu$ | Order ($q$) | MSE (Linear model) | MSE reduced nonlinear model |
|---|---|---|---|---|---|---|---|
| 0.005 | 1 | 0.2390 | 0.0871 | 0.1 | 1 | 0.0068 | 0.0148 |
| | 2 | 0.2390 | 0.0293 | | 2 | 0.0068 | 0.0015 |
| | 3 | 0.2390 | 0.0138 | | 3 | 0.0068 | 0.0003 |
| | 4 | 0.2390 | 0.0080 | | 4 | 0.0068 | 0.0000 |
| | 5 | 0.2390 | 0.0051 | | 5 | 0.0068 | 0.0000 |
| 0.01 | 1 | 0.1857 | 0.0742 | 0.5 | 1 | 0.0001 | 0.0031 |
| | 2 | 0.1857 | 0.0213 | | 2 | 0.0001 | 0.0004 |
| | 3 | 0.1857 | 0.0081 | | 3 | 0.0001 | 0.0000 |
| | 4 | 0.1857 | 0.0036 | | 4 | 0.0001 | 0.0000 |
| | 5 | 0.1857 | 0.0018 | | 5 | 0.0001 | 0.0000 |
| 0.05 | 1 | 0.0335 | 0.0292 | 1 | 1 | 0.0000 | 0.0019 |
| | 2 | 0.0335 | 0.0038 | | 2 | 0.0000 | 0.0002 |
| | 3 | 0.0335 | 0.0007 | | 3 | 0.0000 | 0.0000 |
| | 4 | 0.0335 | 0.0002 | | 4 | 0.0000 | 0.0000 |
| | 5 | 0.0335 | 0.0000 | | 5 | 0.0000 | 0.0000 |

**Table 2** Performance index—Relative error $L_2(\Omega)$

| Parametric Value ($\nu$) | MSE (ANN) | Relative error $L_2(\Omega)$ |
|---|---|---|
| 0.01 | 0.0403 | 0.4223 |
| 0.06 | 0.0010 | 0.3337 |
| 0.1 | 0.0079 | 0.3110 |

**Table 3** Execution time* (In Seconds)

| Order ($q$) | ET (NL model) | ET Lin model | ET ROM |
|---|---|---|---|
| 1 | 0.05012 | 0.04784 | 0.04862 |
| 2 | 0.05012 | 0.04784 | 0.04642 |
| 3 | 0.05012 | 0.04784 | 0.04162 |
| 4 | 0.05012 | 0.04784 | 0.03612 |
| 5 | 0.05012 | 0.04784 | 0.03312 |

*Note* For recording time, Intel Core i5, 2.5 GHz, DDR3 RAM 1600 MHz, 16 GB, MacBook Pro is used state variable which is not accounted during training. The results correlated well with the untrained target. The former ability plays a major role in practical scenarios. The Neural network approximation is nearly accurate and faster than the generic methods. The network is a simple feedforward network with no complex mathematical calculations. This paves way for a novel methodology to model generation through neural network using machine learning.

## References

1. Antoulas AC (2004) Approximation of large-scale dynamical systems: an overview. IFAC Proc Vol 37(11):19–28; 10th IFAC/IFORS/IMACS/IFIP symposium on large scale systems 2004: theory and applications. Osaka, Japan
2. Bailer-Jones CA, MacKay DJ, Withers PJ (1998) A recurrent neural network for modelling dynamical systems. Netw Comput Neural Syst 9(4):531–547. PMID: 10221578
3. Chen S, Billings SA (1992) Neural networks for nonlinear dynamic system modelling and identification. Int J Control 56(2):319–346
4. Foresee FD, Hagan MT (1997) Gauss-newton approximation to bayesian learning. In: Proceedings of international conference on neural networks (ICNN'97), vol 3, pp 1930–1935
5. Demuth HB, Beale MH, De Jess O, Hagan MT (2014) Neural network design, 2nd edn. Martin Hagan, USA
6. Fang Y, Yagoub MC, Wang F, Zhang QJ (2000) A new macromodeling approach for nonlinear microwave circuits based on recurrent neural networks. IEEE Trans Microwave Theory Tech 48(12):2335–2344
7. Goodfellow I, Bengio Y, Courville A (2016) Deep learning. MIT Press
8. Henneron Thomas, Clenet Stephane (2014) Model order reduction of non-linear magnetostatic problems based on POD and DEI methods. IEEE Trans Mag 50(2):33–36
9. Kabir H, Yu M, Zhang QJ (2010) Recent advances of neural network-based EM-CAD. Int J RF Microwave Comput Aided Eng 20(5):502–511
10. Kanaan M, Khazaka R (2011) Model order reduction for nonlinear macromodeling of RF circuits. In: 2011 IEEE 9th International on new circuits and systems conference (NEWCAS). IEEE, pp 217–220
11. Kanaan M, Khazaka R (2017) Nonlinear time-domain macromodeling using proper orthogonal decomposition and feedforward neural networks. In: 30th Canadian conference on electrical and computer engineering (CCECE). IEEE, pp 1–4
12. Kerschen G, Golinval JC, Vakakis AF, Bergman LA (2005) The method of proper orthogonal decomposition for dynamical characterization and order reduction of mechanical systems: an overview. Nonlinear Dyn 41(1–3):147–169
13. Levenberg K (1944) A method for the solution of certain nonlinear problems in least squares. Q Appl Math 2(2):164–168
14. MacKay DJC (1991) Bayesian interpolation. Neural Comput 4:415–447
15. Marquardt D (1963) An algorithm for least-squares estimation of nonlinear parameters. J Soc Ind Appl Math 11(2):431–441
16. Nagaraj S, Seshachalam D (2018) Neural network macromodeling of nonlinear electrical circuit for variable frequency inputs using karhunen-loeve decomposition. In: Third IEEE international conference on computational systems and information technology for sustainable solutions (CSITSS-2018). Bengaluru, India
17. Nagaraj S, Seshachalam D, Hucharaddi S (2018) Model order reduction of nonlinear circuit using proper orthogonal decomposition and nonlinear autoregressive with exogenous input (narx) neural network. In: 2018 16th ACM/IEEE international conference on formal methods and models for system design (MEMOCODE), pp 1–4

18. Nguyen T, Schutt-Aine JE (2018) A pseudo-supervised machine learning approach to broadband lti macro-modeling. In: 2018 IEEE international symposium on electromagnetic compatibility and 2018 IEEE Asia-Pacific symposium on electromagnetic compatibility (EMC/APEMC), pp 1018–1021
19. Norgaard M, Ravn O, Poulsen NK, Hansen LK (2000) Neural networks for modelling and control of dynamic systems: a practitioner's handbook, 1st edn. Springer, Berlin
20. Schilders WH, Van der Vorst HA, Rommes J (2008) Model order reduction: theory, research aspects and applications. Mathematics in industry. Springer, Germany
21. Striebel M, Rommes J (2018) Model order reduction of nonlinear systems: status, open issues, and applications
22. Widrow Bernard, Rumelhart David E, Lehr Michael A (1994) Neural networks: applications in industry, business and science. Commun. ACM 37:93–105
23. Wilson TG (1988) Life after the schematic: the impact of circuit operation on the physical realization of electronic power supplies. Proc IEEE 76(4):325–334
24. Zhang G, Patuwo BE, Hu MY (1998) Forecasting with artificial neural networks:: The state of the art. Int J Forecasting 14(1):35–62
25. Zhang QJ, Gupta KC, Devabhaktuni VK (2003) Artificial neural networks for rf and microwave design-from theory to practice. IEEE Trans Microwave Theory Tech 51(4):1339–1350

# Stability Analysis of Salt Fingers for Different Non-uniform Temperature Profiles in a Micropolar Liquid

**Nisha Mary Daniel, Vennela Lingamneni, Tessy Tom, and N. Arun Kumar**

## 1 Introduction

When heat is supplied to the fluid from above, the temperature makes the fluid molecules lighter at the upper layer, and hence it stabilizes the system. But, when we supply solute from above, the concentration of the solute molecules slowly enters the system and settles at the bottom, and due to this movement, the system becomes unstable and hence the concentration of solute destabilizes the system. In this case, we can observe that temperature stabilizes the system and the concentration destabilizes the system. The double diffusive instability induced by this case in literature is known as salt fingers. Double diffusive convection is a fluid dynamics phenomenon that occurs due to the difference in diffusivity rates of two different density gradients in the fluid. The difference in the thermal and salt diffusivity rates gives rise to instabilities called "salt fingers". Stern [1], Jevons [2], Ekman [3], and Stommel et al. [4] were among the first to explain the physical mechanism of double diffusive convection. Instability at the interface between a layer of denser fluid and an underlying layer of temperature stratified water was observed. Later researches verified that salt fingers develop at the thin interface where two layers of the fluid with constant concentrations of its two distinct density gradients, with different molecular diffusivities, meet. In nature, double diffusive convection is more often observed in large water bodies like seas and oceans. Other occurrences in nature include convection in molten-rock

N. Mary Daniel (✉)
Research Scholar, Department of Mathematics, CHRIST(Deemed to be University), Bengaluru, India
e-mail: nisha.daniel@res.christuniversity.in

V. Lingamneni
Department of Mathematics, Sir MV PU College, Karnataka, India

T. Tom
Department of Commerce and Management, CHRIST(Deemed to be University), Lavasa, India

N. Arun Kumar
School of Engineering and Technology, CHRIST(Deemed to be University), Bengaluru, India

 
S. Srinivas et al. (eds.), *Recent Advances in Applied Mathematics and Applications to the Dynamics of Fluid Flows*, Lecture Notes in Mechanical Engineering,
https://doi.org/10.1007/978-981-19-1929-9_32

chambers and sea-wind formations [4]. The study of this convection has found its way into many researches due to its wide range of applications in nature and industry.

First introduced by Eringen [5], micropolar fluid theory establishes a class of fluids which demonstrates microrotational effects and can sustain body couples. In terms of its physical aspect, micropolar fluids comprise arbitrarily oriented rigid particles which can deform and rotate independently of the motion of the fluid, where the deformation of the particles is not considered. Micropolar fluids can model anisotropic fluids, liquid crystals with inflexible molecules, magnetic fluids, dusty clouds, muddy fluids, and a few biological fluids [5]. The theory of micropolar fluids is discussed in detail in the books of Eringen [5] and Lukasazewicz [6].
The goal of this paper is to investigate the stability of the system on the onset of salt finger convection under the influence of different non-uniform temperature profiles and to analyze the effect of certain micropolar parameters.

## 2 Mathematical Formulation

A layer of Boussinesquian micropolar liquid is considered between two horizontal plates of infinite length. These plates are kept at a distance "d" as shown in Fig. 1. Let $T_0$, $C_0$, and $T$, $C$ be the temperature and solute concentration of the fluid at the lower and upper plates, respectively. Also, $\Delta T$ and $\Delta C$ be the temperature and solute concentration deviation of the fluid between the lower and upper plates. Suitable transport equations for both temperature and solute concentration are chosen considering effective heat capacity ratio and effective thermal diffusivity. The diagrammatic representation is examined under a Cartesian coordinate system where the origin and the x-axis coincide with the lower boundary and the z-axis is vertically upward [7]. Given below are the governing equations for the Boussinesquian micropolar liquid under double diffusive convection:

**Fig. 1** Schematic diagram

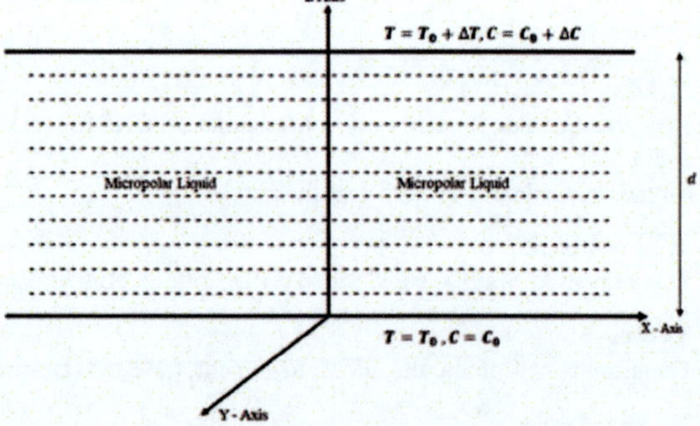

$$\nabla.\mathbf{q} = 0, \tag{1}$$

$$\rho_0\left[\frac{\partial \mathbf{q}}{\partial t} + (\mathbf{q}.\nabla)\mathbf{q}\right] = -\nabla p - \rho g\hat{k} + (2\xi + \eta)\nabla^2\mathbf{q} + \xi\nabla \times \boldsymbol{\omega}, \tag{2}$$

$$\rho_0 I\left[\frac{\partial \boldsymbol{\omega}}{\partial t} + (\mathbf{q}.\nabla)\boldsymbol{\omega}\right] = (\lambda + \eta)\nabla(\nabla.\boldsymbol{\omega}) + \eta\nabla^2\boldsymbol{\omega} + \xi(\nabla \times \mathbf{q} - 2\boldsymbol{\omega}), \tag{3}$$

$$\left[\frac{\partial T}{\partial t} + (\mathbf{q}.\nabla)T\right] = \chi\nabla^2 T + \frac{\beta}{\rho_0 C_v}(\nabla \times \boldsymbol{\omega}).\nabla T, \tag{4}$$

$$\rho = \rho_0[1 - \alpha_t(T - T_0) + \alpha_s(C - C_0)], \tag{5}$$

$$\left[\frac{\partial C}{\partial t} + (\mathbf{q}.\nabla)C\right] = \chi_s\nabla^2 C. \tag{6}$$

Here, $q$, $\omega$, and $p$ are the velocity, angular velocity, and pressure, respectively. $\rho$ is the density of the liquid at temperature $T$ and $\rho_0$ is the density at temperature $T_0$. $g$ and $I$ are the acceleration due to gravity and moment of inertia, respectively. $\xi$ is the coefficient of coupling viscosity, $\lambda$, $\eta$, and $\lambda'$, $\eta'$ are the coefficients of bulk and shear spin viscosity, $\chi$ and $\chi_s$ are the thermal and solute conductivity, $\beta$ is the coefficient of micropolar heat conduction, $\alpha_t$ and $\alpha_s$ are the coefficients of thermal and solutal expansion, respectively, $\sigma$ is the electrical conductivity, and $C_v$ is the specific heat at a constant volume.

## 3 Basic State

The basic state of the liquid in its quiescent condition is depicted by

$$\mathbf{q} = \mathbf{q}_b = (0, 0, 0), \boldsymbol{\omega} = \boldsymbol{\omega}_b = (0, 0, 0), p = p_b(z) = \rho = \rho_b(z), \tag{7}$$

$$\frac{dT_b}{dz} = \frac{\nabla T}{d} f(z), \frac{dC_b}{dz} = \frac{\nabla C}{d} g(z), \tag{8}$$

where the subscript "$b$" denotes the basic state. The basic temperature gradient $f(z)$ and concentration gradient $g(z)$ are non-dimensional and non-negative and satisfy the condition $\int_{d=0}^{d=1} f(z)\,dz$ and $\int_{d=0}^{d=1} g(z)\,dz$, respectively. The concerned basic state variable satisfies Eqs. (1) and (3) equivalently. This paper considers four different temperature profiles to analyze the onset of convection which is given below in Table 1.

## 4 Linear Stability Analysis

To examine the instability, an infinitesimal thermal perturbation is now introduced to the quiescent basic state of the liquid. We now have

**Table 1** Non-uniform basic temperature profiles

| S.No. | Profile | $f(z)$ |
|---|---|---|
| CM1 | Linear | $f(z) = 1$ |
| CM2 | Piecewise linear | $f(z) = \frac{1}{\epsilon} \quad \forall\, 0 \le z \le \epsilon$<br>$0 \quad \forall\, \epsilon \le z \le 1$ |
| CM3 | Inverted parabolic | $f(z) = 2(1 - z)$ |
| CM4 | Parabolic | $f(z) = 2z$ |

$$\mathbf{q} = \mathbf{q}_b + \mathbf{q}', \boldsymbol{\omega} = \boldsymbol{\omega}_b + \boldsymbol{\omega}', \rho = \rho_b + \rho', p = p_b + p', \tag{9}$$

$$T = T_b + T', C = C_b + C'. \tag{10}$$

The subscript "$b$" indicates the basic state of the quantity and the primes denote the infinitesimal perturbations. By substituting Eq. (9) and (10) in Eqs. (1)–(6), we get the following linearized governing equations with respect to the infinitesimal perturbations:

$$\nabla.\mathbf{q}' = 0, \tag{11}$$

$$\rho_0\left[\frac{\partial \mathbf{q}'}{\partial t} + (\mathbf{q}'.\nabla)\mathbf{q}'\right] = -\nabla p' - \rho' g\hat{k} + (2\xi + \eta)\nabla^2\mathbf{q}' + \xi\nabla\times\boldsymbol{\omega}', \tag{12}$$

$$\rho_0 I\left[\frac{\partial \boldsymbol{\omega}'}{\partial t} + (\mathbf{q}'.\nabla)\boldsymbol{\omega}'\right] = (\lambda' + \eta')\nabla(\nabla.\boldsymbol{\omega}') + \eta'\nabla^2\boldsymbol{\omega}' + \xi(\nabla\times\mathbf{q}' - 2\boldsymbol{\omega}'), \tag{13}$$

$$\left[\frac{\partial T'}{\partial t} + w\frac{\Delta T}{d}f(z)\right] = \chi\nabla^2 T' + \frac{\beta}{\rho_0 C_v}\left[(\nabla\times\boldsymbol{\omega}').\left(\frac{\Delta T}{d}f(z)\right)\hat{k}\right], \tag{14}$$

$$\rho' = -\rho_0\alpha_t T' + \rho_0\alpha_s C', \tag{15}$$

$$\left[\frac{\partial C'}{\partial t} + (w\frac{\Delta C}{d}g(z)\right] = \chi_s\nabla^2 C'. \tag{16}$$

The perturbation Eqs. (11)–(16) are non-dimensionalized using the following definitions:

$$(x^*, y^*, z*) = \left(\frac{x}{d}, \frac{y}{d}, \frac{z}{d}\right), \nabla^{*2} = d^2\nabla^2, \mathbf{q}^* = \frac{\mathbf{q}}{\frac{k}{d}}, t^* = \frac{t}{\frac{d^2}{\chi}}, \tag{17}$$

$$p^* = \frac{p}{p_0}, T^* = \frac{T}{\Delta T}, C^* = \frac{C}{\Delta C}, W^* = \frac{W}{\frac{\chi}{d}}. \tag{18}$$

Using Eq. (15) in Eq. (12), curl is applied twice on the following equation. Curl is applied once on Eq. (13) as well and the derived two equations are then non-dimensionalized along with Eqs. (14) and (16). After ignoring the asterisks, the following equations are obtained:

$$\frac{1}{Pr}\frac{\partial}{\partial x}(\nabla^2 W) = R\nabla_1^2 T - R_s\nabla_1^2 C + (1+N_1)\nabla^4 W + N_1\nabla^2\Omega_z, \quad (19)$$

$$\frac{N_2}{Pr}\frac{\partial \Omega_z}{\partial t} = N_3\nabla^2\Omega_z - N_1\nabla_2 W - 2N_1\Omega_z, \quad (20)$$

$$\frac{\partial T}{\partial t} = \nabla^2 T + (-W + N_5\Omega_z)f(z), \quad (21)$$

$$\frac{\partial C}{\partial t} = \tau\nabla^2 C - Wg(z), \quad (22)$$

where Rayleigh Number $\left[R = \frac{\rho_0\alpha g\Delta T d^3}{(\eta+\zeta)\chi}\right]$, solutal Rayleigh Number $\left[R_s = \frac{\rho_0\alpha_s g\Delta C d^3}{(\eta+\zeta)\chi}\right]$, coupling Parameter $\left[N_1 = \frac{\zeta}{\zeta+\eta}\right]$, couple stress parameter $\left[N_3 = \frac{\eta'}{(\zeta+\eta)d^2}\right]$, micropolar heat conduction parameter $\left[N_5 = \frac{\beta}{\rho_0 C_v d^2}\right]$, and ratio of diffusivity $\left[\Gamma = \frac{\chi_s}{\chi}\right]$ are the non-dimensional parameters used.

A normal mode solution for the stationary convection is obtained as the infinitesimal perturbations $W, \Omega_z, T$ and $C$ are assumed to be periodic waves [8]. The solution is represented as

$$[W, \Omega_z, T, C] = [W(z).e^{i(lx+my)}, G(z).e^{i(lx+my)}, T(z).e^{i(lx+my)}, C(z).e^{i(lx+my)}]. \quad (23)$$

Here, the wave number a has horizontal components $l$ and $m$. Equation (23) is substituted in Eqs. (19)–(22) to obtain

$$-Ra_2T + R_s a^2 C + (1+N_1)(D^2-a^2)^2 W + N_1(D^2-a^2)^2 G = 0, \quad (24)$$

$$\left[2N_1 - N_3(D^2-a^2)\right]G + N_1(D^2-a^2)^2 W = 0, \quad (25)$$

$$(D^2-a^2)T + (-W + N_5 G)f(z) = 0, \quad (26)$$

$$\left[\tau(D^2-a^2)\right]C - Wg(z) = 0, \quad (27)$$

where $a^2 = l^2 + m^2$.

Applying the Galerkin procedure to Eqs. (24)–(27), we obtain general results on the eigenvalue for different temperature gradients under the given boundary conditions by considering trial functions for velocity $W(z,t)$, microrotation $G(z,t)$, and temperature $T(z,t)$, concentration perturbations $C(z,t)$.

$$W(z,t) = \sum A_i(t)W_i(z), \quad G(z,t) = \sum E_i(t)G_i(z),$$
$$T(z,t) = \sum B_i(t)T_i(z), \quad C(z,t) = \sum F_i(t)C_i(z),$$

where $A_i(t)$, $E_i(t)$, $B_i(t)$, and $F_i(t)$ are constant functions and $W_i(z)$, $G_i(z)$, $T_i(z)$, and $C_i(z)$ are polynomials in z, that generally satisfy the given boundary conditions.

Now taking $i = j = 1$, Eqs. (24), (25), (26), and (27) are multiplied by $W, G, T$, and $C$, respectively. The resulting equation is integrated by parts with respect to

$z$ from 0 to 1. The equation for the Rayleigh number is generated by substituting $W = AW_1; G = EG_1; T = BT_1; C = FC_1$, where $A, B, E$, and $F$ are constants and $W_1, G_1, T_1$, and $C_1$ are the trial functions:

$$R = \left[\frac{X_1 X_3 L_7}{a^2 L_1(\langle W_1 T_1 f(z)\rangle X_1 - N_1 N_5 \langle T_1 G_1 f(z)\rangle L_5)}\right], \tag{28}$$

where

$$X_1 = N_3 L_2 - 2N_1 L_3, \quad X_3 = \frac{R_s a^2 L_8 \langle W_1 C_1\rangle}{-\tau L_{10}} + L_{14}$$

$$L_1 = \langle W_1 T_1\rangle, \quad L_2 = \langle G_1(D^2 - a^2)G_1\rangle, \quad L_3 = \langle G_1^2\rangle, \quad L_4 = \langle T_1 G_1\rangle,$$

$$L_5 = \langle G_1(D^2 - a^2)W_1\rangle, \quad L_6 = \langle T_1^2\rangle, \quad L_7 = \langle T_1(D^2 - a^2)T_1\rangle,$$

$$L_8 = \langle W_1 C_1\rangle, \quad L_9 = \langle C_1^2\rangle, \quad L_{10} = \langle C_1(D^2 - a^2)C_1\rangle,$$

$$L_{11} = \langle W_1(D^2 - a^2)^2 W_1\rangle, \quad L_{12} = \langle W_1(D^2 - a^2)W_1\rangle, \quad L_{13} = \langle W_1(D^2 - a^2)G_1\rangle,$$

$$L_{14} = (1 + N_1)L_{11} + \frac{N_1^2 L_5 L_{13}}{X_1}, \quad L_{15} = \langle W_1 T_1 f(z)\rangle, \quad L_{16} = \langle T_1 G_1 f(z)\rangle.$$

Here, integral with respect to $z$ under the limits $z = 0$ and $z = 1$ (for calculation purpose, we have taken the distance between the plates as $d = 1$) is denoted by $\langle \ldots \rangle$. The value of the concentration gradient $g(z)$ is taken as 1 in Eq. (27) to examine the effect of non-uniform temperature profiles.

The combinations of boundary conditions considered in this problem are Free–Free Isothermal-Permeable No-Spin condition, Rigid–Free Isothermal-Permeable No-Spin condition, and Rigid–Rigid Isothermal-Permeable No-Spin condition. The critical Rayleigh number is subjected to the three boundary conditions and the value of the trial functions that satisfy them are taken as $W_1 = 3z - 2z^2 + z^4, 4z - 6z^3 + 3z^4$ and $5z^4 - 6z^3 - z^2$, respectively, and $G_1 = T_1 = C_1 = 2z(z - 1)$.

# 5 Results and Discussion

## 5.1 *Effects of Non-uniform Temperature Gradient Profiles*

The effects of one uniform and three non-uniform temperature gradient profiles on the onset of salt finger convection in a micropolar liquid are examined in this paper. The values of these profiles are discussed in Table 1. To analyze the effect of the temperature gradient, the value of the concentration gradient $g(z)$ is taken as 1 and the following observations are made:

For Free–Free and Rigid–Free boundary condition,
$$R_{CM3} < R_{CM1} < R_{CM4} < R_{CM2}.$$

For Rigid–Rigid boundary condition, $R_{CM4} < R_{CM1} < R_{CM2} < R_{CM3}$.

Figures 2 and 3 represents the graphs of critical Rayleigh number $R_C$ plotted against micropolar liquid parameters $N_1$, $N_3$, and $N_5$ taken one at a time while keeping the other two fixed for various solutal Rayleigh number $R_S$ and ratio of diffusivity under free–free and rigid–rigid boundary conditions, respectively. The graphs are obtained for the non-uniform temperature profiles mentioned in Table 1.

Under free–free boundary conditions, the inverted parabolic function appears to be the temperature profile that destabilizes the system considerably compared to the other profiles and the piecewise function profile comparatively stabilizes the system. (The same effect can be observed for rigid–free boundary conditions), whereas, for rigid–rigid boundary conditions, the parabolic function is the temperature profile that destabilizes the system the most, and the inverted parabolic function profile comparatively stabilizes the system.

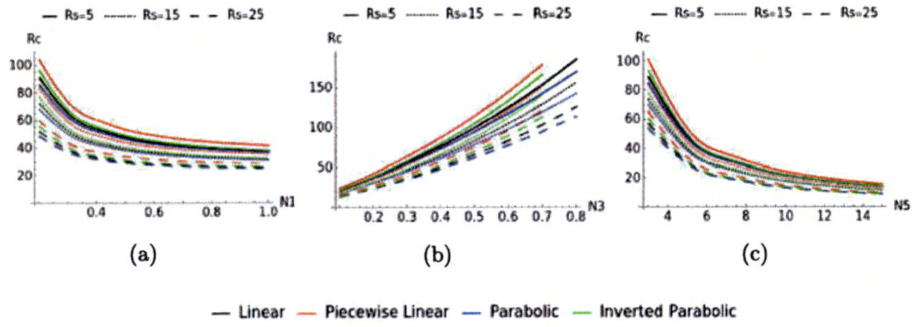


**Fig. 2** **a** $R_c$ is plotted against $N_1$ by fixing $N_3 = 0.1$, $N_5 = 5$, and $\tau = 0.2$, **b** $R_c$ is plotted against $N_3$ by fixing $N_1 = 0.5$, $N_5 = 10$, and $\tau = 0.2$, **c** $R_c$ is plotted against $N_5$ by fixing $N_1 = 0.5$, $N_3 = 0.1$, and $\tau = 0.2$ for free–free boundary condition

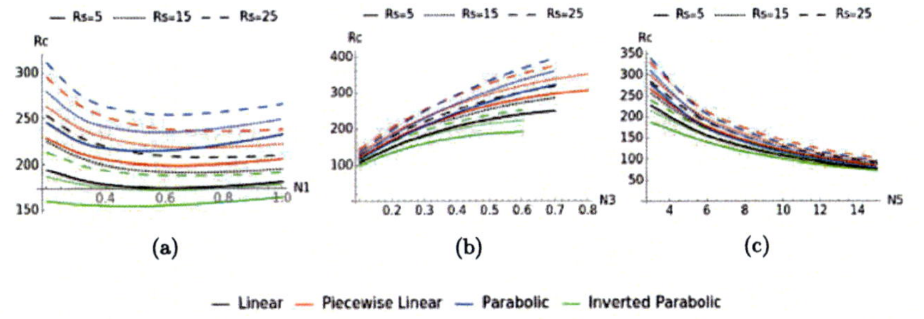


**Fig. 3** **a** $R_c$ is plotted against $N_1$ by fixing $N_3 = 0.1$, $N_5 = 5$, and $\tau = 0.2$, **b** $R_c$ is plotted against $N_3$ by fixing $N_1 = 0.5$, $N_5 = 10$, and $\tau = 0.2$, **c** $R_c$ is plotted against $N_5$ by fixing $N_1 = 0.5$, $N_3 = 0.1$, and $\tau = 0.2$ for rigid–rigid boundary condition

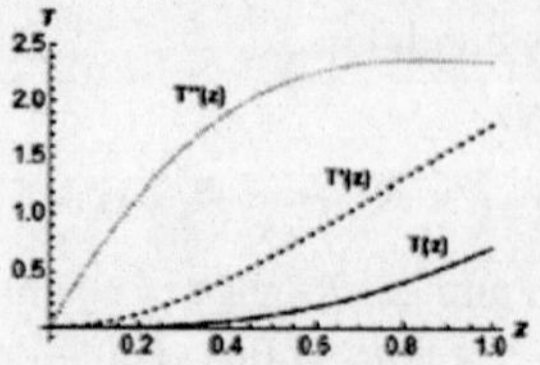


(a) Free-Free boundary condition

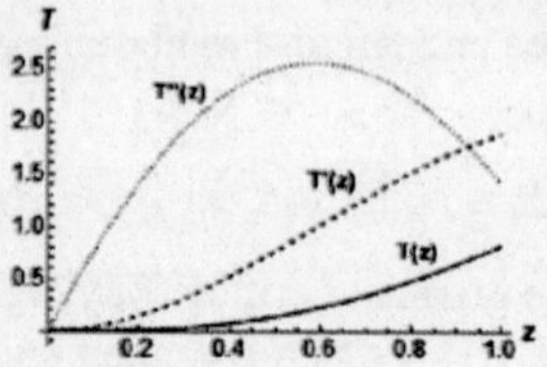


(b) Rigid-Free boundary condition

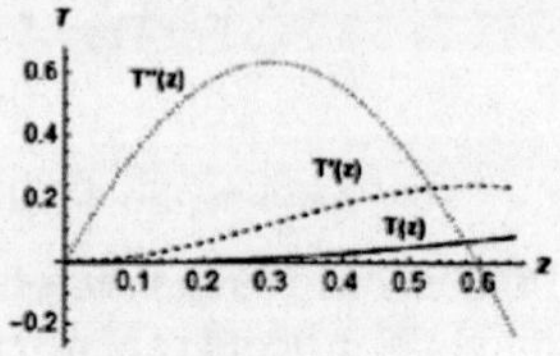


(c) Rigid-Rigid boundary condition

**Fig. 4** Phase of temperature flow

## 5.2 *Phase of Temperature Flow*

The graphs for the phase of temperature flow are plotted for free–free, rigid–free, and rigid–rigid boundary conditions (Fig. 4a, b, c). They are obtained by solving the conservation of energy equation (temperature equation) numerically for different cases as mentioned. The $x$-axis and the $y$-axis represent the distance between the parallel plates and the amount of the temperature supplied from above into the system, respectively. These plots are basically the graphical representation of the solution of $T(z)$ and its first and second derivatives. From Fig. 4a, b, c, the following observations can be made: (i) Temperature is more at $z = 1$ (upper plate) and decreases gradually as we move towards $z = 0$ (lower plate ). This is because the temperature is supplied from above and the diffusion rate will be higher or faster near the upper plate. Later, it slows down as it comes in contact with the colder molecules in the system (it is also because heat diffuses much faster than solute in the beginning). (ii) The phase of temperature flow of $T(z)$ is lesser than its derivatives. (iii) For rigid–rigid boundary conditions, the temperature flow effect will be seen only towards the center of the system and near the upper plate. This is mainly because the plates are rigid, hence more amount of temperature is required for the onset of convection.

# 6 Conclusion

- The inverted parabolic function appears to be the temperature profile that destabilizes the system in free–free and rigid–free boundary conditions.
- The piecewise function profile comparatively stabilizes the system in free–free and rigid–free boundary conditions.
- The parabolic function is the temperature profile that destabilizes the system the most in rigid–rigid boundary conditions.
- The inverted parabolic function profile comparatively stabilizes the system in rigid–rigid boundary conditions.
- The Micropolar parameters $N_1$ and $N_5$ destabilize the system with respect to $R_c$ in all the boundary conditions.

- The Micropolar parameters $N_3$ stabilize the system with respect to $R_c$ in all the boundary conditions.
- The phase of temperature flow will explain the variation of the temperature between the parallel plate boundary (i.e., at $z = 0$ and at $z = d$). Finally, it is observed that the temperature distribution is more at the upper plate boundary and this is due to the nature of salt fingers (as shown in the schematic diagram).

## References


1. Stern ME (1960) The 'Salt-Fountain' and thermohaline convection. Tellus 12(2): 172–175
2. Jevons WS (1858) On clouds; their various forms, and protruding causes. Syd Mag Sci Art 1:163–176
3. Ekman VW (1906) "On dead water", Scientific results of the Norwegian North Polar. Expedition 5:1–152
4. Stommel H, Arons AB, Blanchard D (1956) An oceanographical curiosity: the perpetual salt fountain. Deep Sea Res 3(2):152–153. https://doi.org/10.1016/0146-6313(56)90095-8
5. Eringen AC (1965) Theory of micropolar fluids
6. Lukaszewicz G (1999) Micropolar fluids-theory and applications. Birkhauser, Boston
7. Pranesh S, Narayanappa A (2012) Effect of non-uniform basic concentration gradient on the onset of double-diffusive convection in micropolar fluid. Appl Math 3(5):417–424. https://doi.org/10.4236/am.2012.35064
8. Chandrasekhar S (1981) Hydrodynamic and hydromagnetic stability. Dover Publications, New York

# Hydrodynamic Analysis of Submarine Profiles for Drag Reduction

**E. Lalith Sairam and Amit Kumar**

## 1 Introduction

In the fourth era of the industrial revolution, man and machine are becoming increasingly linked with automation and ultimate precision and accuracy in manufacturing. As a result, more conventional and ancient techniques of manufacturing have suffered a precipitous fall, particularly in the twenty-first century. This is not to say that more archaic systems have been extinct, but they are also constantly changing. In this modern era of technology and engineering, simulations have become a great tool for estimating, experimenting and calculating results virtually for large-scale projects, it has been used to provide an accurate and clear understanding of actual real-time procedure [1, 2].

Hydrofoils are specially CAD engineered structures that generate lift force as water passes over them. The study of hydrofoils in the field of mechanical engineering made it better propagate through water and also allowed us to explore vast areas of ocean, generally lift is created when an object passes through water. At the same time, the passage of water over the body creates resistance called drag [3]. To calculate drag, certain parameters like lift and drag coefficients are used. In this project, the water flow is stimulated with different constraints and submarine profiles to compute the lift and drag coefficients generated with various angles of attack, e.g. like (0°, 5°, 10° and 15°).

The objective of this investigation was to understand the parameters for selecting the submarine profile and to analyse its performance in the preferred conditions using

E. L. Sairam (✉) · A. Kumar
School of Mechanical Engineering, VIT-Andhra Pradesh, Amaravati 522237, India
e-mail: elalith.sairam@vitap.ac.in



S. Srinivas et al. (eds.), *Recent Advances in Applied Mathematics and Applications to the Dynamics of Fluid Flows*, Lecture Notes in Mechanical Engineering,
https://doi.org/10.1007/978-981-19-1929-9_33

simulations. The object's displayed surface determines how much skin friction drag there is. When a solid moves through a fluid, friction was formed, which inhibits the solid from moving ahead. This is more applicable to turbulent flows because it increases the quantum of contact between the solid and the fluid. By reducing the exposed surface area and delaying the transition from laminar to turbulence, skin friction drag was decreased. Further skin friction drag can be caused by a boundary layer existing on the surface. Form drag was defined as the drag caused by the separation of the boundary layer from the wake. It is more reliant on the solid's arrangement than on the surface area shown. A smooth surface equals in a smaller wake zone, which results in less form drag [4, 5].

## 2 Methods

In the present investigation, first, a hydrofoil geometrical profile was constructed using CAD Software and the profile was simulated under steady-state and transient conditions. By simulating the hydrofoil, it was used to understand how the drag, and lift coefficients were generally affected by the different Angle of attack (AoA). When a profile was subjected to water flow, the top and bottom sections encounter most of the force, while calculating the lift coefficient. Therefore, it was considered because the lift force works in that direction, an area may be seen from the bottom view.

Software used: COVERGE CFD

By enabling the bounding box in stimuli, the dimensions of the hydrofoil were modified. The bounding box was utilised to calculate the dimensions of any shape. The geometry must then be exported as a.stl file and loaded into AutoCAD. It then checked for defects such as intersection errors, open edges, and non-manifolding.

**Table 1** Angles of attack versus drag coefficient and lift coefficient in steady state and transient state (Coverage CFD)

| Angle of attack (AoA) | Steady state | | Transient state | |
|---|---|---|---|---|
| | Drag coefficient | Lift coefficient | Drag coefficient | Lift coefficient |
| 1° | 0.10837 | 0.2266 | 0.0100 | 0.0320 |
| 5° | 0.1382 | 0.7209 | 0.0019 | 0.0735 |
| 10° | 0.2998 | 0.1639 | 0.0327 | 0.1122 |
| 15° | 1.6048 | 2.4869 | 0.0648 | 0.8980 |

## 3 Optimization Technique

In real conditions, steady state does not exist since complete laminarity is impossible, and there will always some turbulence. However, steady flow state was still stimulated for an industrial approach like in wind tunnel. When process data were posted in stimulus, the flow was turned spontaneous and irregular [5–7].

**Transient state stimulation**
On stimulating transient water flow over the profile in simulation, it appeared to coverage, and when the AoA was raised further, the pressure and velocity fluctuated greatly. It was also observed that creating a flow separation area lowered pressure.

**Hydrodynamic analysis**
The profile is selected on the basis of the $C_L/C_D$ ratio and the attack angle is taken as 0 degree.

The selected profiles (naca16015, naca2418) were preferred based on the streamline structure to increase the efficiency [8].

The following dimension inputs are given to the stimulus ()

- Submarine length = 25.1 m
- Speed = 12 m/s
- Hull diameter = 3 m
- Water temperature = 20 °C

If one simply views that lift = weight in order to establish a relationship between the Coefficient of lift and object weight. Then it is used to calculate the relationship between coefficients of lift and coefficient of drag, such that $C_l/C_d$ is maximum according to the submersible mission that is to be completed.

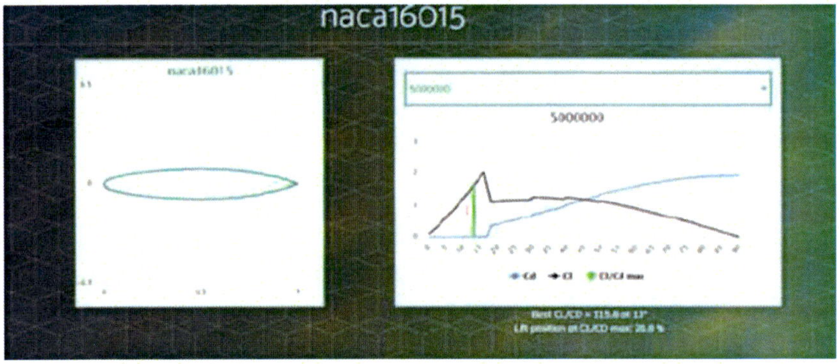


**Fig. 1** Performance analysis of naca16015

One can then establish a value for the drag force experienced (for different $C_l/C_d$ ratios if desired) for a given weight of the submersible, $C_l/C_d$ provided us with a measure of the Lift/drag ratio. The particular design point of interest, in that the higher the Lift to drag ratio, the better the profiles climb performance, fuel/energy economy, range and endurance. Basically the method might be used to compare the relationship between drag and L/D max $(C_l/C_d)$ max per unit weight of the submersible—for UAVs and manned submersibles and probably only indicates the relative power required in order for surveillance missions (long endurance or furthest range—take your pick—for the type of propulsion system). Keep in mind however that the $C_l/C_d$ data chosen for the UAV was arbitrary in nature and thus meaningless except in terms of being a desired target value [9].

The Reynolds number we got from the given stimulus input is approximately same for both the profiles, hence the Reynolds matching is fixed same for both the profiles. While comparing the performances the $C_l/C_d$ ratio is high for profile naca2418 at a smaller angle when compared to naca16015. As Naca16015 at 5,00,000 Reynolds number $C_l/C_d$ was 76.7 at AOA 8° while Naca2418 −5,00,000 Reynolds number $C_l/C_d$ was 81.7 at AOA 7°. Hence, NACA2418 was selected for the further analysis [10, 11]

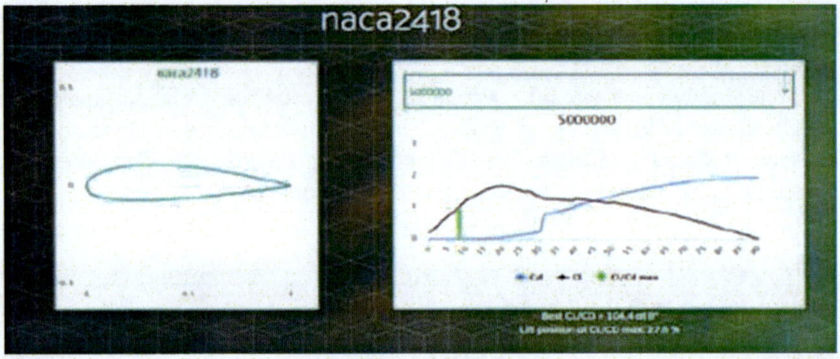


**Fig. 2** Performance analysis of naca2418

**Fig. 3** CAD generated models: **a** varient 1, **b** varient 2, and **c** varient 3

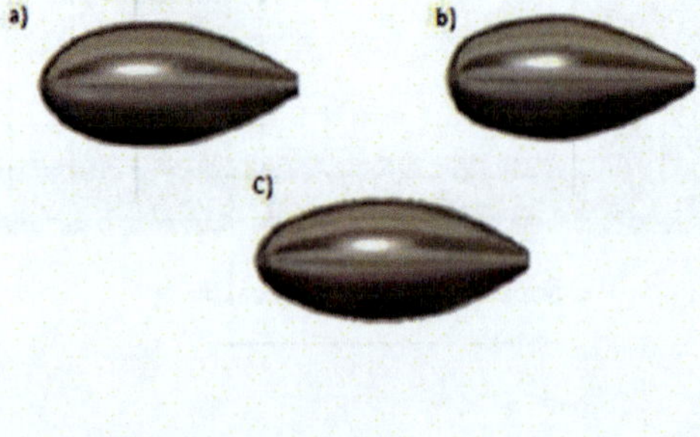

**Effect of attack angle and streamlining of the submarine in drag reduction**

- Here in this analysis, we are finding the effect of streamlining the profile
- NACA2418 in drag reduction, pressure and velocity
- To find this, three variants with little difference in AOA of NACA2418 are taken as shown in Fig. 3.

The submarine variants were stimulated in real conditions, similar to wind tunnel where the profile is static and water is moving. The dimensions of variants were same for all variants and only the frontal area is varied, with a difference of around 10° in $AOA_2$ (here $AOA_2$ means the angle between the tangent of water to the tip of the submarine with the axis of the submarine [variant (a) 76.8° , variant (b) 82.3°, variant (c) 90.2°] shown in Fig. 3.

Dimensions of variants and boundary conditions used were:

- Length = 20 cm and width = 7 cm
- Water is taken seawater
- Water velocity = 1.5 m/s in $x$ direction, $y = 0$, $z = 0$.
- Pressure = 3000 Pa
- Depth = 3

## 4 Results and Discussion

- Table 1 shows the effect of Angle of attack on the $C_l$ and $C_d$ in transient and steady conditions and how it is varying with AoA. Further, the data were utilized in selecting the submarine profiles naca16015 and naca2418 for the analysis, by comparing the performances on desired dimensions and conditions, the naca2418 has shown better performance. The $C_l/C_d$ ratio produced is 5% more than the other corresponding variant (Figs. 1 and 2) [15–17].

The selected profile (naca2418) was processed to generate the variants with varying geometrics, this significantly contributes in analysing the effect of streamlining the profiles on the parameters like velocity, pressure and nut [12]. When the submarine velocity increases the total resistance increases, in contrast the normal wake decreases with increment in velocity affects the wake length (Figs. 4, 5 and 6). The corresponding reduction in the parameters and enhanced performance can be seen in Table 2.

Analysing pattern in the above table a slight variation of AOA can reduce significant amount of pressure on the profiles. The reduction in pressure and velocity in percentage are shown in Table 3. The effect of streamlining can be understood by the reduction in the pressure and velocity components between the variants. It was understood how to use the parameter for maximizing the submersible ability to perform in different conditions and investigate an underwater profile numerically in commercial code and knowledge of the fluid dynamics has increased.

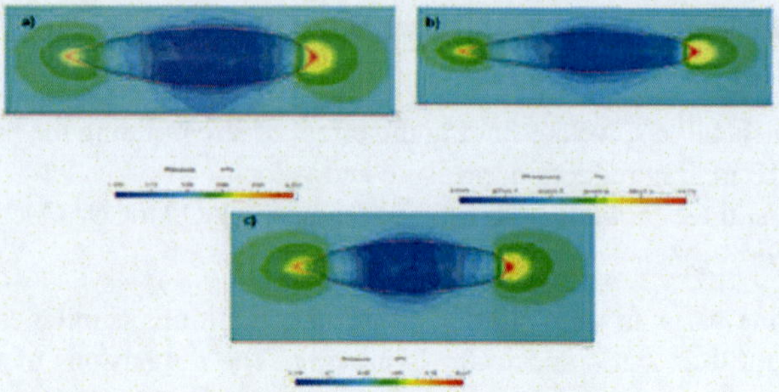

**Fig. 4** Pressure component and its zones for **a** variant 1, **b** variant 2, and **c** variant 3

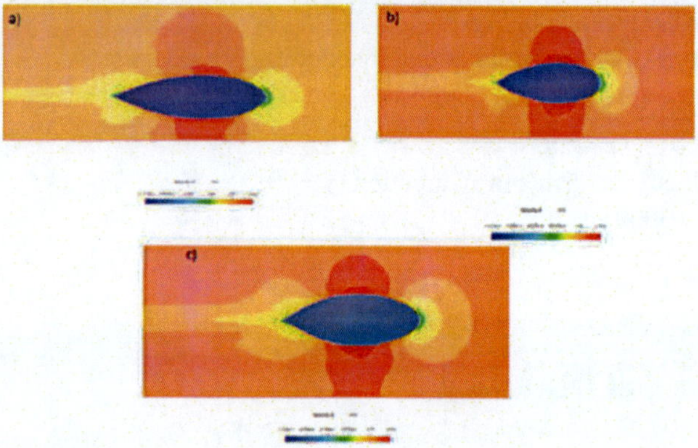

**Fig. 5** Velocity component and zones for **a** variant 1, **b** variant 2, and **c** Variant 3

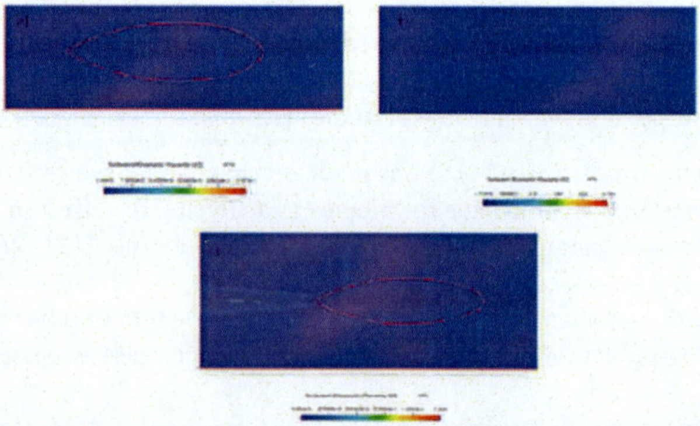

**Fig. 6** NUT component and its zones for **a** variant 1, **b** variant 2, and **c** Variant 3

**Table 2** The generated results for different Variants' Pressure

| Variant | Pressure | Velocity | Nut |
|---|---|---|---|
| 1 | The pressure is acting on a wider region in front with Max. 4203 Pa and the sides are 2336 Pa | The velocity is noted at max of 1.824 m/s at sides and 6.5e–1 m/sat front the velocity of the fluid continues to decrease as it gets streamline because it gets easy for the object to penetrate but the side regions get darker as the skin friction increases | The nut generated was 1.44e–5 |
| 2 | The pressure gets slightly reduced to 4175 Pa at the frontal area and increased to 2390 Pa in sides and also the pressure affected region is also reduced and increased in the sides with darker shades in sides | The velocity is 1.744 m/s at front and 0.6 m/s at sides with darker shade the wake created decreases, which reduces the separation | The nut generated is 1.712e–6 |
| 3 | The pressure gets even more reduced to 4157 Pa in the frontal area and with much less affected area in the front but there is constant increase in the sides | The velocity generated to 1.763 at front and 5.8e–1 m/s at sides with darker shades | The nut generated is 3.286e–6 |

**Table 3** The effect of streamlining

| Condition | Pressure | Velocity (%) |
|---|---|---|
| Between 1 and 2 variants | 0.67% reduction | Front velocity 4.38 |
| Between 2 and 3 variants | 0.43% reduction | Front velocity −1.1 |
| Between 1 and 3 variants | 1.1% reduction | Front velocity 3.34 |

## 5 Conclusion

The conclusions that can be taken in to consideration from the study are:

1. The angle of attack affects the lift performance of submarine. As AoA increases, lift and drag coefficient increase.
2. Naca2418 has shown better performance naca16015 as it has shown more $C_l/C_d$.

3. The more the streamlined an object is, the smaller the wake it creates, which was analysed and discussed. The three variants of Naca2418 have the same diameter but the more streamlined has better performance, which penetrate and slip through the water smoothly and help drag reduction.

## References


1. Versteeg HK, Malalasekera W (1995) An introduction to computational fluid dynamics
2. Clement EP, Blount DL (1963) Resistance tests of systematic series of planing hull forms. Trans SNAME 71:491–579
3. Renilson M, Submarine hydrodynamics. https://continuum-dynamics.com/lib-pro-svlt.html
4. Date JC, Turnock SR, A study into the techniques needed to accurately predict skin friction using RANS solver with validation against Froude's Historical flat plate experimental data 49
5. Li F, Zhao G, Liu WX, Sun ZZ (2014) Simulation on flow control and drag reduction with bionic jet surface. J Basic Sci Eng 22(3):574–583
6. Perzon S, Davidson L (2000) On transient modeling of the flow around vehicles using the Reynolds equation. In: Proceedings of the ACFD 2000. Beijing, China, pp 720–727. Available online: https://www.tfd.chalmers.se/lada/postscript_files/sven_acfd_paper_2000.pdf
7. Versteeg HK, Malalasekera W (1995) An introduction to computational fluid dynamics
8. Yang HW, Gao G (1997) Experimental study for turbulent drag reduction using a novel boundary control technique. Acta Aeronautica Et Astronautica Sinica. 18(4):455–457
9. Katayama T, Hayashita S, Suzuki K, Ikeda Y (2002) Development of resistance test for high-speed planing craft using very small model scale effects on drag force. In: Asia Pacific workshop on hydrodynamics
10. Stavisky D (1964) Hydrodynamic analysis of planing hulls. Marine Technol 1:71–95
11. Jahanbakhsh E, Panahi R, Seif MS (2009) Catamaran motion simulation based on moving grid technique. J Marine Sci Technol 17:9
12. Yildiz technical university naval architecture and maritime faculty department of naval architecture numerical evaluation of submarine hydrodynamics for different appendage configurations
13. Fluid mechanics submarine design, Australian Maritime College (2011)
14. Sarkar S, Thummar K, Shah N, Vagrech V, A review paper on aerodynamic drag reduction and CFD analysis of vehicles
15. Evaluation of submarine model test in towing tank and comparison with CFD and experimental formulas for fully submerged flows. Indian J Geo Marine Sci (2013)
16. Pathan KA, Dabeer PS, Khan SA (2017) CFD analysis of effect of area ratio on suddenly expanded flows. In: 2nd international conference on convergence in technology (I2CT-2017) Organized by Siddhant College of Engineering, Pune, India and Sponsored by IEEE, USA
17. Moonesun M, Korol YM, Tahvildarzade D (2014) Optimum L/D for Submarine Shape. In: The 16th marine industries conference (MIC 2014)
18. Khan SA, Fatepurwala MA, Pathan KN, Dabeer PS, Baig MAA, CFD analysis of human powered submarine to minimize drag
19. Moonesun M, Korol YM, Nikrasov VA, Ursalov A, Brajhko A, CFD analysis of the bow shapes of submarines
20. Baker C (2004) Estimating drag forces on submarine hulls. Rep DRDC Atlant 125
21. Praveen PC, Krishnankutty P (2013) Study on the effect of body length on the hydrodynamic performance of an axi-symmetric underwater vehicle. Indian J Geo-Marine Sci 42(8):1013–1022
22. Abinesh J, Arunkumar J (2014) CFD analysis of aerodynamic drag reduction and improve fuel economy. Int J Mech Eng Robot Res 3(4)
23. Submarine pressure hull design and the application of gained experience to the design of other pressure structures